Current Developments in Nanowires

Edited by **Rich Falcon**

NY RESEARCH PRESS

New York

Published by NY Research Press,
23 West, 55th Street, Suite 816,
New York, NY 10019, USA
www.nyresearchpress.com

Current Developments in Nanowires
Edited by Rich Falcon

International Standard Book Number: 978-1-63238-109-5 (Hardback)

Printed in China.

Contents

Preface

This book has been an outcome of determined endeavour from a group of educationists in the field. The primary objective was to involve a broad spectrum of professionals from diverse cultural background involved in the field for developing new researches. The book not only targets students but also scholars pursuing higher research for further enhancement of the theoretical and practical applications of the subject.

Structures having diameter of the orders of nanometers are referred as nanostructures. Significant amount of research has been carried out on single dimensional nanoscale structures like nanowires. Miniature sizes bring exclusive properties to nanowires because of quantum captivity. Large surface-to-volume ratios enable high sensitivity to surface effects in nanowires. The distinctive geometrical benefits and properties assist the utilization of nanowires in nanoelectronics. This book is an attempt to provide an advanced assessment of the various uses of different nanowires and related progress in creation and properties characterization. It comprises of discourses on recent developments in metal oxide nanowires, silicon nanowires, carbon based nanotubes and nanowires.

It was an honour to edit such a profound book and also a challenging task to compile and examine all the relevant data for accuracy and originality. I wish to acknowledge the efforts of the contributors for submitting such brilliant and diverse chapters in the field and for endlessly working for the completion of the book. Last, but not the least; I thank my family for being a constant source of support in all my research endeavours.

Editor

Section 1

Oxide Nanowires

Chapter 1

Metal-Oxide Nanowires for Gas Sensors

Supab Choopun, Niyom Hongsith and
Ekasiddh Wongrat

Additional information is available at the end of the chapter

1. Introduction

In past decades, gas sensors based on the metal oxide semiconductors (MOSs) have been studied in diverse field for wide applications. A gas sensor is a device that can be used to detect various gas such as ethanol, LPG, CO_2 and CO gases etc. The gas sensors based on MOSs such as SnO_2, TiO_2, WO_3, ZnO, Fe_2O_3, and In_2O_3 have an important role in environmental monitoring, chemical process controlling, personal safety (Q. Wan et al., 2004), industrial process controls, for the detection of toxic environmental pollutants in human health, and for the prevention of hazardous gas leaks, which comes from the manufacturing processes (K. Arshak&I. Gaidan, 2005), wine quality monitoring, and traffic safety (X.F. Song et al., 2009).

The first generation of MOS gas sensors were based on thick films of SnO_2 since 1960s which was firstly reported by Taguchi (E. Comini et al., 2009). The MOS gas sensors have some advantages such as small size, low-power-consumption (E. Comoni et al., 2009), simple construction, good sensing properties (K. Arshak&I. Gaidan, 2005), and high compatibility with microelectronic processing (E. Comini et al., 2002). So, they have rapidly gained attention over the years.

Recently, various morphologies of MOS nanostrcutures such as wire-like, belt-like, rod-like, and tetrapods have been widely investigated for gas sensor applications. It is well-known that the sensitivity characteristics of these sensors strongly depend on the morphology of MOS. Especially, one-dimensional nanostructures such as nanowires, nanobelts, nanoneedles have gained a lot of interest for nanodevice design and fabrication (Wang et al., 2008). The sensors based on MOS nanowires are promising due to feasibility for ultrahigh sensitive sensors or ppb-level sensors. These nanowires can be prepared by various techniques such as pulse laser deposition (PLD), chemical vapor deposition (CVD), thermal evaporation, metal-catalyzed molecular beam epitaxy (CBE) and thermal oxidation technique. Moreover, there are many reports on gas sensor based on the nanowires.

The fundamental mechanism of gas sensing based on MOS depend on the reaction between the surface complexes such as O^-, O^{2-}, H^+, and OH^- reactive chemical species and the gas molecules (reducing/oxidizing gas) to be detected (E. Comini et al., 2009). Thus, it is important to understand the surface reactions between semiconductor surface and target gas for improving the sensing characteristics. Typically, the important parameters in sensor development are sensitivity, selectivity, and stability that called "3S" (E. Comini et al., 2009). However, in this chapter we will mostly discuss on only the sensitivity parameter.

The sensitivity of sensors based on bulk and thin film MOS is typically low. Thus, sensitivity improvement has been extensively studied by using several techniques. The two techniques that commonly used for sensitivity improvement are in the following:

1. *Using nanostructures*

The MOS sensors based on various nanostructures such as nanowires, nanobelts, nanoparticles, nanrods, and nanotubes etc. have been demonstrated to be excellent candidates for ultrahigh sensitivity due to their high surface-to-volume ratio. A large surface-to-volume ratio means that a significant fraction of atoms (or molecules) are much quantity on the surface. So, the reaction between target gas and reactive chemical species (O^-, O^{2-}, H^+, and OH^-) on the surface can extremely occur. A list of MOS sensors with the sensitivity for the different morphologies of ZnO nanostructures is summarized in Table 1. It can be seen that the sensor sensitivity strongly depends on size and morphology of ZnO nanostructures.

2. *Adding noble metal*

The noble metals such as Au, Pt, Pd, and Ag on the surface of MOSs can act as a catalyst to modify surface reactions of MOSs toward sensing gas and result in high sensitivity. The ZnO sensors with noble metal additive are also listed in Table 1.

Usually, the gas sensors based on MOS nanostructures exhibit high sensitivity and sometimes up to a few hundred folds over a conventional MOS sensor at a moderate concentration. On the other hand, the sensors based on a larger size of MOS such as in the form of thin film or micro-tetrapod shows lower sensitivity. Several models have been proposed to explain sensitivity characteristics of MOS sensors and still be a subject of discussion.

There are many reports about gas sensor model that are used to explain sensor response characteristics. Wang and co-workers have proposed a surface-depletion model and a contact-controlled model that are used to describe the sensing mechanism of resistance-type metal-oxide semiconductor sensors (Feng et al., 2005). The surface-depletion controlled model is used to explain the sensing mechanism of semiconducting oxide sensors based on nanocrystal/ nanowire/nanobelt structures, while the contact-controlled model is proposed to explain the contact between the outer ends of the rods. This leads to the formation of many junctions in the sensors that significantly modify the potential barrier of contact between rods. These barriers can control the transport of electrons between the rods resulting in the change of the sensor resistance. Chen and co-workers (Chen et al., 2006) have used space-charge model to explain the sensitivity improvement when the grain size is close to or smaller than the Debye length ($2L_d$).

In this chapter, we investigate the sensing characteristics of the MOS nanowire sensors and present the sensitivity formulas that are developed in order to explain all circumstances of gas sensors based on MOS nanowires. The size and morphology dependences on the sensitivity are explained in terms of the two important parameters including surface-to-volume ratio and depletion layer width. The developed formulas will be discussed and related to the experimental sensing characteristics of ZnO sensors.

Materials	Morphology	Diameter (nm)	Target gas	Sensitivity (ppm)										
				1	10	25	50	100	150	200	300	500	1000	2000
ZnO (Son et al., 2008)	nanowire	20	Ethanol	~16		~40	~54	~62	~70	~70		-	-	
ZnO (Xu et al., 2008)	nanorod	40-80	Ethanol				7.3	-				-	-	
ZnO (Chen et al., 2006)	nanorod	<15	Ethanol		20.5		104.9	176.8		224.2		267.7	-	
ZnO (Bie et al., 2007)	nanorod	10-30	Ethanol		~5		~10	18.29				~32	~42	
ZnO (Li et al., 2009)	nanoneedle	5-10	Ethanol	11	56		116	176		~300		~650	-	
ZnO (Wan et al., 2004)	nanowire	25	Ethanol	~2.5		~8	~16	~33		~47		-	-	
ZnO (Feng et al., 2005)	flowerlike	150	Ethanol	2.2	5.8		11.4	14.6				25.2	30.1	
ZnO (Li et al., 2007)	nanorod	15	Ethanol	4.1	10.7		18.1	29.7				~72	100	
ZnO (Yang et al., 2008)	nanorod	50	Ethanol	10	18		60	100				-	-	
ZnO (Chen et al., 2008)	nanotube	250	Ethanol	2.6			-	24.1		34.8		59.3	-	
ZnO (Choopun et al., 2007)	nanobelt	50-150	Ethanol				7.3		12			21.1	23.2	
ZnO (Hongsith et al., 2008)	nanowire	60-180	Ethanol				-	5.07				9.79	14	14
ZnO (Wongrat et al., 2009)	nanowire	100-500	Ethanol				2	3				5	8	
SnO_2 (Neri et al., 2006)	nanopowder	6-100	Ethanol				~7	~8.6	~10			-	-	
SnO_2 (Lee et al., 2008)	nanorod	<100	Ethanol				-	-				-	~40	
ZnO (Liu et al., 2010)	nanotube	200	H_2				-	2.3				-	-	
ZnO (Bie et al., 2007)	nanorod	10-30	H_2		~5		~7	10.41				~22	~24	
SnO_2 (Zhang et al., 2010)	nanofiber	80-120	H_2				~2	~5				~8	~10	
SnO_2 (Lee et al., 2008)	nanorod	<100	H_2				-	-				-	~10	
SnO_2-Pd (Zhang et al., 2010)	nanofiber	80-120	H_2				~5	~8				~15	~26	

Materials	Morphology	Diameter (nm)	Target gas	Sensitivity (ppm)										
				1	10	25	50	100	150	200	300	500	1000	2000
SnO_2-Pd (Lee et al., 2008)	nanorod	<100	H_2				~33	-				-	~700	
ZnO-Au (Hongsith et al., 2008)	nanowire	60-180	Ethanol				-	6				12	24	37
SnO_2-Pt (Neri et al., 2006)	nanopowder	6-100	Ethanol				~18	26.5	~31			-	-	
SnO_2-Pd (Lee et al., 2008)	nanorod	<100	Ethanol				~15	-				-	~125	
ZnO-Au (Li et al., 2007)	nanorod	15	Ethanol		20.1		41.8	89.5			193.6	-	-	
ZnO-Au (Wongrat et al., 2009)	nanowire	100-500	Ethanol				7	10				20	32	

Table 1. List of MOS sensors with the sensitivity for the different morphologies of ZnO nanostructures.

2. Sensing mechanism and sensitivity parameters

Normally, the gas sensor based on MOS has an optimum operating temperature at high temperature about 250-350°C. When the MOS is heated at lower temperature about 100-200°C, oxygen molecules in the atmosphere are adsorbed on its surface and form oxygen ion molecules by attracting an electron from the conduction band of MOS as shown in the Eq. (1).

$$O_2(gas) + e^- \Leftrightarrow O_2^-(adsorb) \tag{1}$$

At higher temperature, the oxygen ion molecules are dissociated into oxygen ion atoms with singly or doubly negative electric charges by attracting an electron again from the conduction band as shown in Eq. (2) and (3)

$$\frac{1}{2}O_2 + e^- \overset{k_{Oxy}}{\Leftrightarrow} O_{ads}^- \tag{2}$$

$$\frac{1}{2}O_2 + 2e^- \overset{k_{Oxy}}{\Leftrightarrow} O_{ads}^{2-} \tag{3}$$

where k_{Oxy} is the reaction rate constant. The oxygen ions on the surface of MOS are extremely active with the target gas molecule and give up the electrons from the surface back to the conduction band of MOS. The generally chemical reaction between gas molecule and oxygen ions is shown in Eq. (4)

$$X + O_{ads}^{b-} \xrightarrow{k_{gas}} X' + be^{-} \tag{4}$$

where X and $X^{'}$ is target gas and out gas, respectively. The b value is the number of electron and k_{gas} is the reaction rate constant of the gas reaction.

The chemical reaction causes change of the carrier concentration in the conductivity and thus, change of sensor resistance. The change of sensor resistance depends on a type of MOSs. A schematic diagram for change of the sensor resistance upon exposure to the target gas (reducing gas) in the cases of n-type and p-type MOS sensors is illustrated in Fig 1 and will be discussed in the following.

2.1. -type metal oxide gas sensor

Since majority carriers in n-type MOSs are electron, the resistance of n-type MOS sensor decreases when the temperature increases due to their semiconducting properties. However under oxygen ambient, from Eq. (1)-(3), the electrons in the conduction band of n-type MOS are removed by the adsorbed oxygen ions. This causes a decrease of the carrier concentration and thus, an increase of resistance of n-type MOS sensor at operating temperature. When the n-type MOS sensor is under the target gas ambient (reducing gas), the electrons obtained from the chemical reaction as in Eq. (5) are given back to the conduction band leading to a decrease of the sensor resistance.

2.2. p-type metal oxide gas sensor

In another hand, the majority carriers in p-type MOSs are holes. Similar to n-type MOSs, the sensor resistance of p-type MOS decreases when the temperature increases. However, under oxygen ambient, p-type MOS generates holes when the oxygen ions are adsorbed on the surface via the excited electrons from valence band. This process results in raising the number of charge carriers, which leads to a decrease of the sensor resistance (opposite to n-type). When the p-type MOS sensor is under the target gas ambient (reducing gas), the electrons inject into the valence band and recombine with the holes and this method resulting in reducing the number of holes, which leads to an increase of the sensor resistance (opposite to n-type).

It should be noted that in the case of an oxidizing gas the change of resistance will be just opposite to the above discussions.

In addition, from Eq. (4), a rate equation of an electron density can be written as

$$\frac{dn}{dt} = k_{gas}[O_{ads}]^{b}[X]^{b} \tag{5}$$

where n is the electron density or the electron concentration under the gas atmosphere, and k_{gas} is the reaction rate constant or reaction rate coefficient described as

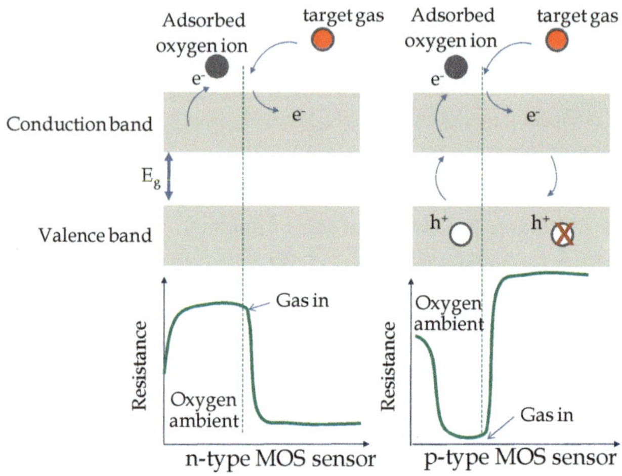

Figure 1. Schematic diagram for change of the sensor resistance upon exposure to the target gas (reducing gas) in the cases of n-type and p-type MOS sensors.

$$k_{gas} = A\exp\left(-E_a / k_B T\right) \tag{6}$$

where E_a is the activation energy of a reaction, k_B is the Boltzmann constant and T is absolute temperature. By integrating Eq. (5) at an equilibrium state under gas ambient and air ambient and using the carrier concentration defined as $n=\alpha/R$ (where R is a resistance and α is a proportional constant), a sensitivity relation can be obtained as (Hongsith et. al, 2010)

$$n = \Gamma_t k_{gas} [O_{ads}^{ion}]^b [X]^b + n_0 \tag{7}$$

$$S_g = \frac{R_a}{R_g} = \frac{\Gamma_t k_{gas} [O_{ads}^{ion}]^b [X]^b}{n_0} + 1 \tag{8}$$

Sometimes, a compact form of the sensitivity relation on gas concentration, X, can be rewritten as

$$S_g = a\mathrm{X}^b + 1 \tag{9}$$

where a is a controllable parameter.

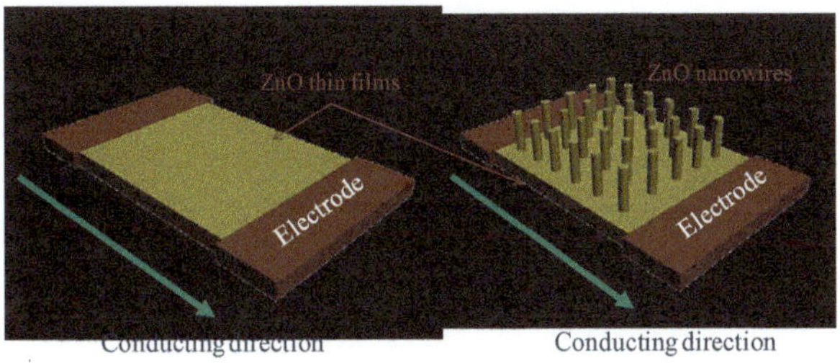

Figure 2. a) sensors based on thin film structrue and (b) based on thin films with perpendicular spike nanowires. The sensitivity enhancement of nanowire sensors can be explained by Eq. (11) which is due to an increase of surface-to-volume ratio.

The sensitivity formula in Eq. (8) can be applied to explain very well in the case of thin film and bulk. In the case of nanostructure, however, two important parameters including surface-to-volume ratio and depletion layer width need to consider in order to explain the sensing characteristics.

2.2.1. *Surface-to-volume ratio*

Generally, this surface-to-volume ratio can be related to the density of the adsorbed oxygen ions. Thus, we have proposed that the density of adsorbed oxygen ions can be written in term of surface-to-volume ratio as

$$[O_{ads}^{ion}] = \frac{\sigma_0 \Phi V_m}{V_s} \tag{10}$$

where σ_0 is a number of oxygen ions per unit area, Φ is a ratio of surface area per volume of material V_m, and V_s is a system volume. By substituting Eq. (10) onto Eq. (8) gives

$$S_g = \frac{\Gamma_t k_{gas} \left(\frac{\sigma_0 \Phi V_m}{V_s} \right)^b}{n_0} X^b + 1 \tag{11}$$

The Eq. (11) can be applied to explain very well in the case of sensors based on MOS nanowires. At the begining, it is quite amazed that sensors based on MOS nanowires exhibited the sensitivity higher than that of thin film or bulk even though the nanowires were not aligned along the conductive direction. For example, sensors based on thin films with perpendicular spike nanowires as shown in Fig.2 exhibit the sensitivity of higher than sensor based on thin film structure. The explanation of sensitivity enhancement is clear with using the Eq. (11) which is due to an increase of surface-to-volume ratio.

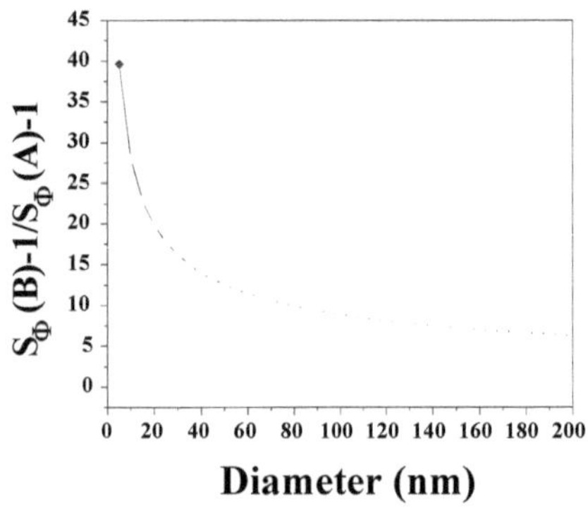

Figure 3. The sensitivity ratio as a function of diameter where L = 10 μm of a gas sensor based on vertical alignment ZnO nanorods with diamerter D and length L and a gas sensor based on ZnO thin films with an area of 1 cm^2.

From Eq. (11), the sensitivity strongly depends on the surface-to-volume ratio (Φ) as discussed earlier. For example, let consider a thin film with an area of 1 cm^2 as shown in Fig. 2 (a). Then, given cylindrical nanostructures of 10 μm in length with various diameters are grown on this area, as shown in Fig. 2 (b). The surface-to-volume ratio can be calculated and put in Eq. (11) for the sensitivity ratio as

$$\frac{S_{\Phi}(B)-1}{S_{\Phi}(A)-1}=\left(\frac{\Phi_B}{\Phi_A}\right)^b \tag{12}$$

The sensitivity ratio as a function of diameter is plotted as shown in Fig. 3. It can be seen that the sensor sensitivity enhances as decreasing the diameter due to an increase of surface-to-volume ratio and rapidly enhances at small diameter.

2.2.2. Depletion layer width

Typically, the sensor resistance is contributed from two parts including the resistance along a cylindrical nanowire, and the resistance between nanowires. The resistance along the nanowire is due to surface depletion layer and conductive channel. The resistance between nanowires is due to band bending (potential barrier between wires) which depend on a size of nano-structure.

First, let consider for only the resistance along the nanowire by neglecting the resistance between nanowires. According to the depletion layer or the space charge model, L_d (Debye length), can be expressed by

$$L_d = \left(\frac{\varepsilon k_B T}{q^2 n} \right)^{1/2} \tag{13}$$

where ε is a static dielectric constant, q is an electrical charge of a carrier, and n is a carrier concentration. It can be seen that at steady operating temperature, the Debye length depends only on the carrier concentration. Let consider in Eq. (13), at the optimum operating temperature based on pure or undoped MOS nanowire sensor, the L_d can be regarded as a constant value and equals to a value of the depletion layer width.

In this model, a cylinder, which is one of the most basic geometric shapes in one-dimension, is considered and a conductive channel is assumed to be along the axis of the cylinder. At an operating temperature, the oxygen ions are adsorbed by attaching an electron on the surface of the cylinder. Therefore, the depletion layer is formed on the surface of cylinder with a thickness of L_d, and then a size of conductive channel is reduced along the radial direction as shown in Fig. 4.

When exposed to the reducing gas atmosphere, gas reacts with oxygen ions on the surface and gives back electrons to MOS sensors resulting in increasing conductive channel (decreasing depletion layer width). The conductive channel can be related to the carrier concentration, and can be written in term of the depletion layer width (L_d) as

$$n' = n_0 \frac{\pi \left(D - 2L_d \right)^2}{\pi D^2} \tag{14}$$

where n_0 is a carrier concentration of a intrinsic material, n' represents a carrier concentrationof the Debye length, and D is a diameter of the cylinder. Thus, the effect of the depletion layer on sensitivity based on cylindrical MOS nanowire are given by inserting Eq. (14) in Eq. (11) and obtained

$$S_{L_d} = \left(\frac{\Gamma_t k_{gas} (\sigma_0 \Phi (V_m / V_s))^b}{n_0} \right) \frac{D^2 X^b}{\left(D - 2L_d \right)^2} + 1 \tag{15}$$

Let compare a diameter of cylinder D to the Debye length (L_d). Since the Debye length is in the order of nanometer, it can be divided into three conditions.

Under condition D >> 2L_d, Eq. (15) turns in to Eq. (11)

In this condition, a diameter of cylinder is much larger than micrometer which is the case of microstructure or bulk materials. The depletion layer width is very small compared with the cylindrical diameter (D >> $2L_d$) and Eq. (15) can be approximated to Eq. (11) which is an equation that has no depletion layer effect.

D> 2L$_d$, Eq. (15) can be approximated to Eq. (11)

When a diameter of cylinder is in the order of nanometer but still larger than Debye length ($D > L_d$), Eq. (15) again can be approximated to Eq. (11) with no depletion layer effect. However, the sensor sensitivity strongly depends on the oxygen ion density due to the surface-to-volume ratio, Φ, parameter.

D~2L$_d$, sensor sensitivity strongly depends on D

When a cylindrical diameter decreases down to the order of nanometer and is comparable to the Debye length ($D \sim 2L_d$), the depletion layer has strong effect and the sensor sensitivity is strongly dependent on a cylindrical diameter. Thus, Eq. (15) can be used to explain sensitivity of all structural size ranging from nanometer to bulk and can be regarded as a general form of sensitivity formula in order to explain the sensing characteristics.

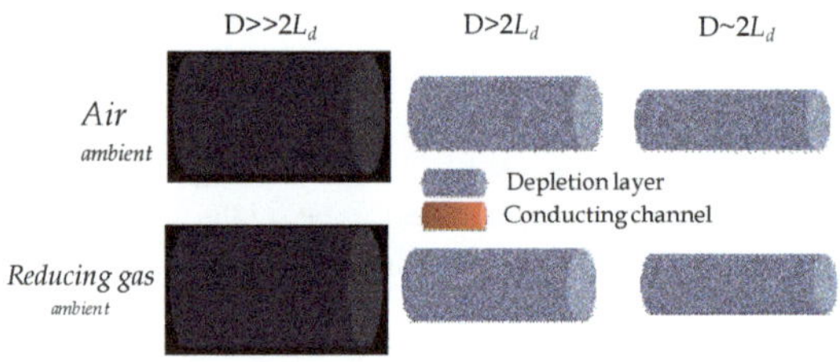

Figure 4. Schematic diagram of the depletion layer on the surface of cylinder with a thickness (or width) of L$_d$, under condition of D>>2L$_d$, D>>2L$_d$, and D~2L$_d$ (not in scale).

Second, let consider for a case that sensor resistance is contributed from both along nanowire and between nanowires. So, the sensor resistance is due to both effects of surface deletion layer and band bending (potential barrier). It should be noted that band bending has also effect on the surface depletion layer. Since the sensor resistance strongly depend on a size of nanowire, let compare a diameter of a cylindrical nanowire D with a depletion layer width (w). It can be divided into three cases.

Under D >>2 w (large size)

In this case, the depletion layer width is much narrow compared to a diameter of the cylinder so the depletion layer has small effect on the electron density in the wires. However, it does greatly affect the potential barrier of the contacts between the wires. Thus, the sensor resistance is dominantly controlled by the potential barrier in this case.

From the oxygen adsorption reaction as given in Eq. (1)-(3) which oxygen ion specie is O^{2-}, at equilibrium condition it can be deduced by using the mass action law that

$$k_a\left[e\right]^2 P^{1/2} = k_d\sigma. \tag{16}$$

$$\sigma = k_{Oxy}\left[e\right]^2 P^{1/2} \tag{17}$$

where k_{Oxy} is an equilibrium constant of oxygen adsorption reaction which is defined as $k_{Oxy} = k_a/k_d$, wherek_a and $k_d$is a reaction constant for oxygen adsorption and desorption

reactions, respectively. σ is a density of O^{2-} per unit area, $[e]$ is a density of conductionelectron at the surface, and P is a partial pressure of oxygen molecule. The density of conduction electrons at surface [e] can be expressed in terms of a surface potential as (Yamazoe & Shimanoe, 2008)

$$\left[e\right] = n_0 \exp\left(-\frac{qV_B}{k_B T}\right) \tag{18}$$

where n_0 is a density of donor electron, which is assumed to be constant throughout MOS and the relation between V_B and w is shown as (Yamazoe & Shimanoe, 2008):

$$w = \left(\frac{2\varepsilon}{qn_d} V_B\right)^{1/2} = L_d\left(2\beta V_B\right)^{1/2}, \quad V_B = \frac{\left(w / L_d\right)^2}{2\beta}. \tag{19}$$

Here L_d is the Debye length defined by $L_D = (\varepsilon kT/q^2N_d)^{1/2}$, and $\beta = q/kT$, where k and T are Boltzmann constant and temperature, respectively. In addition, n_0 is the density of conduction electrons and apparently is equal to the density of donors, n_d. By using Eq. (18), we can rewrite it as (Yamazoe & Shimanoe, 2008):

$$\left[e\right] = n_d \exp\left(-\frac{qV_B}{k_B T}\right). \tag{20}$$

Since the sensor resistance is dominantly controlled by the potential barrier in this case, the sensor resistance R, which is inversely proportional to the electron density is given as (Yamazoe & Shimanoe, 2008)

$$R = R_0 \exp(m^2 / 2), \qquad R \propto \frac{1}{\left[e\right]} \tag{21}$$

where R_0 is a resistance under the flat band condition, m is the reduced depletion depth which is defined by $m=w/L_d$. Moreover, if there are no electron-trapping sites other than the adsorbed oxygen ion O^{2-} on the surface, the depletion layer width (w) can be related to density of O^{2-} per unit area (σ) as (Yamazoe & Shimanoe, 2008):

$$\sigma = n_d w \tag{22}$$

From Eq. (22), the larger depletion layer width results in higher density of O^{2-} per unit area (σ). Finally, the density of O^{2-} per unit area (σ) and [*e*]([*e*]=n_0 in this case) can be inserted into Eq. (15) and give (Wongrat et al., 2012)

$$S_g = \left(\frac{\Gamma_t k_{gas} \left(\Phi n_d w (V_m / V_s) \right)^b \exp(m^2/2)}{n_d} \right) X^b + 1 \tag{23}$$

n_d is density of donor electron, which is assumed to be constant, w is the depletion layer width. From Eq. (23) (under $D \gg 2w$), it can be seen that the gas sensing mechanism is controlled by the potential barrier of the contact between wires. Furthermore, the potential barriers are independent of the cylinder size and therefore, the sensitivity is independent of the diameter of nanowire, D.

Under $D>2w$

In this case, the depletion layer has effect on both electron density in the wires and the potential barrier of the contacts between the wires. Thus, the sensor resistance is controlled by surface depletion layer and the potential barrier in this case.The conductance related to the nanowire after oxygen adsorption is given as:

$$G = \frac{e\mu_0 [e] \pi (D - 2w)^2}{4l} \tag{24}$$

where μ_0 is the crystal electron mobility, l is the distance between electrodes. It can be seen that the gas-sensing mechanism depends on both the width and the height of the contact potential barriers (Sysoev et al., 2009, Rothschild&Komem, 2004). Thus, the sensor resistance due to surface depletion layer and potential barrier is represented by

$$R = R_0 \left(\frac{D}{D - 2w} \right)^2 \exp\left(m^2/2 \right). \tag{25}$$

From Eq. (12), (8) and using $[e]=n_0$, the sensor response formula can be given as (Wongrat et al., 2012)

$$S_g = \left(\frac{\Gamma_t k_{gas} \left(\Phi n_d w \left(V_m / V_s \right) \right)^b \exp\left(m^2/2 \right)}{n_d} \right) \left(\frac{D}{D - 2w} \right)^2 X^b + 1 \tag{26}$$

From this equation, it can be seen that the sensor response depend not only on the potential barrier between wires but also on the size of nanowire.

Under $D \leq 2w$

In this case, the depletion layer has effect only on the electron density in the wires and can be regarded as fully depleted cylinders. The fully depleted cylinders are occurred when the Fermi level is totally controlled by surface states under nearly flat energy bands and so potential barrier can be neglected. Therefore, the sensor resistance is controlled only by the surface depletion in this case.

3. Experimental, results and discussions

3.1. Sensor fabrication

The MOS nanowires have been synthesized by thermal oxidation technique. This technique has been successfully used for synthesizing ZnO or CuO by simply heating pure Zn and Cu material source, respectively. The process is usually conducted in a cylindrical furnace. For the synthesis of ZnO nanowire as shown in Fig. 5, Zn thin films or thick films were screened or evaporated on alumina substrate. Then, they were sintered at temperature in ranging of 500-700ºC to form the nanostructure. For the synthesis of CuO nanowire, the commercial grade copper plate with thickness of 0.1 mm was cut and cleaned by alcohol in an ultrasonic bath for 2 min and dried at room temperature. The copper plate was loaded into a center of a tube furnace at 600°C at normal atmosphere for oxidation reaction.

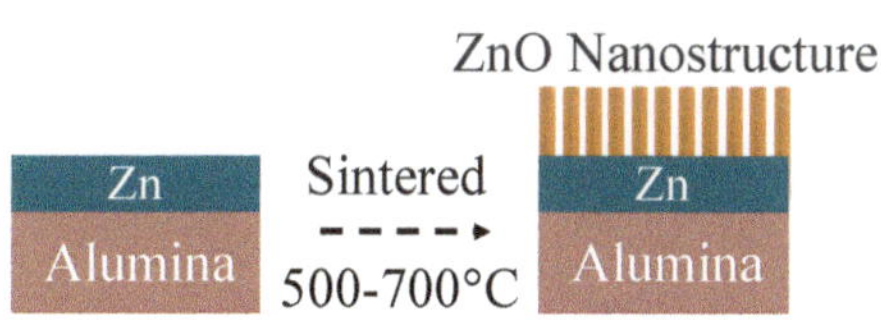

Figure 5. Schematic diagram of the synthesis of ZnO nanowire on alumina substrate by thermal oxidation technique.

The sensors were fabricated by putting gold paste as inter-digital electrodes on the top of the sensors surface and putting a heater underneath the alumina substrate as shown in Fig 6. The heater for the sensor was made from nickel-chromium which could control temperature in the range 280-380 ºC. The operating temperature was measured by a thermocouple placed on the middle of the sensor.

3.2. Setup for ethanol sensing characteristic measurement

A set up for the ethanol sensing characteristic measurement was shown in Fig. 7. The sensing response and recovery characteristics of MOS sensors were studied in a gas flow chamber under a dynamic equilibrium. Electrical measurements were performed using a volt-amperometric

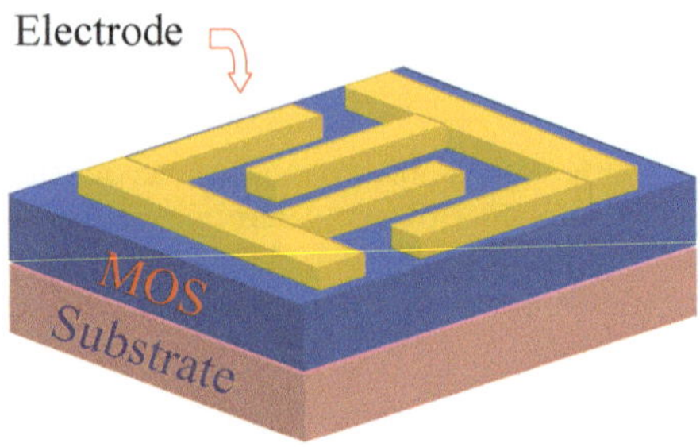

Figure 6. Schematic diagram of the fabricated sensor structures.

technique. A constant bias voltage of 5 V was applied to the sensor. Another voltage was applied to the heater coil used for heating the sensor and maintaining at a desired operating temperature. A thermocouple was employed to monitor the sensor temperature. The ammeter, voltmeter, and thermocouple signals were monitored and recorded via an interfaced personal computer. The ethanol sensing properties of the device were observed by the resistance change under an ethanol vapor atmosphere at different operating temperatures. Since our goal was to apply as an alcohol breath analyzer, the ethanol vapor at various concentrations was generated from ethanol solutions using alcohol simulator (GUTH Laboratories, Inc., Harrisburg, PA). The function of this alcohol simulator was to simulate alcohol concentration at conditions similar to exhaled human breath, being varied for ethanol concentration of 50–1,000 ppm for this experiment. A set up for the ethanol sensing characteristic measurement was illustrated in Fig. 7. The sensor sensitivity is defined by the ratio of the electrical resistance of the sensor in air and the electrical resistance of the sensor in ethanol vapor.

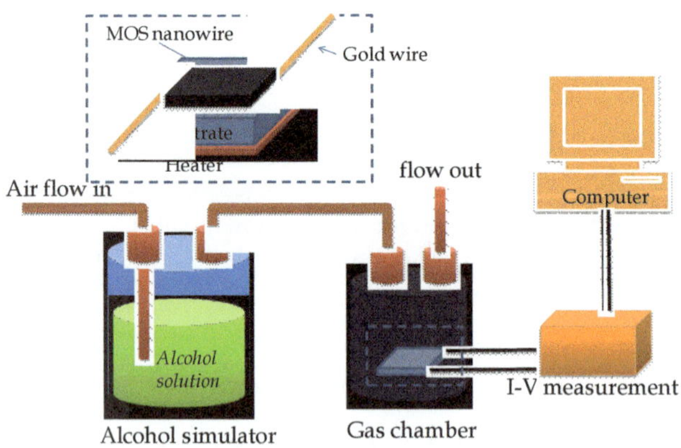

Figure 7. Schematic diagram of a set up for the ethanol sensing characteristic measurement.

3.3. ZnO and CuO nanowires

For the synthesis of ZnO nanowires, it was found that the nanowires exhibit the belt-like or wire-like structures with sharp tip. The diameter and length were in the range of 100-500 nm and 2-7 μm, respectively. For the synthesis of CuO nanowires, the nanowires having a diameter of 100–400 nm and the length of around several micrometers were observed. The different morphologies of ZnO and CuO nanowires were shown in Fig 8. It can be seen that Fig 8 (a) and (b) were for the thick film morphologies. No nanowires were observed in this case representing as the bulk case. Fig 8 (d) and (f) were for vertical aligned and non-vertical aligned ZnO nanowires, respectively. The vertical aligned ZnO nanowires were prepared by zinc screening technique but non-vertical alignment of ZnO nanowires were prepared by zinc evaporation technique. And finally Fig 8 (g) was for the CuO nanowires (Raksa et al., 2005).

Figure 8. FE-SEM images of bulk ZnO (a, b), ZnO nanowires (c-f), and CuO nanowires (g).

3.4. Response and recovery curve

The resistance response and recovery characteristics of ZnO and CuO gas sensor that exposed to ethanol concentration in range of 50-1,000 ppm at the optimum operating are shown in Fig. 9 (a) and (b), respectively. At the beginning, the measured resistance is steady in air atmosphere and then, the resistance of ZnO sensor drops sharply in ethanol atmosphere and recovers to the initial value after removing ethanol vapor. While the resistance of CuO sensor rapidly rises in ethanol atmosphere and falls back after removing ethanol. The decrease and increase of resistance under the ethanol atmosphere is due to the n-type and p-type conducting of ZnO and CuO, respectively as discussed earlier.

The sensitivity of ZnO sensors with different sizes and morphologies at different ethanol concentration are summarized in Table 2. The sensitivity of sensor is defined as R_a/R_g (reverse for CuO case) when R_a was the resistance of sensor in air and R_g was the steady resistance of sensor in ethanol. The sensor sensitivity strongly depends on size and morphology of the sensors. This size and morphology dependence will be discussed in a later section.

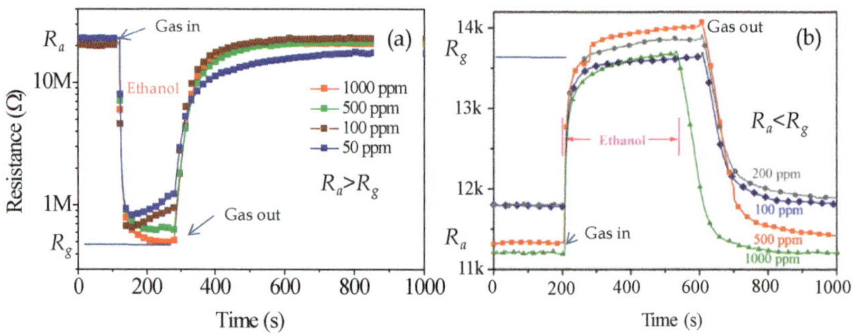

Figure 9. The response and recovery curve of n-type and p-type MOS sensor for ZnO nanowires (a) and CuO nanowires (Raksa et al., 2005) (b), respectively.

Morphology	*Diameter (nm)*	Sensitivity (50 ppm)	Sensitivity (100 ppm)	Sensitivity (500 ppm)	Sensitivity (1000 ppm)
bulk	-	1.3	1.6	2.0	2.2
Nanowire (screening)	100-500	3.1	3.9	6.4	8.7
Nanowire(evaporation)	11-104	17	24	30	40

Table 2. Sensor sensitivity of sensor with different ZnO sizes and morphologies.

3.5. Linear relation and Sensitivity enhancement based on metal oxide nanowires

Sometimes, a compact form of the sensitivity relation on gas concentration can be rewritten as

$$S_g = aX^b + 1 \tag{27}$$

$$a = \frac{\Gamma_t k_{gas}\left(\frac{\sigma_0 \Phi V_m}{V_s}\right)^b}{n_0} \tag{28}$$

where a is a controllable parameter and X is a concentration of gas. At the optimum operating temperature condition, the relation between the sensitivity and gas concentration can be rewritten as:

$$\log(S_g - 1) = \log a + b \log X. \quad (29)$$

It can be seen that $\log(S_g\text{-}1)$ has a linear relation with log X having a slope of b value. Thus, b value which represents oxygen ion species on the surface of MOS sensors can be obtained from a slope of a plot between $\log(S_g\text{-}1)$ and log X.The value of the constant b is normally around either 0.5or 1, depending on the charge state of the surface ion. The adsorbed oxygen ion is $O^{2-}$for b of 0.5, the adsorbed oxygen ion is O^- for b of 1.

The plots between log $(S_g\text{-}1)$ and log X of some works as listed in Table 1 was displayed in Fig 10. It was found that the value b of all sensors is close to 0.5 suggesting that the adsorbed surface oxygen species on ZnO sensor is O^{2-}. In addition, this suggests that the oxygen adsorption species on the surface is independent on the size and morphology of ZnO regardless of bulk, microstructure or nanostructure.

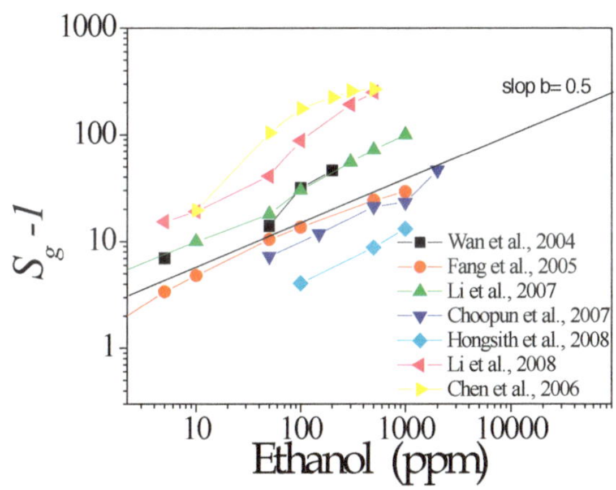

Figure 10. Plot of sensor response (*Sg-1*) and ethanol concentration in log scale for ethanol sensor based on different ZnO morphologies. The linear line in the graph has a slope of 0.5.

3.6. Sensor sensitivity

In the section 2 we have explained the sensing formula for n-type and p-type MOS sensor. In this section, we will explain the sensing parameters for understanding in the mechanism of resistance change under air atmosphere and gas atmosphere and will discuss in term of the size and morphology dependence on the sensor sensitivity.

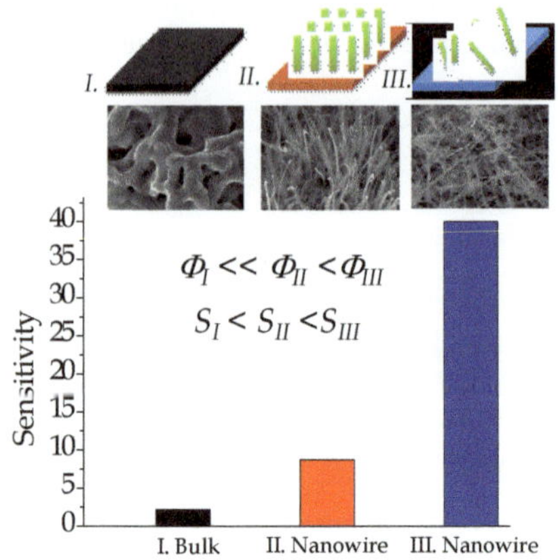

Figure 11. Sensor sensitivity for different ZnO sizes and morphologies with ethanol concentration of 1,000 ppm.

Let consider a sensitivity of MOS gas sensor based on the different sizes and morphologies as seen in Fig. 11, it can be divided into three cases of MOS ethanol gas sensor.

3.6.1. Bulk ZnO

From Table 2, the sensitivity of a bulk ZnO sensor shows the lowest value compare with that of nanostructures. It can be simply explained by using Eq. (8). The sensing parameter of the surface-to-volume ratio is low for the bulk sensor and thus, the oxygen ion surface density is low resulting in low sensitivity.

3.6.2. Large size nanowires

From Table 2, it can be seen that the sensitivity of MOS sensor based on vertical aligned nanowires (zinc screening nanowires) is higher than that of bulk. The sensitivity is 8.7 for vertical aligned nanowires sensor at 1,000 ppm of ethanol compared to that of about 2 for bulk ZnO gas sensor. The higher sensitivity of MOS sensor based on vertical aligned nanowires can be explained in terms of surface-to-volume ratio parameter (Φ) by using Eq. (11) or (15). Since Φ is proportional to sensitivity, the larger Φ due to nanostructure gives the higher sensitivity for the case of vertical aligned nanowires even though the nanowires are not aligned along the conducting direction.

3.6.3. Small size nanowires

It can be seen in Table 2 that the sensitivity of MOS sensor based on non-vertical aligned nanowires (zinc evaporation nanowires) is even higher than that of zinc screening nanowires (the sensitivity increases from 8.7 to 40). This is due to the combination effects of the surface-to-volume ratio and the depletion layer and can be explained by using Eq. (15). The non-vertical aligned nanowires exhibit smaller diameter (11-104 nm) indicating higher surface-to-volume ratio. Also, the diameter of nanowires is comparable to the depletion layer width (two times of Debye length; about 10 nm for ZnO) indicating strong effect of the depletion layer. Thus, the surface-to-volume ratio parameters (Φ) increases and the difference between diameter and the depletion layer width (D-$2L_d$) decreases resulting in the increase of the sensor sensitivity as indicated in Eq. (15).

3.7. Ultra high gas sensing by adding novel metal

In addition, the sensor sensitivity can be further enhanced up to several folds by using adding novel metal technique.There have been several reports on sensitivity enhancement due to metal doping or metal adding effect. We have also reported on the sensitivity enhancement due to gold nanoparticles (AuNPs) on the surface of ZnO (Wongrat et al., 2012).As shown in Fig 12, the sensitivity of ZnO:AuNPs exhibits maximum value of 478 at ethanol concentration of 1,000 ppm while the sensitivity of pure ZnO nanostructure sensor is about 40.

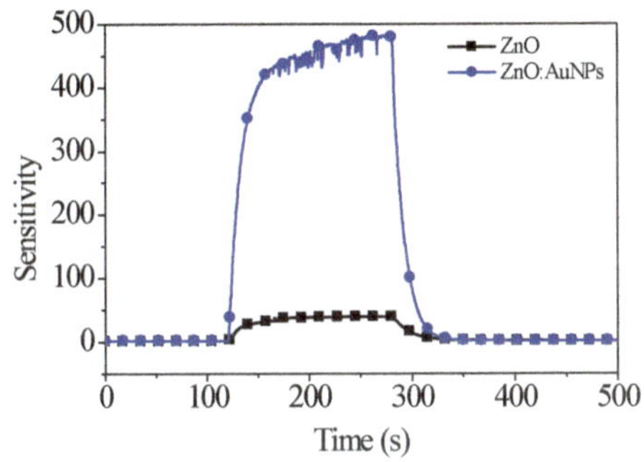

Figure 12. Ultra high sensitivity of ZnO sensor withAuNPs on the surface at ethanol concentration of 1,000 ppm (Wongrat et al., 2012).

The metal adding effect can be simply explained in our model by using Eq. (15), (17) and (26). It can be seen that the sensitivity is proportional to the reaction rate constant, k_{gas} and k_{Oxy}, through the oxygen density. Basically, the reaction rate constant can be affected by noble metals (such as Au, Pd, Pt etc.) in MOS sensors due to the catalytic effect. Thus, metal adding causes an increasing of reaction rate constant and resulting in enhancement of the sensor sensitivity.

The evidence for having higher k_{gas} can be observed from the sensor resistance in the ethanol ambient. The sensor resistance under ethanol ambient of ZnO:AuNPs sensor is about 200 kΩ which is less than of pure ZnO nanostructure sensor(500 kΩ) suggesting higher k_{gas}. In addition, the evidence for having higher k_{Oxy}, can be observed from the sensor resistance in air. The highest resistance in air of ZnO:AuNPs sensor is about 100 MΩ which is more than of resistance of pure ZnO nanostructure sensor (20 MΩ).The increase of resistance in air suggests that the gold metal catalytically activates the dissociation of molecular oxygen due to higher reaction rate that results in increasing an amount of the oxygen adsorption molecules. Therefore, electrons in ZnO nanowire are captured by oxygen adsorption to form oxygen ion and hence ZnO nanowire loses more electrons and causes the larger depletion layer at ZnO surface resulting in higher resistance.

Therefore, the sensitivity enhancement of ZnO sensors with noble metal is due to catalytic effect in both ethanol adsorption reaction k_{gas} and oxygen adsorption reaction k_{Oxy} through the oxygen density. In the other words, the ultra-high sensitivity of ZnO sensors with noble metal can be explained by an increase of both parameters of k_{gas} and w (depletion layer width) as described in Eq. (26).

4. Summary

The MOS nanostructures, especially nanowires have been demonstrated to be an excellent candidate as a gas sensing device. The gas sensors based on MOS nanowires are promising due to feasibility for ultrahigh sensitive sensors or ppb-level sensors. The simple and low-cost technique of thermal oxidation technique is effectively used to prepare the MOSs nanowires. The sensing characteristics of the MOS sensors strongly depend on sizes and morphologies of ZnO nanostructures. The sensitivity formulas have been developed in order to explain all circumstances of gas sensors based on MOS nanowires. The size and morphology dependences on the sensitivity have been explained in terms of the two important parameters including surface-to-volume ratio and depletion layer width. The experimental sensing characteristics of ZnO sensors are in good accordance to the developed formulas. This suggests that the sensing formulas are a powerful tool in sensor design and also can be applied for sensors based on other MOSs such as SnO_2, TiO_2, MoO_3 or WO_3, etc. In addition, the sensor sensitivity can be further enhanced up to several folds by using adding novel metal technique and can be simply used the developed sensitivity formula for explanation in term of the depletion layer parameter.

Symbols used

k Reaction rate

X Density, Concentration, of X

n, n_0 Electron carrier density

Γ Time constant

R_a Resistance in air

R_g Resistance in active gas

S_g Sensitivity

Φ Surface to volume ratio

σ, σ_0 Oxygen ion surface density

V_m Volume of material

V_s Volume of system

C_g Ethanol concentration

L_d Debye length

ε Static dielectric constant

q Electrical charge of carrier

k_B Boltzmann constant

T Temperature (K)

Acknowledgements

This work was supported by Thailand Research Fund (TRF). Also, we would like to thank P. Raksa for useful discussion.

Author details

Supab Choopun[1,3], Niyom Hongsith[2,3] and Ekasiddh Wongrat[2]

1 Department of Physics and Materials Science, Faculty of Science, Chiang Mai University Chiang Mai, Thailand

2 School of Science, University of Phayao, Phayao, Thailand

3 ThEP Center, CHE, Bangkok, Thailand

References

[1] Rothschild, A, Komem, Y, & Appl, J. Phys., (2004).

[2] Li, C. C, Du, Z. F, Li, L. M, Yu, H. C, Wan, Q, & Wang, T. H. Appl. Phys.Lett., (2007).

[3] Li, C. C, Du, Z. F, Yu, H. C, & Wang, T. H. Thin Solid Films, (2009).

[4] Comini, E, Faglia, G, & Sberveglieri, G. Zhengwei Pan and Zhong L. Wang, Appl. Phys.Lett., (2002).

[5] Comini, E, Faglia, G, & Sberveglieri, G. Electrical-Based Gas Sensing." in Solid State Gas Sensing, E. Comini, G. Faglia, and G. Sberveglieri (Ed), Springer Science+Business Media, LLC, New York, (2009). , 47.

[6] Wongrat, E, Pimpang, P, & Choopun, S. Appl. Surr. Sci., (2009).

[7] Wongrat, E, Hongsith, N, Wongratanaphisan, D, Gardchareon, A, & Choopun, S. Sens. Actuators B, (2012).

[8] Neri, G, Bonavita, A, Micali, G, Donato, N, Deorsola, F. A, Mossino, P, Amato, I, & Benedetti, B. D. Sensor. Actuator B, (2006).

[9] Zhang, H. N, Li, Z. Y, Liu, L, Xu, X. R, Wang, Z. J, Wang, W, Zheng, W, Dong, B, & Wang, C. Sensor. Actuator B, (2010).

[10] Son, J. Y, Lim, S. J, Cho, J. H, & Seong, W. K.. and Hyungjun Kim,Appl. Phys.Lett., (2008).

[11] Xu, J. Q, Zhang, Y, Chen, Y. P, Xiang, Q, Pan, Q. Y, & Shi, L. Y. Mater. Sci. Eng. B, (2008).

[12] Arshak, K, & Gaidan, I. Mater. Sci. Eng. B, (2005).

[13] Bie, L. J, Yan, X. N, Yin, J, Duan, Y. Q, & Yuan, Z. h. Sens. Actuators B, (2007).

[14] Hongsith, N, Viriyaworasakul, C, Mangkorntong, P, Mangkorntong, N, & Choopun, S. (2008).

[15] Hongsith, N, Wongrat, E, Kerdcharoen, T, & Choopun, S. Sens. Actuators B, (2010).

[16] Wang, N, Cai, Y, & Zhang, R. Q. Mater. Sci. Eng. R-Rep., (2008).

[17] Yamazoe, N, & Shimanoe, K. Sens. Actuators B, (2008).

[18] Feng, P, Wan, Q, & Wang, T. H. Appl. Phys.Lett., (2005).

[19] Raksa, P, Gardchareon, A, Chairuangsri, T, Mangkorntong, P, Mangkorntong, N, & Choopun, S. Ceramics International, (2009).

[20] Wan, Q, Li, Q. H, Chen, Y. J, & Wang, T. H. Appl. Phys.Lett., (2004).

[21] Choopun, S, Hongsith, N, Mangkorntong, P, Mangkorntong, N, & Physica, E. (2007).

[22] Sysoev, V. V, Schneider, T, Goschnick, J, Kiselev, I, Habicht, W, Hahn, H, Strelcov, E, & Kolmakov, A. Sens. Actuators B, (2009).

[23] Lee, Y. C. Hui Huang, O.K. Tan, and M.S. Tse, Sens. Actuators B, (2008).

[24] Chen, Y. J, Zhu, C. L, & Xiao, G. Sens. Actuators B, (2008).

[25] Chen, Y. J, Zhu, C. L, & Xiao, G. Nanotechnology, (2006).

[26] Liu, Y. X, Gao, C. T, Pan, X. J, An, X. Y, Xie, Y. Z, Zhou, M, Song, J, & Zhang, H. L. Z.Y.

[27] Liu, Q, Zhao, Y. H, & Zhang, E. Q. Xie, Appl. Surr. Sci., (2010).

[28] Song, X. F, Wang, Z. J, Liu, Y. B, Wang, C, & Li, L. J. Nanotechnology, (2009).

[29] Yang, Z, Li, L. M, Wan, Q, Liu, Q. H, & Wang, T. H. Sens. Actuators B, (2008).

Chapter 2

Ferromagnetic ZnO Nanowires for Spintronic Applications

Usha Philipose and Gopal Sapkota

Additional information is available at the end of the chapter

1. Introduction

Dilute Magnetic Semiconductors (DMS) are expected to be the main building blocks for the realization of semiconductor spin based electronics–spintronics. The giant magnetoresistance of these materials, i.e. large resistivity changes in response to an applied magnetic field, make them suitable for applications in magnetic sensors, magnetic random access memories and spintronic devices. DMS materials have been experimentally synthesized [1, 2] and theoretically modeled [3–5] for several years, focused on the transition metal (TM) doped group III-V and II-VI semiconductors. In order to develop spintronics as a practical technology, it is necessary to explore materials whose magnetic property can be controlled by changing either the external field or the carrier concentration. From an industrial point of view, it is essential to fabricate ferromagnetic DMS with Curie temperature (Tc) above room temperature and a ferromagnetism that is intrinsic, rather than due to the presence of dopant clusters. The synthesis of DMS as nanowires is particularly attractive for the fabrication of low dimensional, spin-based electronic and optoelectronic devices. Transition metals with partially filled d states (Sc, Ti, V, Cr, Mn, Fe, Co, Ni, and Cu) have been used as magnetic atoms in DMS. The electron spin of the partially filled d states are responsible for their magnetic behavior.

The field of spintronics gained prominence when $Ga_{1-x}Mn_xAs$ was found to be ferromagnetic up to 110 K [1] and up to 173 K with a Mn concentration of 6.8% [6]. Ferromagnetism in this DMS was explained in terms of carrier-mediated exchange coupling between the localized spins of the transition metal ions. It is based on the theory that doping GaAs with Mn introduces shallow acceptors and hence holes into the valence band of GaAs. These holes are anti-ferromagnetically coupled to the local Mn moments. Dietl calculated the Curie temperature of various DMS doped with Mn [3] and found this temperature to be proportional to the concentration of impurities and to the square root of the p-type charge carrier concentration. Ferromagnetism has been predicted and confirmed in several III-V semiconductors doped with transition metal ions such as Mn, Cr, V, and Co. However, III-V semiconductors require complex processing steps and are expensive to synthesize. Moreover, obtaining T_c higher than room temperature poses a major challenge for development of III-V-based DMS.

The quest for materials with a high magnetic transition temperature continues and in this regard, TM doped II-VI semiconductors like ZnO has emerged as an attractive candidate as it is abundant, non-toxic and inexpensive to synthesize. ZnO is also a transparent, semiconducting oxide exhibiting piezoelectric and possibly ferromagnetic behavior. Dietl predicted room temperature ferromagnetism for 5% Mn in p-type wide band gap semiconductors like ZnO and GaN [3]. Sharma and coworkers reported room temperature ferromagnetism in Mn-doped ZnO thin films [2]. The author reported the first observation of room temperature ferromagnetism in Mn doped ZnO nanowires [7], and since then there have been works presenting evidence for intrinsic ferromagnetism in individual ZnO nanoparticles doped with TM ions [8]. However, recent experiments reveal inconsistent ferromagnetic properties for TM doped ZnO [9–13], leaving the real mechanism for ferromagnetism in ZnO unclear. There are several controversial and conflicting reports on TM doped ZnO; including claims that the magnetic signature arises from clustering, segregated phases [13, 14], or defect mediated ferromagnetic coupling [15, 16]. The controversy in understanding the origin of ferromagnetism in TM doped II-VI semiconductors like ZnO is because in II-VI materials, dopants like Mn behave as an isovalent impurity. When Mn^{2+} substitutes for Zn^{2+} in the ZnO lattice, it provides no additional carriers. Therefore, charge carriers would need to be provided by additional doping in order to render Mn-doped ZnO ferromagnetic. Based on the predictions by Dietl et al. [3], room temperature ferromagnetism is expected in Mn-doped ZnO in the presence of a fairly high hole concentration of $\approx 10^{20} cm^{-3}$. Such high hole concentrations seem rather unrealistic, especially, since it is difficult to achieve any reasonable concentration of holes in ZnO. The origin of experimentally observed ferromagnetism in TM doped II-VI semiconductors thus remains unclear. Kolesnik and coworkers [14] argue that the magnetic signature may arise from segregated phases of spinel oxides such as Mn_3O_4 that is ferromagnetic. Theoretical calculations predict that $Zn_{1-x}Mn_xO$ nanowires favor the ferromagnetic state if Zn vacancies are introduced, while they are anti-ferromagnetic unless additional carriers are incorporated [17].

In this chapter, we will review experimental results of observed room temperature ferromagnetism in TM doped II-VI semiconductors. Studies drawn primarily from the author's research along with similar experimental findings will be presented. The choice of nanowires (as opposed to thin films) for studying the origin of ferromagnetism in II-VI semiconductors is justified because they allow: (a) clean and smooth surfaces with no precipitates or attachments on the nanowire surface, and (b) a unique geometry for realizing spintronic devices with possible one dimensional charge transport.

2. Theory of ferromagnetism in ZnO based DMS

ZnO, a II-VI semiconductor, is a wide band gap (3.37 eV at room temperature) semi-conducting metal oxide with a large exciton binding energy of 60 meV, which makes it a promising material for optoelectronic applications. It has attracted increasing attention and stimulated intense research efforts since theoretical calculations predict Mn-doped ZnO can show ferromagnetism with a T_c value above room temperature making it a promising candidate for spintronic applications too. ZnO crystallizes in the wurtzite structure with tetrahedral sp^3 bonding. When doped with Mn to form $Zn_{1-x}Mn_xO$, Mn atoms substitute the Zn atoms in the crystal lattice. The electronic configuration of Mn is $1s^2, 2s^2, 2p^6, 3s^2, 3p^6, 3d^5, 4s^2$ and during substitution, Mn donates its $4s^2$ electrons to the sp^3 bonding and exists as Mn^{2+} ions. The half filled 3d band of Mn^{2+} has 5 electrons and it hybridizes with the p-bands of oxygen in ZnO. This results in an exchange interaction between the localized magnetic moments (3d spins) and the spins of itinerant carriers in the host ZnO. Several mechanisms have been proposed to explain ferromagnetism in DMS. These include the Zener model [18], carrier mediated exchange [3], double exchange [5, 19] and coupling via a bound magnetic polaron [20, 21].

According to the Zener model, since the Mn-d shell is only half filled, the direct interaction between d shells of adjacent Mn atoms leads to an antiferromagnetic state. However, the indirect coupling of spins through itinerant carriers tends to align the spins of the d shell electrons in a ferromagnetic manner. Ferromagnetism is therefore assumed to occur due to interactions between the local moments of the Mn atoms, mediated by free holes in the material. Based on this model, TM-doped p-type ZnO is predicted to be a promising candidates for ferromagnetic DMS with high Curie temperature. However, these predictions are based on the incorporation of $\approx 5\%$ TM element into the host and hole concentrations of above $10^{20} cm^{-3}$. Considering that p-type doping has always been a challenge in II-VI semiconductors, the achievement of such high hole concentrations is not realistic.

The Zener double exchange was used to explain ferromagnetism in TM doped InAs [19] and is an indirect coupling mechanism between two neighboring TM ions. According to this theory, if neighboring TM magnetic moments are in the same direction, there is a widening of the TM-d band due to hybridization of the spin states. The band energy of the ferromagnetic state can be lowered below that of the antiferromagnetic state if a sufficient number of holes (or electrons) exist. The 3d electrons of the TM ions hop to the 3d orbitals of neighboring TM ions with parallel magnetic moments, thus lowering its energy by hopping in the ferromagnetic state.

In all of the proposed mechanisms, the impact of charge state and the involved electronic 3d-shell configuration of TM ions are crucial points in understanding the magnetic mechanism within the diluted magnetic semiconductors. Experimental investigations have shown that Mn-substituted ZnO samples exhibit distinct magnetic properties including spin-glass behavior [22]; paramagnetism [23]; ferromagnetism with T_c of 45 K [24]; ferromagnetism with T_c of 425 K in thin films of Mn doped ZnO [2] and $T_c > 400$ K for Mn doped ZnO nanowires [7]. Structural defects have also been reported to play an important role in the occurrence and stability of ferromagnetism [25]. To interpret the origin of the magnetic properties, many magnetic interaction mechanisms have been proposed, including direct super-exchange, indirect superexchange, and carrier-mediated exchange. In spite of all these works, the local structure and microscopic origins of the high-T_c ferromagnetism in ZnO, GaN, and Si-based DMSs are still under debate. This debate has emerged as one of the most important challenges in understanding the science of DMS.

3. Ferromagnetic II-VI nanowires

The attractive features in using II-VI semiconductors for fabricating DMS are several: interesting opto-electronic properties, high solubility of transition metal atoms in the host lattice (implies that the amount of injected spins can be large), and also because II-VI materials like ZnO are cheap, abundant in nature and are also environmentally friendly. The ternary nature of the $A^{II}_{1-x}Mn_xB^{VI}$ alloys, allows the possibility of 'tuning' the lattice constant and band parameters by varying the composition of the material. The substitutional Mn atoms in the $A^{II}B^{VI}$ lattice are also characterized by highly efficient electroluminescence, which makes dilute $A^{II}_{1-x}Mn_xB^{VI}$ alloys important in optical flat panel display applications. However, since Mn atoms do not provide any carriers when they are substituted in the insulating host $A^{II}_{1-x}Mn_xB^{VI}$ lattice, it was believed for a long time that the magnetic interaction in II-VI DMS is dominated by antiferromagnetic exchange among the Mn spins, which would result in paramagnetic or antiferromagnetic behavior of the material. The prediction of room temperature ferromagnetism in TM doped II-VI semiconductors initiated intense theoretical and experimental work on II-VI materials, with several contradicting results. Of the II-VI semiconductors, ZnO is particularly interesting because it has a smaller lattice constant compared to other IIŰVI semiconductors, which will serve to increase the magnetic coupling strength among the magnetic ions.

3.1. Synthesis and structural characterization of $Zn_{1-x}Mn_xO$ nanowires

The $Zn_{1-x}Mn_xO$ nanowires were grown on Si substrates by the vapor phase growth, using the VLS mechanism [7]. The growth temperature was $700°C$. Mn doping was achieved by using MnO_2 as the dopant material, which was evaporated at $700°C$. A mixture of ZnO and graphite was used as the source material, evaporated at $900°C$. After growth for an hour, the Si surface was found coated with a grayish colored substance. The morphologies of this product was investigated by a scanning electron microscope (SEM) using Hitachi SEM S-5200, and a high resolution transmission electron microscope (HRTEM) using Hitachi STEM HD-2000. Magnetic properties of these samples were studied by superconducting quantum interference device (SQUID) magnetometer (MPMS MultiVu Application). The Mn content was determined by Energy Dispersive X-ray (EDX) spectroscopy and Ion Coupled Plasma Mass Spectrometry (ICPMS) using Perkin Elmer Sciex ELAN DRC II. The morphology of the $Zn_{1-x}Mn_xO$ nanowires was observed using SEM, which revealed that the nanowires have lengths in the range of 8–10 μm and diameter between 40–150 nm. The crystallinity of the $Zn_{1-x}Mn_xO$ nanowires as evidenced by HRTEM is shown in Figure 2. The image clearly reveals the lattice fringes of ZnO $\{0001\}$ planes with an interplanar spacing of about 0.52 nm, indicating that the nanowire is a single crystal and grows along the [001] direction, as shown in Figure 2. No observable change in structure was noted between pure and Mn doped nanowires. All the wires were straight and had a smooth surface, which also confirms that Mn is doped into the lattice structure instead of being precipitated as a second phase.

The composition of the nanowires was characterized by EDX at nanometer resolution. All samples show uniform distribution of Zn, O and no other impurities other than Mn (Figure 3. The signal from Si is due to the substrate. Due to the limited sensitivity of EDX at high resolution, Mn atoms were detected only in samples with more than 2% Mn concentration although ferromagnetism was observed in all Mn-doped samples as discussed below. The elemental mapping (EDX) spectra show a highly uniform distribution of Mn atoms with

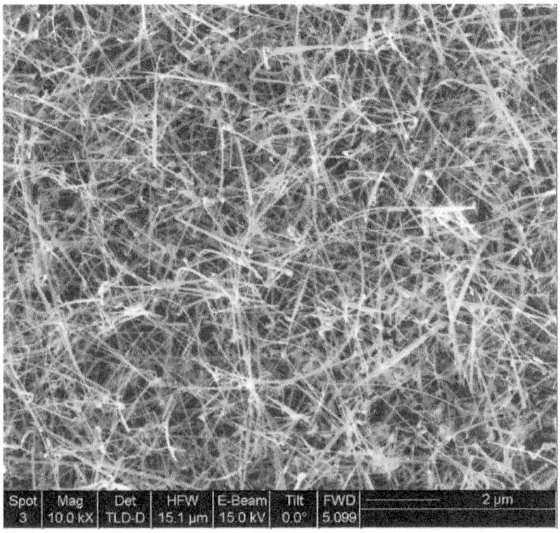

Figure 1. SEM image of $Zn_{1-x}Mn_xO$ nanowires growing in a 'weed-like' manner on Si substrate.

Figure 2. HRTEM image of a crystalline $Zn_{1-x}Mn_xO$ nanowire. The nanowire has smooth surface and there is no evidence of any secondary phases or defects on these nanowires.

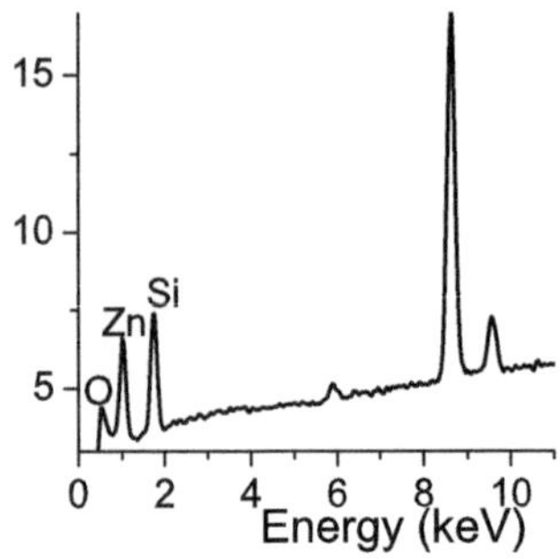

Figure 3. EDX spectra for $Zn_{1-x}Mn_xO$ nanowires, with $\sim 2\ at\%$ Mn.

no clustering or segregated Mn-rich phases. As the growth method does not allow us to precisely control the Mn concentration and the sensitivity of EDX is limited, ICPMS (a destructive chemical analysis method) was done to estimate the Mn concentration. The nanowires were first dissolved in HNO_3 and subsequently introduced into an extremely hot (7,000 to 10,000 K) radio frequency induced argon plasma. The plume of the argon plasma is sampled directly into the entrance orifice of a quadrupole mass spectrometer. Analyte ions produced by the plasma are accelerated, magnetically separated and counted using an electron multiplier. In all samples the Mn concentration was measured to be in the range of 1% to 4%.

3.1.1. *SQUID measurements of magnetization on $Zn_{1-x}Mn_xO$ nanowires*

Magnetization measurements, using a SQUID magnetometer (MPMS MultiVu Application) showed a distinct hysteresis loop superimposed on a diamagnetic contribution from the substrate (Figure 4).

Figure 5 shows the variation of magnetization (M) with temperature T, measured during cooling in an applied magnetic field of 100 Oe. The magnetization decays very slowly with T over the range studied. T_c is well above room temperature, but it is hard to determine the exact value, since the value of T_c is rather high, exceeding the range of measurements. The same conclusion can be reached from the dependence of remnant magnetization (M_r) and coercive field (H_c) on temperature (Figure 5). M_r and H_c are 2.44×10^{-6} e.m.u and 37.05 Oe respectively at 300 K.

3.1.2. *Estimation of magnetic moment per Mn atom in $Zn_{1-x}Mn_xO$ nanowires*

The $Zn_{1-x}Mn_xO$ nanowires were grown on Si substrates of area 9 mm^2. The nanowires had an average length of 10 μm and hence the maximum volume of nanowires is estimated to be $9 \times 10^{-5}\ cm^3$. For Mn concentration varying from 1 $at\%$ to 4 $at\%$, the number of Mn atoms (N_{Mn}) is estimated to vary from $2 \pm 0.8 \times 10^{15}\ atoms/cm^3$ to $8 \pm 3.2 \times 10^{15}\ atoms/cm^3$. The magnetic moment per Mn atom (in terms of the Bohr magneton) was determined by the equation:

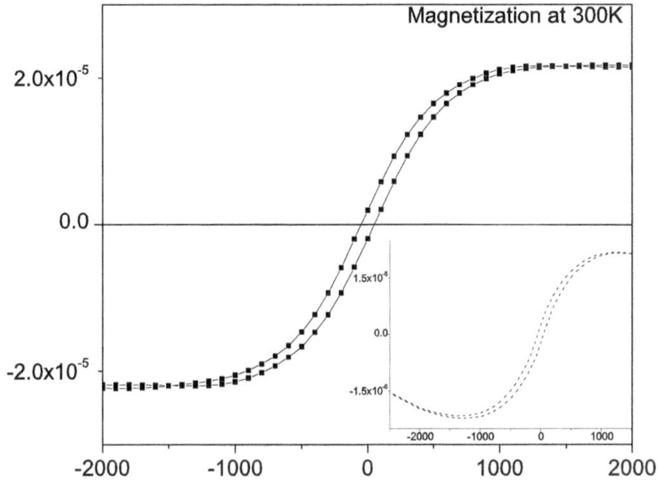

Figure 4. Hysteresis loop (M-H curve) of $Zn_{1-x}Mn_xO$ nanowires at a temperature of 300 K; inset: the as-obtained data from the SQUID measurements including the diamagnetic contribution arising from the Si substrate.

$$\frac{m}{\mu_B} = \frac{M_s}{N_{Mn}(9.27 \times 10^{-24})} \tag{1}$$

where M_s is the saturation magnetization. From the observed saturation magnetization (Figure 4), M_s is of the order of $2.0 - 3.0 \times 10^{-5}$ e.m.u ($2.0 - 3.0 \times 10^{-8}$ $A.m^2$), and the magnetic moment per Mn atom is estimated to be in the range of 0.30 μ_B to 1.2 μ_B, which is smaller than the maximum spin moment (5 μ_B) per Mn^{2+} with S = 5/2 and g=2; but is larger than those reported in previous reports [2].

3.1.3. Optical characterization of Mn doped and undoped ZnO nanowires

The luminescence spectra of ZnO exhibits strong mid-gap green emission, which has been occasionally attributed to Zn interstitials [26] and Zn vacancies [27], the most widely accepted origin of this emission is oxygen vacancies [28, 29]. The $Zn_{1-x}Mn_xO$ and undoped ZnO nanowires were characterized by room temperature photoluminescence (PL) measurements.

The wires were excited by a 355 nm wavelength laser at room temperature. The photoluminescence spectrum shown in Figure 6 shows a very weak band edge luminescence from exciton (UV emission) at about 380 nm and a strong mid-gap emission (green emission) at about 525 nm. The ZnO band edge emission is extremely weak possibly because of strong

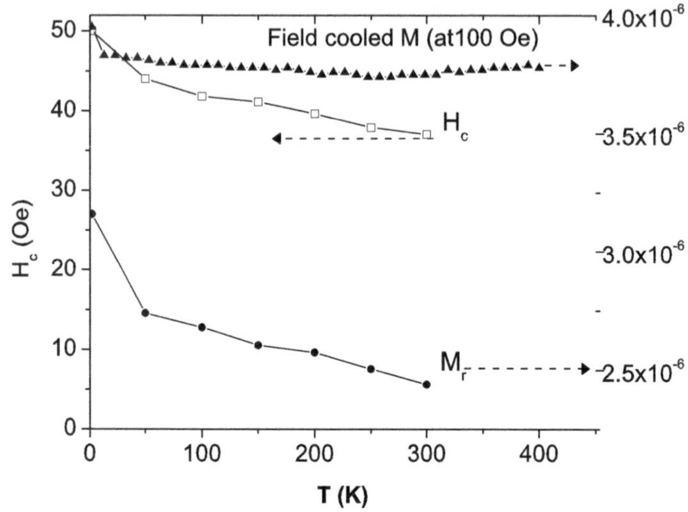

Figure 5. Dependence of remnant magnetization (M_r), coercive field (H_c) and field cooled magnetization (M-T at 100 Oe) of $Zn_{1-x}Mn_xO$ nanowires on temperature.

non-radiative processes due to the large density of surface states in a nanowire. When doped with Mn, a strong suppression of the mid-gap emission is observed followed by a substantial increase in the band-edge luminescence. Although most Mn-doped II-VI materials show a manganese specific orange emission no such luminescence was observed in Mn doped ZnO nanowires. This is consistent with previous reports on Mn-doped ZnO [30, 31], but the reason for this absence of Mn-related emission is not clear. Although, the strong suppression of the oxygen vacancy related emission could be due to Forester type dipole-dipole energy transfer, the small oscillator strength of Mn absorption would make such a process very inefficient. The suppression of mid-gap emission in Mn-doped II-VI materials implies a more efficient direct capture of photo excited carriers before getting trapped by states responsible for mid-gap emission. However, such a process would lead to suppression of band-edge luminescence as well, contrary to the observation of enhanced band-edge emission. These results therefore indicate that there is an interaction between Mn doping and native defects like oxygen vacancies. The mechanism of such an interaction is not clear and deserves further study.

3.1.4. Discussion

Unlike previous reports, the observed ferromagnetism can be attributed only to the Mn doped into the ZnO lattice structure since HRTEM revealed a clean and smooth surface of the nanowire and no precipitates or attachments to the wire surface were observed. No ferromagnetic behaviour was observed for pure ZnO nanowires. Moreover, ferromagnetic impurity phases responsible for the observed magnetic ordering can be ruled out since there

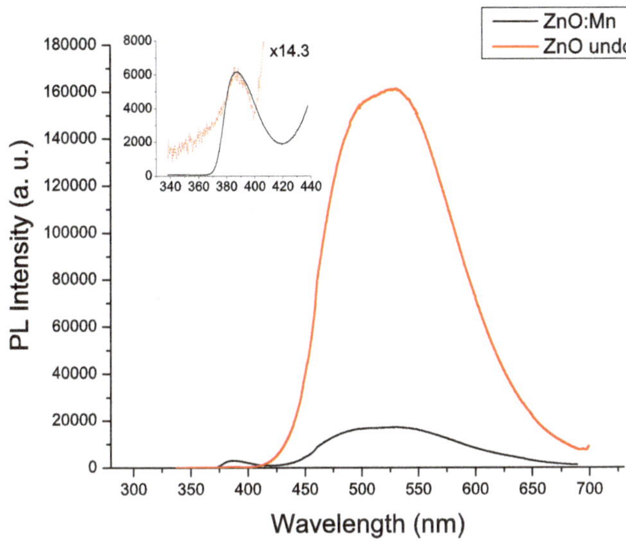

Figure 6. Room temperature photoluminescence spectra of undoped and $Zn_{1-x}Mn_xO$ nanowires.

is no known phase of manganese oxide with above room temperature T_c. All the possible MnO_x phases that could be forming in our growth process are antiferromagnetic, with the exception of Mn_3O_4, which is ferromagnetic below 45 K. In contrast to the bulk and thin film samples studied in the past we have single crystal nanowires and no evidence for multiple phases. The nanowires studied here are not intentionally doped. In general, undoped ZnO tends to be n-type due to intrinsic defects. Although the large density of surface states and the VLS mechanism of nanowire growth can potentially make them p-type, we do not have any evidence for substantial hole concentration in these samples. In order to estimate the conductivity of the nanowire array, a sample for conductivity measurements was prepared following the procedure described in Reference [32]. Conductivity measurements on ZnO nanowire arrays show that these wires are conducting but the observed resistance is more than 1 GΩ per wire implying very low mobile carrier concentration. However, carrier mediated exchange interaction only requires itinerant carriers that hybridize with the Mn 3-d band so that hopping of localized spins between Mn ions can stabilize the parallel spin configuration. The material may still remain semi-insulating or show hopping conductivity [33]. Further, conduction through nanowires are known to be very sensitive to charge inhomogeneities along the length of the wire [32]. Even though ferromagnetism is predicted and expected in p-type Mn doped ZnO through strong p-d hybridization [3], Sato et al. [5] predicted that the ferromagnetic state of Co^{2+}(d7) in Co-doped ZnO can be stabilized by $s-d$ hybridization, thus making it possible to obtain high Curie temperature ferromagnetic materials in n-type ZnO.

3.2. Synthesis and characterization of $Zn_{1-x}Fe_xO$ nanotubes

Sato et al.,[34] used first principles calculations to predict that the ferromagnetic state is stable in V-, Cr-, Fe-, Co- or Ni-doped ZnO without any additional carrier doping treatments. It is believed that (Zn,Fe)O is a promising candidates for high T_c ferromagnets, and there are reports of ferromagnetism in Fe-doped ZnO nanocrystals [35], and with co-doping [36, 37]. In this section, a recent work in the author's research group[38] on the synthesis of $Zn_{1-x}Fe_xO$ nanotubes with clear evidence of Fe ions substituting for Zn^{2+} ions in the ZnO crystal lattice is presented. The Fe-doped ZnO nanotubes were synthesized by a simple electrochemical process at 75°C using $ZnCl_2$ and KCl mixed with Fe as an electrolyte. The undoped ZnO compound has near-zero or negligible magnetic moment, since all its electrons are paired. However, when ZnO is doped with Fe (which has an incomplete d subshell), it exhibits a magnetic moment that is determined by the number of unpaired electrons in the doped compound. The determination of the magnetic moment was done using a technique similar to the Gouy method [39]. In this technique, an Evans balance was employed to measure the change in current required to balance suspended magnets when their magnetic field interacts with a magnetic sample. This value was then used to determine the effective magnetic moment. Magnetization measurements on the $Zn_{1-x}Fe_xO$ nanotubes have not been completed and so no claims can be made on the ferromagnetic property of these nanostructures.

Synthesis of ZnO nanotubes was done in a three electrode standard electrochemical cell with a Si(111) substrate coated with 300 nm Au film as the cathode and the working electrode. The counter electrode was made of Pt. The electrolyte comprised of a mixture of $FeCl_3$ (0.1 mM) + $ZnCl_2$ (6 mM) + KCl (0.2 M)]. Synthesis was carried out in a temperature-controlled water bath maintained at 75°C . Oxygen (20 sccm) was continuously bubbled through the electrolyte during the growth process. A constant bias of -1 V was maintained between the working and reference electrodes during the growth process. Post growth, the nanostructures were characterized by SEM (NOVA dual-beam SEM/FIB, FEI Nova 200 Nano-Lab), HRTEM (FEI Tecnai G2 F20 S-Twin 200 keV), and selected-area electron microscopy (SAED) pattern analysis. Composition and structural analysis were further done by EDX and x-ray diffraction (XRD) (Rigaku Ultima III). The hexagonal ZnO nanotubes had wall thickness of 20 nm and pore diameter in the range of 80 nm to 120 nm, as shown in Figure 7(a) and (b). EDX analysis of the nanotubes showed a Fe peak, confirming a dilute concentration of Fe ($< 4wt.\%$) in the nanotubes. Electrolyte concentration, growth time, and growth temperature play a vital role in determining the morphology of the nanostructures. The first stage of the synthesis process was the growth of nanorods, which was followed by etching of the nanorods along a specific plane. The etching process was facilitated by the high concentration of KCl in the electrolyte.

3.2.1. XRD measurements on $Zn_{1-x}Fe_xO$ nanotubes

Figure 8 shows results of XRD measurements on undoped and Fe doped ZnO nanotubes. The diffraction peaks have been indexed to the hexagonal ZnO wurtzite structure. The strong peak at $2\theta = 34.36^0$ corresponds to the (0002) c-axis of ZnO, which indicates that the nanotubes preferentially grow along (0001) direction. No additional peak attributed to Fe or its oxide could be detected in the XRD spectrum of Fe-doped ZnO nanotubes, though a small shift of the (0002) peak was observed. This shift of the (0002) peak position for

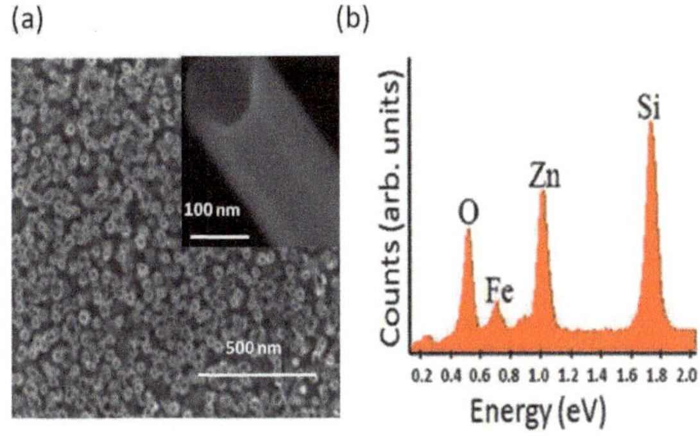

Figure 7. SEM image of nanotubes: (a) $Zn_{1-x}Fe_xO$ nanotubes with wall thickness of about 20 nm and pore diameter of about 80 nm to 120 nm. Inset shows a magnified SEM image of a single Fe-doped ZnO nanotube. (b) EDX spectrum confirming the presence of Fe, Zn, and O. The Fe concentration is estimated to be about 3-4 wt.%.

the Fe-doped ZnO indicates that there is an increase in the c-axis lattice parameter for the Fe-doped sample, indicative of the fact that Fe ions substitute for Zn^{2+} ions in the ZnO lattice, without changing its wurtzite crystal structure. The ionic radius of Fe^{2+} is larger than that of Zn^{2+} by about 5%. Hence, when Fe^{2+} ions occupy the Zn^{2+} sites of ZnO, the in-plane atomic arrangement with closed atomic packing will be strained due to the substitution of Zn by the larger Fe atom. This cause an increase in the c-lattice constant and a shift of the (0002) peak to smaller angle in the XRD pattern of the Fe-doped ZnO nanotubes. It is also possible that Fe^{2+} and Fe^{3+} ions are both present and occur as substitutional as well as interstitial dopants. In this case, the atomic arrangement is disturbed and the number of interstitials will increase, which will result in lattice disorder and strain. A comparison of the XRD spectrum of the undoped and doped nanotubes (inset of Figure 8) shows that the full-width at half maximum (FWHM) of the (002) diffraction peak is broadened (about 3% to 4%) and its intensity decreased in the Fe-doped ZnO nanotubes. This indicates that the dilute (<4 wt.%) Fe doping influences the crystalline quality of the nanotubes by causing lattice defects. This result was further validated by comparison of the PL spectra before and after doping.

3.2.2. *Optical characterization of $Zn_{1-x}Fe_xO$ nanotubes*

The nanostructures were optically characterized by photoluminescence (PL) spectroscopy and Raman spectroscopy. PL spectra were obtained using a TRIAX 320 spectrometer with excitation provided by a Kimmon continuous-wave HeCd laser with spot size of about 1 mm^2. The Raman spectrum was obtained using an Almega XR 532 nm (green) laser.

Figure 9 shows the room-temperature PL spectra for the undoped and Fe-doped ZnO nanotubes. The curves with solid lines and symbols represent the experimentally obtained data for the undoped and Fe-doped ZnO nanotubes.

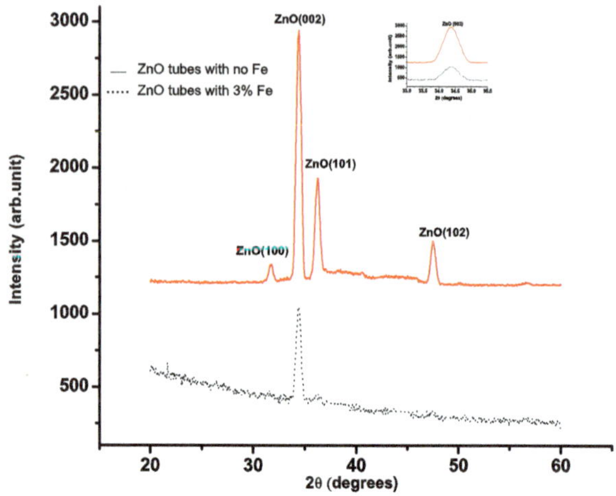

Figure 8. XRD spectrum of undoped (solid line) and Fe-doped (dotted line) ZnO nanotubes. Inset shows the broadening of (0002) peak for Fe-doped nanotubes.

The broad defect peak in the Fe-doped sample was de-convoluted using a Gaussian distribution (shown by dashed curves). The PL spectra for undoped ZnO tubes showed strong band-edge emission at 370.5 nm (3.35 eV) and no other peaks, attesting to the high crystalline quality of the undoped nanotubes. However, when doped with Fe, the PL spectrum showed an ultraviolet (UV) emission peak at $\approx 381nm$ (red-shifted) and a

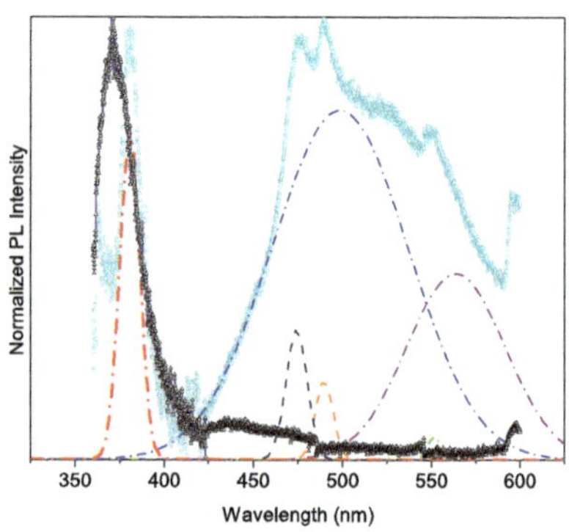

Figure 9. Room-temperature PL spectra of undoped and Fe-doped ZnO nanotubes. The normalized curves with solid lines and symbols represent the experimentally obtained data. The curves shown by dotted lines represent the Gaussian fit of PL spectra for Fe-doped nanotubes.

broad emission peak in the visible region. The red-shift of the band edge in Fe-doped II-VI semiconductors has been attributed to the $sp-d$ spin exchange interactions between the band electrons of the host and localized d electrons of the Fe ion substituting the group II ion [40]. A Fe^{2+} cation is in $3d^6$ configuration. The sŰp and pŰd exchange interactions between ZnO as host and Fe as dopant give rise to a negative and positive correction to the conduction- and valence-band edges, leading to narrowing of the band gap. Thus, the red-shift and quenching of the UV emission peak in the doped sample provide further evidence of the fact that Fe substitutes for Zn in the ZnO crystal [41]. The broad defect-related emission peak was deconvoluted into Gaussian peaks at 474 nm, 489 nm, 499 nm, 551 nm, and 563 nm. The peak observed at 474 nm is attributed to interstitial Zn and oxygen. The peaks at 489 nm and 499 nm are attributed to singly ionized oxygen vacancies, whereas the peaks at 551 nm and 563 nm are attributed to oxygen interstitials.

Figure 10(a) and (b) compare the Raman spectrum for the undoped and Fe-doped ZnO nanotubes. Both samples show a Raman shift at 521 cm^{-1} corresponding to the Si substrate used for the growth of ZnO nanotubes. The undoped sample shows a characteristic Raman peak at 438 cm^{-1}, related to the hexagonal wurtzite phase of ZnO. The Fe-doped sample shows additional peaks at 336 cm^{-1}, 384 cm^{-1}, and 644 cm^{-1}.

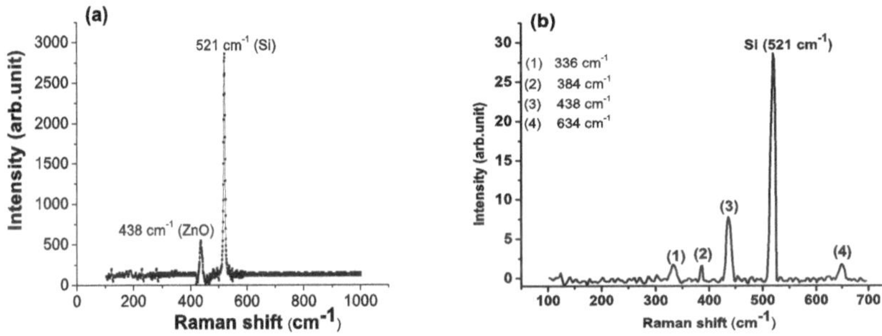

Figure 10. Raman scattering spectrum for: (a) undoped ZnO nanotubes and (b) Fe-doped ZnO nanotubes

The peaks at 336 cm^{-1} and 384 cm^{-1} are attributed to intrinsic defects in ZnO such as oxygen vacancies and zinc interstitials. The additional vibrational mode at 644 cm^{-1} in the Fe-doped ZnO nanotubes is also related to intrinsic ZnO lattice defects, which either become activated as vibrating complexes upon addition of Fe, or their concentration increases upon Fe incorporation into the ZnO lattice.

3.2.3. *Determination of effective magnetic moment in $Zn_{1-x}Fe_xO$ nanotubes*

Further evidence of the successful incorporation of Fe into ZnO was obtained through determination of the effective magnetic moment (μ_{eff}) for the undoped and Fe-doped ZnO nanotubes. This was done using a magnetic susceptibility balance (Evans balance), which measures the force experienced by a substance when it is placed in a magnetic field. The nanotubes were placed in a cylindrical tube, which was placed in a uniform magnetic field.

The effective magnetic moment was measured in units of Bohr magnetron (μ_B)using the equation:

$$\mu_{eff} = 2.828(\chi_A T)^{1/2} \tag{2}$$

where χ_A is the corrected molar susceptibility of the sample at temperature T (room temperature). χ_A for $Zn_{1-x}Fe_xO$ and undoped ZnO nanotubes was determined to be $\approx 1.31 \times 10^{-2}$ cgs units and $\approx 1.68 \times 10^{-4}$cgs units respectively. The effective magnetic moment was then calculated for both samples and found to be $\approx 5.41\mu_B$ for Fe-doped and $\approx 0.62\mu_B$ for undoped ZnO nanotubes. These values of μ_{eff} correspond to four unpaired electrons in the $Zn_{1-x}Fe_xO$ nanotubes and zero unpaired electrons in the undoped ZnO nanotubes. The four unpaired electrons found in the Fe-doped material is consistent with the electron configuration of Fe^{2+} in a high spin tetrahedral geometry, and is expected from the incorporation of Fe^{2+} as a substitutional dopant in ZnO. The undoped ZnO nanotubes showed a near zero effective magnetic moment.

4. Conclusion

There are several works in favor of and against ferromagnetism in TM doped ZnO. Experimentalists have frequently observed ferromagnetism in this material though the theory to support this claim does not exist. The promise of obtaining materials with high Curie temperature and thus develop spin-based semiconductor devices keeps the field of II-VI DMS very active. The electronic properties of nanoscale DMS are strongly affected by quantum confinement effect and large surface to volume ratios; hence the magnetic coupling between TM ions in low-dimensional systems may be distinctively different from that of bulk materials. Ferromagnetism in Mn-doped ZnO nanowires is yet to be understood, as Mn in its +2 valence state is an isovalent dopant and does not provide any carrier to ZnO. Hence, by first-principles studies, it must lead to an antiferromagnetic ground state. However, experimental observation of room temperature ferromagnetism in Mn-doped ZnO thin films and nanostructures have posed a challenge to the theoretical understanding of ferromagnetism for such systems. To complicate matters further, recent experiments reveal inconsistent ferromagnetic properties for TM doped ZnO, leaving the real mechanism for ferromagnetism in ZnO unclear. Experimental studies have revealed the importance of defects such as zinc and oxygen interstitials and vacancies on the magnetic ordering in such systems. It is evident that the observed ferromagnetic ordering as well as the measured magnetic moment and T_c depends largely on sample preparation techniques. Even in low dimensional systems, there could be several ferromagnetic sources, such as (a) contaminated substrate, (b) clusters of intrinsic defects with ferromagnetically aligned spins, (c) clusters of TM ions, and (d) uncompensated surface spins.

The first step in the synthesis of DMS is the successful incorporation of the TM ion into the host (ZnO) lattice. In this regard, we present our recent work on the synthesis of $Zn_{1-x}Fe_xO$ nanotubes with clear evidence of Fe ions substituting for Zn as 2+ ions in the ZnO crystal lattice. Nanotubes are interesting since the pores in them provide an additional dimension for band gap engineering and tuning of properties.

There is no clear understanding on the origin of ferromagnetism in low dimensional II-VI systems such as ZnO when doped with TM ions. There are several important issues that need to be resolved. It needs to be ascertained if the existence of room temperature ferromagnetism is related to some intrinsic origin and not to the presence of defects. If defect induced ferromagnetism is implicated, then the application of this DMS becomes very limited.

Author details

Usha Philipose and Gopal Sapkota

Department of Physics, University of North Texas, USA

References

[1] H. Ohno(2001). Toward Functional Spintronics. *Science*, Vol.291[5505]: 840-841

[2] P. Sharma et al.,(2003). Ferromagnetism above room temperature in bulk and transparent thin films of Mn doped ZnO. *Nature Materials*, Vol.2: 673-677

[3] T. Dietl,H. Ohno,F. Matsukura,J. Cibert,D. Ferrand (2000). Zener model description of ferromagnetism in zinc-blende magnetic semiconductors. *Science*, Vol.287[5455]: 1019-1022

[4] K. Sato and H. Katayama-Yoshida (2000). Material design for transparent ferromagnets with ZnO-based magnetic semiconductors. *Japanese Journal Of Applied Physics*, Vol.39[6B]: L555-558

[5] K. Sato and H. Katayama-Yoshida (2001). Stabilization of ferromagnetic states by electron doping in Fe-, Co- or Ni-doped ZnO. *Japanese Journal Of Applied Physics*, Vol.40:L334-336

[6] T. Jungwirth et al.,(2005). Prospects for high temperature ferromagnetism in (Ga,Mn)As semiconductors. *Physical Review B*, Vol.72[16]: 165204:1-13

[7] U. Philipose, Selvakumar V. Nair, Simon Trudel, C. F. de Souza, S. Aouba, Ross H. Hill, Harry E. Ruda (2006). High temperature ferromagnetism in Mn doped ZnO nanowires. *Applied Physics Letters*, Vol.88[26]: 263101:1-3

[8] Z. H. Zhang, X. Wang, J. B. Xu, S. Muller, C. Ronning and Quan Li (2009). Evidence of intrinsic ferromagnetism in individual dilute magnetic semiconducting nanostructures. *Nature Nanotechnology*, Vol.4: 523-527

[9] C. Liu, F. Yun and H. Morkoc (2005). Ferromagnetism of ZnO and GaN: A review. *Journal of Materials Science: Materials in Electronics*, Vol.16: 555-597

[10] S.J. Pearton, W.H. Heo, M. Ivill, D.P. Norton and T. Steiner (2004). Dilute magnetic semiconducting oxides. *Semiconductor Science and Technology*, Vol.19[10]: R59-74

[11] F. Pan, C. Song, X. J. Liu, Y. C. Yang and F. Zeng (2008). Ferromagnetism and possible application in spintronics of transition-metal-doped ZnO films. *Materials Science and Engineering: R: Reports*, Vol.62[1]: 1-35

[12] S.J. Pearton et al., (2007). Ferromagnetism in transition-metal doped ZnO. *Journal of Electronic Materials*, Vol.36[4]: 462-471

[13] S. J. Han, T. H. Jang, Y. B. Kim, B. G. Park, J. H. Park and Y. H. Jeong (2003). Magnetism in Mn-doped ZnO bulk samples prepared by solid state reaction. *Applied Physics Letters*, Vol.83[5]: 920-922

[14] S. Kolesnik and B. Dabrowski (2004). Absence of room temperature ferromagnetism in bulk Mn-doped ZnO. *Journal of Applied Physics*, Vol.96[9]: 5379-5381

[15] H. S. Hsu et al.,(2006). Evidence of oxygen vacancy enhanced room-temperature ferromagnetism in Co-doped ZnO. *Applied Physics Letters*, Vol.88[24]: 242507:1-3

[16] K. R. Kittilstved, D. A. Schwartz, A. C. Tuan, S. M. Heald, S. A. Chambers and D. R. Gamelin (2001). Direct Kinetic Correlation of Carriers and Ferromagnetism in Co^{2+} : Zn. *Physical Review Letters*, Vol.97[3]: 037203:1-4

[17] H. S. Hsu, et al.,(2008). Magnetic coupling properties of Mn-doped ZnO nanowires: First-principles calculations. *Journal of Applied Physics*, Vol.103[7]: 073903:1-5

[18] C. Zener (1951). Interaction Between the d Shells in the Transition Metals. *Physical Review B*, Vol.81[3]: 440-444

[19] H. Akai (1998). Ferromagnetism and Its Stability in the Diluted Magnetic Semiconductor (In, Mn)As. *Physical Review Letters*, Vol.81[14]: 3002-3005

[20] J.M.D Coey, M. Venkatesan and C.B Fitzgerald (2005). Donor impurity band exchange in dilute ferromagnetic oxides. *Nature Materials*, Vol.4: 173-179

[21] M. Herbich, A. Twardowski, D. Scalbert and A. Petrou (1998). Bound magnetic polaron in Cr-based diluted magnetic semiconductors. *Physical Review B*, Vol.58[11]: 7024-7034

[22] T. Fukumura et al., (2001). Magnetic properties of Mn-doped ZnO. *Applied Physics Letters*, Vol.78[7]: 958-960

[23] A. Tiwari, C. Jin, A. Kvit, D. Kumar, J.F Muth and J. Narayan (2002). Structural, optical and magnetic properties of diluted magnetic semiconducting $Zn_{1-x}Mn_xO$ films. *Solid State Communications*, Vol.121[6]: 371-373

[24] S.W Jung, S.J An, G. Yi, C.U Jung, S. Lee and S. Cho (2002). Ferromagnetic properties of $Zn_{1-x}Mn_xO$ epitaxial thin films. *Applied Physics Letters*, Vol.80[24]: 4561-4563

[25] K.R Kittilstved, W.K Liu and D.R Gameln (2006). Electronic structure origins of polarity-dependent high T_C ferromagnetism in oxide-diluted magnetic semiconductors. *Nature Materials*, Vol.5: 291-297

[26] L. V. Korsunska et al., (2003). The influence of defect drift in external electric field on green luminescence of ZnO single crystals. *Journal of Luminescence*, Vol.102-103: 733-736

[27] A. F. Kohan, G. Ceder, D. Morgan and C.G. Van de Walle (2000). First-principles study of native point defects in ZnO. *Physical Review B*, Vol.61[22]: 15019-15027

[28] N.E. Hsu, W.K. Hung and Y.F. Chena (2004). Origin of defect emission identified by polarized luminescence from aligned ZnO nanorods. *Journal of Applied Physics*, Vol.96[8]: 4671-4673

[29] F. H. Leiter, H. R. Alves, A. Hofstaetter, D. M. Hofmann and B. K. Meyer (2001). The Oxygen Vacancy as the Origin of a Green Emission in Undoped ZnO. *Physica Status Solidi (b)*, Vol.226[1]: 073903:1-5

[30] X. T. Zhang, Y. C. Liu, J. Y. Zhang, Y. M. Lu, D. Z. Shen, X. W. Fan and X. G. Kong (2003). Structure and photoluminescence of Mn-passivated nanocrystalline ZnO thin film. *Journal of Crystal Growth*, Vol.254: 80-85

[31] M. Liu, A. H. Kitai and P. Mascher (1992). Point defects and luminescence centres in zinc oxide and zinc oxide doped with manganese. *Journal of Luminescence*, Vol.54[1]: 35-42

[32] U. Philipose, H. E. Ruda, A. Shik, C. F. de Souza and P. Sun (2006). Conductivity and photoconductivity in undoped ZnSe nanowire array. *Journal of Applied Physics*, Vol.99: 066106:1-3

[33] Y-J. Ma, Z. Zhang, F. Zhou, Li Lu, A. Jin and C. Gu (2005). Hopping conduction in single ZnO nanowires. *Nanotechnology*, Vol.16[6]: 746-749

[34] K. Sato and H. Katayama-Yoshida (2001). Electronic structure and ferromagnetism of transition-metal-impurity-doped zinc oxide. *Physica B*, Vol.308-310: 904-907

[35] D. Karmakar, S. K. Mandal, R. M. Kadam, P. L. Paulose, A. K. Rajarajan, T. K. Nath, A. K. Das, I. Dasgupta and G. P. Das (2007). Ferromagnetism in Fe-doped ZnO nanocrystals: Experiment and theory. *Physical Review B*, Vol. 75[14]: 144404:1-14

[36] S. W. Yoon et al., (2003). Magnetic properties of ZnO-based diluted magnetic semiconductors. *Journal of Applied Physics*, Vol.93[10]: 7879-7881

[37] M. S. Park and B. I. Min (2003). Ferromagnetism in ZnO codoped with transition metals: $Zn_{1-x}(FeCo)_xO$ and $Zn_{1-x}(FeCu)_xO$. *Physical Review B*, Vol.68[22]: 224436:1-6

[38] G. Sapkota, K. Gryczynski, R. McDougald, A. Neogi and U. Philipose (2012). Low-Temperature Synthesis of Fe-Doped ZnO Nanotubes. *Journal of Electronic Materials*, Vol.41[8]: 2155-2161

[39] Z. Szafran, R.M. Pike, and M.M. Singh, *Micro-Scale Inorganic Chemistry*, Wiley, ISBN 0-471-61996-5, New York.

[40] K.J. Kim and Y.R (2004). Optical investigation of $Zn_{1-x}Fe_xO$ films grown on Al_2O_3(0001) by radio-frequency sputtering. *Journal of Applied Physics,* Vol.96[8]: 4150-4153

[41] A.J. Chen, X.M. Wu, Z.D. Sha, L.T. Zhuge and Y.D. Meng (2006). Structure and photoluminescence properties of Fe-doped ZnO thin films. *Journal of Physics D: Applied Physics,* Vol.39: 4762-4765

Chapter 3

Self-Assembly of Copper Oxide Core-Shell Nanowires Through Ethyl Alcohol

Sheng Yun Wu

Additional information is available at the end of the chapter

1. Introduction

The control of the core-shell structure at the nanometer is one of the most fundamental challenges in condensed matter science [Deville et al., 2009; Woodley et al., 2008; Bannin, 2007; Ji et al., 2007; Lauhon et al., 2002]. One of the fundamental concepts regarding the synthesis of nanowires via the spontaneous self-organization of individual nanoparticles, called spontaneous organization nanocrystals, refers to the formation of two-dimensional (2-D) and 3-D arrays of nanoparticles or nanowires [Tang et al., 2002; Lu et al., 2002; Balazs et al., 2006; Tang et al., 2006]. The copper-oxide based system, such as Cu_2O [Huang et al., 2012; Hong et al., 2011; Yao et al., 2010], CuO [Wu, 2012; Cheng et al., 2007; Chou et al., 2008], and Cu_2O@CuO [Yec et al., 2012], has been known to facilitate oxidation reactions in the bulk material, which may allow it to be a cost effective substitute for noble metals in various catalytic systems. With recent developments in nanostructures synthesis leading to the ability to control size, reproducibility and structural complexity [Lauhon et al., 2002; Fan et al., 2006; Lu et al., 2005], it becomes worthwhile and possibly paramount to define specific target structures for the nanoparticles based on an understanding of the mechanism of the self-organization mechanism. Upto now, many complex procedures, such as vapor-liquid-solid, chemical vapor deposition, thermal evaporation, and chemical reactions, have been developed for the synthesis of one-dimensional materials of Copper-based nano compounds [Huang et al., 2012; Hong et al., 2011; Yao et al., 2010]. In addition to these methods, solution chemical route including solvothermal, hydrothermal, self-assembly, and template-assisted chemical vapor deposition has become a promising option for large –scale production of nanomaterials, due to the simple, fast, and less expensive virtues [Xin et al., 2002; Li et al., 1999; Lu et al., 2000; Hung et al., 2004; Roy et al., 2003; Baxter et al., 2003]. However, a new type of immersing nanoparticles into liquid alcohol to form core-shell nanowires at room temperature has been difficult, primarily because control of nucleation and growth is still a challenge.

In the present work, we report on a simple method for bottom-up fabrication by means of template-free growth of high density Cu_2O/CuO core/shell nanowires. These can be spontaneously self-assembled by the oxidation of ultra-small copper nanoparticles simply immersed in ethyl alcohol. This leads to the formation of Cu_2O/CuO core/shell nanowires. Then nanosized growth effects of these nanowires are observed experimentally. We find that the formation of nanowires may be far more due to chemical reactions than previously realized.

2. Synthesis method

The initial Cu nanoparticles used as a precursor in the synthesis of Cu_2O/CuO core/shell nanowires, are fabricated by employing the gas condensation method, as shown in Fig. 1. High purity Cu ingots (~ 0.5 g, 99.999% pure and 3 mm in diameter) are heated by a current source of 90 A. They are evaporated at a rate of 0.01 Å/s in an Ar pressure of 0.1 Torr. The evaporated Cu nanoparticles are collected on a non-magnetic quartz plates (1 cm x 1 cm), maintained at liquid nitrogen temperature. After restoration to room temperature, the nanoparticles, which are only loosely attracted to the collector, are stripped off. The resultant samples were in the form of dried powder, consisting of a macroscopic amount of individual Cu nanoparticles. The combined effect of high gas flow rate, low deposition pressure and relatively low substrate temperature resulted in the deposition of copper nanoparticles. The moving Cu nanoparticles were washed and stripped off from quartz plates using ethyl alcohol(99.5%) in argon gas environment (chemical hood), as shown in Fig. 2(a). They were then kept in a glass container and in a sealed quartz capillary (Fig. 2(b) and 2(c) for few days) for *in situ* x-ray scattering examination. The sample is apparently not sensitive to being exposed to air for a short period of time, which is supported by the fact that the x-ray diffraction patterns of the as-grown sample and sample exposed to air for few hours are indistinguishable. However, the sample used in the study is kept in glass container with liquid alcohol. The main key method for sample syntheses in this study was that the ethyl alcohol solution-phase was important tofostering spontaneous growth of Cu nanoparticles which then led to the formation of Cu_2O/CuO core/shell nanowires at room temperature. In general, metal-oxide nanowires can be synthesized using a solid-state method involving high temperature oxidation of the metal [Cheng et al., 2007; Devan et al., 2008] and a solution-phase under hydrothermal conditions [Tan et al., 2007]. However, a typical reaction requires very high temperature or pressure and therefore would seem to be incompatible with our system [Yang et al., 1998]. Copper oxide nanoparticles in a limited size range have been prepared with other techniques using chemical routes, and some of these techniques have resulted in alimited growth of rodlikenanocrystals reported by Fang group [Fang et al., 2004]. Ethyl alcohol (CH_3-CH_2OH)is chosen as the reactive solution for the spontaneous organization method, because of its dipole moment (1.684 Debye), which is comparable to that of water (1.69 Debye), and its polar character, which allows the formation of Cu_2O/CuO core/shell nanowires with an alcohol bridge. In the presence of a rich OH^- solution, spontaneous oxidation can be drastically accelerated. This procedure has been reported by Tello and coauthors [Tello et al., 2003] for the oxidation of *p*-silicon in water and ethyl alcohol. They demonstrated that local oxidation in liquid ethyl alcohol environments is different from that in aqueous solutions, with a significant improvement in the growth rate. The Liu group [Liu et al., 2009] also reported that the dosage rate can influence the morphologies of Cu_2O, so that the shape of the nanostructures can be controlled through adjusting the reduction rate of Cu^{2+} ions. In this study, Cu^{2+} ions are released continuously from the cooper nanoparticles into the ethyl alcohol under ambient conditions. These can be immediately captured through coordination with OH^-, to form $Cu(OH)_2$. After 4-days (D=4) of immersion, the $Cu(OH)_2$ loses water and CuO or Cu_2O is formed by the release of O_2. The resultant copper-based phases can be either cuprous oxide or cuprite oxide depending on the temperature and surrounding environment. At room temperature, however, the cuprous phase is the dominant product of oxidation. The cuprous surface can be terminated by a few mono layers of cuprite oxide product (crystalline or amorphous phase) at the solution interface, forming Cu_2O/CuO core-shell nanowires. The prepared samples were mounted in a capillary rotor sample holder and the process of the formation of nanowires was systematically studied using an *in situ* x-ray diffractometer (Philips X'Pert PRO)with CuK_α radiation (λ=1.541 Å) at times ranging from one to seven days. The morphology of sample was characterized by a field emission scanning electron microscope (FE-SEM, JEOL JSM-6500F) equipped with anenergy dispersive x-ray spectroscope (EDS, Oxford Instrument INCAx-sight 7557). Atomic-resolution transmission electron microscopic (TEM) analysisand high-resolution transmission electron microscopy (HRTEM) images were taken with the CCD-camera of an electron microscope (JEOL JEM-2100) at 200 kV. Analysis software (Digital Micrograph) was employed to digitalize and analyze the obtained images.

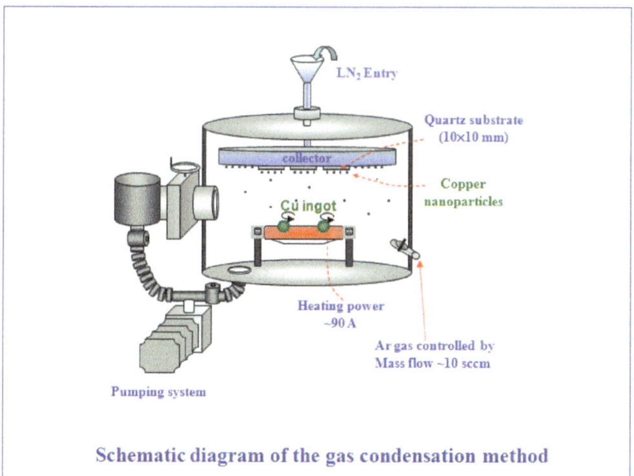

Figure 1. Schematic representation of the gas condensation method. Reproduced with permission [4]; copyrights 2010, Institute of Physics.

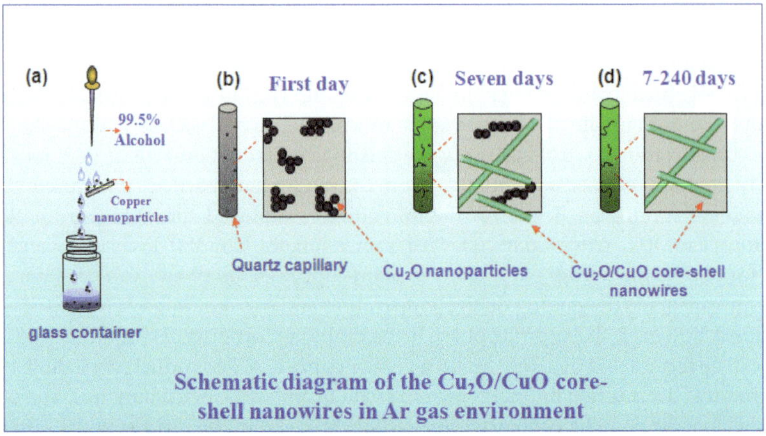

Figure 2. The processing steps for Cu nanoparticles include: (a) washing of the quartz plate with liquid ethyl alcohol and then collecting the liquid in a glass container; (b)-(c) quartz capillary for the experiments of time dependency of the *in situ* x-ray scattering at D=4 and 7, and (d) at D=7-240. Reproduced with permission [4]; copyrights 2010, Institute of Physics.

3. Results and discussion

The *in situ* x-ray diffraction patterns at various time periods were shown in Fig. 3. As indicated on the bottom of Fig. 3, there are three nuclear peaks at the {1 1 1}, {2 0 0} and {2 2 0} positions, as indexed based on a cubic Cu -*Fm-3m* structure, with a lattice expansion of ~1.5 % compare with Cu bulk. According to the Scherrer diffraction formula, the average size of initial Cu nanoparticles can be described as:

$$\frac{(\Delta 2\theta_B)^2}{\tan^2\theta_B} = \frac{\lambda}{<d>} \times \frac{\Delta 2\theta_B}{\tan\theta_B \sin\theta_B} + 16e^2, \tag{1}$$

where <d> is the mean size, λ (=1.541 Å) is the incident x-ray and $\Delta 2\theta_B$ is FWHM at the scattering angle of $2\theta_B$.The obtained mean size of Cu nanoparticles is <d>=5.4 nm, estimating from these peaks in the diffraction pattern. After immersion in an ethyl alcohol solution four days, we observe a significant broader peak around 2θ=36°, which is associated with the Cu_2O structure of the Miller index {1 1 1}, and exhibits the coexistence of core/shell nanocrystals. This may be understood by assuming the existence of the Cu_2O phase near the surface, with the oxidation contribution coming from the Cu atoms in the core. The pattern in the upper part of Fig. 3 should contain, in principle, contributions from the Cu_2O phase, after 7-days (D=7) of immersion. X-ray diffraction patterns are known as the fingerprints of crystalline materials. They reveal details of the crystalline structure and their formation during synthesis, and even the crystalline phase transitions or separation at various temperatures. An example of x-ray and Rietveld refined diffraction patterns of the sample, taken at D=7, is shown in Fig. 4. Diffraction patterns were utilized to characterize the crystalline structure in the prepared samples. The diffraction peaks appeared to be much broader than the instrumental resolution, reflecting the nano-size effects. The analysis was performed using the program package of the General Structure Analysis System (GSAS) [Larson et al., 1990] following the Rietveld method [Rietveld, 1969]. Several models with different symmetries were assumed during the preliminary analysis. In our structural analysis we then pay special attention to searching for the possible symmetries that candescribe the observed diffraction pattern well. All the structural and lattice parameters were allowed to vary simultaneously, and refining processes were carried out until R_p, the weighted R_{wp} factor, differed by less than one part in a thousand within two successive cycles. Figure 4 shows the diffraction pattern (red crosses) taken at room temperature, where the solid curve (green curve) indicates the fitted pattern and the difference (pink curve) between the observed and the fitted patterns is plotted at the bottom of Fig. 4. The refined lattice parameters are a=b=c=4.2821(5) Å with a space group of Cu_2O-*Pn-3m* structure. The formation of Cu_2O requires an expansion of only 0.026 % of the metallic Cu nanoparticles to make room for the O atoms, comparing with Cu_2O bulk (a=4.281 Å). The decrease in surface energy of the nanowires will tend to cause an increase in their size by elastic distortion of their crystal lattice. Of course, this kind of distortion is very small compared to the whole particle size. It is worth notingthat the contribution of the intensity of the CuO shell (as *can not* be seen in the x-ray patterns) is undetectable or amorphous in state. These characteristics (i.e., the amorphous CuO shell) agree with previously reported results for Cu_2O/CuO nanoparticles obtained by Yin et al. [Yin et al., 2005]. In their comprehensive analysis, they found that Cu nanocrystals formed initially, but were subsequently oxidized to form a highly crystalline Cu_2O coated with an amorphous epitaxial CuO layer.

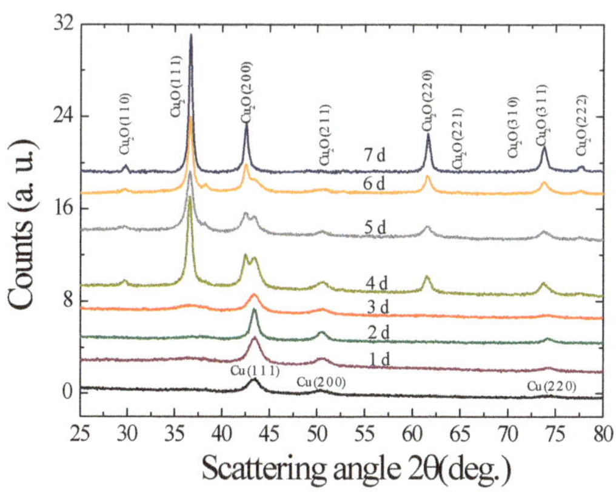

Figure 3. In situtime dependence of x-ray diffraction patterns taken at room temperature.

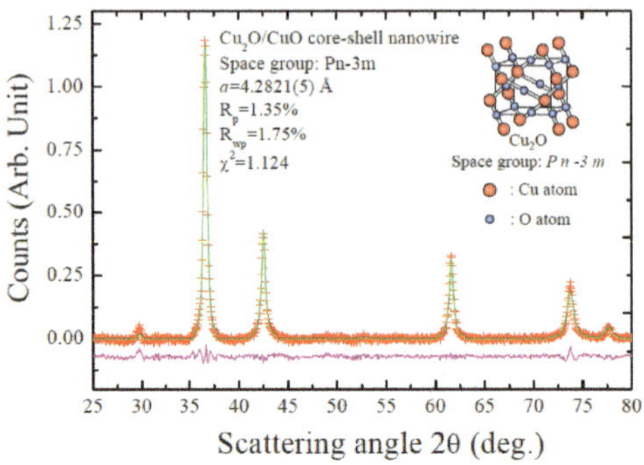

Figure 4. The observed (crosses) and Rietveld refined (solid lines) x-ray diffraction patterns of Cu_2O/CuO core-shell nanowires. The contribution of the intensity of the CuO shell *can not* be seen in the x-ray patterns due to amorphous state.

The morphology of selected samples at D=1, 4 and 7, was characterized by FE-SEM, as shown in Figs. 5(a) to (c). As can be seen in Fig. 5(a), it is clearly evident that the nanoparticles are spherical and stick together due to an electrostatic effect as well as an artifact of the drying of aqueous suspensions. It can be seen that the Cu_2O/CuO nanowires grew homogeneously in a large area to form straight nanowires, as shown in Fig. 5(b). The diameters ranged from 100-500 nm and the lengths were up to several tens of microns. As can be seen in Fig. 5(c), the Cu_2O/CuO core-shell nanowires are long and straight, ranging from tens of micrometers to more than a hundred micrometers. The growth of the nanowires is clear. An evaluation of the diameter distribution can be obtained and calculated from a portion of the SEM image, as shown in the Figs. 5(d) to 5(f). They are quite asymmetric and can be described using a log-normal distribution function defined as follows:

$$f(d) = \frac{1}{d\sigma\sqrt{2\pi}} \exp\left(-\frac{(\ln d - \ln \langle d \rangle)^2}{2\sigma^2}\right), \quad (2)$$

where <d> is the mean value and σ is the standard deviation of the function. The mean diameters and standard deviations for the nanowires obtained from the fits are <d>=201(5) to 225(5) nm, σ=0.42 and 0.31 nm at D=4 and 7, indicating increasing diameter as the reaction time increases. The small standard deviation (σ< 0.5) of the function indicates that the distribution is confined to a limited range. It can be seen that the width of the distribution profile broadens due to the crystalline structure. The stability of growth during the formation of core-shell nanowires is confirmed by examination of the SEM images even at D=240 (not shown). This result indicates that increasing the reaction time by immersion in ethyl alcohol *cannot* make the growth of diameter of the Cu_2O core more intensive, due to the termination of amorphous CuO at D=7.

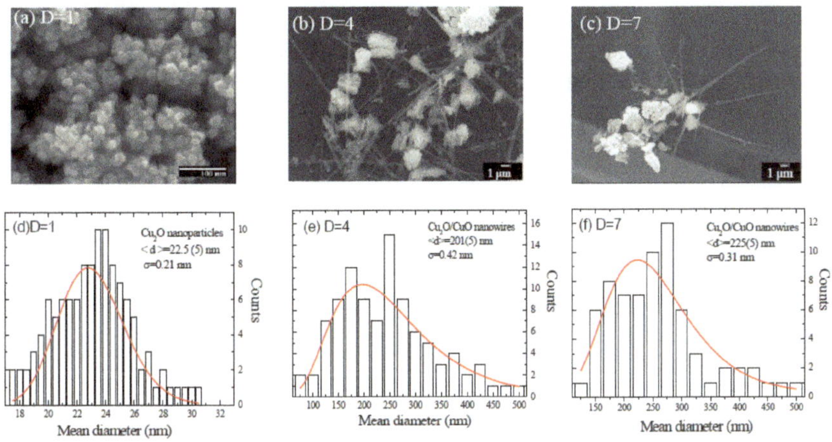

Figure 5. SEM images of Cu_2O/CuO core/shell after various reaction time with (a) D=1, (b) d=4 and (c) D=7.(d)-(f) The distribution of mean diameter <d> at D=1, 4, and 7,respectively.

The results of an early study (x-ray patterns) show that the prepared core-shell nanowires are crystalline. We further characterized the structure of sample using TEM microscopy. Figure 6 shows an example of the Cu_2O/CuO core-shell nanowires taken at D=4, which has a uniform coating of amorphous CuO (light contrast) surrounding a single-crystal Cu_2O core (dark contrast), forming core-shell Cu_2O-CuO nanowires. The chain-like configuration of the Cu_2O nanoparticles intertwined with the nanowires can also be observed. The results are in good agreement with the NVT Monte Carlo (MC) method [Phillies, 1974] results for the intermediate state of pearl necklace-like aggregates. In the MC simulation gives the formation of nanowires involves a process where Au nanoparticles could self-assemble into chain-like intermediate states then recrystallize into one-dimensional nanowire. The dipole-dipole-interaction in the Cu_2O nanoparticles is strong and long-range, leading to the formation of the pearl necklace-like nanostructures. A significant surface CuO layer (a few nms in thickness) was observed, indicating the existence of an amorphous CuO shell. In order to determine the amorphous CuO and the surface characteristics of the core-shell nanowires, we have utilized XANES technique to examine the existence of surface CuO thickness, simulating through an intensity ratio of Cu_2O-core and CuO-shell by using a proposed core-shell nanowire model.

For a cylindrically multilayered structure without consider the effect of diameter distribution. An electron being photoionized into the surface CuO has a probability of escaping through the surface without suffering any inelastic processes. Therefore, it can be detected as a photoelectron. In generally, the probability is conveniently expressed in terms of the mean free path for surface layer λ_{CuO} and core λ_{Cu2O}, respectively. The intensity of ionized electron is proportional to $I_o\ exp(-f/\lambda)$, where f is the distance between the surface and the point of origin of the photoelectron, I_o is the normalization constant depends on the nanocrystals. Thus, integrating the portion of contribution intensity over the cylindrical volume, then we can easily obtain the total intensity from each of the two regions of the nanocrystals. The total intensity from each region in terms of cylindrical polar coordinates can be expressed as [Nanda et al., 1999]:

$$I = I_0 \int_{-L/2}^{L/2} \int_0^{\pi} \int_{R_1}^{R_2} \exp\left[\frac{-f(r,\phi,z)}{\lambda}\right] r dr d\phi dz, \tag{3}$$

where L is the length of the nanowire, λ is the mean free path of the photoionized electron,$R_1$is the radius of the core, R_2 is the radius including the surface layer. The distance between the surface and the point of origin of the photoelectron then can be schematically plotted in Fig. 7 and defined as:

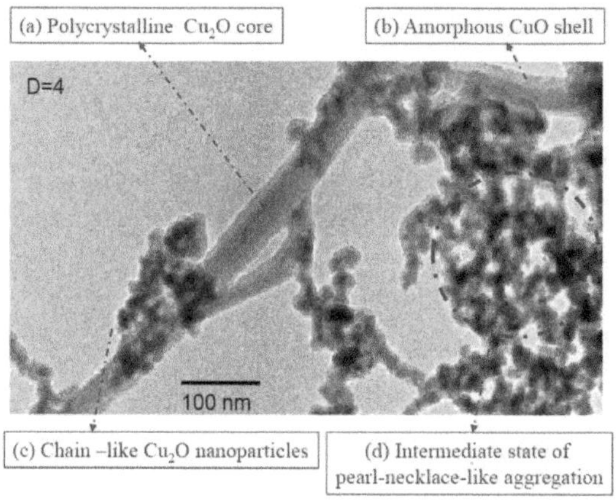

Figure 6. Representative TEM image Cu_2O/CuO core/shell nanowires, revealing the core-shell structure and a number of interesting portions. (a) The formation of a single Cu_2O/CuO core/shell nanowire reveals a straight polycrystalline Cu_2O core; and (b) an amorphous CuO layer. (c) The chain-like configuration of Cu_2O nanoparticles and (d) pearl-necklace-like aggregation intertwined with the nanowires can be observed.

$$f(r,\phi,z) = \left[(\sqrt{R^2 - r^2\cos^2\phi} - r\sin\phi)^2 + z^2 \right]^{\frac{1}{2}} \tag{4}$$

The mean free paths of λ_{CuO} and λ_{Cu2O} used in this study are obtained from previous reports [36]. The mean free paths of λ_{CuO} and λ_{Cu2O} used in the simulation are defined as

$$\lambda_{CuO} = \frac{0.0062E_p(Cu^{2+})}{\log(0.0501E_p(Cu^{2+}))} and\, \lambda_{Cu_2O} = \frac{0.00767E_p(Cu^{+})}{\log(0.0501E_p(Cu^{+}))} nm \tag{5}$$

In order to clarify the contribution from Cu_2O- or CuO- intensity, we have defined a theoretical integrated intensity ratio γ taken as:

$$\gamma = \frac{I_{Cu_2O}}{I_{CuO}} = \frac{I_0 \int_{-L/2}^{L/2} \int_0^{\pi} \int_0^{R_1} \exp\left[\frac{-f(r,\phi,z)}{\lambda_{Cu_2O}}\right] r dr d\phi dz}{I_0 \int_{-L/2}^{L/2} \int_0^{\pi} \int_{R_1}^{R_2} \exp\left[\frac{-f(r,\phi,z)}{\lambda_{CuO}}\right] r dr d\phi dz} \tag{6}$$

The above ratio cannot be evaluated analytically, and we can compute numerically by varying these parameters of R_1 and R_2, respectively, to match the experimental integrated intensity ratio of $\gamma=I_{Cu2O}/I_{CuO}$. The electronic geometric structure around Cu in the as-synthesized material was characterized with X-ray absorption near edges(XANES) measurements. XANES spectroscopy of Cu $L_{2,3}$–edge was performed using the 6-m HSGM beam-line (BL20A1) of the National Synchrotron Radiation Research Center in Hsinchu, Taiwan. The recorded spectra were corrected for the energy dependent incident photon intensity as well as for self-absorption effects and normalized to the tabulated standard absorption cross sections in the energy range of 900-1020 eV for the Cu L-edge. The radiation was monochromatized using a 6m SGM monochromator and the photon energy was calibrated by measuring X-ray absorption spectrum of Cu metal foil by using the Cu L_3 white line. Figure 8 shows XANES spectra of as-synthesized material and copper based references such as Cu_2O (red curve), CuO (green curve) bulk. The main peak of Cu $2P_{3/2}$ at 932.3 eV reveals the presence of Cu_2O phase. It is accompanied by a second peak at 935.1 eV related to the CuO phase, and can be regarded as characteristic of Cu^+ and Cu^{2+} ($2p_{3/2} \rightarrow 3d$), respectively [Yin et al., 2005; Xu et al. 2007; Wen et al., 2005]. It is worth noting that the occurrence of a weak broad satellite feature on the higher energy (~ 943.5 eV) side of the Cu 2p main peak indicates the slight presence of CuO on the surface. The observed amount of CuO is very tiny due to the slightly oxidized surface of Cu2O nanowires. These results are well consistent with the XRD and TEM observations shown in Figs. 3 and 6. This result can be understood as due to the occurrence of amorphous CuO and crystalline Cu_2O on the surface and core, respectively. Based on the proposed core-shell model, we use the proposed model of a single core-shell nanowire with an inner crystalline Cu_2O core coated by a thin amorphous CuO surface layer to simulate the ratio of integrated intensity. These can be obtained from the data. The intensity ratio of the Cu_2O core and CuO shell is thus determined by integrating the peak intensity I_{Cu_2O} /I_{CuO}. Then we can tune the value of R_1 and R_2 in the model to fit the ratio of I_{Cu_2O}/I_{CuO}. All the parameters are allowed to vary simultaneously. Refining processes are carried out until the results meet the ratio of I_{Cu_2O} /I_{CuO} =3.302 obtained from XANES. The refined parameters, as calculated from the model are listed in Table 1. From XANES analysis, we can conclude that the core-shell nanowires are mostly of Cu_2O phase. However, CuO phase in the as-grown sample could be explained either by the surface oxidation in liquid ethyl alcohol or by the core-shell-type of structure of Cu_2O nanowires with CuO as the capping layer [Sahoo et al., 2009; Yu et al., 2007].

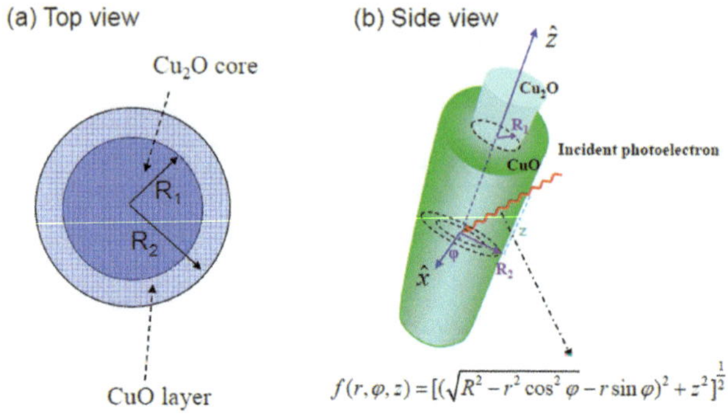

Figure 7. Schematic models of core-shell nanowires based on the photoemission core-level analysis. (a) Top view of this model, where R_1 is the radius of core and R_2 is the radius including surface layer. (b) Side view of the proposed model, where the penetrated distance is f(r, φ, z) and the r integration is over suitable limits.

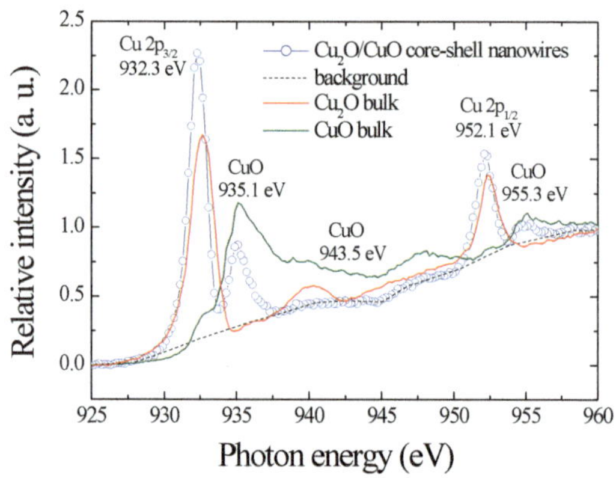

Figure 8. X-ray absorption spectra of Cu_2O/CuO core-shell nanowires at D=4. The experimental spectra and fitted curve are indicated by the solid circles and the line, respectively.

$E_p(Cu^+)$	I_1	$E_p(Cu^{2+})$	I_2	Observed I_{Cu2O}/I_{CuO}	Simulated I_{Cu2O}/I_{CuO}	R_1 (nm)	R_2 (nm)	CuO thickness$_2$ (nm)
932.3	3.423	935.1	1.037	3.302	3.36	99	101	2

Table 1. Summary of the simulation of XANES results for Cu_2O/CuO core-shell nanowires at D=4. The XANES of finite size core-shell model is convoluted with the Gaussian instrumental resolution function $G(w_1, \bar{E}_p)$,where w and I are the FWHM and amplitudes of peak $\bar{E}_p$.

4. Conclusion

A new type of spontaneous organization method was successfully utilized to grow Cu_2O/CuO core/shell nanowires by immersing copper nanoparticles into liquid alcohol at room temperature. The microstructure of the nanowires during the nucleation and growth process from CuO nanoparticles into Cu_2O/CuO core-shell nanowires can be examined using time dependent *in situ* x-ray diffraction, SEM, and TEM. We show that the reactions of Cu nanoparticles oxide to form Cu_2O/CuO core-shell nanowires can occur at room temperature with unusually fast reaction rates. An intermediate state of pearl necklace-like aggregates formed by a chain-like configuration of Cu_2O nanoparticles was observed which then intertwined to form the nanowires. The results agreed with the prediction of NVT Monte Carlo simulation[Phillies, 1974]. A significant thin surface layer was observed on Cu_2O nanowires, indicating the existence of an amorphous CuO shell, as also confirmed by the x-ray absorption near-edge structure technique.

Acknowledgment

The authors would like to thank the National Science Council of the Republic of China for financially supporting this research under Contract No. NSC-100-2112-M-259-003-MY3. We would also like to thank Professor Yuan-Ron Ma of National Dong Hwa University and Dr. Ting Shan Chan of National Synchrotron Radiation Research Center (Taiwan) for their valuable contributions and discussions in this work

Author details

Sheng Yun Wu

Department of Physics, National Dong Hwa University, Hualien, Taiwan, Republic of China

References

[1] Deville, S.; Marie, E.; Bernard-Granger, G.; Lasalle, A.; Bogner, A.; Gauthier, C.; Leloup, J. & Guizard, C. (2009) Metastable and Unstable Cellular Solidification of Colloidal Suspensions, *Nat. Mater.* Vol. 8: 966-972.

[2] Woodley, S. M.&Catlow, R. (2008) Crystal structure prediction from first principles,*Nat. Mater.*Vol. 7:937–946.

[3] Bannin, U. (2007) Nanocrystals: Tiny seeds make a big difference,*Nat. Mater.* Vol. 6: 625-626.

[4] Ji, J.-Y.; Shih, P.-H.; Yang, C. C.; Chan, T.-S.; Ma, Y.-R.& Wu, S. Y. (2010) Spontaneous self-organization of Cu_2O/CuO core-shell nanowires from copper nanoparticles, *Nanotechnology* Vol. 21: 045603.

[5] Lauhon, L. J.; Gudlksen, M. S.; Wang, D. & Lieber, C. M. (2002) Epitaxial core–shell and core–multishell nanowire heterostructures, *Nature* Vol. 420: 57-61.

[6] Tang, Z.; Kotov, N. A. & Giersig, M. (2002) Spontaneous Organization of Single CdTe Nanoparticles into Luminescent Nanowires, *Science* Vol. 297: 237.

[7] Lu, Q.; Gao, F. & Zhao, D. (2002) One step synthesis and assembly of copper sulfide nanoparticles to nanowires, nanotubes and nanovesicles through an organic amine-assisted hydrothermal process,*Nano Lett.* Vol. 2, 725-728.

[8] Balazs, A. C.; Emrick, T.& Russell, T. P. (2006) Nanoparticle Polymer Composites: Where Two Small Worlds Meet, *Science* Vol. 314, 1107-1110.

[9] Tang, Z.; Zhang, Z.; Wang, Y.;Glotzer, S. C. &Kotov, N. A. (2006) Self-Assembly of CdTe Nanocrystals into Free-Floating Sheets, *Science* Vol. 314, 274-278.

[10] Huang, W. C.;Lyu, L. M.; Yang, Y. C.& Huang, M. H. (2012) Synthesis of Cu_2O Nanocrystals from Cubic to Rhombic Dodecahedral Structures and Their Comparative Photocatalytic Activity, *J. Am. Chem. Soc.* Vol. 134: 1261-1267.

[11] Hong, F.; Sun, S.; You, H.; Yang, S.; Fang, J.; Guo, S.; Yang, Z.; Ding, B.& Song, X. (2011) Cu_2O Template Strategy for the Synthesis of Structure-Definable Noble Metal Alloy Mesocages, *Crystal Growth & Design* Vol. 11: 3694-3697.

[12] Yao, K. X.; Yin, X. M.; Wang, T. H. & Zeng, H. C. (2010) Synthesis, Self-Assembly, Disassembly, and Reassembly of Two Types of Cu_2O Nanocrystals Unifaceted with {001} or {110} Planes, *J. Am. Chem. Soc.* Vol. 132: 6131–6144.

[13] Wu, S. Y. (2012) Design and Controlling of in-plane CuO Nanowires: An Investigation of Growth Mechanism and Phonon Confinement Effect, *J. Nanosci. Lett.* Vol. 2: 5.

[14] Cheng, C. L.; Ma, Y.R.; Chou, M. H.; Huang, C. Y.;Yeh, V. & Wu, S. Y. (2007) Direct Observation of Short-Circuit Diffusion During the Formation of a Single Cupric Oxide Nanowire, *Nanotechnology* Vol. 24: 245604-245608.

[15] Chou, M. H.; Liu,S. B.; Huang, C. Y.;Wu,S. Y. & Cheng,C.L. (2008) Confocal Raman Spectroscopic Mapping Studies on a Single CuO Nanowire, *Applied Surface Science* Vol. 254: 7539-7543.

[16] Yec, C. C. & Zeng, Hua, C. (2012) Synthetic Architecture of Multiple Core–Shell and Yolk–Shell Structures of $(Cu_2O@)_nCu_2O$ (n = 1–4) with Centricity and Eccentricity, *Chemistry of Materials* Vol.24: 1917-1929.

[17] Fan, H. J.; Knez, M.; Scholz, R.; Nielsch, K.; Pippel, E.; Hesse, D.; Zacharias, M. & Gösele, U. (2006) Monocrystalline spinel nanotube fabrication based on the Kirkendall effect, *Nature Materials* Vol. 5: 627-631.

[18] Lu, C. H.; Qi, L. M.; Yang, J. H.; Wang, X. Y.; Zhang, D. Y.;Xie, J. L. & Ma, J. M. (2005) One-Pot Synthesis of Octahedral Cu_2O Nanocages via a Catalytic Solution Route, *Adv. Mater.* Vol. 17: 2562-2567.

[19] Li, X. Z. and Xia, J. B. (2002) Electronic structure and optical properties of quantum rods with wurtzite structure, *Phys. Rev. B* Vol. 66: 115316.

[20] Li, W. J.; Shi, E. W.;Zhong, W. Z. & Yin, Z. W. (1999) Growth mechanism and growth habit of oxide crystals,*J. Crystal Growth* Vol. 203: 186-196.

[21] Lu, C. & Yeh, C. (2000) Influence of hydrothermal conditions on the morphology and particle size of zinc oxide powder, *Ceram. Int.* Vol. 26: 351.

[22] Hung, C. H. & Whang, W. T. (2004) Low-temperature solution approach toward highly aligned ZnOnanotip arrays,*J. Crystal Growth* Vol. 268: 242-248.

[23] Roy V. A. L., Djurisic A. B., Chan W. K., Gao J., Liu H. F. & Surya C. (2003) Luminescent and structural properties of ZnO nanorods prepared under different conditions, *Appl. Phys. Lett.* Vol. 83: 141.

[24] Baxter, J. B.; Wu, F. &Aydil, E. S. (2003) Growth mechanism and characterization of zinc oxide hexagonal columns, *Appl. Phys. Lett.* Vol. 83: 3797.

[25] Devan, R. S.; Ho, W. D.; Lin, J. H.; Wu, S. Y. & Ma, Y.R. (2008) X-ray diffraction study of a large-scale and high-density array of one-dimensional crystalline tantalum pentoxidenanorods, *Cryst. Growth Des.* Vol. 8: 4465-4468.

[26] Tan, Y.; Xue, X.; Peng, Q.; Zhao, H.; Wang, T. & Li, Y. (2007) Controllable Fabrication and Electrical Performance of Single Crystalline Cu2O Nanowires with High Aspect Ratios, *Nano Lett.* Vol. 7: 3723.

[27] Yang, J. C.;Kolasa, B.; Gibson, J. M.; Yeadon, M. (1998) Self-limiting oxidation of copper, *Appl. Phys. Lett.* Vol. 73: 2841.

[28] Fan, H.; Yang, L.;Hua, W.; Wu, X.; Wu, Z.;Xie, S.&Zou, B. (2004) Controlled synthesis of monodispersed CuO nanocrystals, *Nanotechnology* Vol. 15: 37.

[29] Tello, M. & Garcia, R. (2003) Giant growth rate in nano-oxidation of *p*-silicon surfaces by using ethyl alcohol liquid bridges, *Appl. Phys. Lett.* Vol. 83: 2339.

[30] Liu, H.; Miao, W.; Yang, S.; Zhang, Z. & Chen, J. (2009) Controlled Synthesis of Different Shapes of Cu_2O via γ-Irradiation, *Cryst.Growth Des.* Vol. 9: 1733-1740.

[31] Larson, A. C. & Von Dreele, R. B. (1990). General Structure Analysis System, Report LA-UR-86-748, Los Alamos National Laboratory, Los Alamos, NM.

[32] Rietveld, H. M. (1969). A profile refinement method for nuclear and magnetic structures. *Journal of Applied Crystallography*,Vol. 2: 65-71, ISSN 1095-9203.

[33] Yin, M.; Wu, C. K.; Lou, Y.; Burda, C.; Koberstein, J. T. ; Zhu, Y.; O'Brien, S. (2005) Oxidation in coppernanocrystals*J. Am. Chem. Soc.*Vol. 127,9506.

[34] Phillies, G. D. J. (1974) Excess chemical potential of dilute solutions of spherical polyelectrolytes, *J. Chem. Phys.*Vol.60: 2721-2731.

[35] Nanda, J.; Kuruvilla, B. A.& Sarma, D. D. (1999) Photoelectron spectroscopic study of CdSe nanocrystallites, *Phys. Rev. B* Vol. 59: 7473-7479.

[36] Xu, J. S. & Xue, D. F. (2007) Five branching growth patterns in the cubic crystal system: A direct observation of cuprous oxide microcrystals, *Acta Mater.* Vol. 55: 2397-2406.

[37] Wen, X.; Xie, Y.; Choi, C. L.; Wan, K. C.; Li, X. Y.& Yang, S. (2005) Copper-based nanowire materials-Templated syntheses, characterizations, and applications, *Langmuir* Vol. 21: 4729-4737.

[38] Sahoo, S.; Husale, S.; Colwill, B.; Lu, T. M.; Nayak, S. & Ajayan, P. M. (2009) Electric field directed self-assembly of cuprous oxide nanostructures for photon sensing. *ACS nano* Vol. 3: 3965-3944.

[39] Yu, Z. Q.; Wang, C. M.; Engelhard, M. H.; Nachimuthu, P.; McCready, D. E.; Lyubinetsky, I. V. & Thevuthasan, S. (2007) Epitaxial growth and microstructure of Cu_2O nanoparticle/thinfilms on $SrTiO_3$(100), *Nanotechnology* Vol. 18: 115601-5.

Chapter 4

Synthesis of ZnO Nanowire by MOCVD Technique: Effect of Substrate and Growth Parameter

Sachindra Nath Das, Jyoti Prakash Kar,
Junjie Xiong and Jae-Min Myoung

Additional information is available at the end of the chapter

1. Introduction

ZnO nanostructures have been studied extensively in the past few years due to their fundamental and technological importance [1-5]. In particular, it is a wide-direct-band gap (~3.37 eV), II–VI semiconductor with many potential applications such as nanolaser arrays [2,6], gas sensors [7-10], field emission devices [11-12] and luminescent materials [13-16]. A great deal of attention has been focused on one-dimensional nanostructures because of their superior properties such as high surface-to-volume ratio, high crystalline quality, and quantum confinement effects, which are suitable for various electronics and optoelectronics applications. Up to now, many ZnO configurations have been reported such as nanobelts, nanowires [10–11], nanoneedles [3], nanotetrapods [9], nanocombs [17] and so on. It has also been suggested that ZnO nanostructures are probably the most abundant forms of any known materials. Therefore, proper control of processing parameters is essential to reproduce the desired nanostructures. Synthesis of aligned one-dimensional ZnO nanowire arrays has attracted much attention, which includes the techniques of metal-organic chemical vapour deposition (MOCVD), vapor-liquid-solid (VLS), pulsed laser deposition method (PLD) and solution based chemical technique [21-24]. A solution-based technique is an attractive approach for the growth of ZnO nanostructures because of its simplicity and low cost [25, 26]. In addition, this method allows large-scale growth of aligned ZnO nanowires on an arbitrary substrate, which promotes the integration of ZnO nanomaterials for various applications. But the main drawback of chemical process is the poor quality of nanowire. On the other hand, physical process are of particular interest since they have many advantages such as the ability to fabricate nanostructures of better quality, well controllable configurations, and good reproducibility. From a physical point of view, the synthesis methods for one-dimensional ZnO nanostructures are categorized into two groups, metal catalyst-assisted growth and catalyst-free growth which corresponded to vapor–liquid–solid and vapor–solid mechanisms, respectively. The latter is better than the former in terms of no metal catalytic impurity. On the other hand, MOCVD is superior because, it has abilities to produce large area; catalytic free, long, vertically aligned ZnO NWs, which have excellent morphological, structural and optical properties [27]. As the growth of ZnO nanowires by MOCVD is a bottom–up technique, the nature of substrates has a vital role for the determination of nanowire dimension and alignment. According to lattice misfit, the most suitable substrate for ZnO growth is $ScAlMgO_4$, which is expensive and technologically inconvenient [28]. Interestingly, C-plane sapphire overcomes some of the limitations. On the other hand, ZnO nanostructures on R-plane sapphire have an advantage for the fabrication of the next generation sensors and detectors in the form of surface acoustic wave (SAW) devices due to higher electromechanical coupling [29-31]. However, in the pursuit of next generation ZnO-based nano devices, it would be highly preferred if well-ordered ZnO nanowires with high aspect ratio could be aligned onto cheap and CMOS compatible substrates, such as silicon. In addition, silicon is the most popular substrate for microelectronics and micro electro mechanical systems (MEMS), and it serves as a good integration platform. But, it is difficult to obtain well-aligned ZnO nanowires on silicon substrates, because the formation of an interfacial layer (SiO_2) and large lattice misfit [32,33]. On the other hand, microchannel-based solid-state devices on Si have played an important role for chemical, gas and flow sensors with compact size and superior performance [34-36]. The growth of vertically aligned ZnO microtubes on a silicon wafer without any catalyst are of particular interests for monolithic integration of micro/nano devices.

As far as the growth of ZnO nanowires on glass or SiO_2 layer is concerned, deterioration in the quality and alignment of the nanowires come into picture because of the lack of specific epitaxial arrangements. In such cases the nanowires become quasi-aligned instead of vertical aligned. Since, MOCVD is a bottom up technique, the introduction of an appropriate interfacial layer may also solve the above problem. It is more preferable if the interfacial layer would be transparent and conductive for its integration with optoelectronic devices. Taking these issues into consideration, Ga doped ZnO (GZO) film may be a suitable interfacial layer for the growth of vertical aligned ZnO nanowires. In addition, GZO film can be used as a transparent conducting layer during the integration of the devices. Furthermore, the cost of GZO is much lower than conventionally used indium doped transparent conductive oxide layer. The nature of the substrate with a interfacial layer also take a significant role on the shape of nanowire. The control over size and morphology of nanometer and micrometer-sized semiconductor materials represents a great challenge to realizing the design of novel functional devices. Several efforts have been directed towards the synthesis of nanostructures to control the shape, aspect ratio (length/width), growth site, and growth direction. One salient feature of the ZnO-based nanostructures is a gradual transition between different morphologies with different substrate which leads to the investigation for the effects of substrate and growth parameter on MOCVD technique.

2. Experimental

ZnO nanowires were grown on different substrates (e.g. different types of sapphire, Si, Si microchanel, GZO coated glass, ZnO microcreators) using MOCVD technique with diethylzinc (DEZn) as the zinc precursor. Argon and oxygen (5N) were used as carrier and reactive gases, respectively. The base pressure of the reactor chamber was 10^{-6} Torr. During the growth, the substrate temperature, working pressure and the nozzle-substrate spacing were fixed, but they were varied for different substrates to synthesize desired nanostructure. Morphological investigations of the nanowires were carried out using field-emission scanning electron Microscope (FESEM). The crystallinity of the nanowires was studied by X-Ray diffractometer. In addition, the microstructures of ZnO NWs were observed using high-resolution transmission electron microscope (HRTEM). For the cross-sectional TEM and energy dispersive X-ray (EDX) study, the samples were prepared by focused ion beam (FIB) system.

3. Result and discussion

3.1. Effect of sapphire substrate

The morphology of nanostructures greatly depends on the growth parameters as well as the underlying substrates. c-plane sapphire substrates were used for the growth of ZnO nanowires by MOCVD technique with substrate temperature and substrate to nozzle distance at 630 °C and 3 cm, respectively. The experimental details are given in table 1. Fig. 1(a–c) is typical FESEM images of ZnO nanowires arrays prepared with different growth conditions. The samples constitute uniform and densely packed nanowires arrays with different surface morphologies, namely nanoneedles (Fig. 1(a)), nanonails (Fig. 1(b)) and nanowires with rounded tip (Fig. 1(c)). Cross-sectional view of the samples (inset of Fig 1.) revealed that the nanowires were grown almost perpendicularly on the substrates and their lengths were about 3 μm. It is known that each nucleation site usually supported growth of only one nanowires, but when the nucleation sites were close together, two nanowires combined into one with a larger diameter [37] and different surface morphologies were obtained. From XRD pattern (Fig. 2), a prominent ZnO (0002) peak is observed, which corresponds to the growth of ZnO nanowires along c-axis. No characteristic peaks indicating other impurities are observed in the samples. This is attributed to the hexagonal wurtzite structure of nanowires.

In order to investigate the structural characteristics of ZnO nanowires, TEM study of individual nanowires is a very important characterization technique. The bright field TEM micrographs of the ZnO nanowires and the corresponding HRTEM image (inset) are shown in Fig. 3. The HRTEM images indicated that the nanowires were structurally uniform and did not exhibit any noticeable defects. The HRTEM image showed well resolved lattice with an inter-planar spacing of 0.52 nm, which corresponds to the (0001) plane of wurtzite structure. Hence, one may infer that the nanowires possessed a single-crystal hexagonal structure with good crystalline quality along the [0001] axis.

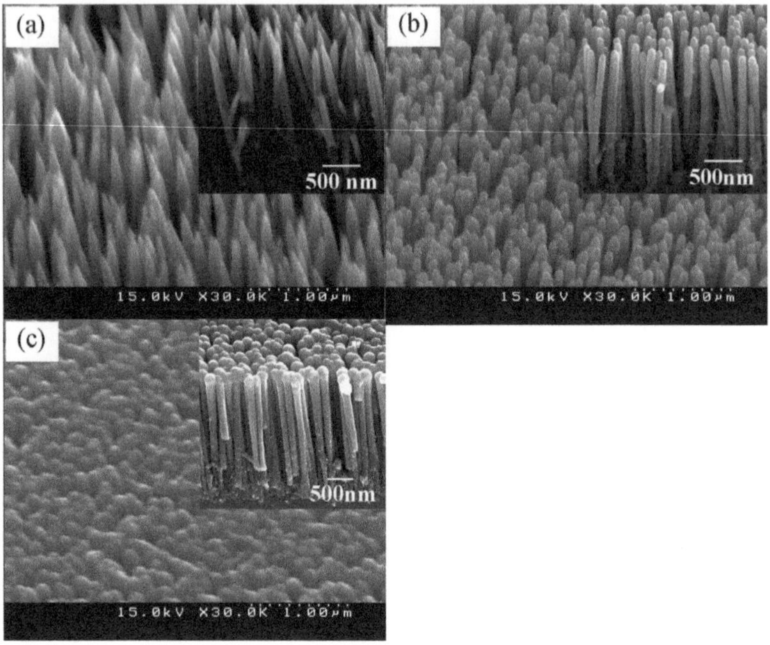

Figure 1. Surface morphologies of vertically aligned ZnO nanowires: (a) nanoneedles, (b) nanonails and (c) nanowires with rounded tip.

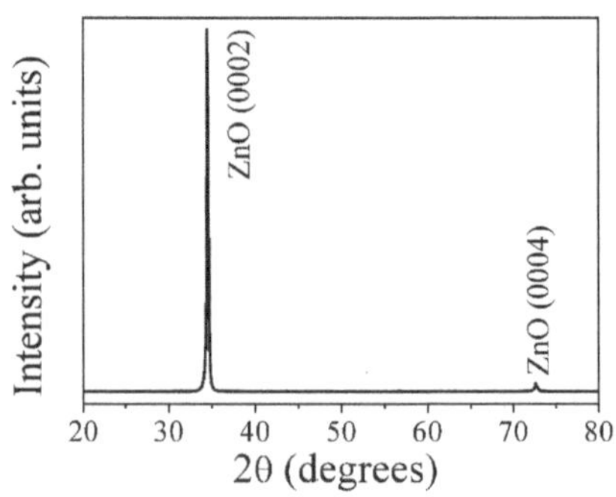

Figure 2. XRD pattern of a representative ZnO nanowire sample

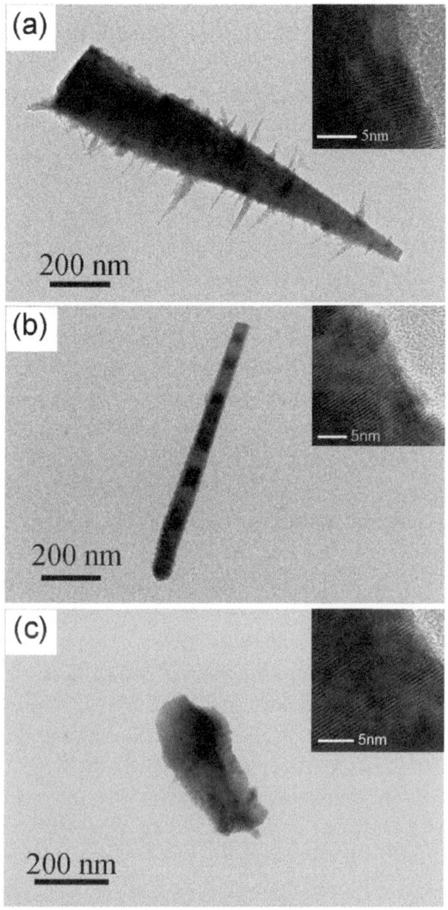

Figure 3. TEM images of ZnO NRs: (a) nanoneedles, (b) nanonails and (c) nanorods with rounded tip. Inset shows the corresponding HRTEM images

Sample	Substrate temp. (°C)	O_2 flow (SCCM)	N_2 flow (SCCM)	Working Pressure (torr).	Time (min)
(a)	630	20	70	3	90
(b)	630	50	50	1	30
(c)	630	50	50	1.5	60

Table 1. Growth parameter of ZnO nanorod arrays.

To investigate the effect of lattice mismatch on the growth mechanism, ZnO nanowires were grown on A (11$\bar{2}$0), C (0001) and R ($\bar{1}$012) plane sapphires using MOCVD system. Here, the working pressure of the reactor was maintained at 6 Torr and the distance between the susceptor and the nozzle was fixed at 5 cm. Fig. 3 depicts the FESEM images of ZnO nanostructures on various sapphire substrates. Vertically aligned nanowires of width 100–150 nm and length 3–4 μm are grown on A- and C-plane sapphires (Fig. 4(a) and (b)), but nanopencil/tip like ZnO nanostructures of length 1.5–2.5 μm is observed for the R-sapphire substrate (Fig. 4(c)). The average basal diameter of the nanopencils is around 250–300 nm and its shape becomes narrow towards top-end. Sharp tip of the nanostructures may be due to the residual amount of precursors at the end of the growth process. For better understanding of the growth process, nanostructures were grown for a longer time (90 min). For both A- and C-plane sapphires, the nanowires retain their vertical nature and are uniform throughout its length, which clearly indicates the growth is along c-axis (Fig. 4(d)). On the other hand, Fig. 4(e) shows a typical image of ZnO nanostructures on R-plane sapphire, in which lateral growth has been taken place indicating a deterioration of c-axis growth. This may be due to the generation of stress during the initial stage of growth and afterwards these nanostructures join together to make thicker structures. The magnified image (inset of Fig. 4(e)) revealed that the thicker nanostructures are randomly oriented and their surface is appeared to be comparatively rough. The bright-field (BF) TEM image of single ZnO nanowire is shown in Fig. 4 (e). This shows the diameter of the nanowires grown on A, C sapphire are around 100 nm and where as it is around 300 nm for the nanowire grown on R sapphire. The tip of the nanowire was gradually tapered down from several hundred of nanometers to 50 nm. This is due to the lack of the fresh reactant supply towards the final stage of growth process. At the initial stage of the growth, the width of the nanowires were thin and the individual thin nanowires laterally merged together to give thicker one with increase in growth time. Generally, the formation mechanism of nanostructures consists of two stages: nucleation and growth. At first, nanosized crystalline nuclei are formed on the substrate and afterwards the nanowires start to grow along a specific direction. The diameter of the nanowires depends on the size and density of the initial ZnO nuclei. In case of C-plane sapphire, ZnO has a partial ionic character, which originates due to the opposite charge on the (0001) and (000$\bar{1}$) planes being Zn and O termination, respectively [38]. Thus there is a net dipole moment when the crystal is terminated by the basal plane, which causes the surface energy to diverge. On the other hand, the (11$\bar{2}$0) plane is nonpolar as it comprises an equal number of O and Zn atoms, and hence has a lower surface energy [39-40]. Therefore the growth rates of ZnO nanostructures on C- and R-plane sapphires are different, which is confirmed from the variation in aspect ratio (length/width) of the nanostructures. Hence, it is expected that the condition of supersaturation is different for the growth of ZnO nanostructures on C and R-plane sapphires. At present, it is believed that there is a formation of an unintentional thin interfacial ZnO layer during the initial stage of growth on R-plane sapphire. Furthermore, the shape of the nanowires was change to nanopencil using an intentional thin ZnO film as an interfacial layer [41]. This indicates that the different shape, size and density of the nanowires can be attributed to the in-plane stress during the initial stage of growth, which comes from thermal and lattice mismatches between ZnO nanostructures and the substrates [42-43]. According to crystallographic point of view, Miyauchi et al. [44] reported that the probability of oxygen vacancies at the terminated surfaces of ZnO nanostructures grown on R-plane sapphire is higher than others. The surface morphology as well as crystallinity of ZnO nanostructures, grown on R-plane sapphire, is quite different from A- and C-plane sapphires. For R-plane sapphire, it is expected that simultaneous existence of O and Zn atoms in $(11\bar{2}0)$ plane plays a vital role on the property of nanostructure.

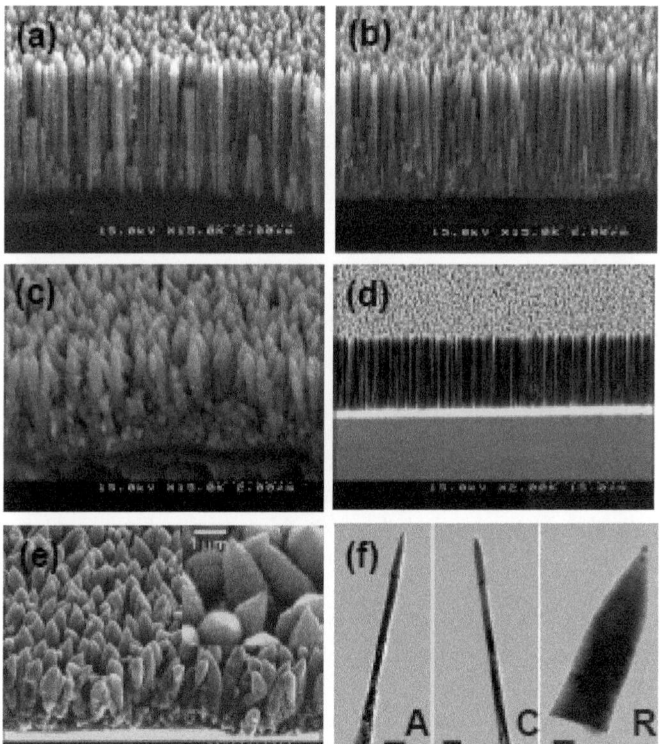

Figure 4. Micrographs of ZnO nanostructures on (a) A-, (b) C-, (c) R-plane sapphires for 15 min growth, (d) C-, (e) R-plane sapphires for 90 min growth, and (f) bright-filed TEM images [J. P. Kar et al. Applied Surface Science 256 (2010) 4995].

3.2. Effect of Si Substrate

To investigate the effect of substrate, ZnO nanostructures were grown on Si substrate and Si with ZnO interfacial layer. Pencil like ZnO nanostructures were obtained on Si substrate with ZnO thin film as an interfacial layer (Fig. 5). Here, the substrate temperature, working pressure and substrate to nozzle distance were kept at 600 °C, 3 Torr and 3 cm, respectively. With the decrease in substrate to nozzle distance (~1 cm) and increase in the substrate temperature 700 ^{O}C, a hallow microtube like structures were grown. FE-SEM images of the vertically aligned ZnO microtubes with a hollow tubular shape are shown in Fig. 6. The magnified SEM images (insets) show that all of the microtubes have a hollow cavity and a faceted hexagonal shape. The tubular wall-thickness is in the range of micrometers towards base and it reduced to nanometers range towards top-end. The average outer basal diameter of the microtubes is around 10 μm and its shape becomes narrow (5 μm) towards top end. The length of the ZnO microtubes is about 40 μm and the diameter is ~4 μm towards the open tip whereas it becomes 1–1.5 μm at the basal part. Typical crystal morphology and a representative cross-sectional view of a single ZnO microtube is shown in Fig. 6(c). Hollow basal part of the tubular structure (inset of Fig. 6(c)) consists of many nanospikes (width ~50–120 nm and length ~500 nm). We believe that these nanospikes grow further during the course of the growth process and merge together to make solid surface, which in fact reduces the inner diameter. From XRD pattern (not shown here), a prominent ZnO (0002) peak is observed, which corresponds to the growth of ZnO microtubes along c-axis. No characteristic peaks indicating other impurities are observed in the samples. This is attributed to the hexagonal wurtzite structure of microtubes. Vertically aligned ZnO microtubes display inerratic or perfect hexagon-shaped cross-sectional structure and open tips. At the initial stage of growth, several vertically aligned nanowires are generated due to the high growth rate of ZnO crystal along [0001] direction, which is thermodynamically stable. As the growth time increases, the ZnO nanowires may grow laterally even though the lateral growth rate is much lower than that of the vertical direction. Moreover, a high reaction temperature may accelerate the lateral growth of the ZnO nanowires. Therefore, neighboring ZnO nanowires gradually start to coalesce together to form hexagonal circles. As the reaction time increases, the complete prismatic ZnO microtubes are finally formed by the coalescence of ZnO nanowires. Furthermore, the substrate was very close (1 cm) to the nozzle of the precursor, which favored the reasonable supply of precursors. It is clear that the morphology changed from the nanowires to microtubes with the in increase growth temperature and decrease in nozzle-substrate spacing. Hence, the higher growth temperature and lower nozzle-substrate spacing favor the "oriented attachment" process of the nanowires to form microtubes. At present, it is believed that substrate temperature has more influence for the quality of the microtubes. On the other hand, it has also been reported that catalysts like carbon and metastable zinc-rich oxide layer played a vital role for the generation of tubular structures [45,46].

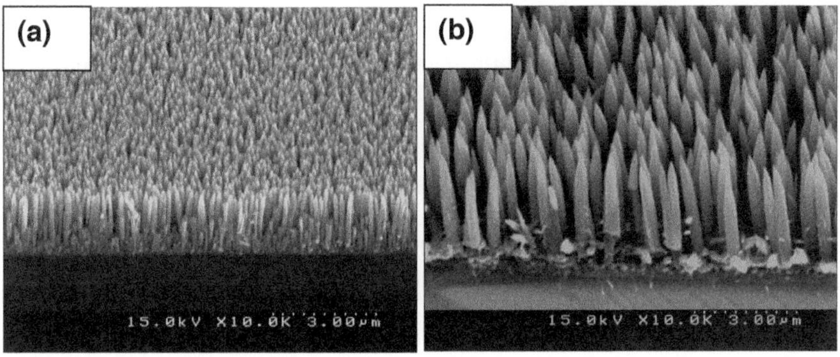

Figure 5. FE-SEM images of ZnO nanowires grown on (a) Si substrate, (b) Si substrate with ZnO interfacial layer.

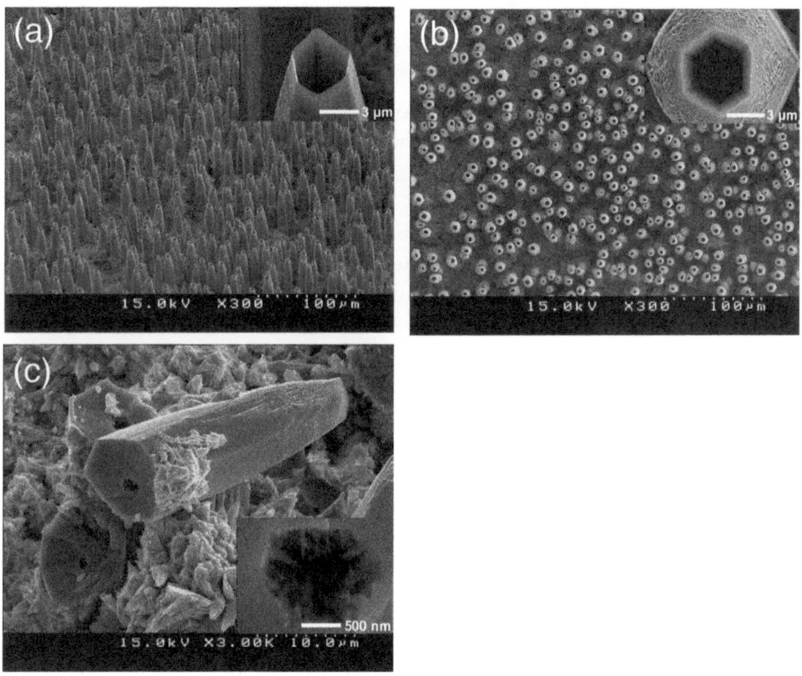

Figure 6. FE-SEM (a) 45° tilted, (b) top, and (c) cross-sectional images of ZnO microtubes grown on Si substrates [J. P. Kar et al. Materials Letters 63 (2009) 2327].

To investigate the effect of the shape of Si substrate, ZnO nanostructures were grown by MOCVD technique on Si microchanel. Before ZnO growth, Si substrates were patterned by wet chemical etching process to form microcavity. At first, 200-nm-thick silicon nitride masking layer was grown by low-pressure chemical vapor deposition (LPCVD) at 800 ^{0}C in the mixture of di-chloro silane (SiH_2Cl_2) and ammonia (NH_3) with 240 mTorr pressure.Thereafter, standard photolithography and etching techniques were adopted in order to fabricate micropatterns. Reactive ion etcher (RIE) was used to remove silicon nitride layer from the micro- patterned windows and the etch rate was 350 nm/min. Afterwards, the samples were immersed in KOH solution for 10 hour to fabricate the truncated pyramid-shaped etch patterns. Before the growth of ZnO nanowires, no passivation layer was intentionally used. The substrate temperature was kept at 700 ^{0}C during the nanowire growth. The working pressure of the reactor was maintained at 3 Torr and the distance between the susceptor and the nozzle was fixed at 1.5cm. Fig. 7 and Fig. 8 show the crosssectional image (FESEM) of the different microchannel in silicon substrate, which contains ZnO nanowires. Three parts (top surface [Fig. 7(a), (b) and Fig. 8(a) and (b)], side wall [Fig. 7(c) and Fig. 8(c)] and bottom surface [Fig. 7(d), (e) and Fig. 8(d) and (e)]) have ZnO nanowires with different dimensions and surface morphologies. On the top of silicon substrate the nanowires were almost uniform along the length (2 mm) and width of 200–250 nm. The aspect ratio, packing fraction and the number density of nanowires on top surface are around 10, 0.8 and 10^7 per mm^2, respectively. The packing fractions of nanowires were determined by measuring the area occupied by the nanowires on a top view micrograph. Interestingly, vertical nanopencil /tip like ZnO nanostructure was observed towards the bottom of the trench. The ZnO nanowires have preferentially grown along a particular direction, which his normal to the substrate surface. It is expected that at the bottom surface the nanowires would have a high perpendicular anisotropy because of the high aspect ratio and the low packing fraction. Figs. 7(e) show the microstructures of ZnO nanowires at the bottom corner of the microchannel. Furthermore, randomly distributed few nanoflowers are also observed at side wall sand bottom of the microchannel. It may be generated from the defects of Si surface, arrived during etching of silicon. The flower has several wings, which have grown from the common origin. These happen due to the different directional growth of ZnO nanowires at the defect points. The possible growth mechanisms of ZnO nanowires are stated below. At the top surface, high-density island were nucleated on the substrate and grew preferentially along the c-axis orientation. The continuous growth led to the coalescence of these islands and as a result wider nanowires, with a low density and a reduced aspect ratio, were grown. But at the inclined surface, the growth direction (c-axis) and the direction of incoming precursor molecules are not same.

So, the width of nanowires is less than the top surface. Now at the bottom surface of the microchannel, number of incoming fresh precursor molecule is less than the top surface due to stagnant of gaseous species at the cavity. It is assumed that the sharpness of the tips originates due to the lack of fresh reactant supply to the bottom of cavity during the growth of nanostructures. Thus, the probability of coalescence of grain is less, which leads to the formation of thin nanowires compared to the top surface.

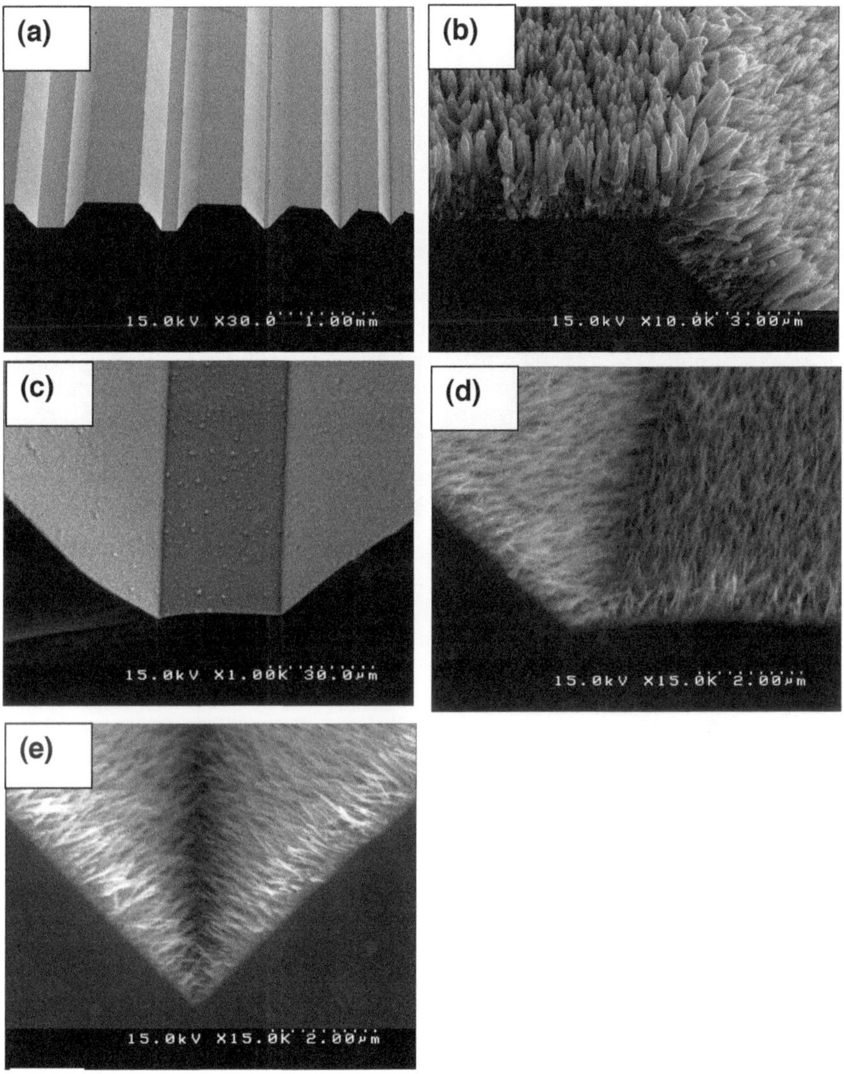

Figure 7. FESEM images of ZnO nanowires on (a) microchannel, (b) Top surface, (c-e) magnified image of bottom surface.

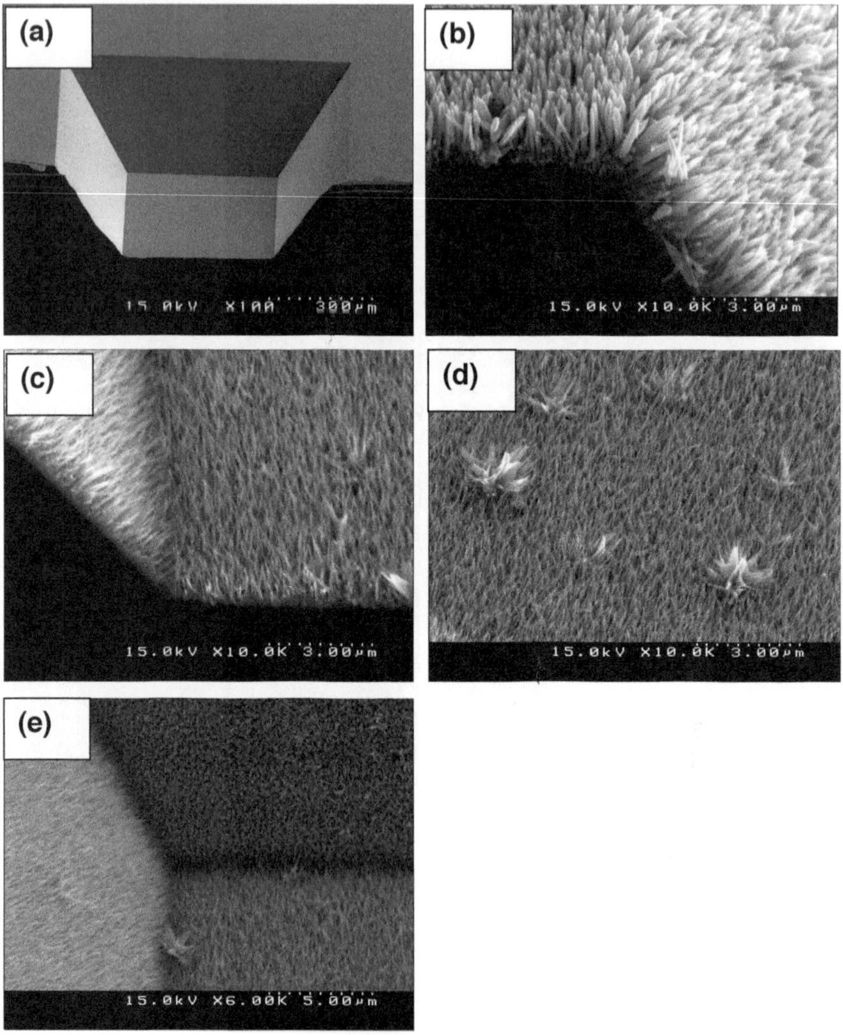

Figure 8. FESEM images of ZnO nanowires on (a) microchannel, (b) Top surface, (c) bottom surface, (d) magnified image of bottom surface, (e) bottom corner.

3.3. Effect of interfacial layer

To investigate the effect of interfacial layer on ZnO growth by MOCVD technique, Ga doped ZnO coated glass and ZnO microcators were used as substrates. Ga-doped ZnO films were deposited by DC sputtering of GZO target on corning glass (1737) substrates. The substrate temperature, working pressure and sputtering power were fixed around 280 ºC, 3×10^{-3} torr and 1.2 kW, respectively. The substrate temperature and working pressure were fixed at 600 ºC, 3 Torr and 1.5 cm, respectively. The evolution of columnar NWs were possible when the flow rate of O_2 (20 sccm) was lower than the carrier Ar gas (50 sccm).

Vertical aligned long ZnO nanowires (width 400-500 nm) were then grown on GZO coated glass substrates and the FESEM image is shown in Fig. 9(b). In order to compare, ZnO nanowires were also grown on c-sapphire substrates keeping all of the parameters same (Fig. 9(a)). This observation depicts the lengths (~ 25 μm) of the nanowires are similar for both the substrates. These kind of long nanowires are suitable for the single nanowire based devices, where the larger length is feasible for various electrical contacts. XRD patterns (not shown here) of ZnO nanowires, grown on GZO/glass substrate shows the presence of (0002) and (0004) peaks. The absence of additional peaks in the XRD pattern excludes the possibility of any extra phases. The nanowires exhibited a prominent (0002) peak with full width at half maximum (FWHM) values around 0.06, indicating high crystallinity of the ZnO nanowires. These data reveal that the ZnO nanowires are all c-axis oriented with wurtzite structure. The bright-field (BF) TEM image of single ZnO nanowire is shown in Fig. 10 (a). This shows the diameter of the nanowires is 460 nm without any side branches. The tip of the nanowire was gradually tapered down from several hundred of nanometers to 50 nm. This is due to the lack of the fresh reactant supply towards the final stage of growth process. The dopant redistribution at high process temperature is a usual phenomenon during the fabrication of semiconductor devices. Herein, the nanowires were grown at 600 ºC and hence there may be a chance of dopant diffusion at the interfacial layer. In this regard, cross-sectional TEM as well as EDX spectroscopic measurements of the ZnO nanowires on GZO film are carried out. Figure 10 (a) shows the enlarged image of the interface between the ZnO nanowires GZO films. This enlarged image indicates that the sputtered film has columnar grains and the nanowires are grown along the columnar grains of the GZO film.

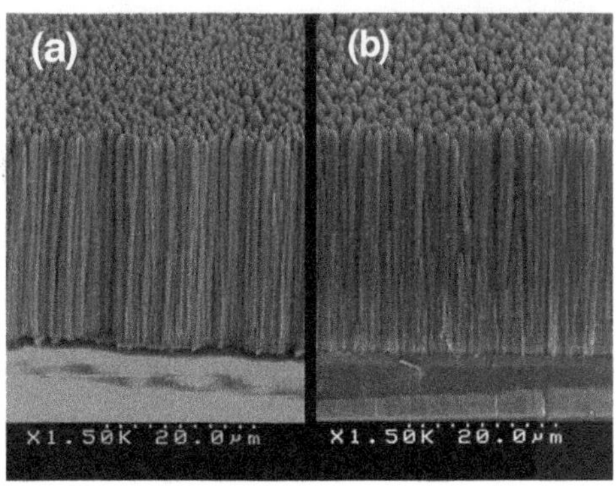

Figure 9. FESEM of ZnO NWs on (a) c-plane sapphire. (b) GZO/glass substrates

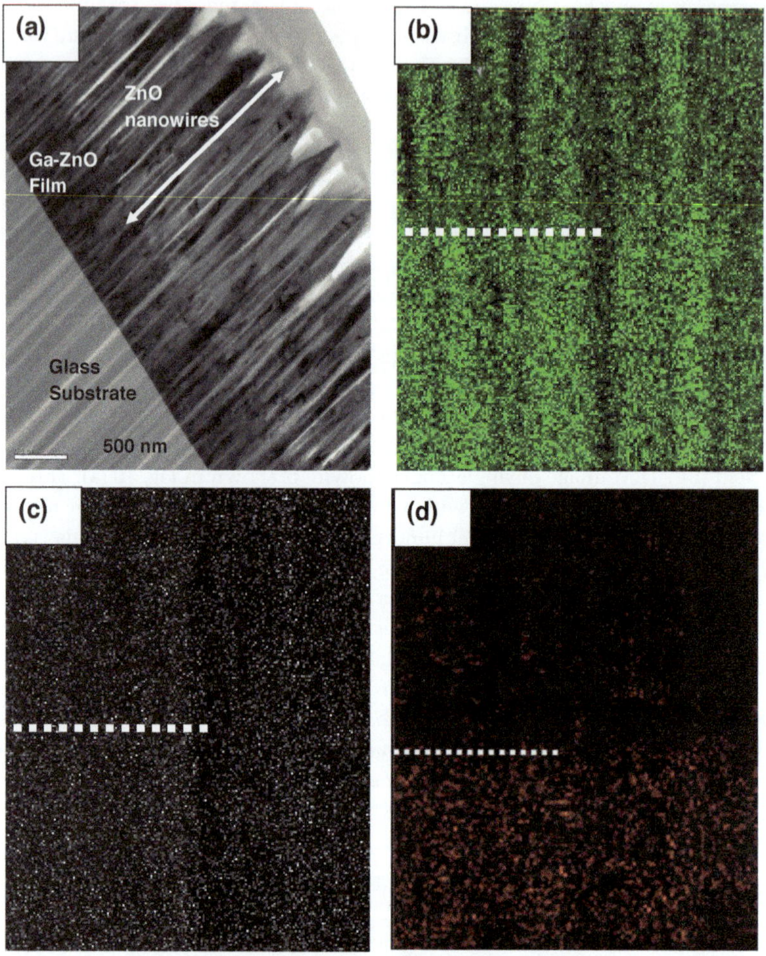

Figure 10. a) Cross-sectional TEM image across the interface of ZnO NWs and GZO film, and EDX spectroscopic mapping images of (a) Zn, (b) O and (c) Ga across the interface.

The mapping images Fig. 10(b-c), of Zn and O atoms were distributed uniformly at the interface, whereas nonuniform distributions of the Ga atoms are clearly observed (Fig.10 (d)). The Ga atoms are appeared not only in the underlying film, but also seen towards the bottom of the nanowires. The Ga concentration in the nanowires decreases with increase in distance from the interface. This observation indicates that the growth temperature provided sufficient thermal energy to Ga atoms in the interfacial film to be diffused into the lower part of nanowires. The concentration of Ga is estimated as 2.95 at.% and 0.32 at.% for GZO film and the bottom of the ZnO nanowires, respectively. The lightly incorporation of Ga towards the bottom of the nanowires, during the initial stage of growth, does not alter the columnar growth due to the similar atomic radius of Ga and Zn as well as the comparable covalent bond lengths of Ga-O and Zn-O. Yan et al. observed the evolution of nanostructures with Ga doping in contrast to undoped ZnO, where Stranski-Krastanov (SK) model was used to explain the phenomena [47]. We have already observed that undoped ZnO film on Si substrate favours the formation of low density, thick nano-pencil structure by MOCVD technique. Thus Ga has an ability to scale down the grain size of the GZO film, which stimulates the formation of large number of homogeneous nucleation sites at the initial stage of nanowires growth. The c-axis orientation in ZnO nanowires can be explained by the "survival of the fastest", where the nuclei were formed with various orientations at the initial stage of the deposition and each nucleus competed to grow but only the nuclei in c-axis orientations could survive due to the fastest growth rates. Once the nanowires have grown, the high-speed laminar gas flow under our specific growth parameters induces turbulent flow between the nanostructures, which results in adsorption of fresh reactant gases only on nanowires tips. Hence the growth of very long nanowires is achieved. It has been reported that beyond a certain doping level ZnO nanowires retains its vertical shape and above this the tip of the nanostructures bends due to the generation of stacking faults and the surface becomes very rough. In the present case, such kind of abnormality has not been seen because of the mild doping of Ga at the bottom of the nanowires during the initial growth.

To study the effect of interfacial layer as well as the shape of the substrate, ZnO micro-craters were fabricated. To do so, AZO thin films were deposited on Corning glass (1737) by pulsed DC sputtering of an AZO ceramic target (Al ~ 2 wt.%) with Ar (5 N) as the ambient gas. The sputtering chamber was evacuated to a base pressure of about 5×10^{-5} Torr. Prior to the sputtering, the target was pre-sputtered to remove any residual contamination from the target surface. The deposition parameters, pulse frequency (2 kHz), DC power (2 KW), substrate temperature (200 ^{0}C), working pressure (3 mTorr) and deposition time (10 min), were optimized and kept constant for all the films. Then the AZO films were anisotropically etched for 90 s by dilute hydrochloric acid (0.5% HCl in H_2O) in order to create textured surfaces. Fig. 11(a–b) shows the 45^0 tilted FESEM images of the as-deposited film and the AZO films etched for 90 s. Before etching, the film is very smooth (Fig. 11(a)). After etching, hexagonal, conical craters with diameters of 0.5–2.5 lm are created on the film surface (Fig. 11(b)). The formation of craters mainly occurs at positions of O-terminated ZnO columns, as it is highly sensitive to acid attack. The exposed facets of the etched surfaces may mainly be (1 0 1) plane [48]. Thereafter, the etched AZO films were put into a metal–organic chemical vapor deposition (MOCVD) chamber for growth of ZnO nanowires. During the growth period, the substrate temperature, working pressure and the nozzle-substrate spacing were fixed at 550 ^{0}C, 3 Torr and 1 cm, respectively. The deposition time was 5 min. ZnO nanowires onto its surface to fabricate the hierarchical structure. Fig. 11(c) shows the SEM images of the hierarchical structure of the ZnO nanowires. All surfaces of the craters are covered by nanowires with cone-shaped tips, 50–100 nm in diameter and 150–300 nm in length. The structure as shown is very similar to that of lotus leaf. The lotus leaf surface is a good example of how, the combination of a micro-/nano-hierarchical structure and low surface energy wax layer, exhibits high water con-

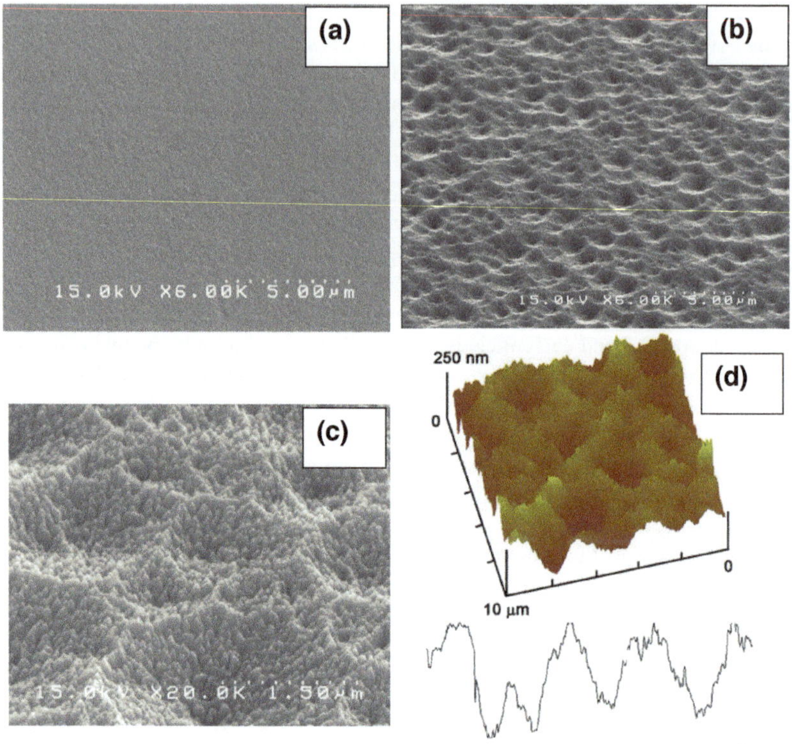

Figure 11. FESEM images (45° view) of the AZO films: (a) as-deposited, (b) etched for 90 seconds (c) nanowires on etched sample. (d) AFM image with cross section profile (J. Xiong et al J. Colloid and Interface Science 350 (2010) 344.).

tact angle (CA) and low CA hysteresis. A superhydrophobic and self-cleaning surface mimics the surface of a lotus leaf. The combination of both self-cleaning and antireflective properties is very desirable for applications in outdoor photovoltaic and display devices, self-cleaning windows and car windshields. A two step method, using a combination of top down and bottom-up approaches, for fabrication of ZnO based hierarchical structures with nanowires on microcraters helps to mimic the lotus leaf structure. Fig. 11(d) shows the AFM image with the corresponding typical cross sectional profile of the etched films after the growth of nanowires. The microcraters can be easily recognized by the hexagonal holes on the surface. Due to the co-existence of micro- and nano-scale features and the non-vertical geometric alignment, the structure can induce multiple scattering of the incident light. The co-existence of micro- and nano-structures can interact with light of different wavelengths, causing reflective suppression over a wide wavelength range [49]. Another advantage of the hierarchical structure over flat surface film having an antireflection property that is less angle dependent. This indicates a potential application in developing Si thin film solar cell, where it can act as a front electrode (replacing ITO, etched AZO or textured FTO film) to achieve improved light trapping. Furthermore, it can also be used in the dye-sensitized solar cell as an electrode for the collection of electrons. The hierarchical structure can trap light, improve dye absorption and introduce a nano-scale pathway for carrier transportation, similar with the flower-like ZnO structure. Similarly, this structure can also be used for enhancing the extraction efficiency of LED, which is usually realized by using photonic crystals, conductive omnidirectional reflectors, or surface roughening of transparent oxide electrodes (normally ITO).

4. Conclusion

Among the different synthesis process, MOCVD is superior to reproduce large area; catalytic free, long, vertically aligned ZnO NWs with excellent morphological, structural and optical properties. As the growth of ZnO nanowires by MOCVD is a bottom–up technique, the nature of substrate, shape of the substrate and interfacial layer on the substrate have vital role for the determination of nanowire dimension and alignment. According to lattice misfit, the most suitable substrate for ZnO growth is $ScAlMgO_4$, which is expensive and technologically inconvenient. Generally, the formation mechanism of nanostructures consists of two stages: nucleation and growth. At first, nanosized crystalline nuclei are formed on the substrate and afterwards the nanowires start to grow along a specific direction. The c-axis orientation in ZnO nanowires can be explained by the "survival of the fastest", where depending on the nature of the substrate, the nuclei were formed with various orientations at the initial stage of the deposition and each nucleus competed to grow but only the nuclei in c-axis orientations could survive due to the fastest growth rates. Once the nanowires have grown, the high-speed laminar gas flow under our specific growth parameters induces turbulent flow between the nanostructures, which results in adsorption of fresh reactant gases only on nanowires tips. Hence the growth of very long nanowires is achieved. At the initial stage of the growth, the width of the nanowires were thin and the individual thin nanowires laterally merged together to give thicker one with increase in growth time. Thus, the diameter of the nanowires depends on the size and density of the initial ZnO nuclei. Also the supply of the fresh reactant during the growth process takes a vital role on the nanostructure. Among different substrates, nanowires on C-plane sapphire show the excellent optical and structural properties. On the other hand, by changing the deposition condition of MOCVD and type of substrate one can synthesize ZnO nanowire of desired structure for different application.

Acknowledgements

This research was supported by WCU (World Class University) program through the National Research Foundation of Korea funded by the Ministry of Education, Science and Technology (R32-20031).

Author details

Sachindra Nath Das[1,3], Jyoti Prakash Kar[2,3], Junjie Xiong[3] and Jae-Min Myoung[3]

1 Department of Physics, Burdwan Raj College, Burdwan, West Bengal, India

2 Department of Physics, National Institute of Technology, Rourkela, Odisha, India

3 Department of Material Science and Engineering, Yonsei University, Seoul, South Korea

References

[1] Pan Z.W., Dai Z.R., Wang Z.L. Science 2001; 291: 1947.

[2] Lee C.J., Lee T.J., Lyu S.C., Zhang Y., Ruh H., Lee H.J. Appl. Phys. Lett 2002; 81: 3648.

[3] Zhao Q., Zhang H.Z., Zhu Y.W., Feng S.Q., Sun X.C., Xu J., Yu D.P. Appl. Phys. Lett. 2005; 86: 203115.

[4] Xu C.X., Sun X.W. Appl. Phys. Lett. 2003; 83: 3806.

[5] Gao P.X., Wang Z.L. Small 2005; 1: 945.

[6] Yan H., He R., Johnson J., Law M., Saykally R.J., Yang P. J. Am. Chem. Soc. 2003; 125: 4728.

[7] Xu J., Pan Q., Shun Y., Tian Z., Sens. Actuators B 2000; 66: 277.

[8] Cheng X.L., Zhao H., Huo L.H., Gao S., Zhao J.G., Sens. Actuators B 2004;102: 248.

[9] Xiangfeng C., Dongli J., Djurišic A.B., Leung Y.H., Chem. Phys. Lett. 2005; 401: 426.

[10] Wan Q., Li Q.H., Chen Y.J., Wang T.H., He X.L., Li J.P., Lin C.L., Appl. Phys. Lett. 2004; 84: 3654.

[11] Liao L., Zhang W.F., Lu H.B., Li J.C., Wang D.F., Liu C., Fu D.J., Nanotechnology, 2007;18: 225.

[12] Huang Y., Yu K., Zhu Z., Curr. Appl. Phys. 2007; 7: 702.

[13] Park S., Kim S., Han S., Nanotechnology 2007; 18: 055608.

[14] Wu L., Wu Y., Pan X., Kong F., Optical Mater. 2006; 28: 418.

[15] Mahalingam T., Lee K.M., Park K.H., Lee S., Ahn Y., Park J.Y., Koh K.H., Nanotechnology 2007; 18: 035606.

[16] Kang H.S., Kang J.S., Kim J.W., Lee S.Y., Appl. Phys. Lett. 2004; 95: 1246.

[17] Leung Y.H., Djurišić A.B., Gao J., Xie M.H., Wei Z.F., Xu S.J., Chan W.K., Chem. Phys. Lett. 2004; 394: 452.

[18] Li Y., Liu F., Sun J., Chem. Commun. 2009; 2730.

[19] Cao M., Song X., Zhai J., Wang J., Wang Y., J. Phys. Chem. B 2006; 110: 13072.

[20] Zhang X., Fujishima A., Jin M., Emeline A.V., Murakami T., J. Phys. Chem. B 2006; 110: 25142.

[21] Lee, W., Jeong, M. C., and Myoung, J. M. Acta Mater. 2004; 52: 3949.

[22] Li, S. Y., Lin, P., Lee, C. Y., and Tseng, T. Y. J. Appl. Phys. 2004; 95(7): 3711.

[23] Sun, Y., Fuge, G. M., and Ashfold, M. N. R. Chem. Phys. Lett. 2004; 396: 21.

[24] Wang, M., Ye, C. H., Zhang, Y., Wang, H. X., Zeng, X. Y., and Zhang, L. D. J. Mater. Sci. Mater. Electron. 2008; 19(3): 211.

[25] Vayssieres, L. Adv. Mater. 2003;15(5): 464.

[26] Gao, X. D., Li, X. M., Zhang, S., Yu, W. D., and Qiu, J. J., J. Mater. Res. 2007;22(7): 1815.

[27] Hsu H.C., Cheng C.S., Chang C.C., Yang S., Chang C.S., Hsieh W.F., Nanotechnology 2005;16: 297.

[28] Ievtushenko A., Karpyna V., Lashkarev G., Lazorenko V., Baturin V., Karpenko A., Lunika M., Danko A., Acta Phys. Pol. A 2008;114: 1131.

[29] Emanetoglu N.W., Patounakis G., Liang S., Gorla C.R., Wittstruck R., Lu Y., IEEE

[30] Trans. Ultrason. Ferroelectr. Freq. Control 2001; 48 (5):1389.

[31] Wang W.S., Wu T.T., Chou T.H., Chen Y.Y., Nanotechnology 2009; 20: 135503.

[32] Sadek A.Z., Wlodarski W., Li Y.X., Yu W., Li X., Yu X., Kalantar-zadeh K., Thin Solid Films 2007; 515: 8705.

[33] Ievtushenko A., Karpyna V., Lashkarev G., Lazorenko V., Baturin V., Karpenko A., Lunika M., Danko A., Acta Phys. Pol. A 2008; 114: 1131.

[34] Liou S.C., Hsiao C.S., Chen S.Y., J. Cryst. Growth 2005; 274:438.

[35] Kim Y.S., Ha S.C., Kim K., Yang H., Choi S.Y., Kim Y.T., Park J.T., Lee C.H., Choi J., Paek J., Lee K., Appl.Phys.Lett. 2005; 86: 213105.

[36] Rasmussen A., Gaitan M., Locascio L.E., Zaghloul M.E., J.Microelectromech. Syst. 2001; 10: 286.

[37] Sanchez J.B., Berger F., Fromm M., Nadal M.H., SensorsActuatorsB 2006; 119: 227.

[38] Fang F., Zhao D.X., Zhang J.Y., Shen D.Z., Lu M., Fan X.W., Li B.H., Wang X.H., Nanotechnology 2007; 18: 235604.

[39] Miyauchi M., Shimai A., Tsuru Y., J. Phys. Chem. B 2005; 109: 13307.

[40] Zhang Y., Du G., Zhu H., Hou C., Huang K., Yang S., Opt. Mater. 2004; 27: 399.

[41] Miyauchi M., Shimai A., Tsuru Y., J. Phys. Chem. B 2005; 109: 13307.

[42] Kar J.P., Lee S.W, Lee W., Myoung J.M., Appl. Surf. Sci. 2008; 254: 6677.

[43] Kar J. P., Das S. N., Choi J. H., Lee T. I., Myoung J. M., Appl. Surf. Sci. 2010; 256: 4995.

[44] Zhu Q.S., Wang Z.G., Appl. Phys. Lett. 2005; 87: 231903

[45] Miyauchi M., Shimai A., Tsuru Y., J. Phys. Chem. B 2005; 109: 13307.

[46] Sun T, Qiu J. Mater Lett 2008; 62:1528.

[47] Xu W.Z, Ye Z.Z, Ma D.W, Lu H.M, Zhu L.P, Zhao B.H, Appl Phys Lett 2005;87:093110.

[48] Li M.K, Wang D.Z, Shi F., Ding S., Hong J. Chin Phys Lett 2007;24:236–9.

[49] Ellmer K., Klein A., Rech B., Transparent Conductive Zinc Oxide, Springer, 2007.

[50] Zhang L., Zhang X., Dai Z., Wu J., Zhao N., Xu J., J. Colloid Interface Sci. 2010; 345: 116.

Chapter 5

Synthesis, Superhydrophobicity, Enhanced Photoluminescence and Gas Sensing Properties of ZnO Nanowires

M.G. Gong, Y.Z. Long, X.L. Xu, H.D. Zhang and B. Sun

Additional information is available at the end of the chapter

1. Introduction

The past decades have witnessed major worldwide interested in the fabrication of nanodevices based on quasi one-dimensional (1D) semiconducting nanostructures (Xu and Wang, 2011; Long *et al.*, 2011; Long *et al.*, 2012). ZnO, an important II-VI group semiconductor with many excellent properties (Xu and Wang, 2011), has been suggested for a diverse range of applications, including optoelectronics (Ko *et al.*, 2011; Wang *et al.*, 2004), ultraviolet laser devices (Liu *et al.*, 2011), gas sensors (Wang *et al.*, 2011; Ra *et al.*, 2008; Wang *et al.*, 2006) and transparent electrodes (Kusinski *et al.*, 2010; Goris *et al.*, 2009), etc. 1D ZnO nanorods, nanowires and nanofibers have been synthesized by a range of techniques, such as top-down approaches by etching (Wu *et al.*, 2004), and wet chemical methods (Vayssieres *et al.*, 2001; Gong *et al.*, 2009), metal-organic chemical vapor deposition (MOCVD) (Yang *et al.*, 2004; Pfüller *et al.*, 2011; Lee *et al.*, 2004), physical vapor deposition (Huang *et al.*, 2001; Jeong and Lee, 2010; Wang *et al.*, 2005; Hsu *et al.*, 2005), molecular beam epitaxy (MBE) (Heo *et al.*, 2002), pulsed laser deposition (Shen *et al.*, 2010; Cao *et al.*, 2007), sputtering (Chiou *et al.*, 2003), flux (Kong and Li, 2003), electrospinning (Sen *et al.*, 2011; Wu *et al.*, 2008; Lin *et al.*, 2007; Sui *et al.*, 2005) methods, etc.

Among those techniques, MOCVD and MBE can yield high quality ZnO nanowires arrays, but usually suffer from low product yield, poor deposition uniformity, and limited choices of substrate. Especially, the fabrication cost is usually very high, so they have been less widely used. The physical vapor deposition and flux methods usually require high temperature, which are less likely to be able to integrate with flexible organic substrates for future foldable and portable electronics, and can easily incorporate catalysts or other impurities into the ZnO nanowires. Top-down, pulsed deposition and sputtering approaches have less reproducibility and controllability compared with other techniques. Electrospinning usually yields polycrystalline nanowires or fibers, which may lower down the mobility of charge carriers in devices. Compared with the above methods, wet chemical methods are lower cost, higher reproducibility, less hazardous, easier controllability, and thus can be able to large area of preparation. Meanwhile, the nanowires growth temperature is relatively low and compatible with most flexible organic substrates. There are a variety of parameters that can be tuned to effectively control the nanowires length, diameter, distribution and properties of the final product (Gong *et al.*, 2009; Govender *et al.*, 2004). In summary, wet chemical methods have been demonstrated as one of very powerful and versatile techniques for growing 1D ZnO nanowires.

In this chapter, we focus on the 1D ZnO nanostructures that have been prepared by simple hydrothermal self-assembly method and electrospinning. The morphology of ZnO nanostructures were characterized by field emission scanning electron microscopy (SEM). The special reticulate structure assembled by ZnO nanowires exhibits superhydrophobicity with a contact angle of 170° and a sliding angle of 2°. Significant photoluminescence enhancement from ZnO nanowires coated with Ag nanoparticles is observed at 387.6 nm, the ratio of the enhancement reaches 6.6 compared with pure ZnO nanowires. Electrospun ZnO nanowires have also been fabricated into field-effect transistors and sensor devices. The sensors exhibit fast and large response to CO even at room temperature due to highly porous, continuous and compact-grains structures of the electrospun ZnO nanowires.

2. Experimental details

2.1. Hydrothermal self-assembly method

Well-aligned single crystalline ZnO nanowires were prepared from zinc acetate dehydrate in a neutral aqueous solution under hydrothermal conditions. In detail, the Si substrates were first immersed in boiling 98% H_2SO_4 for 10 min and then ultrasonically washed with acetone, ethanol and deionized water, respectively. In order to fabricate well-aligned dispersed ZnO nanowires, a two-step method was used (Wang *et al.*, 2008; Gong *et al.*, 2010): (1) A 10 nm ZnO seed layer was first deposited on a Si substrate by a radio-frequency magnetron sputtering method. A commercially supplied ZnO ceramic plate with a purity of 99.99% was used as a target. The background pressure of the vacuum chamber was 1×10^{-4} Pa. A mixed gas of oxygen and argon with a volume ratio of 1:1 was used as the sputtering gas with a total pressure of 3 Pa. The ZnO seeds were annealed in air for 1hr at 550°C. (2) The hydrothermal growth was carried out at 100°C by immersing the substrates vertically in a mixed aqueous solution containing zinc acetate dehydrate (Zn-AD) and hexamethylenetetramine (He-T). The mixed volume ratio of the two solutions is 1:1, in which the concentration of the two solutions is always the same (i.e. 0.01M/L zinc acetate dehydrate mixed with 0.01M/L hexamethylenetetramine, etc.). The growth time is 12hr. After growth, the substrates were dried in the Far Infrared Drying Oven at 80°C for 5 min (samples 1, 3-6) just after they were immersed in deionized water for 10 seconds. As a comparison, sample 1* is the same as sample 1 except it was not immersed in deionized water. Another substrate was immersed in ethanol for 30min, then in n-Hexane for 90min, drying under natural conditions (sample 2). The reason we chose the different temperature is that the evaporation rate of water at 80°C is almost the same as that of ethanol and n-Hexane at room temperature. Hydrothermal growth and treatment process of ZnO samples are shown in Table 1.The results and the formation mechanism will be discussed in section 3. The morphology of the ZnO film was revealed by SEM images.

samples No.	solution composition	length : diameter	nano-wires density	AA/RA	CA/SA	treatment process
sample 1	Zn-AD 0.01M/L He-T 0.01M/L	104 : 1	$(2.0\pm0.5)\times10^8/cm^2$ (papillary nodes density)	171° / 169.5°	170° / 2°	Immersed in deionized water for 10 seconds, drying in the Far Infrared Drying Oven at 80°C
sample 1*	Zn-AD 0.01M/L He-T 0.01M/L	104 : 1	$(2.9\pm0.5)\times10^8/cm^2$ (papillary nodes density)	169° / 166°	167° / 3°	Without immersed in deionized water, drying in the Far Infrared Drying Oven at 80°C
sample 2	Zn-AD 0.01M/L He-T 0.01M/L	104 : 1	$(2.0\pm0.5)\times10^9/cm^2$	167.5° / 162°	165° / 5°	Immersed in ethanol for 30min, then in n-Hexane for 90min, drying under natural conditions
sample 3	Zn-AD 0.02M/L He-T 0.02M/L	57:1	$(4.0\pm0.5)\times10^9/cm^2$	160° / 151°	157° / 10°	same as sample 1
sample 4	Zn-AD 0.025M/L He-T 0.025M/L	47:1	$(3.1\pm0.5)\times10^9/cm^2$	158° / 142.5°	156° / 18°	same as sample 1
sample 5	Zn-AD 0.04M/L He-T 0.04M/L	17:1	$(1.3\pm0.5)\times10^9/cm^2$	156.5° / 132.5°	153° / 25°	same as sample 1
sample 6	Zn-AD 0.05M/L He-T 0.05M/L	6:1	$(2.4\pm0.5)\times10^8/cm^2$	157° / 145°	154° / 15°	same as sample 1

*The advancing and the receding contact angles (AA/RA), and the contact and the sliding angles (CA/SA) of water droplets on the samples surface (error: ±1°).

Table 1. Hydrothermal solution composition (Zn-AD: zinc acetate dehydrate,andHe-T:hexamethylenetetramine) process and superhydrophobicity of the ZnO nanowires structure

2.2. Electrospinning

Electrospinning has been recognized as an efficient and highly versatile method which can be further developed for mass production of uniform, ultrafine and continuous fibers with nanometer-to-micrometer sized diameter. Up to date, a variety of materials such as polymers, metal oxides, ceramics, metals and carbon have been successfully electrospun into ultrafine fibers mostly from solvent solution(Huang *et al.*, 2003; Tan *et al.*, 2008; Li *et al.*, 2010; Li *et al.*, 2011; Sun *et al.*, 2012; Zhang *et al.*, 2012; Zheng et al., 2012). In this work, ZnO nanowires are also fabricated by calcination of the electrospun fibers of polyvinyl acetate (PVA)/ zinc acetate composite. The processing involved three steps: (1) Preparation of a sol with suitable inorganic precursor and polymer content, and achieving an appropriate viscosity for electrospinning. (2) Spinning of the solution to obtain composite fibers. (3) Calcination of the composite fibers to remove the polymer coumpounds and obtain the pure oxide nanofibers. In a typical procedure, aqueous PVA solution (2.0 g PVA and 18.0 g deionized water) is dropped slowly into the solution of zinc acetate (2.0 g zinc acetate and 2.0 g deionized water), and the reaction proceeds in a water bath at 90 °C for 2 h.The PVA/zinc acetate composite is cooled down to room temperature, and then the precursor solution is introduced into a syringe with metal spinneret, which is connected to positive electrode of a high-voltage power supply. The spinneret tip-to-collector distance is usually in the range of 8-12 cm, and a dc high voltage in the range of 10-15 kV is necessary to generate the electrospinning, as shown in Fig. 1a. A dense web of PVA/zinc acetate composite wires is then collected on the aluminium foil. Finally the as-electrospun composite wires are calcinated at elevated temperatures (500-600°C) to decompose the PVA organic components, meanwhile the inorganic precursors oxidize and crystallize to form ZnO nanowires.

Fig.1b-e show the morphology evolution of electrospun ZnOwires following hydrolyzation, electrospinning, and calcination processes by our group. After calcination, the average diameter of nanowires appears to be decreased (80-200 nm) attributed to the loss of polymer compounds from the composite fibers and the crystallization of metal oxide. Moreover, the wires often exhibited continuous and compact-grains structures. It reveals that the electrospun wires have a high surface-to-volume ratio. This peculiar morphology enlarges contact area between the fiber and surrounding gas. It may play a pivotal role in the ultra-high sensitivity gas sensor of metal oxide by electrospinning (Kim *et al.*, 2006; Landau *et al.*, 2009).

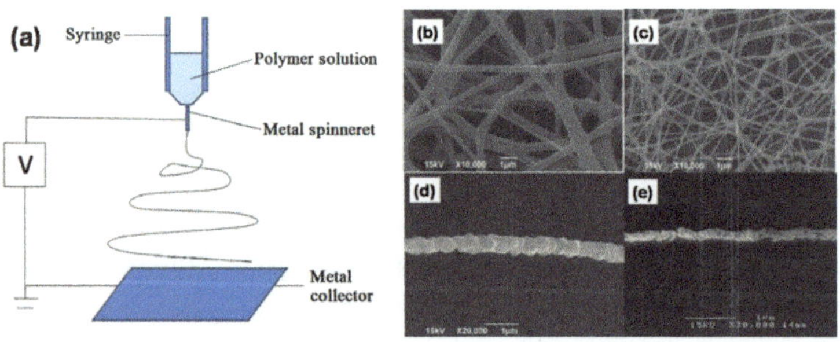

Figure 1. a) Schematic illustration to show polymer nanofibers by electrospinning.SEM images of (b) as-spun PVA/zinc acetate composite fibers and (c) ZnO fibers calcined at 600 °C for 5 h. (d-e) High-magnified SEM images of individual ZnO fibers.

3. Superhydrophobicity of ZnO Nanowires by Hydrothermal Method

3.1. Amalgamation of ZnO Nanowires

Fig. 2 shows the long and slender ZnO nanowires fabricated on the seed layer in the mixed aqueous solution of Zn-AD plus He-T. The ratio of length to diameter is about 104 to 1. Interestingly, the ZnO nanowires in sample 1 self-assembled into reticulate papillary node film (ZnO-RPNF). The growth process of the ZnO-RPNF is illustrated in Fig.3.As shown in Fig.3, during the evaporation and coagulation process of water film on the ZnO nanowires, the ZnO nanowires were aggregated to be a reticulate papillary nodes structure(see Fig.2a). As for the comparison, the ZnO nanowires treated by ethanol and n-Hexane only have a small bending and also remain in their separate condition (see Fig.2c). On the other hand, the nano-wires of the sample 1* soaked in residual water of the reacted solution can also be bended and assembled to a papillary like structure (as shown in Fig.2e), after the water evaporated. Since the solution on the surface of sample 1* was not pure water and the surface tension of the solution is lower, its bending and assembly grade is lower than that of sample 1. The bending and bundling mechanism of vertically aligned ZnO nanowires synthesized by vapor phase deposition method was discussed by other group (Liu *et al.*, 2008). In their report the bending mechanism was attributed to electrostatic interactions between different termination facets with opposite charges. However, the mechanism is different in a liquid phase. Although there are a lot of positive and negative particles in an aqueous solution, a large number of homologous charges would not gather in a given region in a stoichiometric solution. So the EI effect among ZnO nanowires grown by the liquid phase deposition method can be ignored. At 80 °C, the water film on the top of ZnO nanowires evaporates and contracts rapidly to form tiny drops. Therefore, the water has a centripetal surface tension, making the nanowires to be bent and aggregated. On the contrast, the low liquid surface tension (n-Hexane) makes a little bending but not aggregation; the specific reason is explained by the tension junction model.

It is known that the wettability of an ideal surface, expressed with contact angle θ of water droplets, is given by Young's equation:

$$\cos\theta = \frac{\gamma_{SV} - \gamma_{SL}}{\gamma_{LV}} \tag{1}$$

where γ_{SV}, γ_{SL} and γ_{LV} refer to the interfacial surface tensions with solid-vapor, solid-liquid and liquid-vapour, respectively. Young's angle is a result of the thermodynamic equilibrium of the free energy at the solid-liquid-vapor interphase.

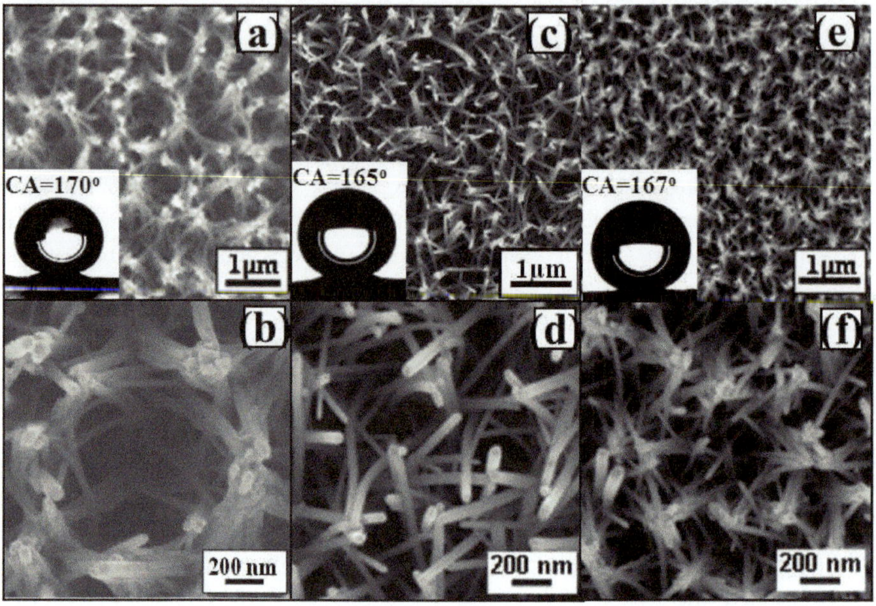

Figure 2. FESEM images of (a) and (b) sample 1, (c) and (d) sample 2, and (e) and (f) sample 1*, where (b), (d) and (f) are the high-magnification for (a), (c) and (e), respectively. The insets show images of droplets that formed on the surface of the different samples during contact angle measurement.

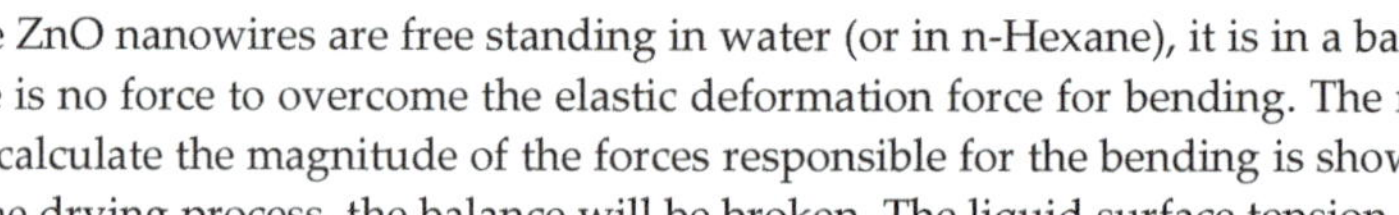

Figure 3. Formation mechanism of the reticulate ZnO film sample 1.

When the ZnO nanowires are free standing in water (or in n-Hexane), it is in a balance state, and there is no force to overcome the elastic deformation force for bending. The model proposed to calculate the magnitude of the forces responsible for the bending is shown in Fig.4. During the drying process, the balance will be broken. The liquid surface tension can be divided parallel to an X-axis factor γ_1 and a Y-axis parallel to the nanowire edge factor γ_2. Note that the γ_2 and the pressure difference $\Delta P=P_L-P_G$ (P_L and P_G are the pressure of the liquid and the atmospheric pressure, respectively) on the droplets arising from the difference in the hollow of the drop cannot let the nanowires be bent. The self-assembling force F_1, parallel to γ_1.F_1 can be obtained from:

$$F_1 = F\sin\theta = \gamma b \sin\theta \tag{2}$$

where $\gamma = \gamma_1 + \gamma_2$ is the liquid tension, $b=\sqrt{3}d/3$ is the hexagonal side length of the ZnO nanowires (d is the diameter of the ZnO nanowires).

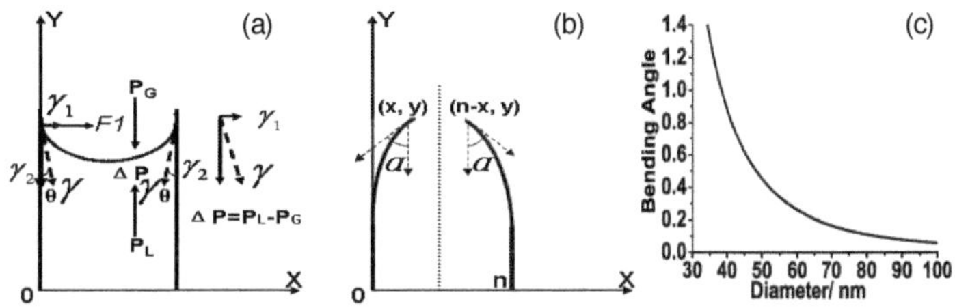

Figure 4. Schematic model of liquid evaporation induced change of the bending angles between two ZnO nanowires, where (a) indicates the initial state of water with two nanowires system; (b) is any intermediate state during the bending process.The bending angle vs. the nanowires diameter is as shown in (c).

To simplify the case, we just consider the liquid surface tension between two nanowires. The initial state for two nanowires with lengths T is shown by Fig.4a, where the two nanowires lengths are defaulted as the same. The self-assembling parallel tension pressed on ZnO nanowires with water F_w and with n-Hexane F_h can be expressed as:

$$\frac{F_w}{b} = \gamma_w \sin\theta_w \tag{3}$$

$$\frac{F_h}{b} = \gamma_h \sin\theta_h \tag{4}$$

where γ_w and γ_h are the surface tension of water and n-Hexane, θ_w and θ_h are the contact angle of water and n-Hexane on a ZnO nanowire, respectively. γ_w=62.61mN/m at 80°C, and γ_h=18.4 mN/m, θ_w =33°, θ_h= 0° at 20 °C, respectively. From the calculation, we observe F_w=34.1bmN/m, and F_h=0. For sample 1, the average diameter of the nanowires d is 42.3 nm. Therefore, the parallel surface tension of water is F_w= 8.33×10^{-10} N.

A tension junction model as shown in Fig.4 is used to estimate the minimum parallel surface tension which is necessary to make two nanowires contact each other. For simplicity, we assumed two symmetric flexible nanowires separated by a distance n. Furthermore, we approximated the parallel surface tension applied on the top of the nanowire. The dependence of the lateral displacement x perpendicular to the nanowire on the elastic bending force applied at the top is given as (Song *et al.*, 2005):

$$F_{elastic} = \frac{3EI}{T^3} x \tag{5}$$

where *T*, *E* and *I* are the length, Young's modulus and moment of inertia of the ZnO nano-wire, respectively. The moment of inertia of a hexagonal-cross-section nano-wire can be expressed as $I = 5\sqrt{3} d^4 / 144$. To make these two symmetric wires a contact at $x=n/2$, the parallel surface tension has to be larger than the bending force for $n/2$ displacement.

Reported values for elastic modulus of ZnO nanostructures vary greatly depending on experimental methods and nanostructure dimensions. For example, Wang *et al.* measured the elastic modulus of vertical ZnO nanowires on sapphire to be 29±8 GPa based on atomic force microscopy (Song *et al.*, 2005). Their nanowires had an average diameter of 45nm and length ranging from 0.277 to 0.638 μm. By manipulating the vertical ZnO nanowires with an atomic force microscopy tip, Hoffmann *et al.* measured an average elastic modulus of about 100 GPa for 250 nm diameter ZnO nanowires (Hoffmann *et al.*,2007). Huang *et al.* used transmission electron microscopy to measure the elastic bending modulus of ZnO nano-wires via electric-field-induced resonant excitation and obtained a mean elastic modulus of ~58 Gpa (Huang *et al.*, 2006). In particular, they measured the elastic modulus of 48.9GPa from a ZnO nanowire with diameter of 43 nm and length of 4.77 μm, which is similar to that (d=42.3 nm, T=4.41 μm) of sample 1. If choosing E=50 Gpa while n = 224 nm based on the nanowires density in Table 1 to calculate the elastic bending force, we therefore obtain

$$F_{elastic} = 3.77 \times 10^{-11} (N) \tag{6}$$

For sample 1 the water's parallel surface tension F_w= 8.33×10^{-10} (N) is about 22 times more than $F_{elastic}$. Therefore, the surface tension of deionized water is big enough to let two or more nanowires bundling together to form the reticulate papillary nodes structure. Note that although the parallel surface tension of n-Hexane is $F_h = 0$, after the substrate was immersed in the n-Hexane, a little bending of the nanowires of sample 2 still happened (see Fig.2d), due to that the residual water in solution cannot be thoroughly removed.

As shown in Fig.4b, the angle α (the angle between the Y-axis and the tangent at the nano-wire tip) is used for defining any intermediate state during the bending process. The relationship between the F_1 (required to overcome the bending-induced elastic energy, for water: $F1= F_w$) and the bending angle is $\alpha=F_1T^2/2EI$ (Hibbeler, 1997). Since there is no other external force in X-axis, the bending forces experienced by the two nanowires must be equal. To simply obtain the relation between the bending angle and the nanowires diameter, we assumed that the parallel surface tension of water, Young's modulus and the length of ZnO nanowires are constant as F_w=8.33×10^{-10}N, E=50 GPa and T=4.41 μm, respectively. The curve in Fig.4c is the bending angle as a function of the nanowires diameter of samples 1, 3-6. The bending angle decreases sharply as the diameter increases. The bending will be faded out as the diameter increases enough.

All obtained ZnO samples are wurtzite structure, as shown by SEM images in Figs.2 and 5. Fig.2a is SEM top image of the reticulate ZnO film, exhibiting the homogeneous micro/nanostructures on a large scale, Fig. 2b is the high magnification. The top inset of Fig.5a shows a selected area electron diffraction pattern of the ZnO-RPNF that confirms the single crystalline property of the ZnO nanowire. The average reticulate width and density, and the factors for the surface roughness of the reticulate ZnO film, are also determined from top-view SEM image. The reticulate ZnO nanowires (sample 1) with the diameters from 20 to 60 nm (the average is 42.3 nm) are even distributed on the whole substrates. The density of surface nodes (top of the nanowire or the papillary) was decreased via assembly of about ten nanowires to one papillary node. In our data, the density was changed from $(2.0 \pm 0.5) \times 10^9/cm^2$ to $(2.0 \pm 0.5) \times 10^8/cm^2$ (see Table1).

It is also found in Table 1 that the nanowires' density decreased as the zinc acetate dehydrate concentration increased, due to the nanowires fusing easily together in their growth process. However, as the concentration is very low, the density of nanowires is also lower which is due to the fact that some seed particles did not grow into the longer nanowires, just like sample 1.

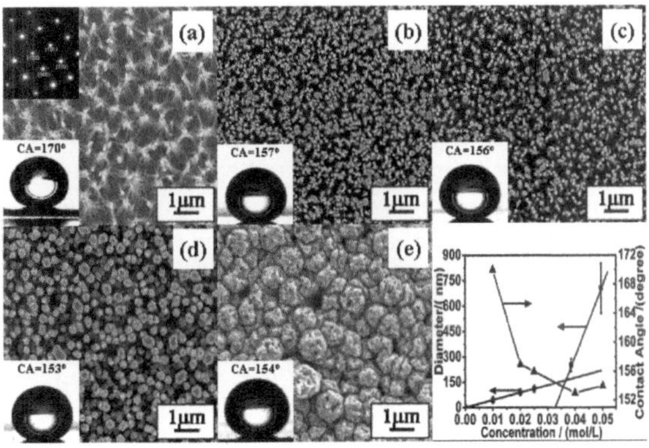

Figure 5. The surface FESEM images of ZnO nanowires of (a) sample 1, (b) sample 3, (c) sample 4, (d) sample 5, and (e) sample 6. The insets show images of the droplets that formed on the surface of the different samples during the contact angle measurement. The top inset of (a) shows the selected area electron diffraction pattern of the ZnO-RPNF that confirms the single crystalline ZnO nanowire. Variation of the ZnO nanowire's diameter and water contact angle measured on the surface of different samples (sample 1, samples 3 to 6) as a function of the concentration of hexamethylenetetramine and zinc acetate dehydrate solution is shown in (f), with the variation range of the nanowires diameter indicated by the error bars.

The general trend was that as ZnO nanowires had a larger aspect ratio and smaller diameter, they were more susceptible to assemble reticulate film with papilla structure. For example, for any given diameter and density, nanowires would not overcome their own elastic modulus and assemble, but assembling by liquid surface tension would become prominent as nanowires grew longer and thinner. The assembling degree of nanowires in Fig.5 was consistent with this general trend. The aspect ratio and diameter in Fig.5a-e were 104:1 and

42.3nm, 57:1 and 88.8nm, 47:1 and 110.3nm, 17:1 and 246.3nm, and 6:1 and 698.5nm, respectively. As expected the degree of assembling was largest in Fig.5a and smallest in Fig.5e.The diameter d variation as functions of the concentration c is shown in Fig.5f. The function is expressed as d=4440c (nm) when the concentration is lower than 0.04M/L, indicating a growth mode for single nano-wire. However, as the concentration increases to 0.04M/L and 0.05M/L, the variation function changes to d=45220c-1562.5(nm). The slope of which is 10 times more than the former.This could be attributed to the merging between the nanowires. As the solution concentration increases, the diameters of the nanowires increase while the distance between them decreases. When the diameters reached a certain value, the spaces between the nanowires became zero, resulting in the amalgamation of nanowires.

3.2. Superhydrophobicity

Wettability of a solid surface has attracted much attention during the past decades, due to its broad technological applications and fundamental researches. A direct expression of the wettability is the contact angle of a water droplet on the surface. Surfaces with very high water contact angles particularly greater than 150° are usually called superhydrophobic surfaces, which have been extended to more applications, such as self-cleaning materials, anti-fog, anti snow, fluid microchips and microreactors (Richard *et al.*, 2002; Erbil *et al.*, 2003; Jiang *et al.*, 2004; Chen *et al.*, 2006).

The obtained ZnO-RPNF was superhydrophobic (see the insert of Fig.2a). The water contact angle is measured to be 170±1°, and the water volume is 10μL. The contact angle hysteresis, which is the difference between the advancing and receding angles, is measured to be less than 2°. The water droplets roll off of the substrate at a sliding angle of less than 2°. The water contact angle on the plane ZnO thin film modified via heptadecafluorodecyltrimethoxysilane (HTMS) was about 114±1°, which is near to the maximum 119° (Nishino *et al.*, 1999). The largest variance in the contact angle is due to the surface roughness. These data suggest that the predominant contribution to the large contact angle is the reticulate papillary nodes structure. Furthermore, the obtained ZnO-RPNF maintains superhydrophobic properties even after ambient environment for three months, showing their favorable stability. This is very important for extending the practical applications of ZnO materials.

The basic effect of the surface structure and chemistry on superhydrophobicity has been elucidated by Wenzel (Wenzel, 1936) and Cassie (Cassie and Baxter, 1944) models. Wenzel's model hypothesizes that the liquid completely penetrates into the troughs of the surface texture. Cassie's model is based on the assumption that the liquid drop doesn't fill the troughs of the rough surface but sits on a composite surface composed of solid material and air. Both the Wenzel and Cassie models have put emphasis on the geometrical structure of a solid surface as an important factor in determining the hydrophobicity. The trapping of air in the surface troughs causes an increase in the water contact angle according to Cassie-Baxter'sequation (Jiang *et al.*, 2004).

$$\cos\theta^* = f_1\cos\theta_1 + f_2\cos\theta_2 \tag{7}$$

$$f_1 + f_2 = 1 \tag{8}$$

where $\theta^*$is the contact angle on HTMS coating the rough surface of reticulate ZnO film with papillary nodes, θ_1 and $\theta_2$are the contact angles on HTMS coating the plane ZnO film and air, respectively; f_1 and $f_2$are the fractions of the papillary nodes' end surface and of the trapped air on reticulate ZnO film. Here, θ_2 is 180°. Substituting θ_2 and (8) into equation (7) will lead to:

$$\cos\theta^* = f_s \cos\theta_e + f_s - 1 \tag{9}$$

here f_s is the fractions of the papillary nodes end surface, and $\theta_e$is the contact angle on HTMS coating ZnO plane film (θ_e=114°).

We notice from eq. (8) that if the reticulate ZnO film with papillary nodes has a very low f_s, a superhydrophobic thin film with a high water contact angle could be fabricated. The value of f_s is estimated by the SEM experiments. As mentioned above, the average diameter of one nanowire is 42.3 nm, and the growth density D is about $(2.0\pm0.5)\times10^9$ per square centimeter by a mass of statistic.Due to the top of ZnO nanowire being hexagonal, the nanowire diameter d and the growth density D is 42.3nm and 2.0×10^9 per square centimeter, respectively. The f_s is obtained as:

$$f_s = \left(2\times\frac{\sqrt{3}}{12}d^2 + \frac{d^2}{2\cos 30^\circ}\right)D = \frac{\sqrt{3}}{2}d^2 D = 0.031 \tag{10}$$

Substituting (10) into eq. (9) will lead to:θ^*=169°. The calculated value is in good agreement with the experimental datum 170°±1° of sample 1. However, the experimental data of sample 1* and sample 2 are 167° and 165° (see Table 1 and Fig.2), respectively. We know from the Cassie model that the air trapped in the surface structure plays an important role for the surface with a larger contact angle. The reticulate structure may serve as a means of trapping sufficient air for larger contact angles to be exhibited. A reticulate papillary node on sample 1 is the assembly of about ten nanowires on average, and the wires bending with a spiral line structure before bundling together at the wires' top. This makes the nano-papillary nodes distribute with circularity and forms a larger column-like micro space (see Fig. 2b). This micro/nano complex structure is closed to the bionic lotus-leaf-structure and much in line with the Cassie model. Nevertheless, the papillary nodes of sample 1* are bended simply by seven nanowires; there is no larger mirco space as that in sample 1. Therefore its superhydrophobicity is lower than that of sample 1. As a comparison, the contact angle of sample 2 is the lowest, because it has fewer or no reticulate papillary nodes.

The contact angle on the surface of different samples formed by nanowires tends to decrease as the surface density of nanowires and the width between them decrease (see Table 1 and Fig.5). As represented in Fig.5f, the contact angle decreases from 170° (sample 1) to 154°

(sample 5). This result is consistent with that of the superhydrophobicity behavior determined by the density of the nanowires and the space between them (Borras *et al.*, 2008). Note that although the nanowires fused together, the contact angle on the surface of sample 6 (Fig.5e-f) is not the minimum, due to the protuberance on the surface.

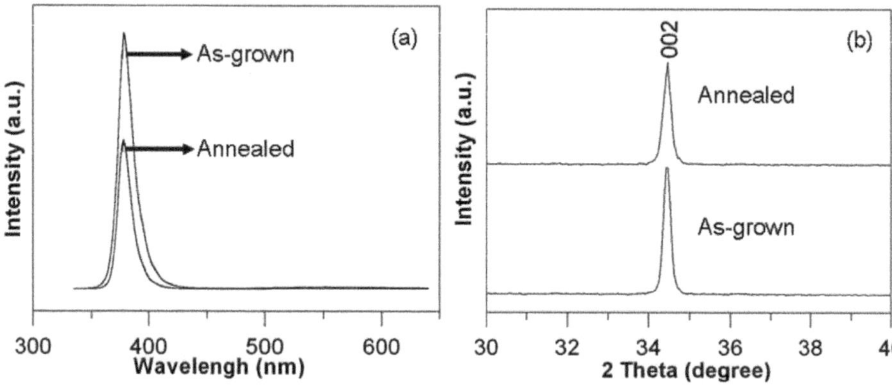

Figure 6. Room-temperature photoluminescence spectra (a), and XRD patterns (b) of ZnO as-grown and annealed at 550 °C for 1h.

4. Photoluminescence of ZnO nanowires via hydrothermal method

ZnO exhibited a wide band gap at room temperature with a large exciton binding energy, which is suitable for effective ultraviolet emission. However, due to the poor crystalline quality of the nano materials, *i.e.* high density of structure defects, the ultraviolet emission of nano-scaled ZnO is liable to be quenched and only defect emission in visible region is detected (Kong *et al.*, 2001). This deficiency hinders process for the application of ZnO in opto-electronic and lasing devices. Therefore, improving the crystal quality of ZnO by synthetic processing and realizing ultraviolet emission and lasing are still major challenges.

The study of photoluminescence is a favorable way to evaluate both ZnO optical properties and its structural defects. Normally, ZnO nanostructures can exhibit ultraviolet emission resulting from the recombination of free excitons and visible emission due to some structural defects(Djurisic and Leung, 2006). Fig.6a gives the room-temperature photoluminescence spectra of ZnO nanowires growing at low temperature of 50 °C by hydrothermal method with as-grown and annealed, in which only ultraviolet emission peaks centered at about 378.5 nm dominate, while the well-known broader emission situated in the blue-red part of the visible spectrum could not be observed (Wang *et al.*, 2007). This result can be attributed to the fine crystallization of the ZnO nanowires. The ultraviolet emission intensity of the as-grown nanowires is better than the annealed one, which indicates the annealing process decreased the crystal quality of ZnO just as the XRD showed Fig.6b. Therefore, ZnO nanowires growing at lower temperature by hydrothermal method have huge potential to fabricate high crystalline quality and good ultraviolet luminescence devices.

The ultraviolet emission intensity of ZnO nanowires is very important in lasing devices. As mentioned above, the intensity can be enhanced by improving crystalline quality of ZnO nanowires. In addition to amending their own quality, surface plasmons or localized surface plasmons of various noble metals also can be used to realize ZnO nanowires photoluminescence enhancement. You's group sputtered ZnO film on Si (001) substrate which had already been coated with 100 nm Ag film previously, and the ultraviolet emission of the composite is found to be greatly enhanced (You *et al.*, 2007). Xiao *et al.* have also found enhancement of near-band-emission of ZnO by coating a layer of Ag implanted silica film(Xiao *et al.*, 2010).

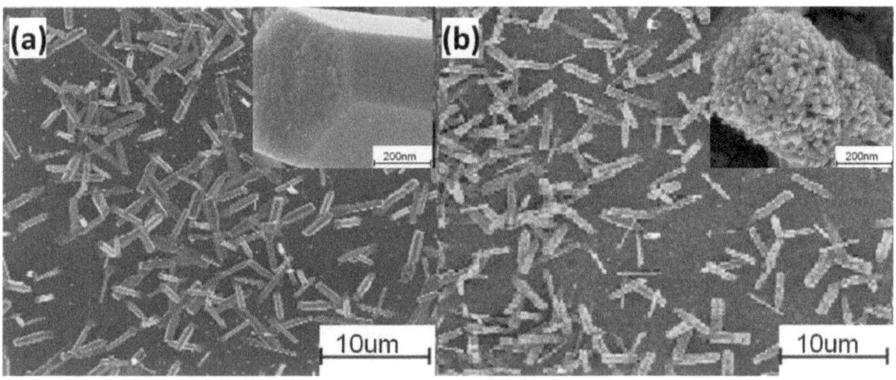

Figure 7. SEM images of (a) ZnO nanowires non-coating and (b) ZnO nanowires coated with separated Ag nanoparticles.

ZnO nanowires grown by the two-step method as mentioned above and annealed in air at 400°C for 1 hr in order to eliminate the defects in ZnO. Then, 40 nm Ag deposited on the ZnO nanowire arrays surface by the radio-frequency magnetron sputtering method at room temperature. Fig.7 shows SEM images of ZnO nanowires non-coating and coated with Ag nanoparticles on the surface. The nanowires, dispersed, lay on the substrate, 4-6 μm in length and 400-600 nm in diameter, and formed a perfect hexagonal structure. It is clear that the nanowires in Fig.7 show almost no difference-either in size or density. There is apparently a layer of 40-50 nm Ag nanoparticles on the nanowires in the inset image of Fig.7b, and the roughness of the surface of the metal is quite high.

The photoluminescence spectra of the ZnO nanowires non-coating and coated Ag nanoparticles are shown in Fig.8. The spectrum of the ZnO nanowires is shown as curve *a*. The peak position at 381.6 nm with 87 meV full width at half maximum, this was attributed to FX-1LO (Shan *et al.*, 2005; Hamby *et al.*, 2003). The emission for the visible light mostly caused by the defect in

ZnO is very week, indicating good quality of the ZnO nanowires. Accordingly, the ZnO nanowires are one of the good optional materials for ultraviolet lasing devices. Curve *b* in Fig.8 is the spectrum of ZnO nanowires coated with 40 nm Ag nanoparticles, showing a greatly enhanced photoemission. The ratio enhancement reaches 6.6 compared with pure ZnO nanowires. Therefore, the photoluminescence of the Ag/ZnO nanowires is apparently stronger than that of ZnO nanowires non-coating because of the intensive coupling between ZnO nanowires and localized surface plasmons of Ag nanoparticles (Lü *et al.*, 2008), which informs us of a good orientation for the composite of metal and semiconductor as a laser source material.

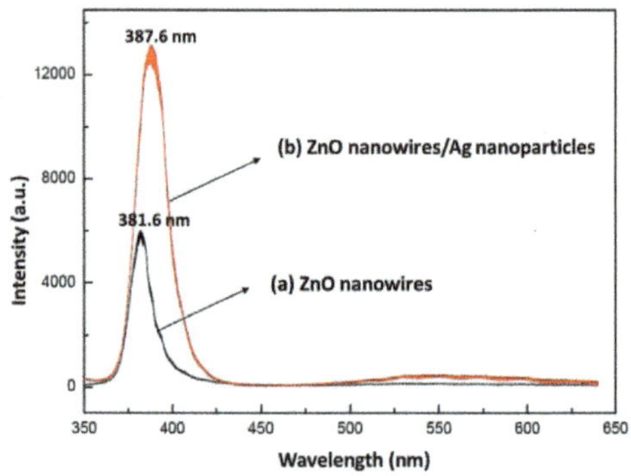

Figure 8. Photoluminescence spectra of (a) ZnO nanowires non-coating and (b) ZnO nanowires coated with 40 nm Ag nanoparticles.

5. Electrical and sensing properties of ZnO nanofibers via electrospinning

Besides hydrothermal self-assembly method, ZnO nanowires were also fabricated by calcination of electrospun poly(vinyl alcohol)/zinc acetate precursor wires at 600 °C for 5 h, as shown in Fig. 1. Raman spectroscopy and XRD were used to confirm the formation of a pure ZnO phase in the nanowires after calcination treatment. Fig.9a shows the clear phonon modes of the wires calcined at 600 °C: the two phonon modes at 439 and 582 cm^{-1} could be assigned to E_2 (high) and A_1 (LO), separately; the phonon mode at 325 cm^{-1} is a second-order phonon, which generally originates from the zone-boundary phonons of $2E_2$ (low). This Raman spectrum is typical for pure, polycrystalline ZnO. Fig.9b gives the XRD pattern for the as-spun fibers after calcinations at 600 °C for 5 h. A clear diffraction pattern is shown, and nine reflection peaks appear at 2θ=31.9° (100), 34.6° (002), 36.5° (101), 47.7° (102), 56.8° (110), 63.1° (103), 66.5° (200), 68.1° (112), and 69.2 (201). It agrees well with the JCPDS card, No. 89-1397 for pure ZnO, confirming the as-prepared product is pure ZnO nanowires after calcination at that temperature.

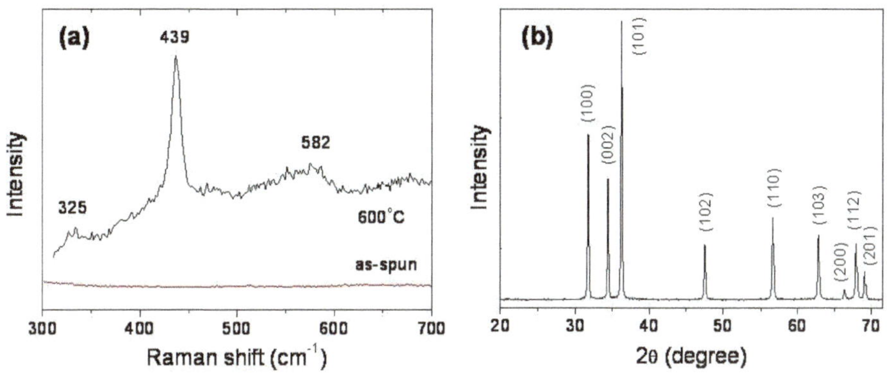

Figure 9. a) Raman spectra of as-spun PVA/zinc acetate nanowires before and after calcinations at 600 °C for 5 h in atmosphere. (b) XRD pattern of the ZnO nanowires.

To fabricate ZnO nanowire-based field-effect transistor or gas sensor device, SiO_2/Si chip with pre-patterned Au electrode arrays via traditional e-beam lithography technology was used as collector to collect PVA/zinc acetate nanowires. The collecting time was about 5 s. The chip was then calcined at 600 °C for 4 h to obtain ZnO nanowires device. As shown in Fig. 10a, the square Au electrodes have a side length of 50 μm, and a long ZnO wire is clearly deposited across the electrodes. In another device, PVA/zinc acetate nanowire yarns were deposited on SiO_2/Si chip by a two-parallel-electrode method or gapping method (Tan *et al.*, 2008). After calcination of the chip at 600 °C for 4 h, two electrodes made of silver paste were attached onto the ZnO nanowire yarns, as shown in Fig. 10b.

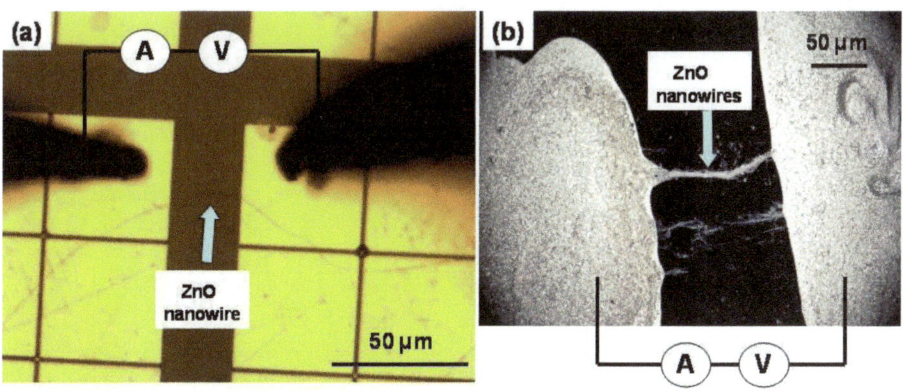

Figure 10. a) Optical image of an individual electrospun ZnO nanowire measured by two-probe method; (b) Optical image of a bundle of electrospun ZnO nanowires and attached Ag paste electrodes.

Current-voltage (I-V) curves were performed on the single ZnO nanowire device by using two-metal-microprobe testing platform. Fig.10a shows the optical picture that two metal microprobes were pressed directly on the square Au electrodes. The scanning voltage was applied from -20 to 20 V with a step of 0.1 V. At the same time the current corresponding to each applied voltage was measured. Fig.11a shows the I_{ds}-V_{ds} characteristic curves of the single ZnO nanowire at dark and under illustration. It is evident that the electrospun ZnO nanowire has a very large resistance (~$1.4 \times 10^{10}\Omega$) at dark: under bias voltage of 20 V, the current is only about 1.4 nA. But the current increases sharply under illustration. The current can reach 19.8 nA (20 V bias) when the color temperature of illustration is 2900 K. In addition, the Si substrate was used as a back gate to measure gate effect of this ZnO nanowire device. As shown in Fig. 11b, higher source-drain currents were measured when positive gate voltages V_g were applied. It indicates that the electrospun ZnO nanowire is an intrinsic *n*-type semiconductor. The above results are in accordance with that obtained from other ZnO nanowire field-effect transistor devices fabricated by the chemical vapour deposition or hydrothermal self-assembly method.

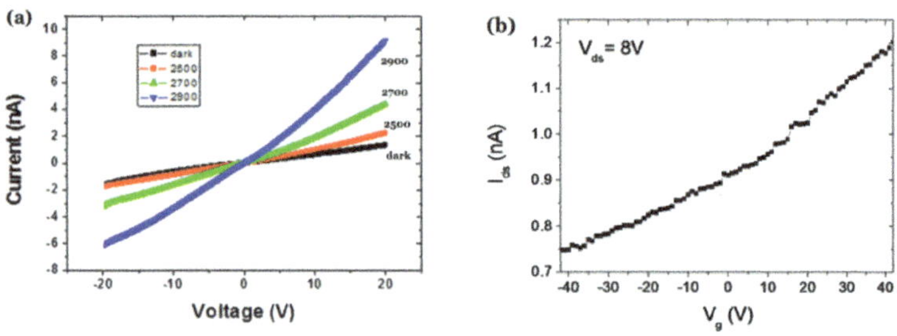

Figure 11. (a) Current-voltage (I_{ds}-V_{ds}) characteristic curves of an individual ZnO nanowire (as shown Fig.10a) at dark and under illustration. The illustration intensity is represented by color temperature of 2500, 2700and 2900 K. (b) Source-drain current (I_{ds}) versus gate voltage (V_g) at V_{ds}=8 V for the same ZnO nanowire device.

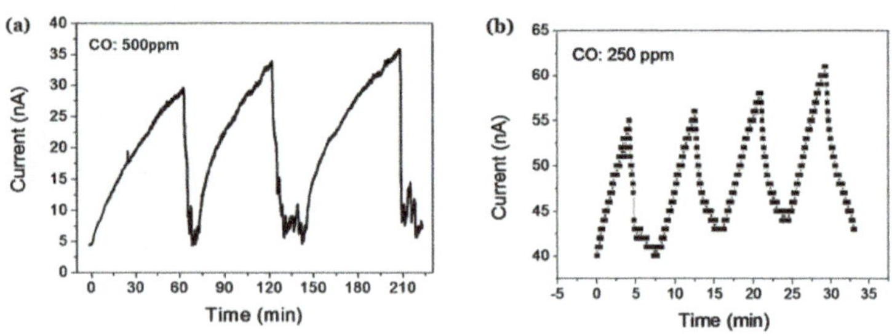

Figure 12. Response of the gas sensor based on (a) an individual ZnO nanowire, and (b) a bundle of ZnO nanowires towards 500 or 250 ppm CO as a function of time at room temperature. The applied voltage is 20 V.

In addition to transistors and photodetectors, ZnO has been proved to be a kind of highly sensitive material for the flammable or toxic gas detection, such as H_2, CO, NO, or ethanol (Zeng *et al.*, 2009; Krishnakumar *et al.*, 2009). To improve the sensing performance of the gas sensors, usually the ZnO nanowires were doped with metal or metal-oxide to form surface depletion layer, and thus enhance the sensing properties (Li *et al.*, 2012; Liu *et al.*, 2009; Pan *et al.*, 2001). Moreover, a heater was necessary for most of the gas sensors to work at high temperatures of about 200-400°C, which leads to the sensor complex and high-cost. In this work, since the electrospun ZnO nanowires exhibit continuous and compact-grains structures and thus have a high surface-to-volume ratio, gas sensing measurements of nanowire devices based on one single ZnO nanowire and one bundle of ZnO nanowires were carried out. As shown in Fig. 12, the ZnO nanowires exhibit rapid, apparent response when exposed to 500ppm or 250 ppm CO even at room temperature. Since the gas response can be defined as the ratio: $S \approx |R_{air}-R_{gas}|/R_{air}$, where R_{air} and R_{gas} are the resistance of the sensor in dry air and in the test gas, separately, Fig. 12 indicates that the gas response can reach 500 % in 60 min towards 500 ppm CO or 26% in 5 min towards 250 ppm CO.

The ZnO carbon monoxide gas sensing can be explained by the chemical reaction between the active gas and oxygen ion adsorption on the surface of ZnO. In air atmosphere, oxygen molecules are adsorbed onto the surface of the ZnO sensor to form O^- or O^{2-} ions by attracting electrons from the conduction band of the ZnO. Under carbon monoxide atmosphere, the CO gas reacts with oxygen ion molecule on the surface and gives back electrons into the conduction band, thereby lowering the resistance of ZnO sensors. ZnO nanostructures exhibit sensor response higher than the bulk; this can be simply explained by the effect of the surface-to-volume ratio. In the present case, the fast and large response to CO at room temperature can be ascribed to small nanocrystal size, high surface area and high porosity of the resultant ZnO nanowires.

6. Summary

ZnO nanostructures have drawn great attention for their wide applications in electronic, optical, optoelectronic, sensing, catalysis, hydrophobic surface modification, and energy harvesting devices (Xu and Wang, 2011). In this chapter, our recent results in synthesis, physical properties and sensor applications of ZnO nanowires are presented. Crystalline ZnO nanowires and polycrystalline ZnO nanofibers are prepared by simple hydrothermal self-assembly method and electrospinning, respectively. The surfaces of ZnO nanowires thin film after surface modification are superhydrophobic with a contact angle of 170° and a sliding angle of 2°. As one of the best optional materials for ultra-violet laser devices, ZnO nanowires grown by hydrothermal method have good photoluminescence properties and photoluminescence enhancement at 387.6 nm by coating a layer of Ag nanoparticles. ZnO nanowires with porous fibrous morphology fabricated via electrospinning have specific surface approximately one to two orders of the magnitude larger than flat films, making them excellent candidates for potential applications in gas sensors. For example, the granular ZnO nanowires exhibit fast and large response to CO even at room temperature. Further studies are in progress to improve the performance of these nanodevices.

Acknowledgments

This work was supported by the National Natural Science Foundation of China (Grant Nos.: 11074138, 50872129 and 11004114), the Taishan Scholars Program of Shandong Province, the Natural Science Foundation of Shandong Province, China for Distinguished Young Scholars (Grant No.: JQ201103), and the National Key Basic Research Development Program of China (973 special preliminary study plan) (Grant No.: 2012CB722705).

Author details

M.G. Gong[1,2], Y.Z. Long[1,3,4], X.L. Xu[2], H.D. Zhang[1] and B. Sun[1]

1 College of Physics, Qingdao University, People's Republic of China

2 Department of Physics, University of Science and Technology of China, People's Republic of China

3 Key Laboratory of Photonics Materials and Technology in Universities of Shandong, Qingdao University, People's Republic of China

4 State Key Laboratory Cultivation Base of New Fiber Materials and Modern Textile, Qingdao University, People's Republic of China

References

[1] Borras, A.; Barranco, A.; Gonzalez-Elipe, A. R. (2008). Reversible superhydrophobic to superhydrophilic conversion of Ag@TiO_2 composite nanofiber surfaces. *Langmuir*, Vol. 24, June 2008, pp. 8021-8026, ISSN: 1520-5827.

[2] Cao, B. Q.; Lorenz, M.; Rahm, A.; von Wenckstern, H.; Czekalla, C.; Lenzner, J.; Benndorf, G.; Grundmann, M. (2007). Phosphorus acceptor doped ZnO nanowires prepared by pulsed-laser deposition. *Nanotechnology*, Vol. 18(45), Oct. 2007, pp. 455707-455711, ISSN: 1361-6528.

[3] Cassie, A. B. D.; Baxter, S. (1944). Wettability of porous surfaces. *Trans. Faraday Soc.*, Vol. 40, June 1944, pp. 546-551, ISSN: 0014-7672.

[4] Chen, Y. B.; Ambade, A. V.; Vutukuri, D. R.; Thayumanavan, S. (2006). Self-assembly of facially amphiphilic dendrimers on surfaces. *J. Am. chem. Soc.*, Vol. 128(46), Nov. 2006, pp. 14760-14761, ISSN: 0002-7863.

[5] Chiou, W. T.; Wu, W. Y.; Ting, J. M. (2003). Growth of single crystal ZnO nanowires using sputter deposition. *Diamond and Related Materials*, Vol. 12(10-11), 2003, pp. 1841-1844, ISSN: 0925-9635.

[6] Djurisic, A. B.; Leung, Y. H. (2006). Optical properties of ZnO nanostructures. *Small*, Vol. 2, Aug. 2006, pp. 944-961, ISSN: 1613-6829.

[7] Erbil, H. Y.; Demirel, A. L.; Avci, Y.; Mert, O. (2003). Transformation of a simple plastic into a superhydrophobic surface. *Science*, Vol. 299, Feb. 2003, pp. 1377-1380, ISSN: 0036-8075.

[8] Gong, M. G.; Xu, X. L.; Yang, Z.; Liu, Y.; Lü, H.; Lü, L. (2009). A reticulate superhydrophobic self-assembly structure prepared by ZnO nanowires. *Nanotechnology*, Vol. 20(16), Apr. 2009, pp. 165602-165608, ISSN: 1361-6528.

[9] Gong, M. G.; Xu, X. L.; Yang, Z.; Liu, Y. S.; Liu, L. (2010). Superhydrophobic surfaces via controlling the morphology of ZnO micro/nano complex structure. *Chin. Phys. B*, Vol. 19(5), May 2010, p. 056701, ISSN: 1674-1056.

[10] Goris, L.; Noriega, R.; Donovan, M.; Jokisaari, J.; Kusinski, G.; Salleo, A. (2009). Intrinsic and doped zinc oxide nanowires for transparent electrode fabrication via low-temperature solution synthesis. *Journal of Electronic Materials*, Vol. 38(4), 2009, pp. 586-595, ISSN: 1543-186X.

[11] Govender, K.; Boyle, D. S.; Kenway, P. B.; O'Brien, P. (2004). Understanding the factors that govern the deposition and morphology of thin films of ZnO from aqueous solution. *Journal of Materials Chemistry*, Vol. 14(16), June 2004, pp. 2575-2591, ISSN: 1364-5501.

[12] Hamby, D. W.; Lucca, D. A.; Klopfstein, M. J.; Cantwell, G. (2003). Temperature dependent exciton photoluminescence of bulk ZnO. *J. Appl. Phys.*, Vol. 93, 2003, p. 3214, ISSN: 1089-7550.

[13] Heo, Y. W.; Varadarajan, V.; Kaufman, M.; Kim, K.; Norton, D. P.; Ren, F.; Fleming, P. H. (2002). Site-specific growth of ZnO nanorods using catalysis driven molecular-beam epitaxy. *Applied Physics Letters*, Vol. 81(16), Aug. 2002, pp. 3046-3048, ISSN: 1616-3028.

[14] Hibbeler, R. C. Mechanics of Materials, 3rd ed. (1997, Prentice-Hall, Engle-wood Cliffs, NJ)

[15] Hoffmann, S.; Ostlund, F.; Michler, J.; Fan, H. J; Zacharias, M.; Christiansen, S. H.; Ballif, C. (2007). Fracture strength and Young's modulus of ZnO nanowires. *Nanotechnology*, Vol. 18(20), 2007, p. 205503, ISSN: 1361-6528.

[16] Hsu, C. L.; Chang, S. J.; Hung, H. C.; Lin, Y. R.; Huang, C. J.; Tseng, Y. K.; Chen, I. C. (2005). Well-aligned, vertically Al-doped ZnO nanowires synthesized on ZnO:Ga/

Glass templates. *Journal of the Electrochemical Society*, Vol. 152(5), 2005, pp. G378-G381, ISSN: 0013-4651.

[17] Huang, M. H.; Wu, Y. Y.; Feick, H.; Tran, N.; Weber, E.; Yang, P. D. (2001). Catalytic growth of znic oxide nanowires by vapor transport. *Advanced Materials*, Vol. 13(2), Jan. 2001, pp. 113-116, ISSN: 1521-4095.

[18] Huang, Z. M.; Zhang, Y. Z.; Kotaki, M.; Ramakrishna, S. (2003). A review on polymer nanofibers by electrospinning and their applications in nanocomposites. *Compos. Sci. Technol.* Vol. 63, 2003; pp. 2223-2253, ISSN: 0266-3538.

[19] Huang, Y. H.; Bai, X. D.; Zhang, Y. (2006). In situ mechanical properties of individual ZnO nanowires and the mass measurement of nanoparticles. *J. Phys.: Condens. Matter*, Vol. 18, Mar. 2006, p. L179, ISSN: 1361-648X.

[20] Jeong, J. S.; Lee, J. Y. (2010). Substrate-facilitated nanoparticles sintering and component interconnection procedure. *Nanotechnology*, Vol. 21(47), Oct. 2010, p. 475204, ISSN: 1361-6528.

[21] Jiang, L.; Zhao, Y.; Zhai, J. (2004). A lotus-leaf-like superhydrophobic surface: a porous microsphere/nanofiber composite film prepared by electrohydrodynamics. *Angew. Chem. Int. Ed.*, Vol. 43, 2004, pp. 4338-4341, ISSN: 1521-3773.

[22] Kim, I. D.; Rothschild, A.; Lee, B. H.; Kim, D. Y.; Jo, S. M.; Tuller, H. L. (2006). Ultrasensitive chemiresistors based on electrospun TiO_2 nanofibers. *Nano. Lett.*, Vol. 6(9), 2006, pp. 2009-2013, ISSN: 1530-6984.

[23] Ko, S. H.; Lee, D.; Kang, H. W.; Nam, K. H.; Yeo, J. Y.; Hong, S. J.; Grigoropoulos, C. P.; Sung, H. J. (2011). Nanoforest of hydrothermally grown hierarchical ZnO nanowires for a high efficiency dye-sensitized solar cells. *Nano Letters*, Vol. 11(2), Jan. 2011, pp. 666-671, ISSN: 1530-6984.

[24] Kong, Y. C.; Yu, D. P.; Zhang, B.; Fang, W.; Feng, S. Q. (2001). Ultraviolet-emitting ZnO nanowires synthesized by a physical vapor deposition approach. *Appl. Phys. Lett.*, Vol. 78, Nov. 2001, p. 407, ISSN: 1616-3028.

[25] Kong, X. H.; Li, Y. D. (2003). Temperature-dependent regulation of antisense activity using a DNA/poly(N-isopropylacrylamide) conjugate. *Chemistry Letters*, Vol. 32(11), 2003, pp. 1062-1063, ISSN: 1348-0715.

[26] Kusinski, G. J.; Jokisaari, J. R.; Noriega, R.; Goris, L.; Donovan, M.; Salleo, A. (2010). Transmission electron microscopy of solution-processe, intrinsic and Al-doped ZnO nanowires for transparent electrode fabrication. *Journal of microscopy*, Vol. 237(3), Mar. 2010, pp. 443-449, ISSN: 1365-2818.

[27] Krishnakumar, T.; Jayaprakash, R.; Pinna, N.; Donato, N.; Bonavita, A.; Micali, G.; Neri, G. (2009). CO gas sensing of ZnO nanostructures synthesized by an assisted microwave wet chemical route. *Sensor and Actuators B: Chemical*, Vol. 143(1), Dec. 2009, pp. 198–204, ISSN: 092-4005.

[28] Landau, O.; Rothschild, A.; Zussman, E. (2009). Processing-microstructure-properties correlation of ultrasensitive gas sensors produced by electrospinning. *Chem. Mater.*, Vol. 21(1), Jan. 2009, pp. 9-11, ISSN: 0897-4756.

[29] Lee, W.; Jeong, M. C.; Myoung, J. M. (2004). Fabrication and application potential of ZnO nanowires grown on GaAs(002) substrates by metal-organic chemical vapour deposition. *Nanotechnol.*, Vol. 15(3), 2004, pp. 254-259, ISSN: 0957-4484.

[30] Li, M. M.; Long, Y. Z.; Tan, J. S.; Yin, H. X.; Sui, W. M.; Zhang, Z. (2010). Dielectric properties of electrospun titanium compound/polymer composite nanofibers. *Chin. Phys. B*, Vol. 19(2), Feb. 2010, p. 028102, ISSN: 1674-1056.

[31] Li, M. M.; Long, Y. Z.; Yang, D. Y.; Sun, J. S.; Yin, H.; Zhao, Z.; Kong, W.; Jiang, X.; Fan, Z. (2011). Fabrication of one dimensional superfine polymer fibers by double-spinning. *J. Mater. Chem.*, Vol. 21(35), 2011, pp. 13159-13162, ISSN: 1364-5501.

[32] Li, Y. J.; Li, K. M.; Wang, C. Y.; Kuo, C. I.; Chen, L. J. (2012). Low-temperature electrodeposited Co-doped ZnO nanorods with enhanced ethanol and CO sensing properties. *Sensor and Actuators B: Chemical*, Vol. 161(1), Jan. 2012, pp. 734-739, ISSN: 092-4005.

[33] Lin, D.; Wu, H.; Pan, W. (2007). Photoswitches and memories assembled by electrospinning aluminum doped zinc oxide single nanowires. *Advanced Materials*, Vol. 19(22), Oct. 2007, pp. 3968-3972, ISSN: 1521-4095.

[34] Liu, J. Z.; Lee, S.; Lee, K.; Ahn, Y. H.; Park, J. Y.; Koh, K. H. (2008). Bending and bundling of metal-free vertically aligned ZnO nanowires due to electrostatic interaction. *Nanotechnology*, Vol. 19(18), April 2008, p. 185607, ISSN: 1361-6528.

[35] Liu, L.; Zhang, T.; Wang, L. Y.; Li, S. C. (2009). Improved ethanol sensing properties of Cu-doped SnO_2 nanofibers. *Mater. Lett.* Vol. 63, Sep. 2009; pp. 2041-2043, ISSN: 0167-577X.

[36] Liu, C. Y.; Xu, H. Y.; Ma, J. G.; Li, X. H.; Zhang, X. T.; Liu, Y. C.; Mu, R. (2011). Electrically pumped near-ultraviolet lasing from ZnO/MgO core/shell nanowires. *Applied Physics Letters*, Vol. 99(6), 2011, p. 63115, ISSN: 1616-3028.

[37] Long, Y. Z.; Li, M. M.; Gu, C. Z.; Wan, M. X.; Duvail, J. L.; Liu, Z. W.; Fan, Z. Y. (2011). Recent advances in synthesis, physical properties and applications of conducting polymer nanotubes and nanofibers. *Progress in Polymer Science*, Vol. 36(10), Oct. 2011, pp. 1415-1442, ISSN: 0079-6700.

[38] Long, Y. Z.; Yu, M.; Sun, B.; Gu, C. Z.; Fan, Z. Y. (2012). Recent advances in large-scale assembly of semiconducting inorganic nanowires and nanofibers for electronics, sensors and photovoltaics. *Chem. Soc. Rev.*, Vol. 41(12), June 2012, pp. 4560-4580, ISSN: 0306-0012.

[39] Lü, H. F.; Xu, X. L.; Lü, L.; Gong, M. G.; Liu, Y. S. (2008). Photoluminescence enhancement of ZnO microrods coated with Ag nanoparticles. *J. Phys.: Condens. Matter*, Vol. 20(47), Dec. 2008, p. 472202, ISSN: 0953-8984.

[40] Nishino, T.; Meguro, M.; Nakamae, K.; Matsushita, M.; Ueda, Y. (1999). The lowest surface free energy based on $-CF_3$ alignment. *Langmuir*, Vol. 15(13), May 1999, p. 4321, ISSN: 1520-5827.

[41] Pan, Z. W.; Dai, Z. R.; Wang, Z. L. (2001). Nanobelts of semiconducting oxides. *Science*, Vol. 291, Feb. 2001; pp. 1947–1949, ISSN: 0036-8075.

[42] Pfüller, C.; Brandt, O.; Flissikowski, T.; Grahn, H. T.; Ive, T.; Speck, J. S.; DenBaars, S. P. (2011). Comparison of the spectral and temporal emission characteristics of homo-epitaxial and hetero-epitaxial ZnO nanowires. *Applied Physics Letters*, Vol. 98(11), Mar. 2011, p. 113113, ISSN: 1616-3028.

[43] Ra, Y.-W.; Choi, K.-S.; Kim, J.-H.; Hahn, Y.-B.; Im, Y.-H. (2008). Fabrication of ZnO nanowires using nanoscale spacer lithography for gas sensor. *Small*, Vol. 4(8), 2008, pp. 1105-1109, ISSN: 1613-6829.

[44] Richard, D.; Clanet, C.; Quere, D. (2002). Surface phenomena: contact time of a bouncing drop. *Nature*, Vol. 417, June 2002, p. 811, ISSN: 1061-4036.

[45] Sen, B.; Stroscio, M.; Dutta, M. (2011). Photoluminescence and Raman spectroscopy of polycrystalline ZnO nanofibers deposited by electrospinning. *Journal of Electronic Materials*, Vol. 40(9), 2011, pp. 2015-2019, ISSN: 0361-5235.

[46] Shan, W.; Walukiewicz, W.; Ager, J. W.; Yu, K. M.; Yuan, H. B.; Xin, H. P.; Cantwell, G.; Song, J. J. (2005). Nature of room-temperature photoluminescene in ZnO. *Appl. Phys. Lett.*, Vol. 86(19), May 2005, p. 191911, ISSN: 1616-3028.

[47] Shen, Y.; Hong, J.-I.; Xu, S.; Lin, S. S.; Fang, H.; Zhang, S.; Ding, Y.; Snyder, R. L.; Wang, Z. L. (2010). A general approach for fabricating arc-shaped composite nanowire arrays by pulsed laser deposition. *Advanced Functional Materials*, Vol. 20(5), Mar. 2010, pp. 703-707, ISSN: 1616-3028.

[48] Song, J. H.; Wang, X. D.; Riedo, E.; Wang, Z. L. (2005). Elastic property of vertically aligned nanowires. *Nano Lett.*, Vol. 5(10), Sep. 2005, pp. 1954-1958, ISSN: 1616-3028.

[49] Sui, X. M.; Shao, C. L.; Liu, Y. C. (2005). White-light emission of polyvinyl alcohol/ZnO hybrid nanofibers prepared by electrospinning. *Applied Physics Letters*, Vol. 87(11), 2005, pp. 113-115, ISSN: 1616-3028.

[50] Sun, B.; Long, Y. Z.; Yu, F.; Li, M. M.; Zhang, H. D.; Li, W. J.; Xu, T. X. (2012). Self-assembly of a three-dimensional fibrous polymer sponge by electrospinning. *Nanoscale*, Vol. 4(6), 2012, pp. 2134-2137, ISSN: 2040-3364.

[51] Tan, J. S.; Long, Y. Z.; Li, M. M. (2008). Preparation of aligned polymer micro/nanofibers by electrospinning. Chin. Phys. Lett., Vol. 25(8), Aug. 2008, pp. 3067-3070, ISSN: 0256-307X.

[52] Vayssieres, L.; Keis, K.; Lindquist, S. E.; Hagfeldt, A. (2001). Purpose-Built anisotropic metal oxide material: 3D highly oriented microrod array of ZnO. *Journal of Physical Chemistry B*, Vol. 105(17), Mar. 2001, pp. 3350-3352, ISSN: 1520-6106.

[53] Wang, X. D.; Summers, C. J.; Wang, Z. L. (2004). Large-scale hexagonal-patterned growth of aligned ZnO nanorods for nano-optoelectronics and nanosensor arrays. *Nano Letters*, Vol. 4(3), Jan. 2004, pp. 423-426, ISSN: 1530-6984.

[54] Wang, L. S.; Zhang, X. Z.; Zhao, S. Q.; Zhou, G. Y.; Zhou, Y. L.; Qi, J. J. (2005). Synthesis of well-aligned ZnO nanowire by simple physical vapor deposition on *c*-oriented ZnO thin films without catalysts or additives. *Applied Physics Letters*, Vol. 86(2), 2005, p. 024108, ISSN: 1616-3028.

[55] Wang, J. X.; Sun, X. W.; Yang, Y.; Huang, H.; Lee, Y. C.; Tan, O. K.; Vayssieres, L. (2006). Hydrothermally grown oriented ZnO nanorod arrays for gas sensing applications. *Nanotechnology*, Vol. 17(19), Sept. 2006, pp. 4995-4998, ISSN: 1361-6528.

[56] Wang, F. F.; Cao, L.; Pan, A. L.; Liu, R. B.; Wang, X.; Zhu, X.; Wang, S. Q.; Zou, B. S. (2007). Synthesis of tower-like ZnO structures and visible photoluminescence origins of varied shaped ZnO nanostructures. *J. Phys. Chem. C*, Vol. 111, 2007, p. 7655, ISSN: 1932-7447.

[57] Wang, Y.; Xu, X. L.; Xie, W. Y.; Wang, Z. B.; Lu, L.; Zhao, Y. L. (2008). Two-step growth of highly oriented ZnO nanorod arrays. *Acta Physica Sinica,* Vol. 57(4), Apr. 2008, pp. 2582-2586, ISSN: 1000-3290.

[58] Wang, J. X.; Sun, X. W.; Yang, Y.; Kyaw, K. K. A.; Huang, X. Y.; Yin, J. Z.; Wei, J.; Demir, H. V. (2011). Free-standing ZnO-CuO composite nanowier array films and their gas sensing properties. *Nanotechnology*, Vol. 22(32), July 2011, p. 325704, ISSN: 1361-6528.

[59] Wenzel, R. N. (1936). Resistance of solid surfaces to wetting by water. *Ind. Eng. Chem.*, Vol. 28(8), Aug. 1936, pp. 988-994, ISSN: 1226-086X.

[60] Wu, J. J.; Wen, H. I.; Tseng, C. H.; Liu, S. C. (2004). Well-aligned ZnO nanorods via hydrogen treatment of ZnO films. *Advanced Functional Materials*, Vol. 14(8), Sept. 2004, pp. 806-810, ISSN: 1616-3028.

[61] Wu, H.; Lin, D.; Zhang, R.; Pan, W. (2008). ZnO nanofiber field-effect transistor assembled by electrospinning. *Journal of the American Ceramic Society* 91 (2), Feb. 2008, 656-659, ISSN: 1551-2916.

[62] Xiao, X. H.; Ren, F.; Zhou, X. D.; Peng, T. C.; Wu, W.; Peng, X. N.; Yu, X. F.; Zhang, C. Z. (2010). Surface plasmon-enhanced light emission using silver nanoparticles embedded in ZnO. *Appl. Phys. Lett.*, Vol. 97, Aug. 2010, p. 071909, ISSN: 0003-6951.

[63] Xu, S.; Wang, Z. L. (2011). One-dimensional ZnO nanostructures: Solution growth and functional properties. *Nano Res.*, Vol. 4(11), 2011, pp. 1013-1098, ISSN: 1998-0124.

[64] Yang, J. L.; An, S. J.; Park, W. I.; Yi, G. C.; Choi, W. (2004). Photocatalysis using ZnO thin films and nanoneedles grown by metal-organic chemical vapor deposition. *Advanced Materials*, Vol. 16(18), Oct. 2004, pp. 1661-1664, ISSN: 1521-4095.

[65] You, J. B.; Zhang, X.W.; Fan, Y. M.; Qu, S.; Chen, N. F. (2007). Surface plasmon enhanced ultraviolet emission from ZnO films deposited on Ag/Si(001) by magnetron sputtering. *Appl. Phys. Lett.*, Vol. 91, Dec. 2007, p. 231907, ISSN: 1616-3028.

[66] Zeng, Y.; Zhang, T.; Wang, L. J.; Kang, M. H.; Fan, H.; Wang, R.; He, Y. (2009). Enhanced toluene sensing characteristics of TiO_2-doped flowerlike ZnO nanostructures. *Sensor and Actuators B*, Vol. 140, 2009, pp. 73–78, ISSN: 092-4005.

[67] Zhang, H. D.; Long, Y. Z.; Li, Z. J.; Sun, B.; Sheng, C. H. (2012). Synthesis, electrical and humidity sensing properties of BaTiO3 nanofibers via electrospinning. *Advanced Materials Research*, Vols. 418-420, 2012, pp. 684-687, ISSN: 1022-6680.

[68] Zheng, J.; Long, Y. Z.; Sun, B.; Zhang, Z. H.; Shao, F.; Zhang, H. D.; Zhang, Z. M.; Huang, J. Y. (2012). Polymer nanofibers prepared by low-voltage near-field electrospinning. *Chin. Phys. B*, Vol. 21(4), April 2012, p. 048102, ISSN: 1674-1056.

Chapter 6

Magnetoresistance of Nanowires Electrodeposited into Anodized Aluminum Oxide Nanochannels

Takeshi Ohgai

Additional information is available at the end of the chapter

1. Introduction

One-dimensional nanowires with a large aspect ratio have received much attention due to their unique shape anisotropy and extremely large surface area. Metallic nanowires such as Au, Ag and Cu nanowires [1-3] and semiconductor nanowires [4-9] would show unique electron transport properties which are not observed in bulk metals. Ferromagnetic nanowires such as Ni, Co and Fe alloy nanowires as well as Co/Cu multilayered nanowires would be best candidate materials for magnetic field sensors with anisotropic magnetoresistance (AMR) and giant magnetoresistance (GMR) effect. Nanowires can be fabricated by manipulating metallic atoms one by one using scanning tunneling microscopy (STM) or catalyst-assisted wet chemical etching technique [10], while they can be also prepared by electrochemically depositing metallic atoms into a nanochannel template with numerous nanochannels [11]. Nanochannel templates such as polycarbonate membrane films or anodized aluminum oxide films with high density of nanochannels (about 10^8~10^{10} cm^{-2}) can be used in the template synthesis technique [12].

Using polymer membrane filters, ferromagnetic metal nanowires have been synthe sized so far. Whitney *et al.* reported that the arrays of Ni and Co nanowires were electrodeposited in polymer templates with the nanometer-sized pores prepared by nuclear track etching technique [13]. They found that the preferred magnetization direction is perpendicular to the film plane and enhanced coercive force as high as 680 Oe. Piraux *et al.* reported that the array of Co/Cu multilayered nanowires with GMR response was electrodeposited in nanoporous polymer template [14]. Blondel *et al.* also reported that Co/Cu and Ni-Fe/Cu multilayered nanowires with GMR response was demonstrated [15]. They synthesized the multilayered nanowires with length of 6 μm, diameter of 80 nm and each layer thickness of 5~10 nm into the nanochannels of ion track-etched polycarbonate membrane filters. In the report, GMR of 14% for Co/Cu and of 10% for Ni-Fe/Cu was demonstrated at ambient temperature in the current perpendicular to the layers. Ohgai *et al.* reported that magneto-sensitive nickel nanowires with AMR response could be fabricated into multi- and single-ion track templates with nanochannels using electrode position technique [16,17].

On the contrary, using anodized aluminum oxide pores on the surface of metallic aluminum substrates, Ni, Co and Fe homogeneous ferromagnetic nanowires have been also electrodeposited and characterized in terms of their magnetization properties. Kawai *et al.* reported that magnetic properties of Ni, Co, Fe, Co-Ni, Fe-Ni and Fe-Co alloy nanowires with the coercive force ranging from 0.5 to 3.2kOe were electrodeposited into the nanopores of an anodic oxide coating films on aluminum [18-20]. Tsuya *et al.* also reported that the alumite films containing Fe nanowires were fabricated as an application to perpendicular magnetic recording medium [21,22]. Huysmans *et al.* reported that the magnetization curling process in perpendicular direction was investigated using Fe nanowire arrays in alumite media [23]. Cheng *et al.* also reported that the magnetic anisotropy of electrodeposited Co nanowires on alumite substrate [24]. Zhang *et al* . reported that the influence of the packing density on the magnetic behaviour was investigated using alumite media containg magnetic nanowires [25]. AlMawlawi *et al.* reported that the coercive force of Fe nanowires electrodeposited into anodic aluminum oxide pores increased up to around 2.1 kOe with increasing the aspect ratio of the nanowires [26]. Nielsch *et al.* reported that the uniform Ni nanowires array was synthesized into the ordered alumina pores by pulsed electrodeposition technique [27]. Evans *et al.* reported that the current perpendicular to plane giant magnetoresistance (CPP-GMR) was found using the multilayered nanowires electrodeposited in a commercially available anodic aluminum oxide membrane filter [28]. Ohgai *et al.* reported that the Co/Cu multilayered nanowires with spin-valve effect and GMR response were synthesized in self-organized anodized aluminum oxide nanopores grown at the surface of bulk aluminum. They also reported that the pore bottom oxide layer was removed by a chemical etching technique using aluminum oxide template with pore-length as short as 2000 nm. In their report, 20% of GMR was demonstrated in Co/Cu multilayered nanowires at room temperature, while the typical resistance switching of spin-valves was also demonstrated in Co/Cu/Co tri-layered nanowires [29-32]. Wu *et al.* reported that the Ni nanodot and nanowire arrays could be prepared using a porous alumina layer on a bulk silicon as a template without a conductive interlayer [33]. Chu *et al.* reported that the integrated ultra high-density Fe-Pt alloy nanowires array was synthesized in a porous alumina layer on an ITO-glass template [34]. Wang *et al.* reported that the Ni nanowires with different diameters were prepared by direct-current electrodeposition into the pores of porous anodic alumina membrane [35]. Friedman *et al.* reported that the optimal parameters were found for synthesis of magnetic nanowires in porous alumina templates [36].

Electrodeposition of metallic nanowires into the as anodized aluminum template can be carried out using an alternating current or a pulsed current in order to reduce the charging up effect of the barrier layer at the interface between the anodized aluminum oxide layer and the metallic aluminum substrate. However, this resistive barrier layer makes it difficult to achieve good electric contacts at the pore bottom and well-controlled layered structure of electrodeposited nanowires. In this chapter, fabrication process of anodized aluminum oxide template without the barrier layer and the magnetoresistance properties of electrodeposited Ni, Co alloy nanowires and Co/Cu multilayered nanowires were investigated to synthesize novel functional ferromagnetic devices with anisotropic magnetoresistance (AMR) effect and giant magnetoresistance (GMR) effect.

2. Fabrication of Anodized Aluminum Oxide Nanochannels Templates

2.1. Barrier Layer Thinning by Chemical Etching Technique

Aluminum sheets with thickness of 500 μm were used as a starting material to prepare anodized aluminum templates. First, the aluminum sheets were electrochemically polished in a ethanol solution containing 25 vol% of perchloric acid to achieve a mirror like surface. During the electrochemical polishing, the cell voltage was kept at 8 V for 10 min at room temperature. Then, the polished aluminum sheets were anodized in an aqueous solution containing 0.3 mol/L oxalic acid to obtain a nanoporous aluminum oxide layer on the surface. During the anodization, the cell voltage was kept at 50 V for 10 min at room temperature. Subsequently, the anodized aluminum sheets were immersed in an aqueous solution containing 5 vol% of phosphoric acid for 50 min to widen the pores, and to thin the oxide layer at the pore bottom. These processing parameters give pores 2 μm long with a diameter of 60 nm.

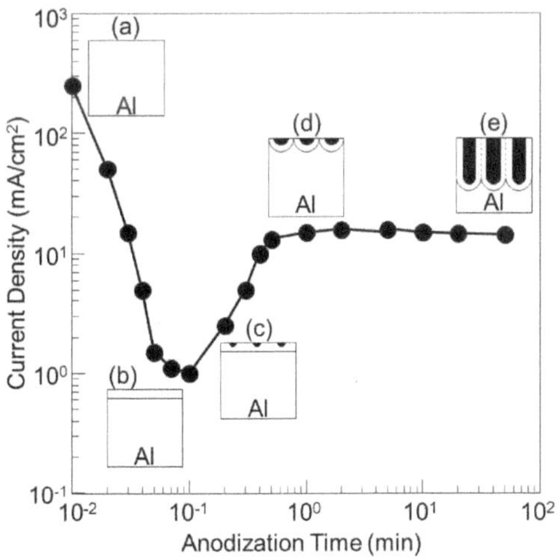

Figure 1. Time dependence of current density on fabrication of self-organized nanoporous anodized aluminum templates. Anodization was conducted under 50 V in aqueous solution containing 0.3 mol/L oxalic acid.

Figure 1 shows the time dependence of current density on the fabrication of self-organized nanoporous anodized aluminum templates. Anodization was conducted under 50 V in aqueous solution containing 0.3 mol/L oxalic acid. At the beginning of anodization, the current density is very high due to oxygen evolution at the surface of aluminum as shown in Figure 1(a). Then, the current density rapidly decreases due to the formation of a highly resistive oxide layer as shown in Figure 1(b). Next, the current density gradually increases due to self-organized formation of nanopores in the oxide layer as shown in Figures 1(c), (d). Finally, the current density reaches a constant value due to a stable growth rate of the porous oxide layer as shown in Figure 1(e).

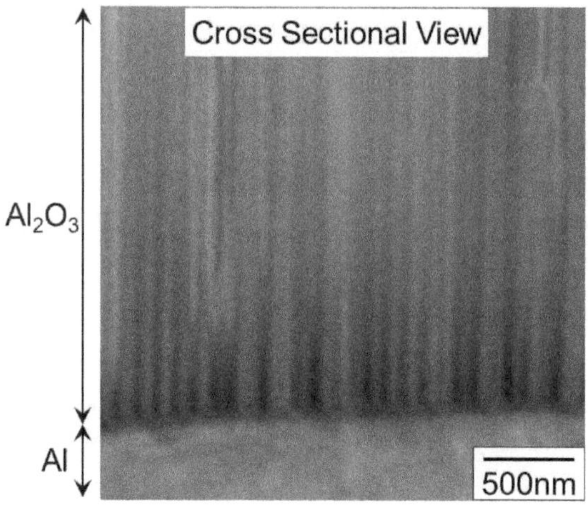

Figure 2. SEM image of a cross section view of an anodized aluminum template. Anodization was conducted under 50 V for 10 min. The anodized aluminum oxide layer has a typical porous columnar structure and the pore length is approximately 2μm. The pore diameter is approximately 60 nm.

Figure 2 shows SEM images of a cross section view of an anodized aluminum template. The oxide layer has a typical porous columnar structure and the pore length is approximately 2μm as shown in Figure 2. The pore diameter is approximately 60 nm and the order of pore density is 10^{10} pores/cm^2 as shown in Figure 2. It is well known that an anodized aluminum template has a resistive oxide layer (barrier layer) at the interface between the anodized aluminum oxide film and the metallic aluminum substrate. Figure 3 shows SEM images of a pore top planar view (a) and a pore bottom back side view (b) of an anodized aluminum template. For observation of pore bottom back side (barrier layer), the massive metallic aluminum substrate was dissolved in hydrochloric acid containing traces of cupric ions. The pore diameter is approximately 60 nm and the order of pore density is 10^{10} pores/cm^2 as shown in Figure 3(a). The existence of a barrier layer is made obvious by looking at the pore bottom of anodized aluminum oxide layer (Figure3(b)).This resistive barrier layer makes well-controlled electrodeposition of layered structures difficult and lowers their AMR and GMR performance. Electrodeposition of metals into the pores of an anodized aluminum surface is usually carried out using alternating current or a pulsed current technique [37–40] in order to reduce the charging up effect of the barrier layer. To remove this barrier layer, the massive metallic aluminum backing is usually dissolved in an aqueous solution containing $HgCl_2$ prior to thinning the barrier layer [41–45]. However, this aluminum substrate dissolving technique can be applied to thick aluminum oxide layers with the thickness of several tens micro meters. On the contrary, when this barrier layer is removed or thinned without removing the aluminum backing, anodized aluminum templates with short pores length can be obtained. Since the thickness of the barrier layer is about several tens of nanometres, this layer can be removed or thinned using a chemical etching technique from the pore side direction [46]. Therefore, after the anodization process, to remove or thin this barrier layer without dissolving the metallic aluminum backing, the anodized aluminum templates were subsequently immersed in an aqueous solution containing phosphoric acid prior to the electrode position process.

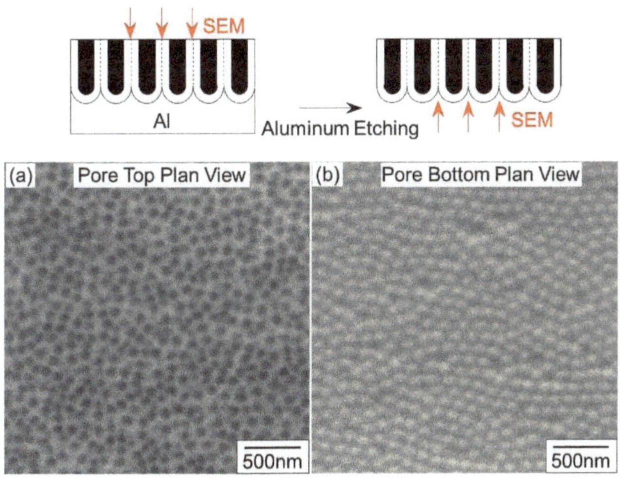

Figure 3. SEM images of pore top plan view (a) and pore bottom plan view (b) of an anodized aluminum template. Anodization was conducted under 50 V for 10 min. The metallic aluminum substrate was dissolved in the hydrochloric acid aqueous solution containing small amount of cupric ions.

2.2. Barrier Layer Thinning by Anodization Voltage Controlling

Figure 4 shows experimental apparatus and time-dependence of applied voltage to synthesize an anodized aluminum oxide membrane filter. Anodized aluminum oxide thick films with numerous nanochannels were exfoliated mechanically by the pressure of hydrogen gas generated at the interface between an oxide layer and a metallic aluminum during the subsequent cathodic reduction process after the growing of anodic aluminum oxide layer. At first,

70 V was applied to growing aluminum oxide long nanochannel. Then, the anodization voltage was decreased gradually down to 0 V for thinning the pore bottom oxide layer (barrier layer). Finally, cathodic voltage was applied to exfoliate an anodized aluminum oxide membrane film from the metallic aluminum rod due to the hydrogen evolution. Figure 5 shows surface appearance of anodized aluminum oxide membrane filters exfoliated from a metallic aluminum rod. These membrane filters were obtained by anodizing at 50 V (Figure 5(a)) and 70 V (Figure 5(b)). The membrane filter anodized at 70 V had round and disc shape with the diameter of 10 mm (Figure 5(b)) while that anodized at 50 V had several cracks (Figure 5(a)).

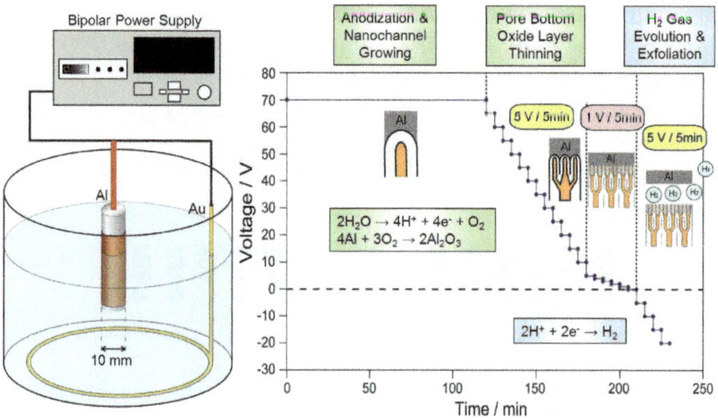

Figure 4. Experimental apparatus and time-dependence of applied voltage to synthesize an anodized aluminum oxide membrane filter.

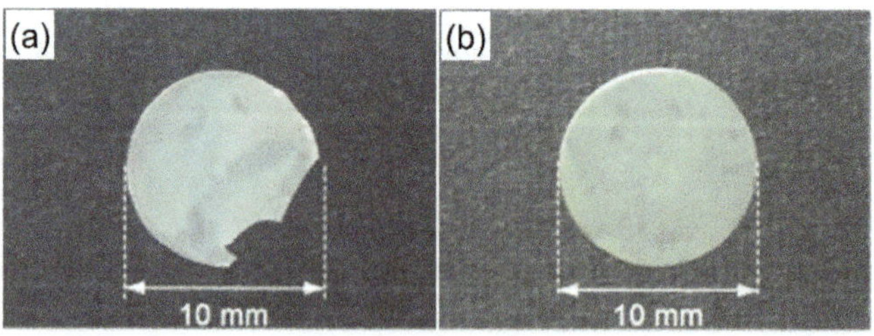

Figure 5. Surface appearance of anodized aluminum oxide membrane filters exfoliated from a metallic aluminum rod. These membrane filters were anodized at 50 V (a) and 70 V (b).

Figure 6 shows SEM images of barrier layer side (plan view (a), cross section (b)) and surface side (plan view (c), cross section (d)) of an anodized aluminum oxide membrane filter exfoliated from a metallic aluminum rod. The anodization was conducted at 70 V. The average channel diameter at the surface side was around 60 nm while that at the barrier layer side was ca. 20 nm. The oxide layer has a typical porous columnar structure and the channel length was ca. 60 μm while the channel density was around 10^8 channels / cm^2.

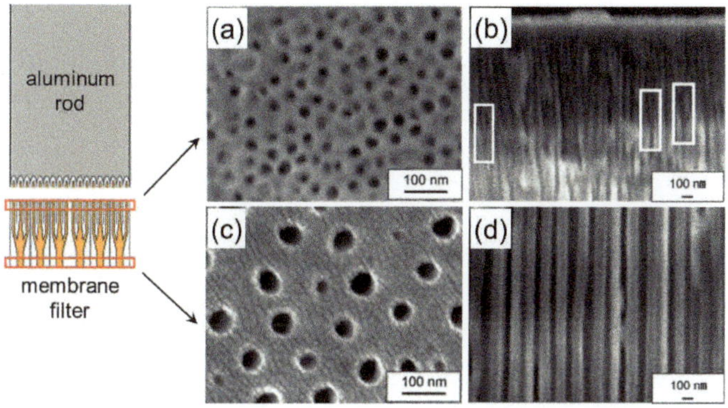

Figure 6. SEM images of barrier layer side (plan view (a), cross section (b)) and surface side (plan view (c), cross section (d)) of an anodized aluminum oxide membrane filter exfoliated from a metallic aluminum rod. The anodization was conducted at 70 V.

Figure 7 shows the effect of anodization voltage on the growth rate of anodized aluminum oxide layer. The oxide layer thickness was determined by observing the cross sectional SEM images. With increasing the anodization voltage V_A(V), the oxide layer thickness was increased exponentially. The average growth rate R_G^{Ave}(nm/sec) of anodized aluminum oxide layer can be expressed by the following equation (1).

$$logR_G^{Ave} = 0.017V_A - 0.34 \quad (1)$$

According to the equation (1), if V_A equal to 70 V, R_G^{Ave} is estimated to be around 7.1 nm/s. Ohgai *et al.* [31] have reported that the average growth rate of the anodized aluminum oxide layer depends on the anodization voltage and the growth rate $R^*{}_G^{Ave}$ (nm/sec) increases exponentially with increasing the anodization voltage V_A (V) as shown in the following equation (2).

$$logR^*{}_G^{Ave} = 0.03V_A - 0.78 \quad (2)$$

According to the equation (2), if V_A equal to 70 V, $R^*{}_G^{Ave}$ is calculated to be around 21 nm/s which is ca. 3 times larger than that obtained from the equation (1). It is well known that the real time growth rate of the anodized aluminum oxide layer R_G^{Real} decreases with increase in the oxide layer thickness. In the equation (1), range of the oxide layer thickness is from 50 to 200 μm while the range is from 1 to 60 μm in the equation (2). Therefore, R_G^{Ave}calculated from the equation (1) would be smaller than that obtained from the equation (2). [1]

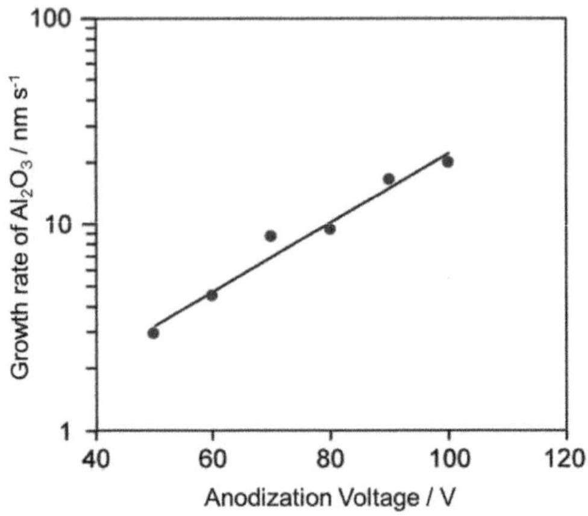

Figure 7. Effect of anodization voltage on the growth rate of anodized aluminum oxide layer.

Figure 8 shows the relationship between film thickness of anodized aluminum oxide layer and anodization time. The anodization was conducted at 70 V. The film thickness y (μm) increased with increasing the anodization time x (hour) according to a parabolic low as shown in the following equation (3).

$$y = 44.7x^{0.47} \tag{3}$$

Therefore, the real time growth rate of the anodized aluminum oxide layer R_G^{Real} can be expressed by the following equation(4).

$$R_G^{Real} = dy / dx = 21.2x^{-0.53}(\mathrm{mm / hour}) = 5.9x^{-0.53}\left(\mathrm{nm / sec}\right) \tag{4}$$

Consequently, R_G^{Real} decreases with increase in the anodization time x and the oxide layer thickness y. Ohgai *et al.* [30] have reported that the film thickness y^*(μm) of anodized aluminum oxide layer increases linearly with increasing the anodization time x (hour) as shown in the following equation (5).

$$y = 50x \tag{5}$$

Therefore, the real time growth rate of the anodized aluminum oxide layer $R^{*}{}_G^{Real}$ would be constant as shown in the following equation (6).

$$R^{*}{}_{G}^{Real} = dy/dx = 50(\text{mm}/\text{hour}) = 14(\text{nm}/\text{sec}) \tag{6}$$

As mentioned previously, the real time growth rate of the anodized aluminum oxide layer R_G^{Real} decreases with increase in the oxide layer thickness. In the equation (4), range of the oxide layer thickness is from 50 to 200 μm while the range is from 0.2 to 50 μm in the equation (6). Therefore, R_G^{Real}calculated from the equation (4) would be smaller than that obtained from the equation (6).

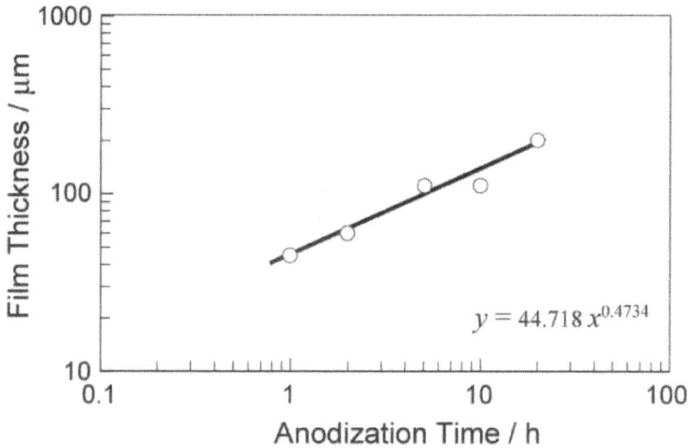

Figure 8. Relationship between film thickness of anodized aluminum oxide layer and anodization time. The anodization was conducted at 70 V.

3. Electrodeposition and Magnetoresistance of Ni and Co Nanowires Array

3.1. Electrodeposition of Ni and Co Nanowires Array

Figure 9 shows experimental apparatus and a nanochannel template for electrodeposition of nanowires array. The exfoliated anodized aluminum oxide nanochannels were used as templates for growing nanowires. On surface of the membrane filter, a gold layer was sputter-deposited to cover the pores and make a cathode. A gold wire and Ag/AgCl electrode were used as an anode and a reference electrode. For example, aqueous solution containing $NiSO_4$ (120 g/L) and H_3BO_3 (45 g/L) was used as electrolyte for electrodeposition of Ni nanowires.

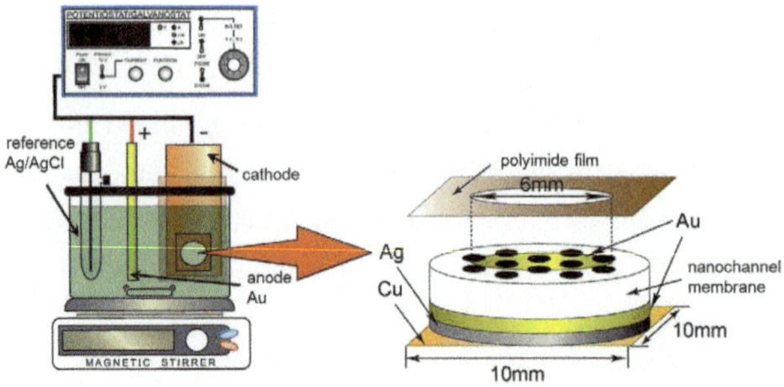

Figure 9. Experimental apparatus and a nanochannel template for electrodeposition of nanowires array.

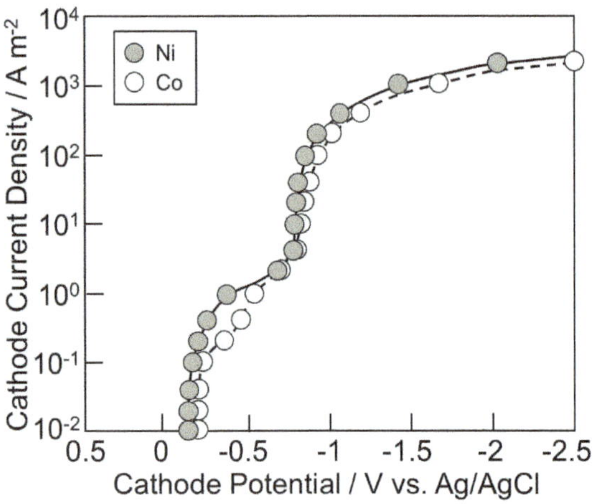

Figure 10. Cathode polarization curves for Ni and Co electrodeposition from aqueous solution containing Ni^{2+} or Co^{2+} ions.

Figure 10 shows cathode polarization curves for Ni and Co electrodeposition from aqueous solution containing Ni^{2+} or Co^{2+}ions. According to Nernst equation, the equilibrium potentials of Ni/Ni^{2+} and Co/Co^{2+} are estimated to be around −0.46 and −0.48V (vs. Ag/AgCl). As shown in Figure 10, the cathodic currents occur at around -0.2V which is more nobler potential region than the equilibrium potential of Ni and Co. Therefore, this cathodic current would be mainly caused by the reduction current of hydrogen ions. With increasing the cathodic current density, at around $2A/m^2$, the potential polarizes significantly to the less noble region. This polarization would be caused by the diffusion limit of hydrogen ions. At around -0.8V which is less nobler potential region than the equilibrium potential of Ni and Co, the cathodic current increases again. It is well-known that the electrodeposition of iron-group metals such as Ni, Co and Fe is accompanied by the over potential due to the rate determination of multi-step reduction process even in the form of their aqua ions. Therefore, in the present work, Ni^{2+} and Co^{2+} ions would electrodeposit with accompanying the over-potential even in simple aqueous solutions containing sulfuric and boric acid. Consequently, this cathodic current would be mainly caused by the reduction current of Ni^{2+} and Co^{2+}. These characteristics are identical to those obtained with the pores in nanochannel polycarbonate templates with metallic gold cathode. This result supports the observation that the barrier layer at the pore bottom of an anodized aluminum template is well removed by the exfoliation process from a metallic aluminum rod and the nanowires growon metallic gold cathode. Using the polarization curves, the optimum electrodeposition potential range for growing Ni and Co nanowires are determined to be from −1.0 to -1.2V (vs. Ag/AgCl) which is the potential region more nobler than the diffusion limit of each metal ions.

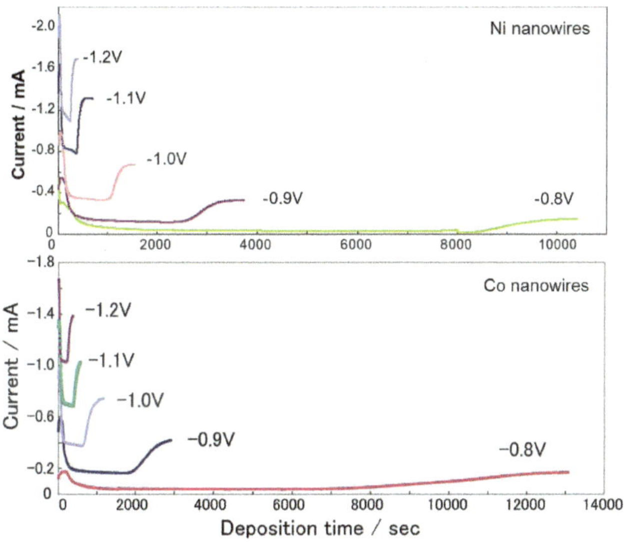

Figure 11. Effect of cathode potential on the time-dependence of cathode current during the electrodeposition of Ni nanowires and Co nanowires.

Time dependence of cathode current was monitored during the electrodeposition to investigate the growing process of nanowires. Ni nanowires and Co nanowires were potentio-statically electrodeposited. Figure 11 shows the effect of cathode potential on the time-dependence of cathode current during the electrodeposition of Ni nanowires and Co nanowires. During the electrodepostion, cathode potentials were fixed to -0.8, –0.9, –1.0, –1.1 and –1.2 V. If the potential was kept to –1.0 V for Ni deposition, the cathode current reached up to ca. 1 mA at the beginning of electrolysis within several tens of minutes. Then, the current rapidly decreased to be ca. 0.4 mA and kept the constant current until around 1000 sec.

During this process, electrodeposition of Ni proceeds in the nanopores. At the initial stage of the electrodepostion, large cathode current was observed in each cathode potential. The concentration of metal ions in the nanopores will decrease with increasing the electrodeposition time due to the reduction of metal ions, while the metal ions will be provided from the bulk solution to the nanopores, where the metal ions are consumed on the cathode due to the electrodeposition process. Finally, as shown in Figure 11, the cathode current rapidly increases at the deposition time more than 1000 sec. At this stage, electrodeposited nanowires reach the surface of the membranes and large hemispheric Ni deposits are formed. Growth rate of Ni nanowires can be estimated from dividing channel length by the filling time. For example, growth rate of Ni nanowires can be estimated as ca. 6 nm/sec at the cathode potential of –1.0 V. Time-dependence of cathode current for Co deposition also showed similar behavior as well as Ni deposition. Growth rate of the nanowires increases up to ca. 30nm/sec with increasing the cathode potential up to -1.2 V.

3.2. Structure of Ni and Co Nanowires Array

After the growing nanowires, anodized aluminum oxide membrane filters were dissolved in aqueous solution containing sodium hydroxide and the remains consisted of nanowires and a gold layer was served as a sample for scanning electron microscope (SEM) and transmission electron microscope (TEM) observation. Figure 12 shows SEM images of Ni nanowires separated from the anodized aluminum oxide templates. Diameter (60 and 300 nm) and length (6 and 30μm) of the nanowires corresponds well to that of nanopores and the cylindrical shape was precisely transferred from the nanopores to the nanowires. Aspect ratio of the nanowires reaches up to around 100.

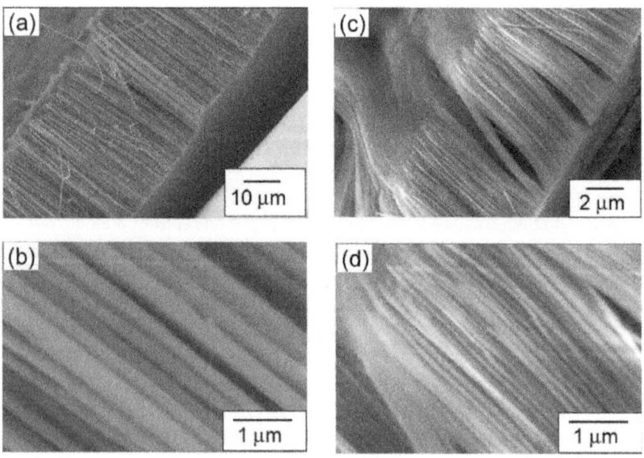

Figure 12. SEM images of electrodeposited Ni nanowires separated from anodized aluminum oxide membrane filters with the pore diameters of ca. 300nm (a),(b) and ca. 60nm (c),(d).

TEM bright field images and electron diffraction patterns of electrodeposited Ni and Co nanowires were also investigated as shown in Figure 13. According to TEM bright field images, shape of the nanowires was almost cylindrical and the electron diffraction patterns are composed of spots, which suggests a nanowire consists of a crystalline phase with preferential orientation.

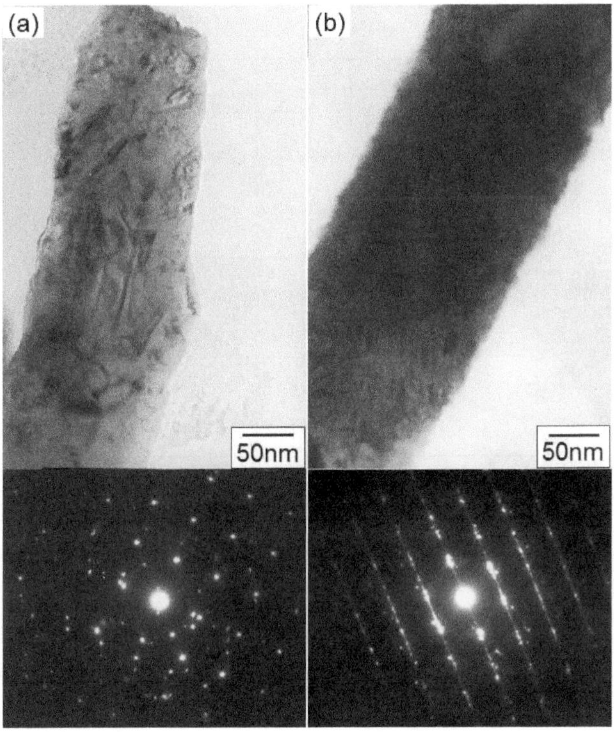

Figure 13. TEM bright field images and electron diffraction patterns of Ni and Co nanowires electrodeposited in nanochannel membrane filters.

3.3. Magnetic Properties of Ni and Co Nanowires Array

Magnetization and anisotropic magnetoresistance (AMR) of Ni alloy and Co alloy nanowires were measured using a vibrating sample magnetometer (VSM) and LCR meter with increasing the magnetic field up to 10 kOe. Figure 14 shows magnetic hysteresis loops of pure Ni, Ni-1.5%Co, Ni-0.8%Fe, pure Co, Co-0.9%Ni and Co-0.1%Fe alloy nanowires electrodeposited into anodized aluminum oxide templates with channel-diameter of 60 nm. Magnetic field was applied to in-plan direction and perpendicular direction to the membrane film plan. The perpendicular direction to the membrane film plan corresponds to the parallel direction to the long axis of nanowires. These nanowires were hardly magnetized in in-plan direction and the magnetization reached to saturation at more than 5 kOe as shown in Figure 14. On the contrary, these nanowires were easily magnetized in perpendicular direction and the coercive force reached up to around 1 kOe. These magnetization curves revealed that the electrodeposited nanowires have a typical perpendicular magnetization behaviour due to the uni-axial shape anisotropy.

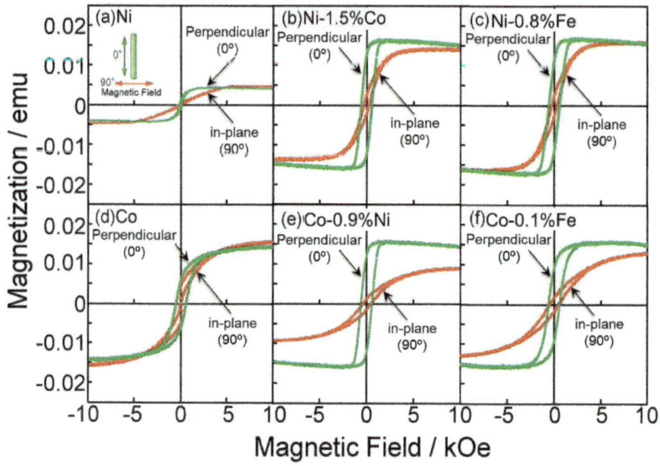

Figure 14. Magnetic hysteresis loops of pure Ni, Ni-1.5%Co, Ni-0.8%Fe, pure Co, Co-0.9%Ni and Co-0.1%Fe alloy nanowires electrodeposited into anodized aluminum oxide templates with channel-diameter of 60 nm. Magnetic field was applied to in-plan direction and perpendicular direction to the membrane film plan. The perpendicular direction to the membrane film plan corresponds to the parallel direction to the long axis of nanowires.

Figure 15 shows magnetoresistance curves of pure Ni, Ni-1.5%Co, Ni-0.8%Fe, pure Co, Co-0.9%Ni and Co-0.1%Fe alloy nanowires electrodeposited into anodized aluminum oxide templates with channel-diameter of 60 nm. Magnetic field was applied to in-plan direction and perpendicular direction to the membrane film plan. The perpendicular direction to the membrane film plan corresponds to the parallel direction to the long axis of nanowires. Here, the anisotropic magnetoresistance (AMR) ratio is defined by the following equation:

$$\mathrm{AMR}\,(\%) = 100 \times (R_{\mathrm{max}} - R_{\mathrm{min}}) \,/\, R_{\mathrm{min}} \tag{7}$$

The magnetoresistive hysteresis of the Ni alloy and Co alloy nanowires depended strongly on the direction of the magnetic field as shown in Figure 15. In the magnetic field direction of 0° (the long axis of nanowires is parallel to the magnetic field), the AMR ratio was almost zero, while a maximum AMR ratio was observed in the magnetic field direction of 90° (the long axis of nanowiresis perpendicular to the magnetic field). Resistance of the nanowires decreased with increase in the magnetic field and the AMR ratio reached 1.0–3.2% with the Ni alloy nanowires. The saturation field of the Ni nanowires was estimated to be about 7 kOe. Ferre *et al.* [47] reported on the magnetoresistance of Ni and Co nanowires electrodeposited in ion-track etched polycarbonate templates. In their report, the magnetoresistance curves obtained from Ni nanowires showed typical AMR behaviour (1.4% of AMR ratio) over a wide range of the diameter in the nanowires, which corresponds well with the results obtained in this study. On the contrary, AMR ratio of the Co alloy nanowires was only 0.6–0.8%. The range of applied magnetic fields was not enough to measure the saturation magnetic field of Co nanowires. Since the magnetic field was insufficient to saturate the magnetization of Co nanowires, the AMR value was not actual full value.

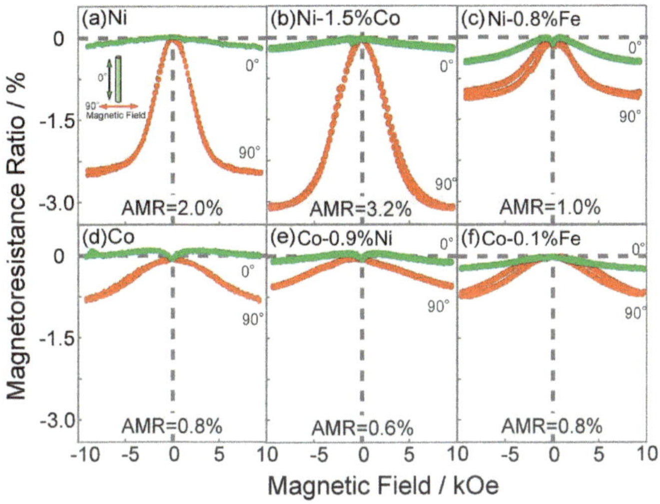

Figure 15. Magnetoresistance curves of pure Ni, Ni-1.5%Co, Ni-0.8%Fe, pure Co, Co-0.9%Ni and Co-0.1%Fe alloy nanowires electrodeposited into anodized aluminum oxide templates with channel-diameter of 60 nm.Magnetic field was applied to in-plan direction and perpendicular direction to the membrane film plan. The perpendicular direction to the membrane film plan corresponds to the parallel direction to the long axis of nanowires.

4. Electrodeposition and Magnetoresistance of Co/Cu Multilayered Nanowires

4.1. Electrodeposition of Co/Cu Multilayered Nanowires

For growing Co/Cu multilayered nanowires, the exfoliated anodized aluminum oxide nanochannels were used as templates. Aqueous solution containing $CoSO_4$ (120 g/L), $CuSO_4$ (1.6 g/L) and H_3BO_3 (45 g/L) was used as electrolyte for electrodeposition of Co/Cu multilayered nanowires. A cathode polarization curve was measured over a wide range of cathode potential to determine the optimum potential for Cu and Co deposition. Figure 16 shows cathode polarization curves for electrodeposition of Cu and Co from the mixed solution (containing Cu^{2+} and Co^{2+} ions) and the solution containing only Co^{2+} ions. The equilibrium potentials of Cu/Cu^{2+} and Co/Co^{2+} are estimated to be around +0.05 V and –0.48 V (vs.Ag/AgCl) on the basis of Nernst equation.The cathode current occurs at the potential region close to the equilibrium potential of Cu as shown in Figure 16. It is well-known that Cu^{2+} ions begin to electrodeposit without accompanying overpotential from the aqueous solution. Therefore, this cathode current corresponds to the deposition current of Cu. With increasing cathode current, at around 10^{-2}A, the potential significantly polarizes to the less-noble region. This polarization would be caused by the diffusion limit of Cu^{2+} ions. In the potential region which is less-nobler than the equilibrium potential of Co, the cathode current increases again at around –0.7 V. It is also well-known that the electrodeposition of iron-group metals such as Ni, Co, and Fe is accompanied by the overpotential due to the rate determination of multistep reduction process. Therefore, this increase in cathode current would be mainly caused by the reduction current of Co^{2+} ions.

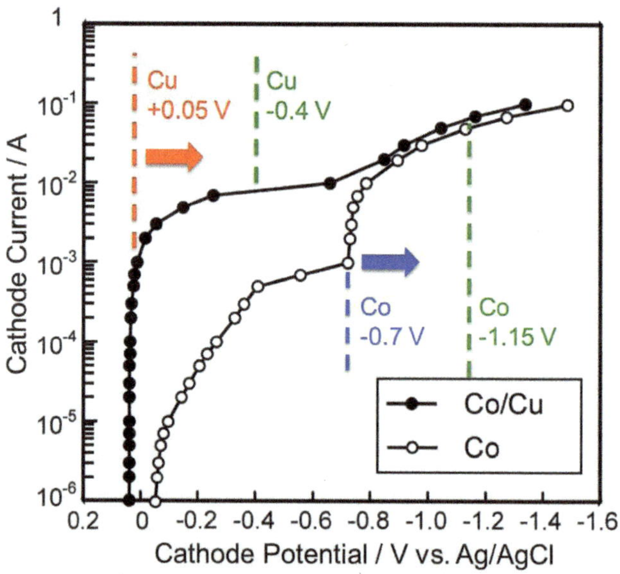

Figure 16. Cathode polarization curves for electrodeposition of Cu and Co from the mixed solution (containing Cu^{2+} and Co^{2+} ions) and the solution containing only Co^{2+} ions.

Growth rates of nanowires were estimated by the channel filling time, which was determined from the time dependence of deposition current at each potential. Figure 17 shows the effect of cathode potential on the time dependence of cathodic current during the electrodeposition of Co nanowires. The cathode potentials were fixed to -0.80,-0.85, -0.90, -0.95 and -1.0 V. To determine the nanowire growth rate, the channel-filling time was estimated by monitoring the deposition current. When the nanowires reach the membrane surface, the current will increase drastically due to the formation of hemispherical caps. The growth rates were calculated dividing the channel length by the channel-filling time. For example, at –1.0 V, the channel-filling time is around 40 s and the deposition rate is estimated to be ca. 150nm/s. On the basis of the results shown in Figure 16 and Figure 17, the optimum deposition potentials of Cu and Co are determined to be about –0.4 and –1.15 V (vs. Ag/AgCl) which is the potential region nobler than the diffusion limit potential of each metal ion. Typical deposition rates of Cu and Co were roughly 10 nm/s (at –0.4 V) and 200 nm s^{-1} (at –1.15 V).

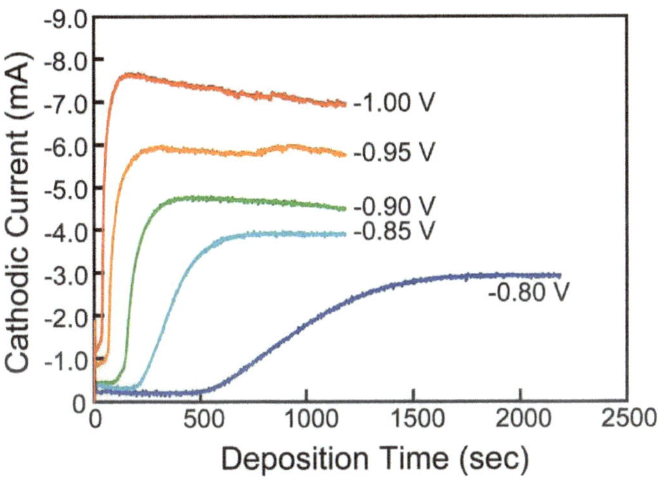

Figure 17. Effect of cathode potential on the time dependence of cathodic current during the electrodeposition of Co nanowires.

Co/Cu multilayered nanowires were synthesized alternatingly switching cathode potential from -0.4 V (for Cu layer) to -1.15 V (for Co layer) as shown in Figure 18. According to this figure, when the potential is switched from -1.15 V to -0.4 V, anodic current is observed. This is resulting from the dissolution of electrodeposited Co, because -0.4V is more nobler than the equilibrium potential of Co. At this potential, it is estimated that the Cu deposition and Co dissolution will proceed simultaneously. According to the time-dependence of cathodic current during electrodeposition of Co/Cu multilayered nanowires as shown in Figure 19, filling time was around 3500 s and the deposition rate was estimated to be about 17 nm/s.

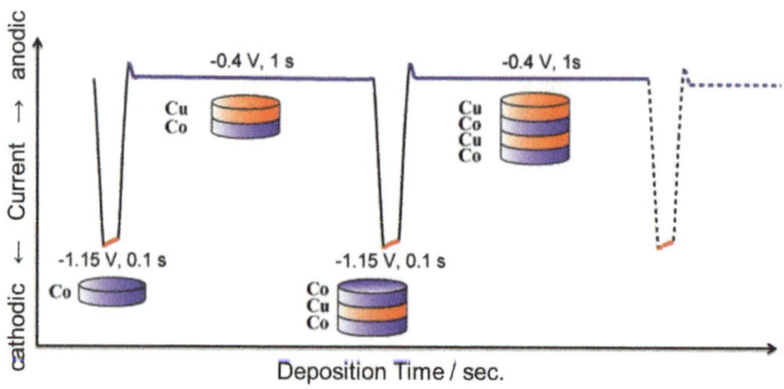

Figure 18. Growing process of Co/Cu multilayered nanowires synthesized alternatingly switching cathode potential from -0.4 V (for Cu layer) to -1.15 V (for Co layer).

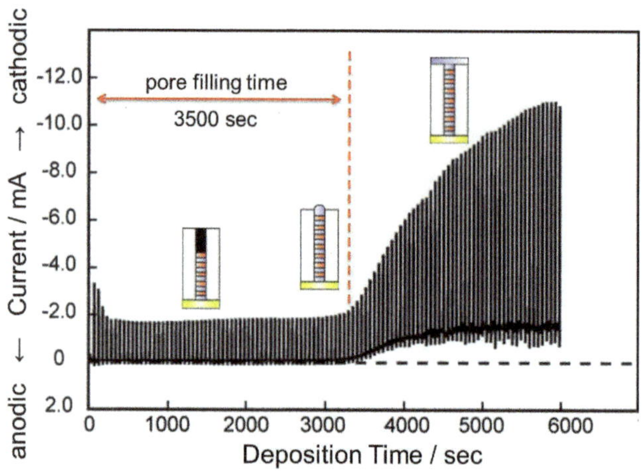

Figure 19. Time dependence of observed-current during the electrodeposition of Co/Cu multilayered nanowires.

4.2. Structure of Co/Cu Multilayered Nanowires

TEM bright field images of electrodeposited Co/Cu multilayered nanowires were investigated as shown in Figure 20. Shape of the nanowires was almost cylindrical and the multilayered structure was also observed as shown in this images. In the structure, several stacking faults were also observed. There would be significant strain energy in the interphase boundary between Co and Cu layer. The strain energy could be stored as the stacking fault energy. It is well known that the stacking fault structure is usually observed in fcc crystals such as Au, Ag, Cu, Ni, Al, etc. Therefore, the stacking fault structure would be introduced to the Cu layer from the interface between Co and Cu layer.

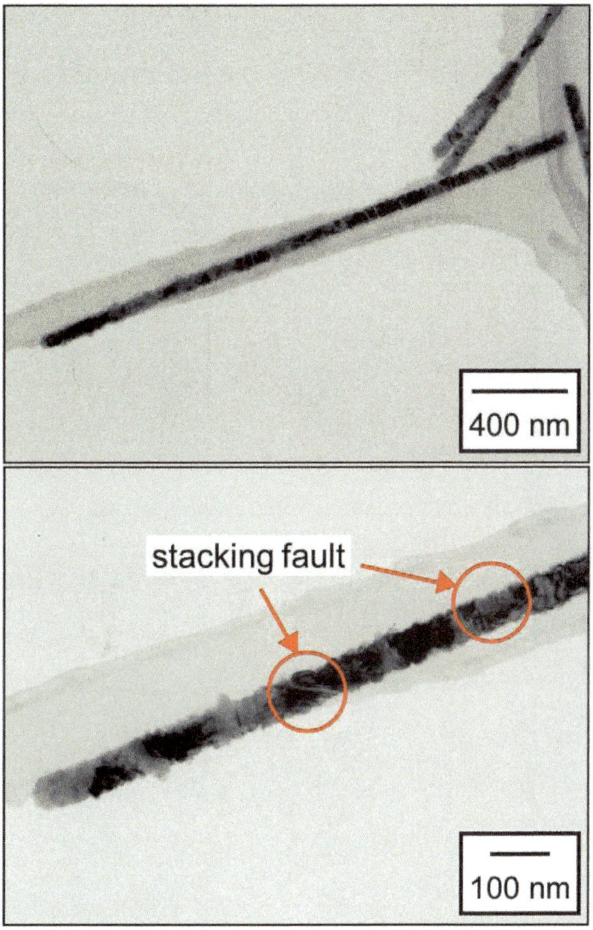

Figure 20. TEM bright field images of Co/Cu multilayered nanowires electrodeposited in nano-channel membrane filters.

4.3. Magnetic Properties of Co/Cu Multilayered Nanowires

Typical perpendicular magnetization behavior was observed from the magnetic hysteresis loops of electrodeposited Co/Cu multilayered nanowires with diameter of 60 nm and length of 60μm as shown in Figure 21-(a). Coercive force of Co/Cu multilayered nanowires was ca.

1 kOe. This is resulting from the shape anisotropy of nanowires with super large aspect ratio of ca. 1000. The magnetoresistance curves revealed that 10.5 % of GMR effect was obtained in the multilayered nanowires with Co layer 10nm, Cu layer 10 nm and 3000 bi-layers as shown in Figure 21-(b).

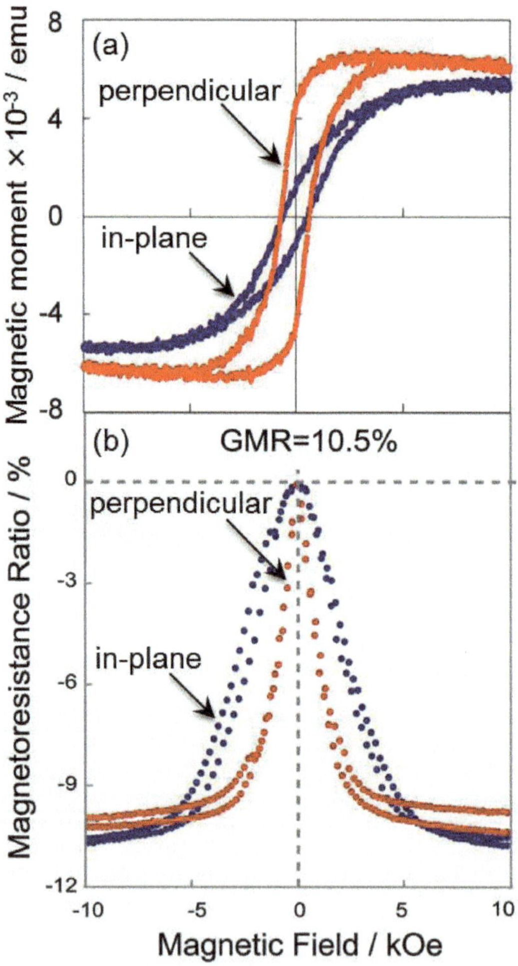

Figure 21. Magnetic hysteresis loops (a) and magnetoresistance curves (b) of Co/Cu multilayered nanowires electrodeposited into anodized aluminum oxide templates with channel-diameter of 60 nm. Magnetic field was applied to in-plan direction and perpendicular direction to the membrane film plan. The perpendicular direction to the membrane film plan corresponds to the parallel direction to the long axis of nanowires.

5. Summary

Anodized aluminum oxide films with the thickness ranging from 2μm to 200 μm were synthesized using a bipolar continuous electrolysis process with anodic oxidation and cathodic exfoliation as well as barrier layer thinning process using chemical etching technique. 3.2 % of AMR effect was observed in Ni-1.5at.% Co alloy nanowires with diameter of 60 nm and length of 60μm. Co/Cu multilayered nanowires with Co-10 nm, Cu-10 nm and 3000 bi-layers were synthesized using a pulsed current electrodeposition technique. 10.5 % of GMR effect was observed in Co/Cu multilayered nanowires electrodeposited into anodized aluminum oxide template with super large aspect ratio of ca. 1000.

Author details

Takeshi Ohgai*

Address all correspondence to: ohgai@nagasaki-u.ac.jp

Division of Chemistry and Materials Science, Nagasaki University, 1-14 Bunkyo-machi, Nagasaki 852-8521, Japan

References

[1] Goad, D. G. W., & Moskovits, M. (1978). Colloidal Metal in Aluminum-oxide. *Journal of Applied Physics*, 49(5), 2929-2934.

[2] Forrer, P., Schlottig, F., Siegenthaler, H., & Textor, M. (2000). Electrochemical Preparation and Surface Properties of Gold Nanowire Arrays Formed by the Template Technique. *Journalof Applied Electrochemistry*, 30(5), 533-541.

[3] Sauer, G., Brehm, G., Schneider, S., Nielsch, K., Wehrspohn, R. B., Choi, J., Hofmeister, H., & Gösele, U. (2002). Highly Ordered Monocrystalline Silver Nanowire Arrays. *Journal of Applied Physics*, 91(5), 3243-3247.

[4] Wang, W., Huang, Q., Jia, F., & Zhu, J. (2004). Electrochemically Assembled p-type Bi_2Te_3Nanowire Arrays. *Journal of Applied Physics*, 96(1), 615-618.

[5] Oh, J., Tak, Y., & Lee, J. (2004). Electrodeposition of Cu_2O Nanowires Using Nanoporous Alumina Template. *Electrochemical and Solid-State Letters* , C 27-C30.

[6] Zhang, X., Hao, Y., Meng, G., & Zhang, L. (2005). Fabrication of Highly Ordered InSb Nanowire Arrays by Electrodeposition in Porous Anodic Alumina Membranes. *Journal ofThe Electrochemical Society*, 152(10), C664-C668.

[7] Ohgai, T., Gravier, L., Hoffer, X., & Ansermet, J. P. (2005). CdTe Semiconductor Nanowires and NiFe Ferro-magnetic Metal Nanowires Electrodeposited into Cylin-

drical Nano-pores on the Surface of Anodized Aluminum. *Journalof Applied Electrochemistry*, 35(5), 479-485.

[8] Katkar, R. A., & Tait, G. B. (2007). The Effect of Stationary Ultraviolet Excitation on the Optical Properties of Electrochemically Self-assembled Semiconductor Nanowires. *Journal of Applied Physics*, 101(5), 053508.

[9] Inguanta, R., Sunseri, C., & Piazza, S. (2007). Photoelectrochemical Characterization of Cu_2O-Nanowire Arrays Electrodeposited into Anodic Alumina Membranes. Electrochemical and Solid-State Letters ., 10(12), K63-K66.

[10] Kim, D. J., Seol, J. K., Lee, M. R., Hyung, J. H., Kim, G. S., Ohgai, T., & Lee, S. K. (2012). Ferromagnetic Nickel Silicide Nanowires for Isolating Primary $CD4^+$ T Lymphocytes. *Applied Physics Letters*.

[11] Martin, C. R. (1991). Template Synthesis of Polymeric and Metal Microtubules. *Advanced Materials*, 3(9), 457-459.

[12] Spohr, R., Zet, C., Fischer, B. E., Kiesewetter, H., Apel, P., Gunko, I., Ohgai, T., & Westerberg, L. (2010). Controlled Fabrication of Ion Track Nanowires and Channels. *Nuclear Instruments and Methods in Physics Research B*, 268(6), 676-686.

[13] Whitney, T. M., Jiang, J. S., Searson, P. C., & Chien, C. L. (1993). Fabrication and Magnetic Properties of Arrays of Metallic Nanowires. *Science*, 261(5126), 1316-1319.

[14] Piraux, L., George, J. M., Despres, J. F., Leroy, C., Ferain, E., Legras, R., Ounadjela, K., & Fert, A. (1994). Giant Magnetoresistance in Magnetic Multilayered Nanowires. *Applied Physics Letters*, 65(19), 484-486.

[15] Blondel, A., Meier, J. P., Doudin, B., & Ansermet, J. P. (1994). Giant Magnetoresistance of Nanowires of Multilayers. *Applied Physics Letters*, 65(23), 3019-3021.

[16] Ohgai, T., Enculescu, I., Zet, C., Westerberg, L., Hjort, K., Spohr, R., & Neumann, R. (2006). Magneto-sensitive Nickel Nanowires Fabricated by Electrodeposition into Multi- and Single-ion Track Templates. *Journalof Applied Electrochemistry*, 36(10), 1157-1162.

[17] Ohgai, T., Hjort, K., Spohr, R., & Neumann, R. (2008). Electrodeposition of Cobalt Based Ferromagnetic Metal Nanowires in Polycarbonate Films with Cylindrical Nanochannels Fabricated by Heavy-ion-track Etching. *Journalof Applied Electrochemistry*, 38(5), 713-719.

[18] Kawai, S., & Ueda, R. (1975). Magnetic Properties of Anodic Oxide Coatings on Aluminum Containing Electrodeposited Co and Co-Ni. *Journal ofThe Electrochemical Society*, 122(1), 32-36.

[19] Kawai, S. (1975). Retentivities of Anodic Oxide Coatings on Aluminum Containing Electrodeposited Cobalt, Nickel, and Cobalt-Nickel. *Journal ofThe Electrochemical Society*, 122(8), 1026-1029.

[20] Kawai, S., & Ishiguro, I. (1976). Recording Characteristics of Anodic Oxide Films on Aluminum Containing Electrodeposited Ferromagnetic Metals and Alloys. *Journal ofThe Electrochemical Society*, 123(7), 1047-1051.

[21] Tsuya, N., Saito, Y., Nakamura, H., Hayano, S., Furugohri, A., Ohta, K., Wakui, Y., & Tokushima, T. A. (1986). Perpendicular Magnetic Recording Medium by Alumite. *Journal of Magnetism and Magnetic Materials*, 54-57(3), 1681-1682.

[22] Tsuya, N., Tokushima, T., Shiraki, M., Wakui, Y., Saito, Y., Nakamura, H., & Katsumata, Y. (1987). Magnetic Alumite Disc for Perpendicular Recording. *IEEE Transactions on Magnetics*, 23(1), 53-55.

[23] Huysmans, G. T. A., Lodder, J. C., & Wakui, J. (1988). Magnetization Curling in Perpendicular Iron Particle Arrays (Alumite Media). *Journal of Applied Physics*, 64(4), 2016-2021.

[24] Cheng, T. J., Jorné, J., & Gau, J. S. (1990). Magnetic Anisotropy of Electrodeposited Cobalt on Alumite Substrate. *Journal ofThe Electrochemical Society*, 137(1), 93-95.

[25] Zhang, L. C., & Lodder, J. C. (1990). The Influence of the Packing Density on the Magnetic Behaviour of Alumite Media. Journal of Magnetism and Magnetic Materials; , 88(1-2), 2236-2246.

[26] Al, Mawlawi. D., Coombs, N., & Moskovits, M. (1991). Magnetic Properties of Fe Deposited into Anodic Aluminum Oxide Pores as a Function of Particle Size. *Journal of Applied Physics*, 70(8), 4421-4425.

[27] Nielsch, K., Müller, F., Li, A. P., & Gösele, U. (2000). Uniform Nickel Deposition into Ordered Alumina Pores by Pulsed Electrodeposition. *Advanced Materials*, 12(8), 582-586.

[28] Evans, P. R., Yi, G., & Schwarzacher, W. (2000). Current Perpendicular to Plane Giant Magnetoresistance of Multilayered Nanowires Electrodeposited in Anodic Aluminum Oxide Membranes. *Applied Physics Letters*, 76(4), 481-483.

[29] Ohgai, T., Hoffer, X., Gravier, L., Wegrowe, J. E., & Ansermet, J. P. (2003). Bridging the Gap Between Template Synthesis and Microelectronics: Spin-Valves and Multilayers in Self-organized Anodized Aluminum Nanopores. *Nanotechnology*, 14(9), 978-982.

[30] Ohgai, T., Hoffer, X., Fabian, A., Gravier, L., & Ansermet, J. P. (2003). Electrochemical Synthesis and Magnetoresistance Properties of Ni, Co and Co/Cu Nanowires in Nano-porous Anodic Oxide Layer on Metallic Aluminum. *Journal of Materials Chemistry*, 13(10), 2530-2534.

[31] Ohgai, T., Gravier, L., Hoffer, X., Lindeberg, M., Hjort, K., Spohr, R., & Ansermet, J. P. (2003). Template Synthesis and Magnetoresistance Property of Ni and Co Single Nanowires Electrodeposited into Nano-pores with Wide Range of Aspect Ratios. *Journal of Physics D: Applied Physics*, 36(24), 3109-3114.

[32] Ohgai, T., Hoffer, X., Gravier, L., & Ansermet, J. P. (2004). Electrochemical Surface Modification of Aluminum Sheets for Application to Nanoelectronic Devices: Anodization Aluminum and Electrodeposition of Cobalt-Copper. *Journalof Applied Electrochemistry*, 34(10), 1007-1012.

[33] Wu, M. T., Leu, I. C., Yen, J. H., & Hon, M. H. (2004). Preparation of Ni Nanodot and Nanowire Arrays Using Porous Alumina on Silicon as a Template without a Conductive Interlayer. *Electrochemical and Solid-State Letters*, 7(5), C61-C63.

[34] Chu, S. Z., Inoue, S., Wada, K., Kanke, Y., & Kurashima, K. (2005). Fabrication and Characterization of Integrated Ultrahigh-density Fe-Pt Alloy Nanowire Arrays on Glass. *Journal ofThe Electrochemical Society*, 152(1), C42-C47.

[35] Wang, X. W., Fei, G. T., Chen, L., X.u, X. J., & Zhang, L. D. (2007). Orientation-controllable Growth of Ni Nanowire Arrays with Different Diameters. *Electrochemical and Solid-State Letters*, 10(4), E 1-E3.

[36] Friedman, A. L., & Menon, L. (2007). Optimal Parameters for Synthesis of Magnetic Nanowires in Porous Alumina Templates. *Journal ofThe Electrochemical Society*, 154(4), E 68-E70.

[37] Li, F., & Metzger, R. M. (1997). Activation Volume of α-Fe Particles in Alumite Films. *Journal of Applied Physics*, 81(8), 3806-3808.

[38] Zeng, H., Zheng, M., Skomski, R., Sellmyer, D. J., Liu, Y., Menon, L., & Bandyopadhyay, S. (2000). Magnetic Properties of Self-assembled Co Nanowires of Varying Length and Diameter. *Journal of Applied Physics*, 87(9), 4718-4720.

[39] Peng, Y., Zhang, H. L., Pan, S. L., & Li, H. L. (2000). Magnetic Properties and Magnetization Reversal of α-Fe Nanowires Deposited in Alumina Film. *Journal of Applied Physics*, 87(10), 7405-7408.

[40] Yin, A. J., Li, J., Jian, W., Bennett, A. J., & Xu, J. M. (2001). Fabrication of Highly Ordered Metallic Nanowire Arrays by Electrodeposition. *Applied Physics Letters*, 79(7), 1039-1041.

[41] Masuda, H., & Fukuda, K. (1995). Ordered Metal Nanohole Arrays Made by a Two-Step Replication of Honeycomb Structures of Anodic Alumina. *Science*, 268(5216), 1466-1468.

[42] Masuda, H., & Satoh, M. (1996). Fabrication of Gold Nanodot Array Using Anodic Porous Alumina as an Evaporation Mask. *Japanese Journal of Applied Physics*, 35, L126-L129.

[43] Masuda, H., Yamada, H., Satoh, M., Asoh, H., Nakao, M., & Tamamura, T. (1997). Highly Ordered Nanochannel-array Architecture in Anodic Alumina. *Applied Physics Letters*, 71(19), 2770-2772.

[44] Jessensky, O., Müller, F., & Gösele, U. (1998). Self-organized Formation of Hexagonal Pore Arrays in Anodic Alumina. *Applied Physics Letters*, 72(10), 1173-1175.

[45] Li, A. P., Müller, F., Birner, A., Nielsch, K., & Gösele, U. (1998). Hexagonal Pore Arrays with a 50-420 nm Interpore Distance Formed by Self-organization in Anodic Alumina. *Journal of Applied Physics*, 84(11), 6023-6026.

[46] Tayaoka, A., Tayaoka, E., & Yamasaki, J. (1996). Preparation of Co-Fe-P Amorphous Fine Needles with Anodization Technique and Measurement of Demagnetizing Factor. *Journal of Applied Physics*, 79(8), 6016-6018.

[47] Ferre, R., Ounadjela, K., George, J. M., Piraux, L., & Dubois, S. (1997). Magnetization Processes in Nickel and Cobalt Electrodeposited Nanowires. *Physical Review B*, 56(21), 14066 .

Chapter 7

Nanowires on a Film for Photoelectrochemical Water Splitting

Miao Zhong, Mario Kurniawan,
Aleksandra Apostoluk, Bertrand Vilquin and
Jean-Jacques Delaunay

Additional information is available at the end of the chapter

1. Introduction

Nanowires-on-a-film structure (NFS) consisting of a dense array of semiconductor nanowires on a conductive film has recently attracted much interest due to their wide range of applications in electronic, optoelectronic, electromechanic and electrochemical devices [15, 5, 25]. Thanks to their 3D geometry, NFSs provide various morphology-induced advantages resulting in improved devices performances. In addition, as the array of semiconductor NFSs can be grown on a conductive film, which realizes electrical connection between the nanowires, the integration of this nanostructured layer into devices is greatly facilitated.

In this chapter, we review the applications of NFSs as efficient photoanodes in photoelectrochemical (PEC) water splitting cells. Recently, the efficiency of the PEC cells has been reported to be enhanced with the use of semiconductor NFSs [23, 10, 30, 32]. The well-aligned geometry of a nanowire array serves as an anti-reflector in order to enhance light absorption in the structure. Also, an effective separation of the photo-generated carriers is achieved along the transverse direction of the nanowires due to the drastically reduced transport distance of minority carriers. Furthermore, the nanowire array possesses large surface areas that greatly enhance the chemical reaction of water splitting as compared to its smooth film counterpart. Lastly, the underneath film realizes electrical connection between the top nanowires. As a consequence, the semiconductor NFS offers a great potential for PEC water splitting applications. However, the fabrication of high quality NFS with functional semiconductor materials still remains a challenge. Single crystalline functional semiconductor materials will ensure the photoanodes to have adequate semiconducting and electrochemical properties, leading to fulfill the PEC water splitting requirements [14, 2], including:

- Efficient light absorption
- Effective carrier separation and transport
- Suitable energy band position for redox reaction of water
- Anti-photocorrosive property in the presence of water.

Furthermore, a controllable approach for the fabrication of the NFSs that enables production of large-area photoanodes at a low cost is not far from being achieved [34].

In the following sections, emphases are placed on the synthesis, property and PEC performance of the NFSs fabricated using various materials. The materials under review include TiO_2, Fe_2O_3 and ZnO, as they have the potential to achieve good performance in PEC water splitting. In addition, surface coating of the ZnO NFS with anti-photocorrosive $ZnGa_2O_4$ photocatalyst layer is discussed as a mean to improve the stability of the ZnO based photoanodes. Finally, the remaining challenges for the NFS to be used as efficient photoelectrodes for PEC water splitting are presented. It should be noted that the NFS described here are based on arrays of one-dimensional (1D) nanowires, however, nanostructured photoanodes based on nanotubes, nanoplates and other nanostructures on a film should also provide improved PEC device performance.

2. TiO_2 NFSs for PEC water splitting

Since the phenomenon of water electrolysis on a titanium dioxide (TiO_2) electrode under UV light illumination was discovered by [9], TiO_2 has been considered as one of the most promising photocatalysts for energy conversion applications. Later in 1977, Frank and Bard achieved the decomposition of cyanide in water using TiO_2 material under UV light illumination [7]. This achievement further confirmed the photocatalytic behavior of TiO_2 material and leaded for the use of TiO_2 in environmental and pollution-degradation applications. Recently, extensive research has been conducted on the TiO_2 nanostructures in order to enhance device performances [3]. Versatile morphologies including TiO_2 nanowires, nanosheets, and nanotubes have been fabricated through different chemical and physical approaches. In many reports, the vertical alignment of the TiO_2 nano-architectures perpendicular to the substrates is found to be a crucial factor necessary to improve device performances [8, 10]. In this section, we focus on the synthesis, characterizations and PEC performances of TiO_2 NFSs.

2.1. Synthesis of the TiO_2 NFSs

Titanium (Ti) is a relatively refractory metal with a high melting temperature of about 1670 °C under air condition. This high melting temperature results in a low Ti vapor pressure of ~ 10^{-3} Torr at 1577 °C, which hinders the TiO_2 nucleation in the direct Ti vapor-phase deposition processes such as the physical vapor deposition (PVD) and the thermal chemical vapor deposition (CVD). In order to increase the Ti precursor concentration for the growth of 1D TiO_2 nanostructures, hydrothermal growth, sol-gel growth and metal organic chemical vapor deposition (MOCVD) using Ti organic compounds as precursors are often employed [3].

The fabrication of TiO_2 NFS via the hydrothermal growth method has been proved to be efficient due to its low-temperature processing (which helps to protect the bottom conductive film during the growth of 1D nanowires), mass production of the nanowires, and their easily controllable alignment. As reported by [10], a densely packed, vertically aligned and single crystalline TiO_2 rutile nanowire array was synthesized on a transparent conductive layer consisting of fluorine-doped tin oxide (FTO) via the hydrothermal method. The fabrication process is described in the following.

The FTO coated glass substrate was first cleaned by sonication in a mixture of ethanol and water for 30 minutes in order to remove the impurities on the surface of the substrate. Then, a seed layer of TiO_2 was synthesized on the cleaned FTO coated glass substrate by a first-step hydrothermal growth. The FTO substrate was soaked in a 0.025 M $TiCl_4$ with n-hexane solution for 30 minutes. The sample was then rinsed with ethanol and finally annealed at 500°C for 30 minutes in air to form a thin layer of TiO_2 on the surface of the substrate. The formation of the TiO_2-seeded layer on the FTO substrate improved the control and the integrity of the later TiO_2 nanowires growth. Moreover, the TiO_2-seeded layer on the FTO layer prevented the formation of electrical short circuits between the electrode and the electrolyte when used as photoanode in the PEC water splitting cell, which guaranteed the photocatalytic function of the TiO_2 photoanode. In the next fabrication step, the TiO_2 nanowires were grown on the TiO_2-seeded FTO substrate with a typical hydrothermal growth procedure. The seeded FTO substrate was placed in sealed Teflon-lined autoclave containing 50 mL of n-hexane, 5 mL of HCL and 2.5-5 mL of titanium isopropoxide. After the reaction was performed at 150°C for a certain amount of time, the autoclave was cooled down to room temperature and finally the TiO_2 NFS was synthesized.

2.2. Characterization of the TiO_2 NFSs

Cross-sectional and top view scanning electron microscope (SEM) images of the synthesized TiO_2 nanowires array are shown in Fig. 1 a and 1b [10]. The TiO_2 nanowires array consisting of vertically aligned and tetragonal shaped nanowires is highly uniform and densely packed on the FTO coated glass substrate. The average diameter of the nanowires is ~5 nm and their length is up to ~4.5 μm. The length of the nanowires can be tuned from ~0.5 to ~5 μm, by modifying the synthesis conditions including the precursors ratio (titanium isopropoxide/n-hexane), the reaction time and the thickness of TiO_2 seed layer. [10]

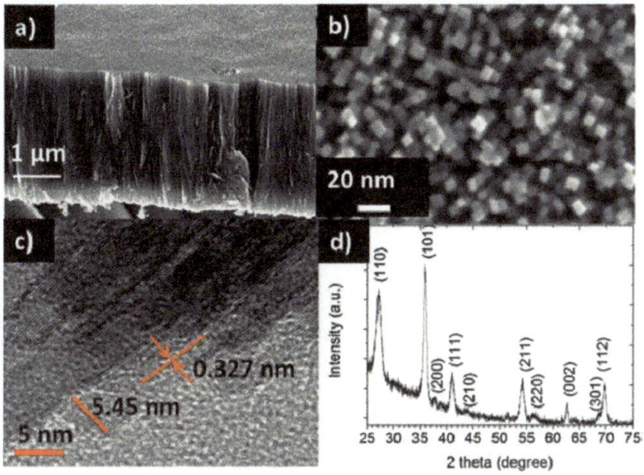

Figure 1. Vertically aligned rutile TiO_2 nanowires. (a) Cross-sectional SEM image, (b) top view SEM image, (c) HRTEM image and (d) GIXRD pattern. Reprinted with permission from [Son 10]. Copyright © 2012, American Chemical Society.

The synthesized TiO_2 nanowires array was characterized by grazing incidence angle X-ray diffraction (GIXRD) technique as shown in Fig. 1 d. The TiO_2 nanowires structure was identified as a rutile crystal from the reference XRD data measured with standard rutile TiO_2 powder. Furthermore, the high resolution transmission electron (HRTEM) image in Fig. 1d showed that the observed rutile phase TiO_2 nanowire is of a single crystal quality, having clear (110) lattice fringes.

2.3. PEC performance of the TiO_2 NFSs

The PEC performance of the synthesized TiO_2 NFS (as previously described) was examined in a three-electrode PEC water splitting cell [10]. The TiO_2 NFS was used as the working electrode, a saturated Ag/AgCl electrode as the reference electrode and a platinum wire as the counter electrode. The electrolyte consisted of one mol/liter KOH solution with pH = 13.5. For the PEC test, TiO_2 nanowires with an average diameter of ~5 nm and average length of ~1.5 μm [10] were studied. The linear sweep voltammetry measurement was performed under a 100 W Xenon lamp illumination (300 nm < wavelength λ < 800 nm) with a bias swept from -0.8 to 0.6 V (vs. Ag/AgCl), as shown in Fig. 2 a. A photocurrent density of ~0.35 mA/cm^2 was obtained at a bias of 1.23 V_{RHE} [10].

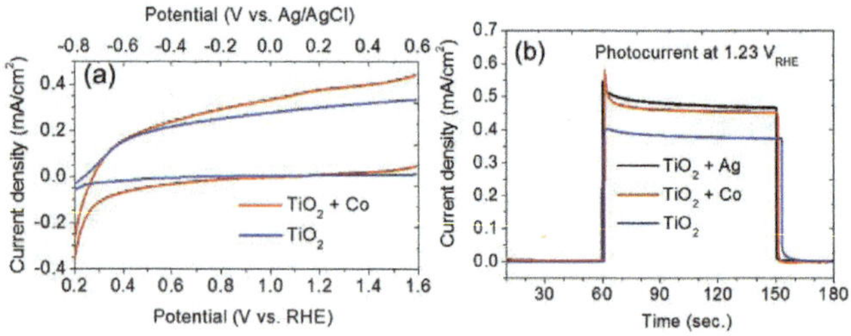

Figure 2. a) Linear sweep voltammetry measurements of the TiO_2 NFS with and without Co treatment, (b) Chronoamperometry measurements of the TiO_2 NFS, Co treated TiO_2 NFS and Ag treated TiO_2 NFS at 1.23 V_{RHE}. Reprinted with permission from [Son 10]. Copyright © 2012, American Chemical Society.

As reported by Tilley et al. [23, 10], decoration of a photoelectrode with metal nanoparticles as co-catalysts exhibited an enhanced yield in hydrogen generation. [10], reported treatments by cobalt (Co) and silver (Ag) performed on the TiO_2 NFS samples. The photocurrent density of ~0.35 mA/cm^2 measured on the undoped TiO_2 NFS increased to ~0.45 mA/cm^2 for the Co and Ag coated TiO_2 NFSs, as shown in Fig. 2 b. The enhancement of the photocurrent after Co and Ag treatments was likely related to the improvement in the extraction of photogenerated holes from the surface of the TiO_2 NFS to the electrolyte, through the Co and Ag nanoparticles [23, 10].

3. Hematite NFS for PEC water splitting

Hematite (α-Fe_2O_3 or iron oxide) has been widely investigated as a photocatalyst candidate for PEC water splitting [13, 14, 30] because of its abundance on Earth, its non-toxicity to environment and its stability in water under light illumination. In addition, a band gap between 2.1 and 2.2 eV of the hematite enables visible light absorption and thus makes this material promising for water splitting under solar light illumination. However, hematite has also some disadvantages such as: small light absorption coefficient, short lifetime of the excited-state carrier (10^{-12} s), short hole diffusion length (2-4 nm), inefficient oxygen evolution reaction kinetics and improper position of the conduction band for hydrogen generation [23, 16], limiting the hematite PEC performance. Different research groups made many efforts to address these issues [23, 14, 16, 10]. To date, improved performances have been realized in the following areas:

1. the synthesis of hematite nanostructures to increase the surface areas;

2. the modification of its electronic structure via elemental doping [23, 14, 16, 10] to enhance the conductivity;

3. the surface decoration with other co-catalysts to reduce the onset water splitting potential.

In the following section, we summarized the results obtained with the silicon-doped hematite nanowire-like array on a conductive film structure (Si-doped NFS-like hematite) decorated with IrO_2 nanoparticles [23] for PEC water splitting.

3.1. Synthesis of the Si-doped NFS-like hematite with IrO_2 surface decoration

Similarly to titanium, iron (Fe) has a relatively high melting temperature of ~1550 °C under air condition and thus hematite nanostructures growth by simple thermal CVD or PVD using solid-phase Fe-compound precursors remains difficult. Instead, solution-phase precursors, such as $Fe(CO)_5$, $Fe(AcAc)_3$ and $FeCl_3$, are used to provide sufficient amount of Fe precursors in the growth phase for the synthesis of hematite nanostructures via atmospheric pressure chemical vapor deposition (APCVD) or hydrothermal growth methods [23, 14, 16, 10].

However, the undoped hematite nanostructure is inactive or very inefficient when used as a photoanode for PEC water splitting, because of its low conductivity and improper conduction band position for hydrogen generation. To improve the conductivity, hematite has been doped by different elements, such as Si, Ti, Sn, Al, Mg and Zn, leading to increase the electron density. To address the issue of improper conduction band position of hematite, hematite surface decoration treatments with co-catalysts nanoparticles, such as IrO_2 nanoparticles [18], were applied to increase the chemical reactions of water oxidation.

As reported by Tilley et al. [23], Si-doped NFS-like hematite sample coated with IrO_2 nanoparticles was fabricated and a large photocurrent density of ~3 mA/cm^2 was obtained at a bias of 1.23 V_{RHE} under AM 1.5 illumination condition. In their synthesis process, a FTO-coated glass was used as the substrate for the NFS growth and put on an aluminum block heater. The substrate was maintained at 415 °C during the APCVD growth process. The precursors $Fe(CO)_5$ and tetraethoxysilane (TEOS) were carried to the FTO coated glass substrate by bubbling argon gas at fixed flow rates, calculated to be 3.5 mg/min for $Fe(CO)_5$ and 0.32 mg/min for TEOS. These two streams were mixed with an air flow of 2 L/min and vertically sprayed to the substrate. After 5 minutes, NFS-like Si-doped hematite was synthesized. According to Tilley's reports, the preparation of homogeneous films on large substrates required a continuously rotating substrate in a rotating belt APCVD reactor. Moreover, the improved PEC performance with Si-doped NFS-like hematite photoanode needed a first growth of an ultrathin SiO_2 interfacial layer of ~1 nm between hematite and FTO. It is very likely that this silica layer worked as an insulator layer to prevent the current leaking from the FTO layer to the electrolyte during the PEC experiments. Finally, the fabricated Si-doped NFS-like hematite was coated with ~2 nm IrO_2 nanoparticles using an electrophoresis deposition process. An increased photocurrent of ~3 mA/cm^2 was achieved with the IrO_2 nanoparticles coating, to be compared with a photocurrent of ~2 mA/cm^2 obtained for the Si-doped NFS-like hematite without co-catalysts [23].

3.2. Characterization of the Si-doped NFS-like hematite with IrO_2 decoration

The cross-sectional SEM images of the synthesized Si-doped NFS-like hematite are shown in Fig. 3 a and 3b [23]. The hematite nanowires with an average height of ~ 500 nm were densely packed on the FTO-coated glass substrate. The diameter of the hematite nanowires at the top area varied from 75 to 250 nm. It is observed that the surfaces of hematite nanowires were very rough and composed of nanoparticles with diameters as small as ~ 10-20 nm. As suggested by Tilley et al., the rough surfaces further increased the nanowire/electrolyte interface areas and also shortened the diffusion length of the photogenerated holes to the interface areas, contributing to the increase of PEC water splitting performance [23].

Figure 3. Cross-sectional SEM images of the dense and rough Si-doped NSF-like hematite. [23] Copyright John Wiley and Sons.

The microstructure and crystal quality of the NFS-like hematite samples with/without Si doping were further studied by XRD 2theta-omega scan technique. The XRD results are shown in Fig. 4. [13]. The undoped and Si-doped NFS-like hematite samples did not contain any other iron oxide phases than hematite, indicating a good experimental control over the hematite synthesis. The fact that the synthesized nanostructures are made of a single-phase hematite is important for the PEC water splitting applications, because other iron oxide phases such as Fe_3O_4 always contain strong electron-hole recombination centers, thus leading to lifetime degradation of the holes used in the oxidation of water [13]. Another important feature shown in the XRD results is that the crystal orientation of the synthesized hematite varied with the synthesis conditions. In the case of the undoped NSF-like hematite sample, there is no dominant crystal orientation. The intensities of the different X-ray diffraction peaks obtained with the undoped NSF-like hematite sample (see Fig. 4b) are similar to the XRD reference data measured on the randomly oriented hematite powders (see Fig. 4c). In contrast, the Si-doped NFS-like hematite sample showed a clear preferential [110] crystal orientation. These [110] oriented hematite nanowires likely have an enhanced conductivity compared to the randomly oriented hematite nanowires. In fact, the conductivity along the hematite [110] orientation was calculated to be of roughly four orders of magnitude higher than the conductivity along the other crystal orientations perpendicular to the [110] in hematite [12]. Therefore, the Si-doped NFS-like hematite with a preferential [110] orientation reported by Tilley et al. achieved a benchmark high photocurrent for PEC water splitting.

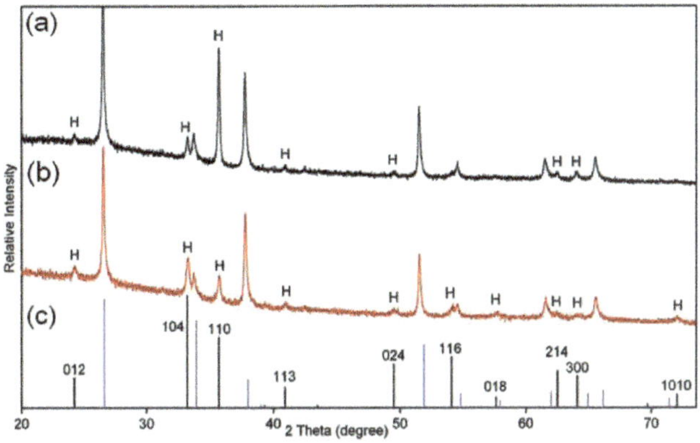

Figure 4. X-ray diffraction pattern of (a) Si-doped NFS-like hematite, (b) undoped NFS-like hematite, (c) standard powders of the hematite (black lines) and the SnO_2 (blue lines). Reprinted with permission from [13]. Copyright © 2006, American Chemical Society.

Finally, the synthesized Si-doped NFS-like hematite was coated with IrO_2 nanoparticles (~2 nm in diameter) by the electrophoresis deposition process [23]. The X-ray photoelectron spectroscopy (XPS) results showed an iridium concentration of ~1% on the surface of the IrO_2 coated sample, confirming the presence of IrO_2 [23].

3.3. PEC performance of the Si-doped NFS-like Hematite with IrO_2 decoration

The PEC performance of the IrO_2 decorated Si-doped NFS-like hematite was studied using a three-electrode PEC cell [23]. In the reported experiment, the Si-doped NFS-like hematite sample was used as the working electrode, an Ag/AgCl in a saturated KCl solution electrode was used as the reference electrode and a platinum wire was used as the counter electrode. One mole/liter NaOH solution with pH = 13.6 was used as the electrolyte. The sample was irradiated with the standard AM1.5 G 100 mW/cm^2 solar simulator.

The chrono-amperometry measurement of the hematite photoanode at a bias of 1.23 V_{RHE} with light on-off cycles shows that an on-off current behavior corresponding to the light on-off cycles is clearly observed (Fig. 5). The low dark current revealed that no/few chemical reactions occurred on the hematite photoanode without light illumination. A high photocurrent density of ~3 mA/cm^2 was obtained at a bias of 1.23 V_{RHE}, as shown in Fig. 5. This photocurrent decreased within the 200-second chrono-amperometry scan. As suggested by Tilley et al., the photocurrent decrease was related to the detachment of the IrO_2 nanoparticles from the surfaces of the Si-doped NFS-like hematite due to a weak chemical bonding between the IrO_2 and the hematite. More robust method to attach the IrO_2 nanoparticles on hematite should further improve the PEC results, as was suggested by Tilley et al. [23]

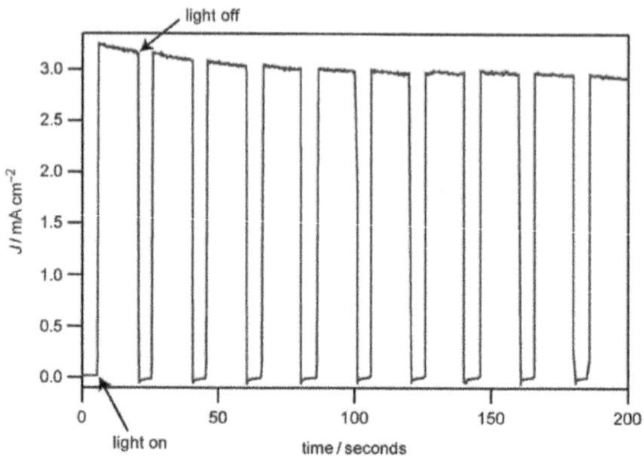

Figure 5. Chrono-amperometry measuremens of the IrO_2 decorated Si-doped NFS-like hematite photoanode at a bias of 1.23 V_{RHE} under AM 1.5 G 100 mW/cm² light illumination [23] Copyright, John Wiley and Sons.

4. ZnO NFS for PEC water splitting

Due to their unique optoelectronic and semiconducting properties, single crystal ZnO densely packed nanowire arrays have been intensively studied in the past few decades for their potential use in electronic, electromechanical, and photonic devices [29, 20, 5, 26, 4, 32]. Recently, much research effort has been devoted to the fabrication of a high quality ZnO nanowire arrays and their application in PEC water splitting with the aim of taking advantage of ZnO material properties such as:

1. suitable band edge positions for PEC water splitting and
2. an increased conductivity when compared to TiO_2 and hematite for efficient charge transport [14, 30, 32].

For PEC water splitting applications, an efficient means permitting a good electrical connection of the nanowires through a conductive film, namely a ZnO NFS, is required. Yang et al. presented a detailed study on the fabrication and PEC performance of ZnO NFS (ZnO nanowire arrays on ITO film) with nitrogen doping [30]. An increased PEC photocurrent was obtained with the nitrogen-doped ZnO NFS sample. Wang et al. reported advanced PEC performances with CdS and CdSe quantum dots loaded ZnO NFSs under the conditions of sacrificial reagents [24]. However promising these results may be, further efforts to increase the stability of the ZnO NFSs in water under light illumination are required.

Since ZnO is chemically unstable in water under light irradiation, surface coating of the ZnO NFS with anti-photocorrosive and photocatalytic shell layers is one of the strategies to improve its chemical stability. However, the search for suitable anti-photocorrosive photocatalysts, which can grow easily on the ZnO nanowire surfaces with a high crystal quality, is a major concern. One possible candidate for the shell material is $ZnGa_2O_4$. Indeed, $ZnGa_2O_4$ is conductive, has anti-photocorrosive ability in water and photocatalytic property for PEC water splitting [11]. In addition, single crystal quality $ZnGa_2O_4$ can be synthesized via high-temperature chemical reactions using ZnO as the precursor material. Thus, surface coating of the ZnO NFS with conductive and anti-photocorrosive $ZnGa_2O_4$ shells to form ZnO-$ZnGa_2O_4$ core-shell NFS has been investigated [33]. With this structure, stable and efficient water splitting was achieved [32]. Further improvements are expected, including enhanced visible light absorption with doping, and efficient charge transfer to the electrolyte with the co-catalyst decoration of the ZnO nanowires.

In this section, we first review the synthesis of high crystal quality ZnO NFSs using different approaches. The realization of high quality ZnO nanowire arrays with efficient electrical connection is crucial for their subsequent use in high performance PEC cells. Then, the surface coating of the anti-photocorrosive $ZnGa_2O_4$ shells on the ZnO NFS is presented. Finally, the electronic properties and a stable PEC performance of the single crystal ZnO-$ZnGa_2O_4$ core-shell NFS are reported in detail.

4.1. Synthesis of the ZnO NFSs

A variety of physical and chemical approaches have been used to synthesize dense and single crystal ZnO nanowire arrays, including laser interference lithography [27], template-assisted wet-chemical growth [30, 4], MOCVD [21] and CVD [29, 5, 34]. As evidenced by these reports, the choice of a suitable substrate is of highest importance to synthesize single crystal ZnO nanowires with an aligned geometry.

Sapphire is a chemically stable and relatively cheap ceramic material and thus has been widely used as substrates for the growth of ZnO nanowire arrays in various fabrication processes. Erdélyi, et al. reported the synthesis of a single crystal quality and mono-oriented ZnO nanowire array on a *c*-plane sapphire by the pulsed laser deposition [4]. In their experiment, the pre-growth of a high quality and size controlled ZnO nanoseeds is needed for the growth of the nanowires. The top-view SEM image of the ZnO nanowire array is shown in Fig. 6 a. Because sapphire is an insulator, the as-grown ZnO nanowires on the sapphire substrate are not electrically connected, preventing their efficient use in optoelectronic applications. Thus, high quality ZnO nanowire arrays grown on conductive films are a prerequisite to the development of advanced nanowire-based materials.

One strategy is to form ZnO NFS on metal substrates. However, the growth of nanowires on metal substrate presents many difficulties. Three major issues should be addressed to realize the functionality of the ZnO NFS using metal substrates. They are as follows:

1. In order to avoid the formation of a Schottky barrier at the interfaces between the ZnO nanowires and the metal substrate, the work function of the metal should be lower than the Fermi level of the ZnO. This requirement severely limits the possible candidates for the metal substrates.
2. Metal substrates usually have large lattice mismatches with ZnO and, therefore, a pre-growth of lattice matched ZnO nanoseeds on the metal substrate is necessary to obtain

vertically aligned ZnO nanowires. As a result, the fabrication process becomes intricate, hard to control and expensive.

3. Another difficult issue related to the metal/ZnO interface is the existence of interface defects between the ZnO nanowires and the metal substrates, or even the formation of an insulator metal oxides layer at the nanowire and substrate interface that will block electron transport.

4. Finally for the purpose of PEC applications, an insulator layer should be deposited over the metal substrate surfaces that are not covered by nanowires in order to prevent a short circuit between the metal substrate and the electrolyte when this structure is used as photoanode.

In summary, the ZnO nanowire array on a metal substrate is still difficult to fabricate with adequate functionalities for use as a photoanode in a PEC cell.

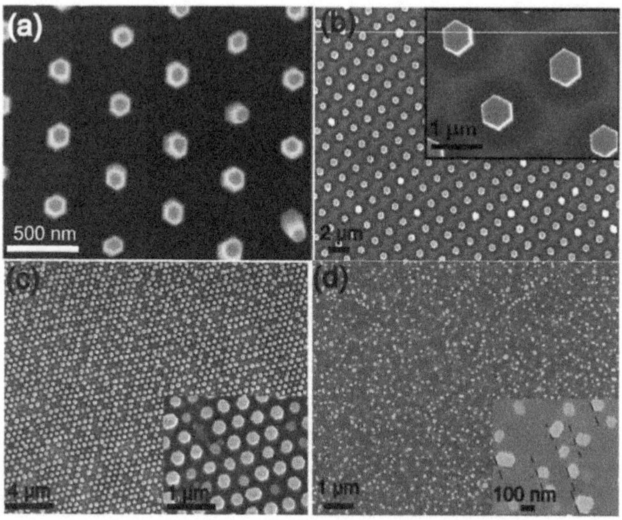

Figure 6. Top-view SEM images of vertically aligned ZnO nanowires on the (a) *c*-plane sapphire. Reprinted with permission from [4]. Copyright © 2011, American Chemical Society. (b) GaN substrate. Reprinted with permission from [27]. Copyright © 20101, American Chemical Society. (c) GaN substrate. [27] Copyright, John Wiley and Sons. (d) high quality ZnO film using *a*-plane sapphire as substrate. [34]

A different strategy is to grow ZnO nanowire arrays on conductive semiconductor substrates. GaN matches the in-plane geometry and lattice constants with ZnO and, therefore, has been employed as a high-quality epitaxial substrate for the growth of ZnO nanowire arrays [5, 27]. Fan et al. and Wei et al. synthesized orderly aligned, size-controlled and single crystalline quality ZnO nanowire arrays using single crystal GaN as substrates via a template-assisted growth process. In Fan's method, a porous anodic alumina membrane was used as a template mask to control the gold catalysts deposition on the GaN substrate, whereas in Wei's method, laser interference lithography technique was used for a pre-growth of high quality ZnO nanoseeds on the GaN substrate. According to their reports, the diameters and distributions of the synthesized ZnO nanowires are governed by the parameters of the porous alumina membrane and the laser interference lithography techniques, respectively. The top-view SEM images of the synthesized ZnO nanowires are shown in Fig 6 b and 6c. Advanced piezo-electric performances have been reported for the fabricated ZnO NFSs. For the use in PEC applications, special attention should be paid to the interface defects and the barrier formation between the ZnO and the GaN in order to realize an efficient electrical charge collection.

The fabrication of a high quality ZnO nanowire array on a ZnO film (ZnO NFS) using a single-step CVD process has been recently reported by our group [34]. The synthesis process enables the growth of a dense ZnO nanowires array on a ZnO film. The nanowires, as well as the film, have a single crystal domain texture quality and the nanowires are perfectly aligned over a large area, which comes as the result of the single crystal domain texture. The top-view SEM image of the fabricated array is shown in Fig 6 d. Efficient electrical connection is realized because nanowires and the underneath film layer are made of the same material ZnO [34]. Compared to other growth processes of ZnO NFSs, our one-step CVD process is easily controlled and cheap. However, the diameters and distributions of the ZnO nanowires cannot be as precisely controlled as in the template-assisted growth processes. The synthesized ZnO NFS offers an ideal template for the subsequent synthesis of a single crystal ZnO-$ZnGa_2O_4$ core-shell NFS, which provides anti-photocorrosive property for the PEC applications.

4.2. Synthesis of the ZnO-$ZnGa_2O_4$ core-shell NFSs

$ZnGa_2O_4$ is a ternary metal oxide compound with a complicated crystal structure: one conventional $ZnGa_2O_4$ unit cell is composed of 8 Zn atoms, 16 Ga atoms and 32 O atoms, arranged into a spinel structure [6]. The direct fabrication of $ZnGa_2O_4$ nanowire array with a single crystal quality is very difficult and has not been reported yet. Thus, a two-step process to grow dense ZnO-$ZnGa_2O_4$ core-shell NFS has been proposed recently [33]:

1. synthesis of the high quality ZnO NFS and
2. surface coating of the ZnO NFS with the $ZnGa_2O_4$ shells (Fig. 7).

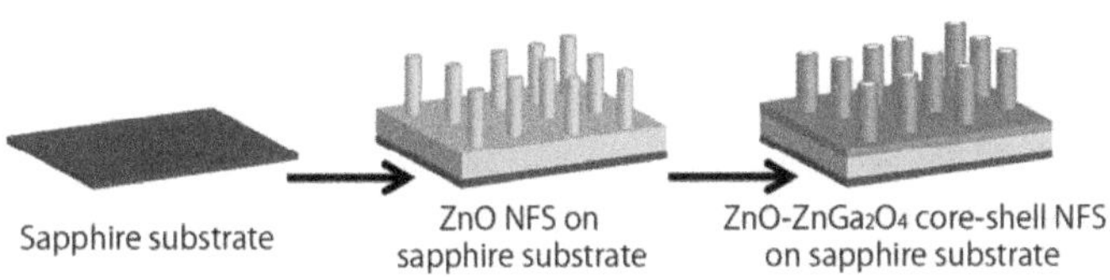

Figure 7. Schematic illustration of the fabrication process of the ZnO-$ZnGa_2O_4$ core-shell NFS. [32]

A high quality ZnO NFS consisting of dense and vertically aligned nanowires on the ZnO film was first fabricated using an *a*-plane sapphire substrate through a simple thermal CVD process [32]. In this process, Au-coated *a*-plane sapphire was selected as the growth substrate. ZnO powder and graphite powder mixed with a 2:1 weight ratio were used as source material and placed at 1 cm upstream position away from the substrate in a tubular furnace. The furnace was operated at 1000 ºC for 30 minutes with the argon and oxygen gas mixture (5:1 in volume ratio) flowing through the tube at the working pressure of 50 mbar. At the end of the synthesis, the furnace cooled down naturally to the room temperature. The synthesized product consisted of a dense nanowires array on a ZnO film (ZnO NFS). Such a structure offers several advantages, namely,

1. the ZnO nanowires grown on the ZnO film are less prone to the formation of defects at the interface between the nanowires and the film and thus an efficient electrical connection of the ZnO nanowires is realized;

2. the underneath ZnO film is stable at high temperature, which enables the subsequent growth of the $ZnGa_2O_4$ shells in the second CVD step, without damaging the NFS.

In a second step, a CVD growth was performed on the ZnO NFS sample. The sample was put in a furnace at a temperature of 900ºC. Mixed precursors of Ga_2O_3, ZnO and graphite were placed at 11 cm upstream position from the sample. The temperature of the precursor zone was set to 1180 ºC. The carrier gases consisted of the argon and oxygen (10:1 in volume ratio) at the working pressure of 50 mbar. After a 30 minutes-long growth, the ZnO-$ZnGa_2O_4$ core-shell NFS was obtained [32].

4.3. Characterization of ZnO-$ZnGa_2O_4$ core-shell NFSs

The morphologies of the synthesized ZnO and ZnO-$ZnGa_2O_4$ core-shell NFSs are presented in Fig. 8. The tilted-angle SEM image of Fig. 8a shows a highly uniform and densely ZnO nanowire array grown over a large area on the Au-coated *a*-plane sapphire substrate. The inset in Fig. 8a is an enlarged SEM image revealing the vertically aligned nanowires. The vertical alignment and smooth sidewalls of the nanowires can be clearly observed. The average diameter of the nanowires is ~70 nm and the average length is about 1 μm. Energy-dispersive X-ray analyses (EDX) performed on the nanowire array gave a Zn/O atomic ratio of ~1/1, indicating the formation of the ZnO nanowires [34].

The tilted-angle SEM image of Fig. 8b shows a densely packed array of ZnO-$ZnGa_2O_4$ core-shell nanowires homogeneously arranged over large areas. The inset in Fig. 8b is an enlarged SEM image of the vertically aligned ZnO-$ZnGa_2O_4$ core-shell nanowires. The initial vertical alignment of the ZnO nanowires is well maintained in the ZnO-$ZnGa_2O_4$ core-shell nanowire sample. The average diameter of the ZnO-$ZnGa_2O_4$ core-shell nanowires is ~100 nm. EDX analysis performed on the surfaces of the core-shell nanowire gave a Zn/Ga/O atomic ratio of ~14/28/58 (~1/2/4), indicating the formation of the $ZnGa_2O_4$ shells. [32] In this work, a nanowires-on-a-film structure

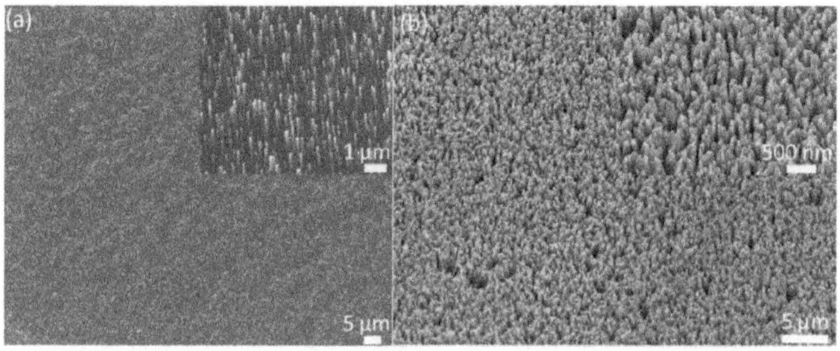

Figure 8. Tilted-angle SEM image of the dense and vertically aligned (a) ZnO and (b) ZnO-$ZnGa_2O_4$ nanowire arrays on the *a*-plane sapphire substrate. Insets in (a) and (b) are close-up images of the vertically aligned ZnO and ZnO-$ZnGa_2O_4$ nanowires. [32 and 34]

TEM and electron diffraction (ED) were used to analyze the structure and the crystalline quality of the ZnO and ZnO-$ZnGa_2O_4$ core-shell NFSs. The cross-sectional high-angle annular dark-field (HAADF) TEM image of the ZnO NFS is shown in Fig. 9a (left image). The average length of the nanowires is about 1 μm and the thickness of the film is about 2.5 μm. Such a thick ZnO bottom layer realized the electrical connection of the nanowire array. The energy dispersive spectroscopy (EDS) mapping images presented in Fig. 9a (the central image) clearly reveal that the NFS is made of ZnO and fabricated on the *a*-plane sapphire substrate.

ED analyses were further performed to characterize the crystal quality and the crystallographic orientation of the ZnO NFS sample. The ED pattern with the incident zone axis of [1-100] (see the right image in Fig. 9a) shows that the ZnO nanowire array and ZnO thick film are all single crystalline and well-aligned along the wurtzite [0001] direction. Such a single crystal and unique [0001] oriented ZnO NFS offers an ideal substrate for the subsequent growth of the ZnO-$ZnGa_2O_4$ core-shell NFS.

The TEM and ED results of the ZnO-$ZnGa_2O_4$ core-shell NFS are shown in Fig. 9b [33]. The bright-field TEM image reveals the nanowire core-shell structure with a different brightness contrast. The ED pattern of the shell region marked as the square A in the fig. 9b presented in the inset A of this figure shows that the shell part of the nanowire consists of a single-crystalline spinel $ZnGa_2O_4$ with a [111] direction parallel to the long axial direction of the nanowire. The ED pattern of the core region of B (square B, the ED pattern is shown in the inset B) confirms that the core part of the nanowire is a single crystal wurtzite ZnO with a [0002] direction parallel to the long axial direction of the nanowire. Furthermore, the ED pattern taken on the interfacial region of C (square C, the ED pattern is shown in the inset C) clearly shows two sets of single-crystalline electron diffraction dot patterns, indexed to the spinel $ZnGa_2O_4$ and the wurtzite ZnO. From the electron diffraction pattern, it is found that the $ZnGa_2O_4$ [111] direction is parallel to the ZnO [0002] direction and the $ZnGa_2O_4$ [2-20] direction is parallel to the ZnO [-2110] direction. An epitaxial relationship between the $ZnGa_2O_4$ (11-2) plane and the ZnO (1-100) plane is therefore predicted. Thus the subsequent growth of $ZnGa_2O_4$ on the ZnO NFS results in a hetero-epitaxial growth of the $ZnGa_2O_4$ shells on the ZnO cores. The final product is well-aligned ZnO-$ZnGa_2O_4$ core-shell nanowires.

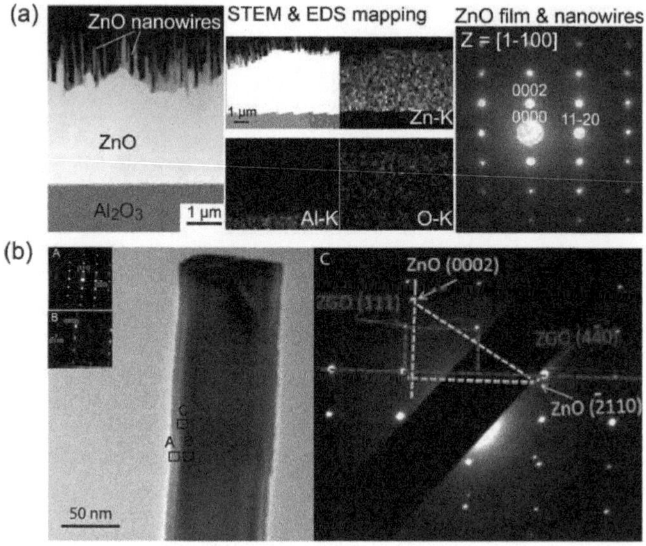

Figure 9. TEM, EDX-mapping and ED analyses of the dense and vertically aligned (a) ZnO and (b) ZnO-$ZnGa_2O_4$ nanowire arrays. [33 and 34]

Finally, standard X-ray diffraction (XRD) 2θ-ω scans were performed to examine the crystal structure and crystal orientations of the ZnO NFS and ZnO-$ZnGa_2O_4$ core-shell NFS over large areas [32 and 34]. Only two main diffraction peaks indexed to the wurtzite ZnO (0002) and (0004) planes were observed on the ZnO NFS sample, indicating a unique out-of-plane crystallographic orientation of the synthesized ZnO NFS over large areas. Furthermore, Zhong et al. performed XRD pole figure and rocking curve measurements. A single crystal domain texture quality (a unique in-plane and out-of-plane orientation over large areas) of the synthesized ZnO NFS was concluded. Such a high quality ZnO NFS was synthesized thanks to a new epitaxial process involving two growth steps. An epitaxial synthesis of the (0001) ZnO on the (11-20) sapphire substrate likely occurred at the beginning of the growth, helping to lock the in-plane orientation of the (0001) ZnO crystal. The Zn atoms then diffused into the Al_2O_3 substrate to form a single crystal, resulting into an intermediate layer of [0-21] oriented $ZnAl_2O_4$. This layer provides an improved in-plane geometry and lattice constant matching over the *a* plane of sapphire and, therefore, helps sustaining the growth of the high quality (0001) ZnO [34]. The fabricated high quality ZnO NFS can be seen as an ideal template to support the synthesis of a single crystal quality ZnO-$ZnGa_2O_4$ core-shell NFS. The XRD 2θ-ω scan result of the core-shell NFS shows only four diffraction peaks. Two diffraction peaks are indexed to the wurtzite ZnO (0002) and (0004) of the cores, and the other two diffraction peaks are indexed to the spinel $ZnGa_2O_4$ (111) and (222) of the shells. This result is strong evidence that the ZnO cores and $ZnGa_2O_4$ shells were both of a single crystal quality and that the crystallographic axes of [0001] ZnO and [111] $ZnGa_2O_4$ are well aligned over large areas.

4.4. Electronic properties of ZnO-$ZnGa_2O_4$ core-shell NFSs

It is of first importance for the development of PEC water splitting cells to understand the electronic properties of the ZnO-$ZnGa_2O_4$ core-shell NFS photoanode in the electrolyte solution. The carrier density and the flatband potential of the ZnO-$ZnGa_2O_4$ core-shell NFS were measured by electrochemical impedance spectroscopy [32]. For this purpose, a solution of 0.5 M $NaClO_4$ buffered (pH of 7.0) was used as an electrolyte, Ag/AgCl electrode in saturated KCl solution as the reference electrode, a Pt wire as the counter electrode and the ZnO-$ZnGa_2O_4$ core-shell NFS sample as a working electrode. Prior to the electrochemical impedance measurement, N_2 gas was bubbled for 10 minutes to get rid of O_2 in the electrolyte solution. The electrochemical impedance analysis was then performed for a bias ranging from - 1 to 1 V (versus Ag/AgCl), at the frequency of 1 kHz under dark condition. The results of this experiment are presented in a $1/C^2$ vs. V plot (Fig. 10a).

The carrier density and the flatband potential of the cylindrical nanowires were quantified using the model proposed by [17]. Solving the Poisson equation with a cylindrical capacitance approximation at the nanowire/electrolyte interfaces, we obtain:

$$\frac{1}{r}\frac{\partial}{\partial r}\left(r\frac{\partial V}{\partial r}\right) = -\frac{q}{\varepsilon\varepsilon_0}N_d \tag{1}$$

$$C_s = \frac{2\varepsilon\varepsilon_0 x^2}{R(R^2 - x^2)} \tag{2}$$

where R is the radius of the nanowires, x the central radius of the quasi-neutral region in the nanowires, C_S the capacitance per unit area of the nanowires, V the applied bias between the electrodes, V_{FB} the flatband potential of the nanowire at the nanowire/electrolyte interface, q the unit electron charge, ε_0 the permittivity of vacuum and ε the relative dielectric constant of ZnO. The results of the fitting of the measured $1/C^2$ vs. V curve according to the Mora-Seró model are shown in Fig. 10 b. The carrier density of the ZnO-$ZnGa_2O_4$ nanowires was estimated to be ~10^{19} cm^{-3}. According to the literature [1, 28, 34], the carrier density of the undoped ZnO nanowires is usually around ~10^{17} – 10^{18} cm^{-3}. The increase in the carrier density of the ZnO-$ZnGa_2O_4$ NFS sample compared to that of the ZnO nanowires can be understood as the formation of conductive $ZnGa_2O_4$ shell layers and the introduction of Ga donors inside the ZnO cores [19, 31]. The increased carrier density in the ZnO-$ZnGa_2O_4$ nanowires decreases their electrical resistance and thus contributes to a reduced energy loss for PEC water splitting applications.

The flatband potential of the ZnO-$ZnGa_2O_4$ NFS sample at the nanowire/electrolyte interface is another important parameter when analyzing the photoanodes properties for water splitting in a PEC cell. Typically, a negative flatband potential and an upward band bending occur at the interface between an *n*-type semiconductor and an electrolyte, due to their Fermi-level difference [22]. A schematic illustration of the band bending with a contact energy barrier at an idealized interface between a n-type semiconductor in equilibrium with an electrolyte is shown in Fig. 10 c. The flat-band potential of the ZnO-$ZnGa_2O_4$ core-shell NFS sample was estimated to be about - 0.4 V (versus NHE). The cathodic flatband potential of the ZnO-$ZnGa_2O_4$ NFS sample provides a suitable energy band position for PEC water splitting.

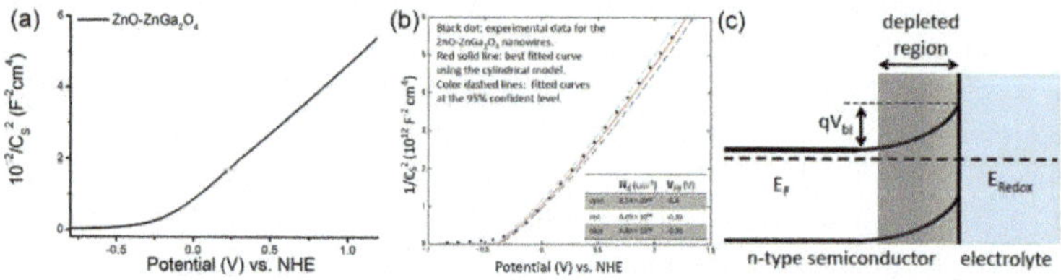

Figure 10. a) Electrochemical impedance plot of the ZnO-$ZnGa_2O_4$ NFS sample at the frequency of 1 kHz in dark condition with a bias varying from -0.8 to 1.2 V (versus NHE); (b) Fitted curves of ZnO-$ZnGa_2O_4$ NFS sample with an assumed ε value of 10 for the ZnO-$ZnGa_2O_4$ nanowires; (c) A schematic of the surface band bending at an idealized interface between an n-type semiconductor and an electrolyte in equilibrium. qV_{bi} is the energy barrier height at the interface. [32] - Reproduced by permission of The Royal Society of Chemistry.

4.5. PEC performances of ZnO-$ZnGa_2O_4$ core-shell NFSs

The PEC performance of the ZnO-$ZnGa_2O_4$ NFS sample used as a photoanode was studied in a 0.1 M Na_2SO_4 electrolyte solution at pH = 6 under illumination with a 300 W Xenon lamp [32]. An Ag/AgCl electrode in a saturated KCl solution was used as the reference electrode and a Pt wire used as the counter electrode. Fig. 11a shows the current-voltage curve for the ZnO-$ZnGa_2O_4$ nanowire array at a bias ranging from -0.8 V to 1 V (versus Ag/AgCl) with light on-off cycles, indicating a good photoresponsibility of the NFS sample used as photoanode. The flatband potential of the NFS sample is equivalent to the onset potential of the anodic photocurrent, as shown to be ~ -0.35 V vs. NHE with an estimated error of ±0.1 V in the inset of Fig. 11a [32]. This value of the onset potential agreed well with the flatband potential of ~ -0.4 V estimated from the fit of the electrochemical impedance data in Fig. 10 b.

The amperometric study of the ZnO-$ZnGa_2O_4$ core-shell nanowire array was performed at a fixed bias of 1.23 V_{RHE} under on-off illumination cycles. The obtained data are plotted in the form of current-time curve (Fig. 11b). The result shows a very low current of < 10^{-4} mA/cm^2 under dark conditions, indicating the absence of any chemical reaction at the ZnO-$ZnGa_2O_4$ NFS anode for an applied bias of 1.23 V_{RHE} under dark conditions. Under illumination, a large photocurrent with a steady value of 1.2 mA/cm^2 was obtained, indicating a stable and efficient photoelectrolysis of water at the surface of the ZnO-$ZnGa_2O_4$ NFS photoanode. As a comparison, the photocurrent measured on ZnO or *n*-doped ZnO nanowire array always deteriorates with time under continuous light illumination, confirming that ZnO is chemically unstable in water under illumination (Li et al. 2009). After switching off the light illuminating the ZnO-$ZnGa_2O_4$ NFS anode, a fast recovery of the current to the dark current value was clearly observed. The on-off current cycles presented a very good reproducibility and anti-photocorrosion stability as shown in Fig. 11 b. Therefore, the core-shell NFS demonstrated an improved stability and a large photocurrent of 1.2 mA/cm^2 when used as the photoanode in the PEC water splitting cell [32].

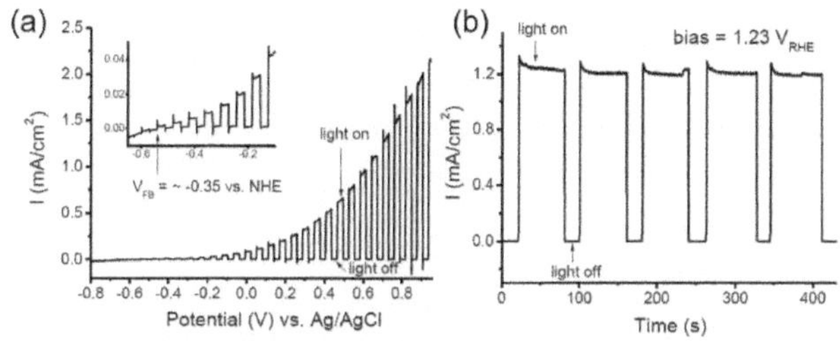

Figure 11. a) Current-voltage curves of the ZnO-$ZnGa_2O_4$ NFS sample in an electrolyte of 0.1 M Na_2SO_4 at pH = 6 under illumination with a 300 W Xenon lamp. Sweep rate of the voltage was 0.5 mV/s. (b) Current-time curve of the ZnO-$ZnGa_2O_4$ NFS when used as a photoanode at an applied bias of 1.23 V_{RHE} with light on-off cycles. [32] - Reproduced by permission of The Royal Society of Chemistry.

5. Conclusions

Recent advances in the development of densely packed and single crystalline quality semiconductor NFSs with a controllable doping level offer promising opportunities for the future PEC water splitting applications. In this type of nanostructured photoanodes, an improved light absorption, enhanced charge separation and an increased surface area for chemical reaction are obtained due to the shell layer of the semiconductor nanowire array of the NFS. Further, the film underneath the nanowires realizes an efficient electrical connection of the semiconductor nanowire array, thus facilitating the direct use of the NFS as the photoanode in PEC water splitting applications. However, many difficulties still remain to be resolved. Two essential issues related to the reliability and the efficiency needs to be addressed for all the PEC water splitting systems. In this Chapter, the NFS-based photoanodes made of different semiconductor photocatalysts have been reviewed and their advantages and the remaining challenges have been discussed in detail. The TiO_2 NFS photoanode has an excellent chemical stability in water, but its large bandgap and inefficient doping limit the efficiency in the PEC applications under solar illumination. The hematite NFS photoanodes have a bandgap which permits a good photoresponsivity under the illumination with the visible part of the solar spectrum and a good chemical stability in water. However, the large resistance and inadequate energy band position of the hematite render this NFS inefficient for the PEC applications without any further efficient doping and the co-catalysts decoration. An efficient doping and the co-catalyst decoration of the hematite NFS have not yet been achieved to improve its performances and obtain a stable operation in water splitting PEC. The ZnO-$ZnGa_2O_4$ core-shell NFS photoanode offers an enhanced conductivity over TiO_2 and hematite, as well as anti-photocorrosive property in water. Thus this material offers stable and efficient operation of water splitting PEC. However, the ZnO-$ZnGa_2O_4$ core-shell NFS only responds to the UV light, which results in a low efficiency under solar light illumination. Further work is now in progress to achieve an efficient and stable solar water splitting with doped ZnO-$ZnGa_2O_4$ core-shell NFSs. Besides the photocatalyst materials already fabricated in the form of NFSs reviewed in this Chapter, it is worth noting that a large variety of alternative photocatalysts exist, which can permit to grow nanostructures for high efficiency PEC applications. Finally, still more progress is expected to be made to realize an efficient and stable solar water splitting system permitting to achieve a new, sustainable and environmentally friendly hydrogen production method, contributing to a better daily life of human beings in the near future.

Author details

Miao Zhong[1], Mario Kurniawan[1], Aleksandra Apostoluk[2*], Bertrand Vilquin[2] and Jean-Jacques Delaunay[1*]

*Address all correspondence to: jean@mech.t.u-tokyo.ac.jp

1 School of Engineering, The University of Tokyo, Japan

2 INL, CNRS UMR-5270, Lyon University, France

References

[1] Allen, M. W., Alkaisi, M. M., & Durbin, S. M. (2006). Metal Schottky diodes on Zn-polar and O-polar bulk ZnO. *Appl. Phys. Lett.*, 89, 103520.

[2] Bak, T., Nowotny, J., Rekas, M., & Sorrell, C. C. (2002). Photo-electrochemical hydrogen generation from water using solar energy. Materials-related aspects. *Inter. J. Hydro. Energy.*, 27, 991-1022.

[3] Chen, X., & Mao, S. S. (2007). Titanium dioxide nanomaterials: synthesis, properties, modifications, and applications. *Chem. Rev.*, 107, 2891-2959.

[4] Erdélyi, R., Nagata, T., Rogers, D. J., Teherani, E. H., Horváth, Z. E., Lábadi, Z., Baji, Z., Wakayama, Y., & Volk, J. (2011). Investigations into the impact of the template layer on ZnO nanowire arrays made using low temperature wet chemical growth. *Cryst. Growth Des.*, 11, 2515-2519.

[5] Fan, H. J., Lee, W., Hauschild, R., Alexe, M., Rhun, G. L., Scholz, R., Dadgar, A., Nielsch, K., Kalt, H., Krost, A., Zacharias, M., & Gösele, U. (2006). Template-assisted large-scale ordered arrays of ZnO pillars for optical and piezoelectric applications. *Small*, 2, 561-568.

[6] Fan, H. J., Yang, Y., & Zacharias, M. (2009). ZnO-based ternary compound nanotubes and nanowires. *J. Mater. Chem.*, 19, 885-900.

[7] Frank, S. N., & Bard, A. J. (1977). Heterogeneous photocatalytic oxidation of cyanide ion in aqueous solutions at titanium dioxide powder. *J. Am. Chem. Soc.*, 99, 303-304.

[8] Feng, X., Shankar, K., Varghese, O. K., Paulose, M., Latempa, T. J., & Grimes, C. A. (2008). Vertically aligned single crystal TiO_2 nanowire arrays grown directly on transparent conducting oxide coated glass: synthesis details and applications. *Nano. Lett.*, 8, 3781-3786.

[9] Fujishima, A., & Honda, K. (1972). Electrochemical photolysis of water at a semiconductor electrode. *Nature*, 283, 37-38.

[10] Hoang, S., Guo, S., Hahn, N. T., Bard, A. J., & Mullins, C. B. (2011). Visible light driven photoelectrochemical water oxidation on nitrogen-modified TiO_2 nanowire. *Nano. Lett.*, 12, 26-32.

[11] Ikarashi, K., Sato, J., Kobayashi, H., Saito, N., Nishiyama, H., & Inoue, Y. (2002). Photocatalysis of water decomposition by RuO_2-Dispersed $ZnGa_2O_4$ with d^{10} configuration. *J. Phys. Chem. B*, 106, 9048-9053.

[12] Iordanova, N., Dupuis, M., & Rosso, K. M. (2005). Charge transport in metal oxides: A theoretical study of hematite α-FeO. *J. Chem. Phys.*, 122, 144305-1-10.

[13] Kay, A., Cesar, I., & Grätzel, M. (2006). New benchmark for water photooxidation by nanostructured α-Fe2O3 films. *J. Am. Chem. Soc.*, 128, 15714-15721.

[14] Krol, R. V. D., Liang, Y., & Schoonman, J. (2008). Solar hydrogen production with nanostructured metal oxide. *J. Mater. Chem.*, 18, 2311-2320.

[15] Law, M., Greene, L. E., Johnson, J. C., Saykally, R., & Yang, P. (2005). Nanowire dye-sensitized solar cells. *Nature*, 4, 455-459.

[16] Ling, Y., Wang, G., Wheeler, A. D., Zhang, J. Z., & Li, Y. (2011). Sn-doped hematite nanostructures for photoelectrochemical water splitting. *Nano. Lett.*, 11, 119-125.

[17] Mora-Seró, I., Fabregat-Santiago, F., Denier, B., & Bisquert, J. (2006). Determination of carrier density of ZnO nanowires by electrochemical techniques. *Appl. Phys. Lett.*, 89, 203117-1-3.

[18] Nakagawa, T., Beasley, C. A., & Murray, R. W. (2009). Efficient electro-oxidation of water near its reversible potential by a mesoporous IrO x nanoparticle film. *J. Phys. Chem. C*, 113, 12958-12961.

[19] Omata, T., Ueda, N., & Ueda, K. (1994). New ultraviolet-transport electroconductive oxide, $ZnGa_2O_4$ spinel. *Appl. Phys. Lett.*, 64, 1077-1078.

[20] Özgür, Ü., Alivov, Ya. I., Liu, C., Teke, A., Reshchikov, M. A., Doğan, S., Avrutin, V., Cho, S. J., & Morkoç, H. (2005). A comprehensive review of ZnO. *Appl. Phys. Lett.*, 98, 041301-1-103.

[21] Park, W. I., Kim, D. H., Jung, S. W., & Yi, G. (2002). Metalorganic vapor-phase epitaxial growth of vertically well-aligned ZnO nanorods. *Appl. Phys. Lett.*, 80, 4232-4234.

[22] Sze, S. M., & Kwok, K. Ng. (2007). Physics of semiconductor devices. *3rd edition. Book, John Wiley & Sons.*

[23] Tilley, S. D., Cornuz, M., Sivula, K., & Grätzel, M. (2010). Light-induced water splitting with hematite: improved nanostructure and iridium oxide catalysis. *Angew. Chem. Int. Ed.*, 49, 6405-6408.

[24] Wang, G., Yang, X., Qian, F., Zhang, J. Z., & Li, Y. (2010). Double-sided CdS and CdSe quantum dot co-sensitized ZnO nanowire arrays for photoelectrochemical hydrogen generation. 10, 1088-1092.

[25] Wang, X., Song, J., Liu, J., & Wang, Z. L. (2007). Direct-Current Nanogenerator driven by ultrasonic waves. *Science*, 316, 102-105.

[26] Wang, Z. L. (2009). Energy Harvesting Using Piezoelectric Nanowires-A Correspondence on "Energy Harvesting Using Nanowires?". *by Alexe et al. Adv. Mater.*, 21, 1311-1315.

[27] Wei, Y., Wu, W., Guo, R., Yuan, D., Das, S., & Wang, Z. L. (2009). Wafer-scale high-throughput ordered growth of vertically aligned ZnO nanowire arrasy. *Nano Lett.*, 10, 3414.

[28] Wolcott, A., Smith, W. A., Kuykendall, T. R., Zhao, Y., & Zhang, J. Z. (2009). Photoelectrochemical study of nanostructured ZnO thin films for hydrogen generation from water splitting. *Adv. Funct. Mater.*, 19, 1849-1856.

[29] Yang, P., Yan, H., Mao, S., Russo, R., Johnson, J., Saykally, R., Morris, N., Pham, J., He, R., & Choi, H. (2002). Controlled Growth of ZnO Nanowires and Their Optical Properties. *Adv. Funct. Mater.*, 12, 323.

[30] Yang, X., Wolcott, A., Wang, G., Sobo, A., Fitzmorris, R. C., Qian, F., Zhang, J. Z., & Li, Y. (2009). Nitrogen-doped ZnO nanowire arrays for photoelectrochemical water splitting. *Nano Lett.*, 9, 2331.

[31] Yuan, G., Zhang, W., Jie, J., Fan, X., Tang, J., Shafiq, I., Ye, Z., Lee, C., & Lee, S. (2008). Tunable n-Type Conductivity and Transport Properties of Ga-doped ZnO Nanowire Arrays. *Adv. Mater.*, 20, 168-173.

[32] Zhong, M., Li, Y., Yamada, I., Delaunay, J., & , J. (2012a). ZnO-$ZnGa_2O_4$ core-shell nanowire array for stable photoelectrochemical water splitting. *Nanoscale*, 77, 1509-1514.

[33] Zhong, M., Li, Y., Tokizono, T., Zheng, M., Yamada, I., Delaunay, J., & , J. (2012b). Vertically aligned ZnO-$ZnGa_2O_4$ core-shell nanowires: from synthesis to optical properties. *J. Nanopart. Res.*, 14, 804-814.

[34] Zhong, M., Sato, Y., Kurniawan, M., Apostoluk, A., Masenelli, B., Maeda, E., Ikuhara, Y., & Delaunay, J. J. (2012c). ZnO dense nanowire array on a film structure in a single crystal domain texture for optical and photoelectrochemical applications. *Nanotechnology*, accepted.

Section 2

Silicon Nanowires

Chapter 8

Silicon Nanowire FinFETs

C. Mukherjee and C. K. Maiti

Additional information is available at the end of the chapter

1. Introduction

Some of the fundamental problems of ultra-small MOSFETs beyond sub-10nm channel length are the electrostatic limits, source-to-drain tunnelling, carrier mobility degradation, process variations, and static leakage. The trend toward ultra-short gate length MOSFETs requires a more and more effective control of the channel by the gate leading to new device architecture. It appears that non-classical device architectures can extend the CMOS lifetime and provide solutions to continue scaling. In case of silicon-based CMOS technologies, planar MOSFETs are limited to scaling beyond 15 nm technology node. As simple scaling of silicon CMOS becomes increasingly complex and expensive, there is considerable interest in increasing performance by using strained channels which can improve carrier mobility and drive current in a device. Multi-gate MOSFETs based on the concept of volume inversion are widely recognized as one of the most promising solutions for meeting the ITRS roadmap requirements. A wide variety of multi-gate architectures, including Double-Gate (DG), Gate-All-Around (GAA), Pi-FET and Fin Field-Effect Transistors (FinFETs), rectangular or cylindrical nanowire MOSFETs has been proposed in the literature. In all cases, these structures exhibit a superior control of short channel effects resulting from an exceptional electrostatic coupling between the conduction channel and the surrounding gate electrode. The nanowire (NW) transistors can be seen as the ultimate integration of the innovative nanodevices and is one of the candidates which have gained significant attention from both the device and circuit developers because of its potential for building highly dense and high performance electronic circuits. Recent advances in nanoscale fabrication techniques have shown that semiconductor nanowires may become the candidate for next generation technologies. Si and Ge nanowire transistors are also important because of their compatibility with the CMOS technology.

Nanowires and FinFETs have attracted considerable attention due to their proven robustness against Short-Channel Effects (SCE) and relatively simple fabrication. Silicon nanowire

FinFET (SNWFT) is being considered as the candidate for CMOS scaling beyond the 32 nm node due to its high performance, excellent gate control and enhanced carrier transportation properties. Thin and multi-gate controlled body provide FinFETs a superior short-channel effect control, electrostatic shielding from the body bias, relaxing channel doping or pocket implants, commonly needed in planar technologies to avoid threshold voltage (V_{TH}) roll-off [1]. Adequate device V_{TH} is now being achieved by using gate stack engineering. Metal-Gates with High-k (HKMG) dielectric stacks provide the desired V_{TH} by work-function tuning as well as preserving low gate leakage [2]. Various process options are being attempted to affect carrier transport in a different manner. The use of metal gates as replacement for Poly-Si stacks eliminates the Poly-depletion effect, benefiting effective carrier mobility by reducing the transverse field [1]. Additionally, use of undoped channels improves low-field μ_{eff} due to the reduction in the substrate impurity scattering [3]. Mobility degradation due to Coulomb scattering in short-channel devices should further be reduced in the absence of pocket implants [4]. However, high-k dielectrics are known to degrade mobility as a result of a combination of Coulomb and phonon scattering mechanisms [5]. In order to account for inversion layer mobility degradation mechanisms for FinFET devices, a robust μ_{eff} extraction algorithm is necessary. The mobility extraction requires accurate measurement of both gate to channel capacitance and channel current, together with reliable estimations for the parasitic Source-Drain series resistance (R_{SD}) and the effective channel length [4]. Unfortunately, the capacitance of short-channel length devices in the presence of large gate leakage is no longer characterized trivially. In multi-gate architectures like the FinFET, the carrier transport occurs in different crystallographic planes. For standard substrates, the current flow occurs in the (100)/(110) for the top surface and (110)/(110) for the sidewalls. The transport in these crystal planes and directions are characterized by different mobility primarily due to the anisotropy of the effective masses.

The most promising among various Si multi-gate MOSFET architectures such as double-gate and tri-gate FinFETs, are nanowire FinFETs due to their superior electrostatic control through gate-all-around structure. Superior gate control, immunity of threshold voltage from substrate bias and excellent carrier transport properties along with more aggressive channel length scaling possibility have made GAA architecture with semiconductor nanowire channel a potential candidate for post-planar transistor design. Two approaches are generally used to fabricate Si NWs as well as other semiconductor NWs: bottom-up and top-down. In the first method, NWs are usually grown using a metallic catalyst on a separate substrate, usually through a Vapor-Liquid-Solid (VLS) growth mechanism. After a chemical or mechanical separation step, the NWs are harvested and transferred to another substrate [6]. In the top down-approach, the NWs are fabricated using a CMOS compatible technology, such as lithography-based patterning and etching [7]. Unlike the bottom-up approach where the NWs are randomly distributed, the top-down method enables accurate positioning of the NWs across the wafer and facilitates the ultra-large-scale-integration (ULSI) for high performance nano-electronic circuits. Moreover, due to processing difficulties related to the length of grown NWs, NW release and gate-etch process, most of the VLS grown NW transistors have omega-shaped gate (Ω-gate) geometry and are thus not full gate-all-around

[6]. Among all the promising post-CMOS non-conventional structures, the silicon nanowire transistor has the most unique advantage – it is based on silicon.

In a MOSFET, the carriers encounter various scattering mechanisms on their way towards the drain terminal. Carrier mobility is a well known benchmark to judge the intrinsic performance of a long channel MOSFET. The state-of-the art short channel devices do not operate in the fully ballistic regime (they are at roughly 60% of the ballistic limit) and mobility is related to velocity through effective mass and ballistic ratio. Therefore, understanding the carrier mobility is beneficial to design and engineer new devices for future CMOS generations. The presence of significant resistive and capacitive parasitics as well as lack of large capacitance due to the small size of the NW channel adds complexities to the extraction of the intrinsic NW device characteristics. Simplified assumptions or incomplete device charge simulations generally lead to inaccurate and unreasonable mobility extraction [9]. In the absence of a well-behaved top-down process (as compared to the bottom-up-fabricated NWs with smooth and near ideal sidewalls) results in significant mobility degradation for smaller NW sizes [8].

Noise is known as a fundamental problem in various fields such as telecommunication, nanoelectronics, and biological systems. The low-frequency noise, or 1/f noise, is the excess noise at low frequencies whose power spectral density (PSD) approximately depends inversely on the frequency and therefore escalates at low frequencies. The 1/f noise originating from the transistors is a major issue in analog circuit design. The 1/f noise is, for example, up converted to undesired phase noise in voltage controlled oscillator (VCO) circuits, which can limit the information capacity of communication systems.

In this chapter we shall present a detailed framework on gate-all-around nanowires and multi-gate FinFETs which will include process and device design and characterization, simulation, and low-frequency noise and tunnelling spectroscopy analyses. Carrier mobility extraction in nanowire and FinFETs using split-CV technique will be described. Channel orientation dependence of the electrical parameters such as current, low-frequency noise will be presented. Self-heating effects in nanowire transistors are also discussed. In Section 2, the growth and fabrication of silicon nanowire transistors and multi-gate FinFET devices will be covered. The detailed fabrication process steps for GAA nanowires as well as tri-gate FinFETs are discussed. 3D device simulation and electrical characterization will also be covered in this Section. Split CV measurement technique is employed for mobility calculation and detailed model equations are formulated for FinFET and nanowire structures. In Section 3, the low-frequency noise characteristics of the nanowire FinFETs will be taken up. Voltage dependence of flicker noise (1/*f*) and power spectral densities and corner frequencies will be discussed based on low-frequency noise measurements. Time-domain random telegraph signal (RTS) and existence of two-level or multi-level RTS noise in nanowire transistors will also be taken up. In section 4, the orientation dependence will be discussed and current-voltage characteristics are compared to show the effect of orientation on the drain current. The effect of channel orientation is also discussed for the nanowire FinFET noise spectral densities for (100) and (110) channel orientations. In section 5, the lattice structures of Si, SiO_2 in the nanowire FinFET are analyzed via inelastic electron tunneling spectroscopy (IETS). The lattice dynamics and other molecular vibrations in silicon nanowire FinFETs are discussed using the IETS spectra.

2. Multi-gate FinFET and Nanowire Transistors

In order to control short channel effects in planar SOI MOSFETs and maintain reasonable electrostatics at a given gate length, L_G, a critical Si thickness, t_{Si} on the order of $L_G/3$ is required [10]. Multi-gate devices have been suggested to significantly relax the critical Si dimension in fully depleted SOI (FD-SOI) device architectures and process technology. For this purpose, planar double-gate, vertical double-gate (FinFET), triple-gate (Tri-gate), omega-shaped gate (Ω-gate), penta-gate, inverted T-channel FET, Φ-FET, and GAA nanowires have been proposed by various research groups over the past decade.

While most of these multi-gate technologies have been experimentally shown to benefit from excellent electrostatics and short-channel effects, they commonly suffer from a complicated fabrication process technology compared to planar bulk or SOI technology. Figure 1. shows the device schematic of

a. double-gate FinFET

b. tri-gate and

c. GAA NW MOSFETs.

The critical Si dimensions to sustain acceptable electrostatics have been reported [11]. It can be seen that the criterion of Si film thickness may be relaxed for the nanowire MOSFET, which is naturally evolved from double-gate FinFET technology.

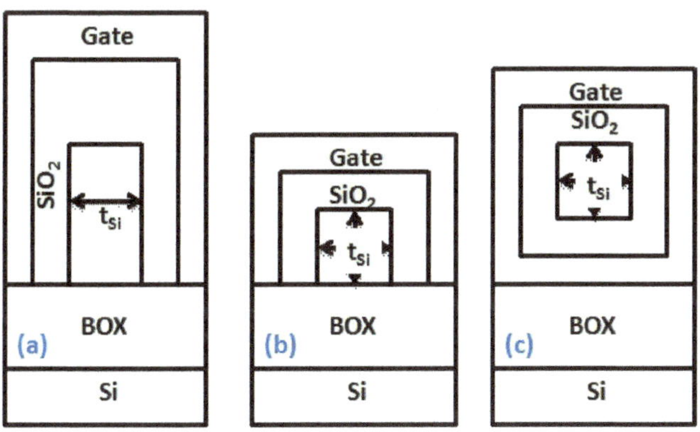

Figure 1. Schematics of device structures of (a) double-gate FinFET (b) tri-gate and (c) GAA NW MOSFETs.

2.1. 3D Quantum Simulation of Multi-Gate FinFETs

The simulation of tri-gate FinFET is performed using SILVACO ATLAS3D quantum simulator framework. The Bohm Quantum Potential (BQP) model is included to model the confinement effects in the simulation. Also, calibration of the BQP model against the 2D Schrödinger-Poisson simulation is done for describing the quantum effects in small geometry Si FinFET and nanowire transistors. To model the effects of quantum confinement, ATLAS allows the solution of Schrödinger's equation along with the fundamental device equations. The solution of Schrödinger's equation gives a quantized description of the density of states in the presence of quantum-mechanical confining potential variations. When the quantum confinement is in one dimension (along y-axis), the calculation of the quantum electron density relies upon a solution of a 1D Schrödinger equation solved for eigen-state energies E_{iv} (x) and wave-functions Ψ_{iv} (x, y) at each slice perpendicular to x-axis and for each electron valley (or hole band) v as described below:

$$-\frac{h^2}{2}\frac{\partial}{\partial y}\left(\frac{1}{m^v{}_y(x,y)}\frac{\partial \psi_{iv}}{\partial y}\right)+E_c(x,y)\psi_{iv}=E_{iv}\Psi_{iv} \quad (1)$$

Here, $m^v_y(x, y)$is a spatially dependent effective mass in y-direction for the v-th valley and E_C (x, y) is a conduction band edge. The equation for holes is obtained by substituting hole effective masses instead of that of electrons and valence band edge -E_V(x,y) instead of E_C (x,y). ATLAS solves the one-dimensional Schrödinger's equation along a series of slices in the y direction relative to the device. The location of the slices in the y direction is developed in two ways. For rectangular ATLAS-defined meshes, the slices will automatically be taken along the existing mesh lines in the ATLAS mesh. If the mesh is non-rectangular or not an ATLAS defined mesh or both, a rectangular mesh must be specified. The potential derived from the solution of Poisson's equation is substituted back into Schrödinger's equation. This solution process (alternating between Schrödinger's and Poisson's equations) continues until convergence and a self-consistent solution of Schrödinger's and Poisson's equations is reached.

For the BQP model, there are two advantages over the density gradient method. First, it has better convergence properties in many situations. Secondly, one can calibrate it against results from the Schrödinger-Poisson equation under conditions of negligible current flow. The model introduces a position dependent Quantum Potential, Q, which is added to the Potential Energy of a given carrier type. This quantum potential is derived using the Bohm interpretation of quantum mechanics and takes the following form [12]

$$Q=-\frac{h^2}{2}\frac{\gamma\Delta(M^{-1}\Delta(n^{\alpha}))}{n^{\alpha}} \quad (2)$$

where γ and α are two adjustable parameters, M^{-1} is the inverse effective mass tensor and n is the electron (or hole) density. The first part of the calibrating the BQP model against the Schrödinger-Poisson (S-P) Model, is to choose a suitable bias for the device. There should be negligible current flow and quantum confinement effects that manifest at the chosen biases. The second part of the calibration is to set the appropriate BQP parameters. The third part of calibration is to choose the quantity to compare with S-P results. The BQP equation is coupled with Poisson's equation using the charge density terms:

$$n = N_C \exp\left(-\frac{E_C + qQ}{kT}\right), p = N_V \exp\left(-\frac{qQ - E_V}{kT}\right) \tag{3}$$

2.2. Tri-gate FinFETs

Typical p-type tri-gate FinFET is fabricated on 1000 Å SOI layer. The FinFET has a fin height of 30 nm. The fin width and fin length are varied from 60-90 nm and 60 nm to 100 μm, respectively. A 50 Å SiO_2 gate oxide is grown. Polysilicon gate was deposited of thickness 1500 Å, followed by pocket implantation and a 300 Å SiO_2 spacer deposition. A deep source/drain implant is followed. Silicidation is done for contact formation by 300 Å Ni/400 Å TiN depositions with RTA at 550-600°C for 1 minute. A 700 Å - 1 μm thick Al-pad is deposited for contact formation. The final device is annealed in forming gas at 420°C for 30 minutes.

3D Device Simulation

The 3D device structure simulated in ATLAS3D framework [13] of the tri-gate bulk Si-FinFET with fin height of 30 nm, fin width and fin length of 60 nm and 160 nm, is shown in Fig. 2.The simulation takes into account the quantum effects using the BQP model and the Schrödinger-Poisson solver. The tri-gate structure includes a poly-Si gate and a Si-Fin as a channel on the buried oxide (BOX). A gate oxide thickness of 5 nm is used for device simulation. The energy band diagram is shown in Fig. 3. depicting the conduction and valence band energies as a function of the distance along the Si-fin channel.

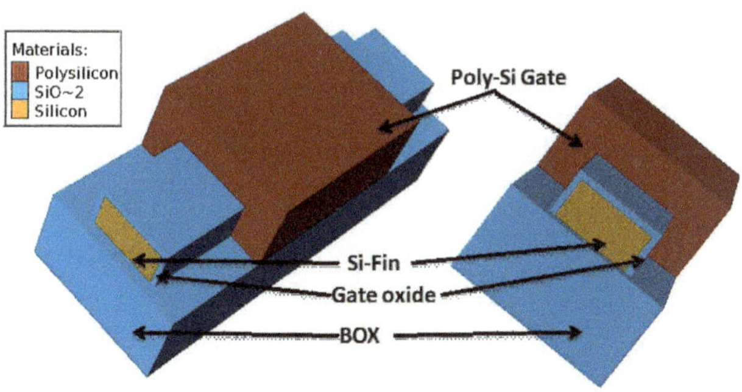

Figure 2. Simulated 3D device structure of a silicon tri-gate FinFET.

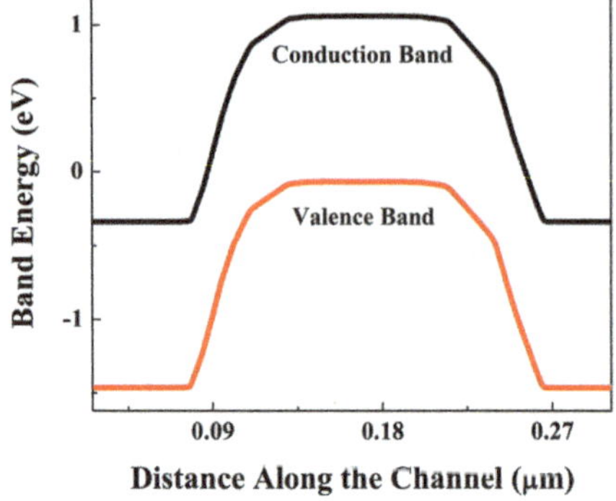

Figure 3. Energy band diagram of a tri-gate Si-FinFET.

Current-Voltage Characteristics

Figure 4. shows the measured I_D-V_{GS} characteristics of the p-type tri-gate FinFET. Typical device dimensions used for measurements are fin length of 160 nm, and width of 60 nm and oxide thickness of ~5 nm. The device displays excellent performance in terms of near ideal subthreshold slope (SS) (~82 mV/dec), and high I_{on}/I_{off} ratios (~10^6). Figure 5. compares the measured I_D-V_{DS} with the TCAD simulation results which shows an excellent agreement showing the need of quantum 3D device simulation.

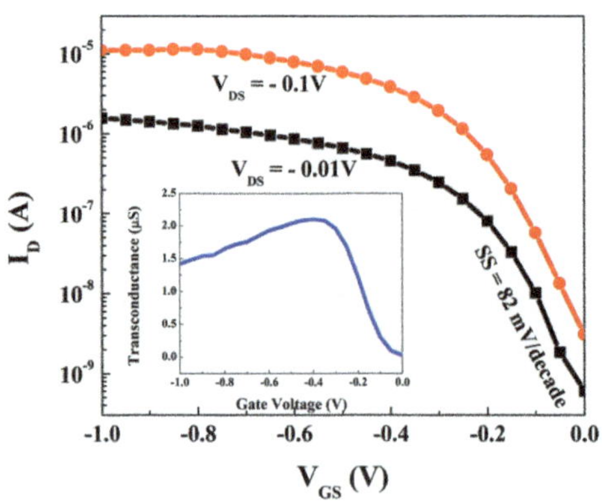

Figure 4. Typical measured I_D-V_{GS} characteristics of a tri-gate p-type Si-FinFET for two drain voltages; the inset shows the transconductance (g_m) plot as a function of gate voltage.

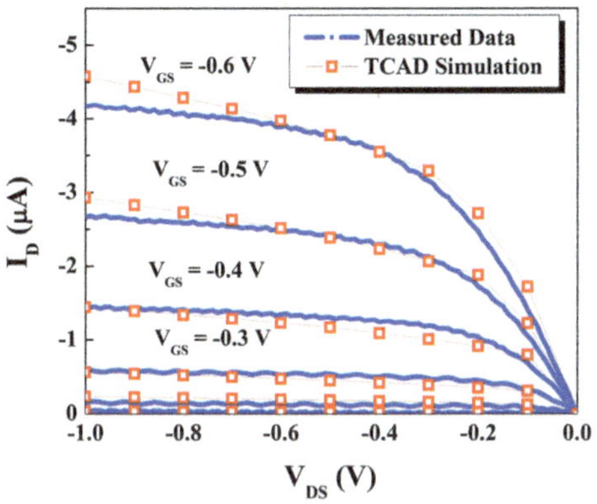

Figure 5. Comparison of I_D-V_{DS} characteristics of a tri-gate p-type Si-FinFET for results of both measured data and TCAD simulation.

2.3. Gate-All-Around Nanowire Transistors

GAA MOSFET is one of the most promising multi-gate structures to extend the scaling of the CMOS devices as it provides the best channel electrostatic control, which improves further with the shrinking of channel thickness. GAA MOSFETs have potential to offer enhanced carrier transport properties compared to planar devices, because of the high carrier mobility on sidewall planes. Careful design and fabrication are critical for GAA MOSFETs for obtaining better transport properties in such structures. The basic nanowire transistor fabrication process flow is illustrated in Fig. 6.

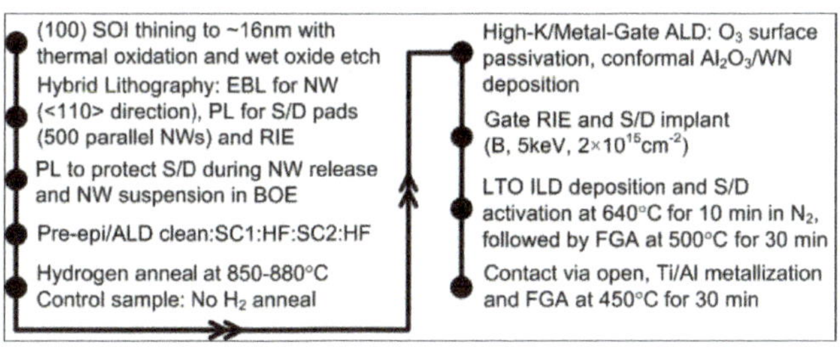

Figure 6. Typical nanowire fabrication process flow (After P. Hashemi, Gate-All-Around Silicon Nanowire MOSFETs: Top-down Fabrication and Transport Enhancement Techniques, PhD dissertation, Massachusetts Institute of Technology, 2010 [14]).

After definition of e-beam lithography (EBL) alignment marks, Si nanowires along <110> direction and S/D pads are defined by hybrid lithography. After reactive-ion etching (RIE) of

Si and stripping the resist, nanowires are locally released using a protective photo resist mask and a buffered-oxide etch bath. The suspended nanowires are then RCA cleaned and are subjected to hydrogen anneal process. Although thermally grown oxide benefits from excellent interface with Si nanowires, the facet dependence of the growth usually results in a non-uniform oxide thickness for various crystallographic planes. In addition, thermally grown oxide can induce a significant amount of stress in the nanowires which can alter their intrinsic carrier transport properties [15]. The ALD dielectric benefits from scalability to thinner effective-oxide-thicknesses (EOT) and its excellent conformality ensures that the entire perimeter of the Si nanowire is uniformly gated. The overlapped gate structures are designed for two reasons. First, it helps to minimize the S/D external resistance as the ion-implanted nanowires may exhibit very large series resistance. Second, it is technologically difficult to remove the metal gate stringers under the nanowires if the gate is under lapped. Moreover, any gate-first process to be used after vertical etching of the gate requires an isotropic etch of the gate metal stringers under the nanowires which also laterally etches the bottom gate and results in high extension resistance. After contact via opening and Ti/Al deposition and patterning, the fabrication is completed by a forming gas annealing at 450°C for 30 minutes.

The 3D device simulation is performed in ATLAS3D framework of SILVACO. The simulated device structure is shown in Fig. 7. depicting a cylindrical GAA FinFET structure with fin diameter ~10 nm and fin length of 100 nm wih a surrounding gate oxide thickness of 10 nm. Gate electrode is defined with 70 nm overlap on source/drain. (100) SOI wafer (p-type, 10^{15} cm^{-3}) was the starting material with 200 nm Si on 150 nm buried oxide (BOX). The gate electrode deposition included 130 nm amorphous silicon (α-Si).

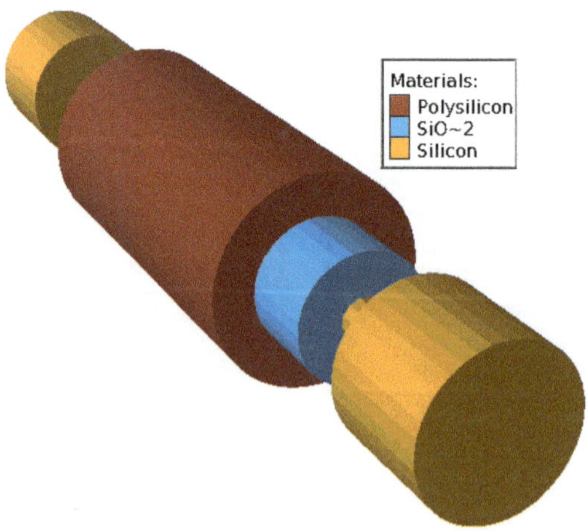

Figure 7. Simulated GAA nanowire FinFET structure.

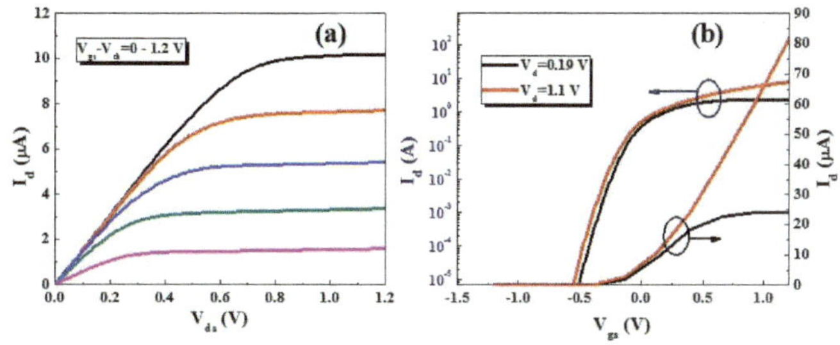

Figure 8. a): I_D-V_{DS} characteristics of a GAA Si-nanowire; (b): I_D-V_{GS} characteristics of a GAA Si-nanowire at two different drain voltages.

Figures 8 (a) and (b)show the typical I_d-V_{ds} and I_d-V_{gs} characteristics of a GAA n-FinFET, respectively. Typical device dimensions used for measurements are fin length of 100 nm, and diameter of 10 nm and oxide thickness of ~10 nm. The excellent subthreshold performance include a high subthreshold slope (SS) of ~90 mV/dec, and I_{on}/I_{off} ratios ~10^6.

2.4. Carrier Mobility in FinFETs

Effective mobility extraction for the FinFET devices was based on an improved split C-V method, similar to planar device mobility extraction [4]. This technique relies on the accurate computation of the inversion charge (Q_i), as well as on the correct calculation of the channel conductance (g_c) along the different operation regimes. Effective channel mobility is defined as [16]:

$$\mu_{eff} = \frac{L_{eff}}{W_{eff}} \cdot \frac{g_c}{Q_i} \quad (4)$$

$W_{eff} = (W_{fin} + 2 * H_{fin}) * N_{fins}$ for a tri-gate structure and $W_{eff} = 2 * (W_{fin} + H_{fin}) * N_{fins}$ for a GAA structure; N_{fins} is the total number of fins, H_{fin} is the fin height. The inversion charge is obtained by integration of the accurate gate to channel capacitance data (C_{gc}).

$$Q_i(V_g) = \int_{V_G<V_T}^{V_G} C_{gc}(V)dV \quad (5)$$

The transfer characteristics and the gate-to-channel capacitance were measured using the split C–V technique to calculate the effective mobility of devices with fin width (W_{fin}) being 60 nm and the height of the fins (H_{fin}) being 30 nm. The effective mobility can be expressed in terms of top, bottom and sidewall mobilities; μ_{top}, μ_{bottom}, $\mu_{sidewall}$, in tri-gate and GAA structures, respectively, as:

$$\mu_{eff}W_{eff} = \mu_{top}W_{fin} + 2*\mu_{sidewall}H_{fin}$$
$$\mu_{eff}W_{eff} = \mu_{top}W_{fin} + \mu_{bottom}W_{fin} + 2*\mu_{sidewall}H_{fin} \quad (6)$$

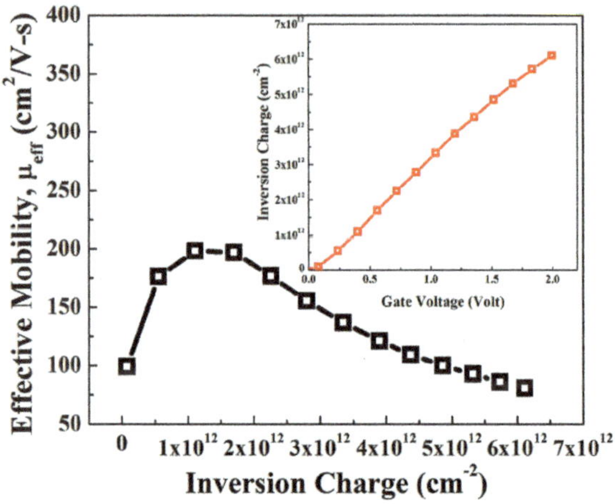

Figure 9. Extracted effective mobility in a p-type tri-gate Si-FinFET with gate length 160 nm, and gate width 60 nm; inset shows the inversion charge as a function of gate voltage.

A linear regression can be used to extract the sidewall and top-surface mobility. Eqs. (4) and (5) are used for calculating the effective mobility in the FinFET device from split C-V measurements. The inset of Fig. 9 shows inversion charge density as a function of gate voltage, as obtained from Eq. (5) using split C-V gate-to-channel capacitance. Figure 9 depicts the effective mobility of the FinFET as calculated using Eq. (4) as a function of inversion charge. The extracted mobility values agree well with the results of [16] for a p-FinFET of with gate length 160 nm.

2.5. Self Heating Effects in Nanowire Transistors

One of the significant challenges in future semiconductor device design is excessive power dissipation and rise in device temperature. With the introduction of new geometrically confined device structures like SOI, FinFETs, nanowires and incorporation of new materials with poor thermal conductivities in the device active region, the device thermal problem is expected to become more challenging in coming years. There is considerable degradation in the ON current due to self-heating effects in silicon nanowire transistors. However, insignificant current degradation is observed in nanowire transistors because of pronounced velocity overshoot effect. Placements of the source and drain contacts also play a significant role on the magnitude of self-heating effect in nanowire transistors. Self-heating and random charge effects simultaneously influence the magnitude of the ON current for both positively and negatively charged single charges. The self-heating affects the ON current in two ways:

1. by lowering the barrier at the source end of the channel, thus allowing more carriers to go through and
2. via the screening effect of the Coulomb potential.

There is larger current degradation because of self-heating due to decreased thermal conductivity. Crystallographic direction dependent thermal conductivity is also an important aspect of self heating effects. Larger degradation is observed in the current along the [100] direction when compared to the [110] direction.

In SOI devices, the low thermal conductivity of the underlying silicon dioxide layer, which is about two orders of magnitude less than that of silicon inhibits cooling in SOI devices and causes severe self-heating, resulting in a higher channel operating temperature. The temperature rise is significant and depends on the buried oxide thickness, silicon thickness, and channel/metal contact separation. The device mobility is reduced as a result of the elevated temperature and results in reduced maximum drain saturation current. Also, high channel temperature leads to increase in interconnect temperature at the silicon-metal contact and make conduction cooling through the source, drain, and interconnects important. It should also be noted that the thermal conductivity of the silicon films decreases as the film thickness is reduced due to boundary scattering of phonons which further exacerbates self-heating and hence device performance. In brief, as semiconductor technology approaches to two-dimensional and three-dimensional transistor structures and are more isolated from the substrate self-heating effects will become increasingly important.

3. Low-Frequency Noise in Nanowire Transistors

Among various options for multi-gate device architecture, such as double-gate, tri-gate, etc., the nanowire (NW) channel with a wrap-around gate, GAA, has the largest advantage in terms of electrostatic integrity. However, several undesired effects become prominent from the miniaturization of the device dimensions. One such unwanted effect is a strong increase in the low-frequency noise generated in the transistor as the size of the device decreases. With dimension-scaling, it is necessary to also scale the power supplies. Noise cannot be completely eliminated and with small signal strength, the accuracy and measurements are limited in electronic circuits in presence of the dominant noise sources. The low-frequency noise, or $1/f$ noise, is the excess noise at low frequencies whose power spectral density approximately depends inversely on the frequency and becomes pronounced at low frequencies. The $1/f$ noise originating from the transistors is a severe obstacle in analog circuit design. For example, it can be up converted to undesired phase noise in voltage controlled oscillator circuits, which can limit the information capacity of communication systems. The problems in down-scaled devices such as the nanowire FinFETs compel one to study noise sources, mechanisms and their physical origins in detail. In this Section, we shall briefly discuss the two fundamental low-frequency noise characteristics, random telegraph signal noise and 1/*f* noise.

3.1. Low-Frequency Noise Measurement Setup

The standard noise measurement setup included an E5263A 2-channel high speed source monitor unit, a SR 570 low noise amplifier (LNA) and a 35670A dynamic signal analyzer. E5263A 2-channel high speed source monitor unit provided the necessary gate-source and drain-source biases as shown in Fig. 10. The minute fluctuations in the drain-source voltage were amplified to the measurable range using low noise amplifier. The output of the amplifier is fed to 35670A dynamic signal analyzer that performs the fast Fourier transform on the time domain signal to yield the voltage noise power spectral density (S_V) in the 1-100 kHz range after correcting for amplifier gain. In order to obtain a stable spectrum, the number of averages was set at 100 and a 90% sampling window overlap was used for optimal real time processing. A computer interface is connected with the measuring system through GPIB connection to control the dynamic signal analyzer and for noise data collection.

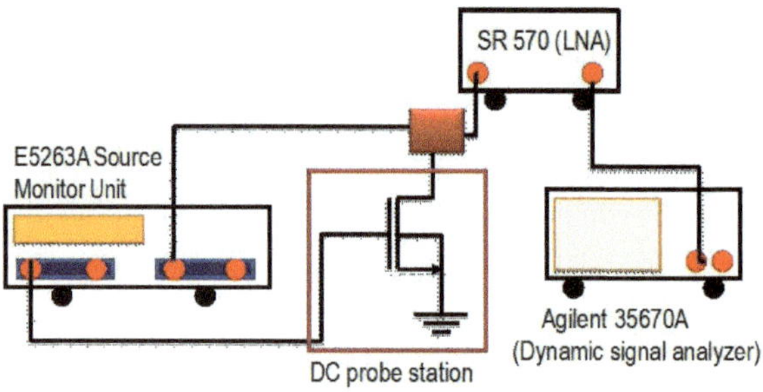

Figure 10. Low-frequency noise measurement setup.

3.2. Random Telegraph Signal (RTS)

RTS noise, also known as burst noise or popcorn noise is observed as random switching events in the time domain voltage or current signal. In a MOSFET, if a carrier is trapped in a single trap or localized defect state, the current or voltage signal displays a random shift in the level denoting a change in the channel resistance. The two-level RTS signal signifies only one active trap. However, when multiple traps are involved, the current (or voltage) can switch between two or more states resembling a RTS waveform due to random trapping and de-trapping of carriers and the phenomenon is much more difficult to explain in order to identify the trapping/de-trapping process. For simple two-level RTS pulses with equal height ΔI and Poisson distributed mean time durations in the lower state τ_l and in the higher state τ_h, the PSD of the current fluctuations is derived as [17].

$$S_I(f) = \frac{4(\Delta I)^2}{(\bar{\tau}_l + \bar{\tau}_h)\left[\left(1/\bar{\tau}_l + 1/\bar{\tau}_h\right)^2 + (2\pi f)^2\right]} \tag{7}$$

Mainly two types of traps are identified depending on the nature of trapping mechanism. They are donor and acceptor traps. The donor trap is charged when it emits and electron (i.e. empty) and is neutral when it captures. The acceptor trap is, contrary to the donor trap, charged when it captures an electron and neutral when empty. In MOSFETs, the channel resistance increases with the charged-trap state changing the current (or voltage) to a high state. Clearly, the donor trap causes high current level after emission of carriers, and the acceptor trap causes the high current level when it captures an electron.

Depending on the values of the mean time constants of the RTS, the traps can be characterized of two types. These are the slow traps with high values of time constants, and the fast traps, with the time constants being very small (order of few 0.1 milliseconds). From RTS noise characterizations, interesting information about the trap energy, capture and emission kinetics and spatial location of the traps inside the semiconductor device can be acquired. The multi-level RTS is due to the activation of multiple traps near the quasi-Fermi level. With smaller area devices, only single traps are active as number of traps is less and the RTS becomes a simple two-level signal, with a Lorentzian PSD ($1/f^2$).

3.3. 1/f Noise

$1/f$ noise, also called flicker noise or pink noise, is the low-frequency noise with fluctuations with a PSD proportional to $1/f^{\gamma}$ with γ close to 1, usually in the range 0.7-1.3. The PSD for $1/f$ noise takes the general form

$$S_I(f) = \frac{KI^{\beta}}{f^{\gamma}} \tag{8}$$

where K is a constant and β is current exponent. There are various theories regarding the $1/f$ noise mechanisms of which most prominent are the carrier number fluctuation (surface phenomenon) and mobility fluctuation theories (bulk phenomenon). The mechanisms behind flicker noise are still a topic of research. The generally accepted origins of the $1/f$ noise are attributed to conductivity fluctuations, damage in crystal structures, and traps due to defects in semiconductors. $1/f$ fluctuations in the conductance have been observed in the low-frequency part of the spectrum (10^{-6} to 10^{6} Hz) in most conducting materials and a wide variety of semiconductor devices [18, 19]. There are essentially two physical mechanisms behind any fluctuations in the current: fluctuations in the mobility or fluctuations in the number of carriers. In 1957, McWorther presented a $1/f$ noise model based on quantum mechanical tunneling transitions of electrons between traps in the gate oxide and the channel [20]. The tunneling time varies exponentially with distance from the trap and the $1/f$

noise is obtained for a trap density that is uniform in both energy and distance from the channel interface. The McWorther model is widely accepted for simplicity and its excellent agreement with experimental data, especially for nMOS transistors. However, the mobility fluctuation noise model explains the 1/*f* noise in pMOS transistors better [21]. It was later explained by the unified flicker noise theory that a trapped carrier also affects the surface mobility through Coulombic interaction. This correlated mobility fluctuation model gave a correction to the number fluctuation noise model which resolves the deviations of the theory for pMOS devices. However, the correction factor was debated for being very high since screening was not accounted for. Also, the carrier mobility at the surface is reduced compared to the bulk mobility due to additional surface scattering (by acoustic phonons and surface roughness), which has an impact on the mobility fluctuations. Moreover, the Hooge mobility noise is sensitive to the crystalline quality, which is deteriorated close to the interface. The most feasible explanation for the higher 1/*f* noise with the carriers being in close proximity of the gate oxide surface is increased mobility fluctuation noise.

3.4. Low-Frequency Noise

Figure 11 shows the RTS of a p-FinFET in time domain in a large span of 60 ms time for V_d = −50 mV and $|V_g - V_{th}|$ = 0.1 V. For this particular RTS, two distinct sets of current switching levels are observed, each of which is consistent with trapping and de-trapping at a single discrete trapping site. A segment of the RTS is enlarged in the inset of Fig. 11 showing a distinct two-level RTS. This shows a fast varying RTS (wit smaller time constants) is superimposed on a slow-varying RTS (with higher time constants) which generates the four-level RTS shown in the figure. This RTS is due to a slow (with higher time constant) and a fast trap (with lower time constant). The corresponding drain current spectral density is plotted in Fig. 12, with a corner frequency f_c and a Lorentzian $1/f^2$ nature. The traps are either donor or acceptor types as described in section 3.2. The high level in current corresponds to the charged trap state. With increasing gate bias, the trap occupancy increases [22]. Also, the time in the high current level increases with gate voltage. So the high level is the occupied charged-trap state or the emission time. The low-current level is the capture time from the same interpretation. The charged trap state is the occupied state which makes the trap to be an acceptor trap. The emission and capture time constants are thus identified in Fig. 12. The mean emission (τ_e) and capture time (τ_c) constants of the RTS in Fig. 11 are shown in Fig. 13 as a function of gate bias. The ratio of capture to emission times (τ_c / τ_e) shows a gradual decrease with gate voltage. The high current time being τ_e and the low current time as τ_c The following can be written [22]:

$$\ln\left(\frac{\tau_c}{\tau_e}\right) = -\frac{1}{kT}\left[\begin{array}{l}(E_{Cox} - E_T) - (E_c - E_F) - \phi_0 \\ +q\psi_s + \dfrac{qx_T}{t_{ox}}(V_{GS} - V_{FB} - \psi_s)\end{array}\right] \tag{9}$$

where E_{Cox} is the conduction band edge of the oxide, E_C is the conduction band edge of the silicon, Φ_0 is the difference between the electron affinities of Si and SiO_2, t_{ox} is the oxide thickness, V_{GS} is the gate bias, V_{FB} is the flat band voltage, ψ_s is the surface potential and x_T is the position of the trap measured from the Si/SiO_2 interface. Differentiating in terms of the gate bias we can obtain the position of traps as:

$$\frac{d}{dV_{GS}}\ln\left(\frac{\tau_c}{\tau_e}\right)=-\frac{1}{kT}\left[\frac{qx_T}{t_{ox}}\right] \tag{10}$$

From Fig. 13, using Eq. (10), the trap depth is calculated as 0.172 nm from the Si/SiO_2 interface. A Lorentzian expression, as expected for an RTS [23, 24], is obtained as

$$WL\frac{S_{I_d}(f)}{I_d^2}=\frac{A}{\left[1+(f/f_c)^2\right]} \tag{11}$$

where A is a constant independent of frequency, $f_c = (1/2\pi)(1/\tau_c + 1/\tau_e)$, and τ_c and τ_e are the capture and emission times, respectively, of a single carrier by an interface trap. It can be seen that the corner frequency $f_c \sim 1$ kHz, which is consistent with the observed switching times in the order of 0.4–1.4 ms.

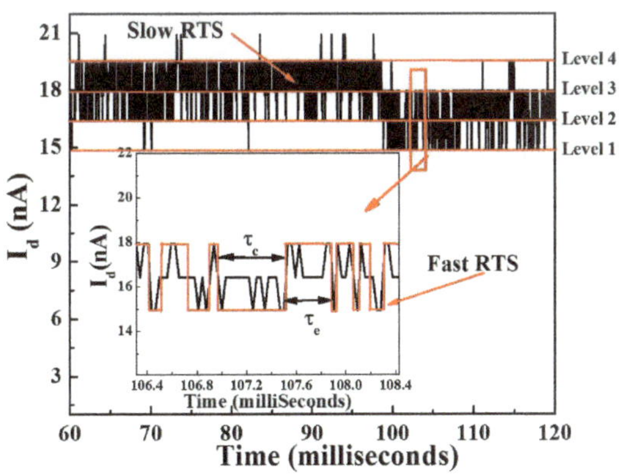

Figure 11. RTS of a p-FinFET in time domain for V_{ds} = −50 mV and $|V_{gs} - V_{th}|$ = 0.1 V, depicting four levels with at least two traps; the inset shows a zoom of the fast RTS with the capture and emission time constants shown in the waveform.

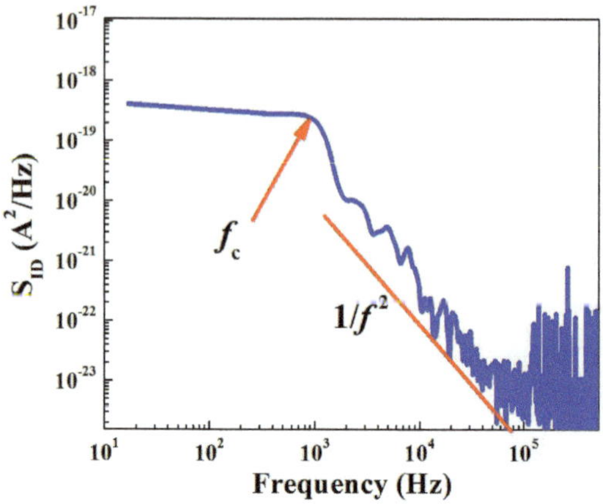

Figure 12. Drain current noise power spectral density with a corner frequency f_c and a Lorentzian nature ($1/f^2$).

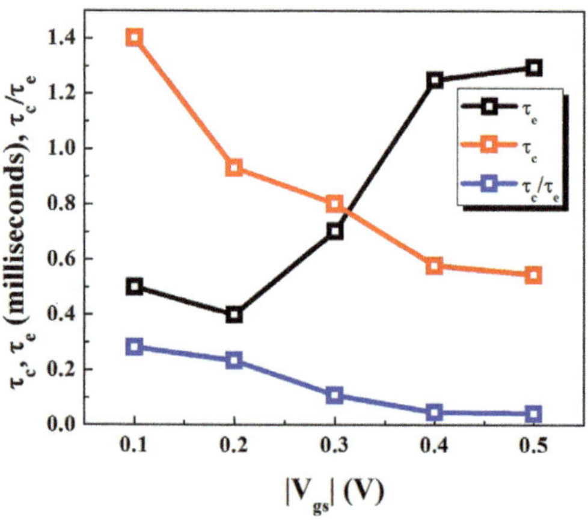

Figure 13. The mean emission (τ_e) and capture time (τ_c) constants and the ratio of τ_c/τ_e of the RTS are shown as a function of gate bias.

The entire FinFET channel is very thin and close to the interface and thus will be highly sensitive to interface defects. The defect potential perturbation, on the order of several nanometres, will have an effect on a significant cross section of the 10-nm-wide fin, hence producing a large current change when the trap is filled. The relative amplitude $\Delta I_d/I_d$ can be written as [25]

$$\frac{\Delta I_d}{I_d} = \frac{1}{WL}\left[\frac{1}{N} \pm \alpha\mu\right] \tag{12}$$

where N is the carrier density, α is the scattering coefficient, and μ is the carrier mobility. Using the double lateral channel FinFET approximation [26], $N = 2\,C_{ox}\,|V_G - V_{TH}|/q$ and $\mu = g_m\,L/(2WC_{ox}\,V_d)$. The mobility limited by Coulomb scattering is calculated as $\mu_C = 1/\alpha N_t$, where N_t is the density of occupied traps [25].

4. Orientation Dependence: Nanowire FinFETs

Theory and measurements on planar transistors show a large difference in mobility for electrons and holes depending on the interface orientation and the direction of current flow in MOSFETs [27, 28, 29]. Figure 14 illustrates the measured drain currents (I_D-V_{DS}) for two different channel orientations; (100) and (110). It is observed that the current is higher in (100) orientation compared to the (110) orientation. The change in channel crystallographic direction (and thereby a change in the direction of current flow) greatly influences the carrier mobilities due to direction dependency of the heavy hole effective mass affecting the low field hole mobility. The effect of channel orientation is dominant in p-type devices due to different band curvatures along the <100> and <110> channel directions.

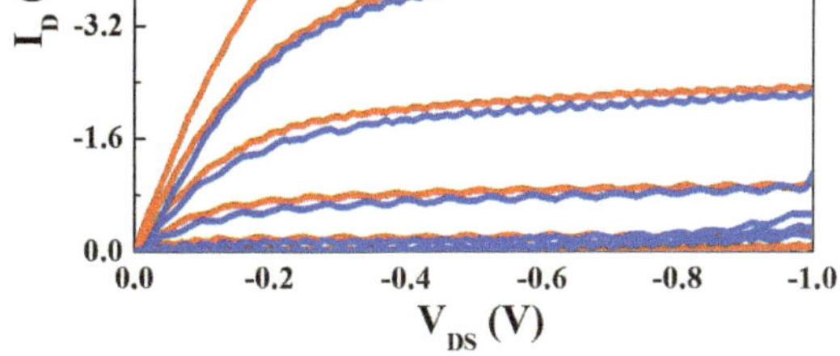

Figure 14. Comparison of the measured drain currents (I_D-V_{DS}) for two different channel orientations; (100) and (110) of a p-type tri-gate Si-FinFET.

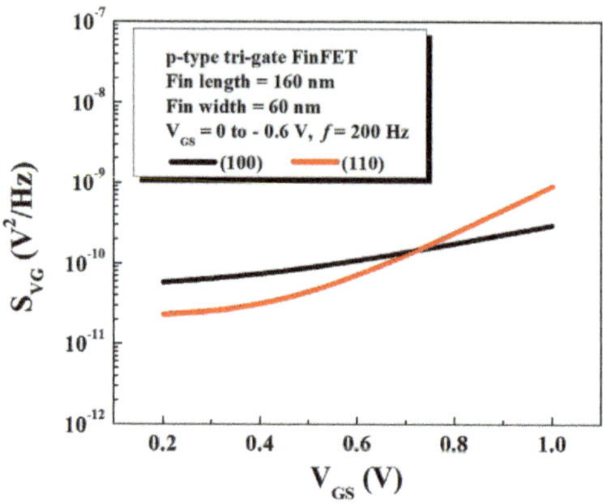

Figure 15. Comparison of the measured gate voltage noise power spectral density for two different channel orientations; (100) and (110) of a p-type tri-gate Si-FinFET.

The input referred gate voltage noise power spectral density (PSD) of the FinFET is shown in Fig. 15 for two different channel orientations ((100) and (110)) as a function of the gate-source voltage. It is observed that the gate voltage PSDs are comparable in magnitude for both orientations. It is well known that technological factors such as gate oxide quality, gate oxide material, channel material, conduction path, annealing, etc. can be of great importance for the low-frequency noise properties [30, 31, 32, 33]. The gate oxide quality and other technological factors can be considered as unchanged by the channel rotation, which explains why the orientation displays a negligible influence on the low-frequency noise. If the mobility fluctutation noise becomes dominant, the channel orientation may play significant part in determining the noise PSD. This is due to the fact that the carrier mobiliy depends on channel orientation. The difference in noise PSD for the two orientation is large in the subthreshold region. This is due to the fact that the mobility-fluctuation noise generation mechanism is dominant in the subthreshold region, since in the number-fluctuation model, there is no dependence of noise PSD on mobility at a fixed drain-current level. The channel orientation dependence of noise is more dominant in pMOS devices as also reported in reference 34. Stronger current and mobility enhancement in certain orientations may also cause a higher LFN level in the corresponding orientation.

5. Lattice Dynamics: Nanowire FinFETs

Lattice vibrations in small geometry devices and its orientation-dependence are important, which however, have not been studied in detail. Inelastic electron tunneling spectroscopy (IETS) has gained importance for characterization of small-geometry devices and ultra-thin-insulators due to its high resolution and sensitivity and has been used by many previous researchers for studying defects and molecular vibrations in metal-insulator-semiconductor (MIS) structures [35, 36, 37, 38, 39, 40, 41].

The principle of IETS depends on the tunneling current through the ultra-thin dielectric layer of the MIS structures and becomes more sensitive as the tunneling current increases. Tunneling spectroscopy detects interaction of tunneling electrons (through the dielectric) with lattice vibrations of the substrate, dielectric, electrodes or other molecular species as inelastic peaks in the second derivative of the current-voltage characteristics. The application of tunneling spectroscopy was first demonstrated by Hall [36]. The technique of inelastic electron tunneling spectroscopy was properly introduced in 1966 by Jaklevic and Lambe [37]. With high sensitivity and resolution, IETS has been used for various applications including the determination of density of states, molecular vibrations, band gap of superconductors and semiconductors, defects in semiconductors and insulators [38, 39]. IETS has also been used for studying the traps and molecular vibrations in ultra thin oxides [40] and high-k gate dielectrics [41]. A tunneling spectrum reveals mainly the vibration energies of the molecules between the electrodes and phonon modes. Yanson *et al.* have shown that the intensity of the inelastic tunneling process is characterized by the relative change in $\Delta g/g$ for conductance g at energy $\hbar\omega$. The quantity $\Delta g/g$ is related to the observed I-V characteristics by the following expression [42]:

$$\frac{\Delta g}{g} = V\int_{V_1}^{V_2} \frac{1}{I}\frac{d^2 I}{dV^2}\, dV \tag{13}$$

where, V_1 and V_2 are the starting and ending voltages of the vibration band under consideration. This relation can be used to quantify the strength of the vibration modes, i.e. the change in the $\Delta g/g$ ratio indicates the strength of the interactions. In the following, we present the experimental tunneling spectra of Si, SiO_2 phonon modes obtained for p-type Al/SiO_2/Si tri-gate FinFETs and study the influence of crystal orientation. A comparison of lattice vibration modes and dispersion characteristics for (100) and (110) silicon orientation in tri-gate FinFET is made. It is observed that the lattice dynamics of silicon and SiO_2 changes for different crystal orientations.

The low temperature (77K) inelastic electron tunneling spectroscopy measurement setup includes an E5263A 2-channel high speed source monitor unit for dc biasing, and a SR830 DSP lock-in amplifier for capturing the tunneling spectra by measuring the second derivatives of the current-voltage characteristics. The setup is shown in Fig. 16. AC modulation signal generated from oscillator output of the lock-in amplifier with f = 890 Hz, peak amplitude of 2 mV, is superimposed on a slowly varying dc gate voltage. A time constant of 3 sec was used. A notch filter in the lock-in amplifier was used to remove the unwanted harmonic frequencies. A computer interface is connected with the measuring system through GPIB connection to control the dynamic signal analyzer and for noise data collection.

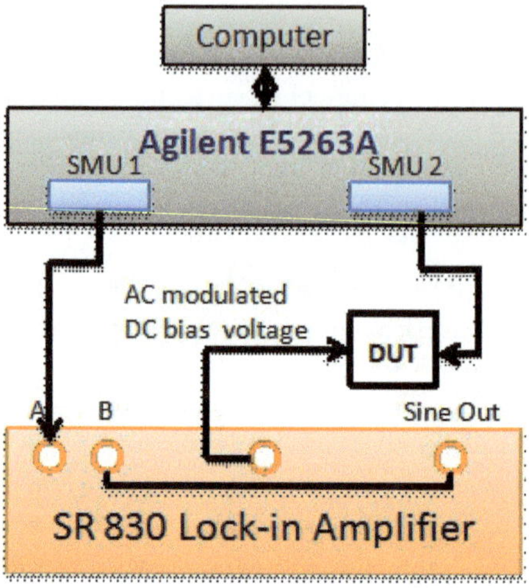

Figure 16. IETS measurement setup.

The phonon spectra of Si are observed in the 0-70 mV range. Figure 17(a)shows the inelastic peaks in the phonon spectra of Si for FinFET on (100) silicon orientation obtained from the average of six scans after de-convolution of the IETS spectra. The most intense peak is found at 60 mV (wave number 483.9 cm^{-1}). The peak is due to optical phonons at the conduction band minima (X-point) of silicon in transverse optical (TO) mode. Other significant Si-phonons include longitudinal acoustical (LA) phonon peak towards the X point at 47.2 mV (380.6 cm^{-1}), the longitudinal optical (LO) phonon towards X point at 53 mV (427.4 cm^{-1}). The transverse acoustic (TA) mode was observed at 19.7 mV (158.9 cm^{-1}). Phonons of Al-electrodes are observed at 26 and 33 mV. These findings are in well agreement with reported results for Si phonons with (100) orientation [35]. The values of phonon frequencies of unstrained silicon were calculated by Sui et al. using a modified Keating model [43]. The experimental values are compared with these theoretical values. We suspect the minor shifts in peak locations are due to the inherent strain in the small geometry structure of the FinFET. The TO and LO mode towards the Γ point, however, was not observed from the IETS. Figure 17(b)shows similar phonon modes of silicon for (110) orientation. From the Brillouin zone of the Si-crystal the dispersion in (110) direction is calculated along the main symmetry directions Γ→X through K point and X→X through W point [44]. The identified peaks include TO mode at Γ and W points, respectively, at 68 (548.4 cm^{-1}) and 58 mV (467.7 cm^{-1}). LO and LA mode towards X point are found at 54 (435.5 cm^{-1}) and 45.6 mV (367.7 cm^{-1}), and LA towards K point is observed at 41 mV (330.6 cm-1). The TA modes at X and W points are observed at 19.5 (157.3 cm^{-1}) and 26 mV (209.7 cm^{-1}), respectively. The peaks at W point or K point are mostly of weaker intensity compared to the peaks at Γ and X points, which are the conduction band minima for the lattice. Also, unlike the (100) orientation, the symmetry directions in (110) orientation reveal inelastic interactions at the K and W points along with the Γ and X points which describes the Brillouin zone for Si in more detail.

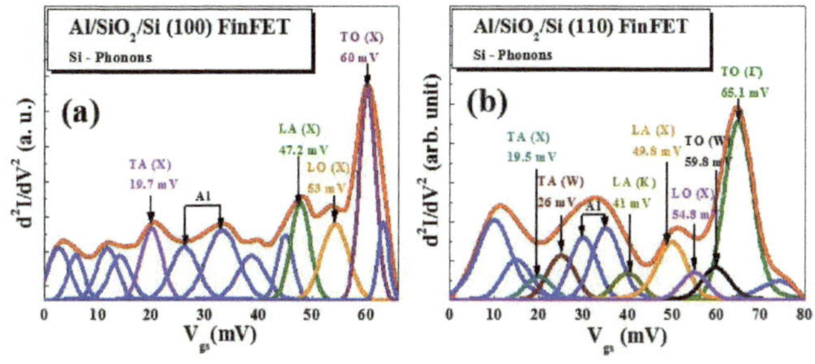

Figure 17. IETS spectra showing Si-phonons in a Al/SiO_2/Si FinFET for (a) (100) channel orientation and (b) (110) channel orientation.

Figures 18 (a) and (b) show the SiO_2 phonon modes (in the 129-170 mV range) for (100) and (110) orientations, respectively. The peaks exhibit almost the same energy position for both the Si (110) and Si (100) devices. These peaks include, TO3 mode at 134 mV, LO4 at 144.2 mV, TO4 at 149.8 and 153.4 mV for (100) device. The corresponding peaks for (110) devices are observed at 132, 144.2 148 and 159 mV, respectively. The main discrepancy between the two orientations comes in the magnitude of the phonons. The asymmetric stretch TO4 mode is theoretically found at 148.8 mV [45] which appears as a strong peak in the Si (110) device, and as a weak peak in the (100) orientation, as also described by [35]. The ratio of intensities of vibration modes may vary for different orientations of the molecular dipoles in the tunnel oxide barrier for different device orientations.

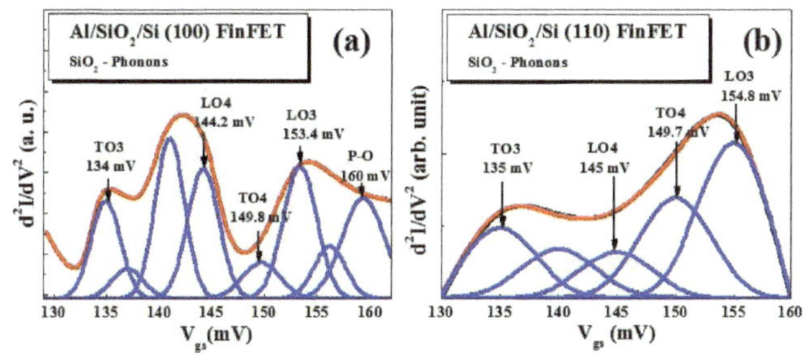

Figure 18. IETS spectra showing SiO_2-phonons in a Al/SiO_2/Si FinFET for (a) (100) channel orientation and (b) (110) channel orientation.

6. Summary

In this chapter, we have discussed silicon nanowire FinFETs in detail. Several important electrical characteristics such as mobility, self-heating, low-frequency noise, impact of channel crystallographic orientation and lattice dynamics are presented for nanowire FinFETs. The fabrication process flow of the nanowire and FinFETs are described. Application of split C-V measurement to calculate effective carrier mobilities is shown and mobility models for FinFETs including side and top-wall mobilities are developed. Low-frequency noise measurements are performed and results of 1/*f* noise and RTS in FinFET devices are presented. Effects of channel orientation on current drive and low-frequency noise are discussed. Lattice vibrations and phonon mode of Si and SiO_2 are studied for (100) and (110) channel orientations and Brillouin zone boundaries of the semiconductor are explored. Crystal dynamics in FinFETs are explained via inelastic electron tunneling spectroscopy. It is expected that the combination of strain effects with NWs can lead to very high performance devices.

Author details

C. Mukherjee and C. K. Maiti

*Address all correspondence to: ckm@ece.iitkgp.ernet.in

Indian Institute of Technology, Kharagpur, India

References

[1] Kedzierski, J., Nowak, E., Kanarsky, T., Zhang, Y., Boyd, D., Carruthers, R., Cabral, C., Amos, R., Lavoie, C., Roy, R., Newbury, J., Sullivan, E., Benedict, J., Saunders, P., Wong, K., Canaperi, D., Krishnan, M., Lee, K. L., Rainey, B. A., Fried, D., Cottrell, P., Wong, H. S. P., Ieong, M., & Haensch, W. (2002). Metal-gate FinFET and fully-depleted SOI devices using total gate silicidation. *Proc. IEEE IEDM Tech. Dig.*, 247-250.

[2] Daewon, H., Takeuchi, H., Choi, Y. K., King, T. J., Bai, W. P., Kwong, D. L., Agarwal, A., & Ameen, M. (2004). Molybdenum gate HfO2 CMOS FinFET technology. *Proc. IEEE IEDM Tech. Dig.*, 643-646.

[3] Takagi, S., Toriumi, A., Iwase, M., & Tango, H. (1994). On the universality of inversion layer mobility in Si MOSFET's: Part II Effects of surface orientation. *IEEE Trans. Electron Dev*, 41(12), 2363-2368.

[4] Romanjek, K., Andrieu, F., Ernst, T., & Ghibaudo, G. (2004). Improved split C-V method for effective mobility extraction in sub-0.1-μm Si MOSFETs. *IEEE Electron Dev. Lett*, 25(8), 583-585.

[5] Zhu, W., Jin-Ping, H., & Ma, T. P. (2004). Mobility measurement and degradation mechanisms of MOSFETs made with ultrathin high-k dielectrics. *IEEE Trans. Electron Dev*, 51(1), 98-105.

[6] Cui, Y., Zhong, Z., Wang, D., Wang, W. U., & Lieber, C. M. (2003). High Performance Silicon Nanowire Field Effect Transistors. *Nano Lett*, 3(2), 149.

[7] Suk, S. D., Lee, S., , Y., Kim, S. M., Yoon, E. J., Kim, M. S., Li, M., Oh, C. W., Yeo, K. H., Kim, S. H., Shin, D. S., Lee, K. H., Park, H. S., Han, J. N., Park, C. J., Park, J. B., Kim, D. W., Park, D., & Ryu, B. I. (2005, Dec). High Performance 5nm radius Twin Silicon Nanowire MOSFET (TSNWFET) Fabrication on Bulk Si Wafer, Characteristics, and Reliability. *Proc. IEEE IEDM Tech. Dig*, 735-738.

[8] Chen, J., Saraya, T., Miyaji, K., Shimizu, K., & Hiramoto, T. (2008). Experimental study of mobility in [110] and [100]-directed multiple silicon nanowire GAA MOSFETs on (100) SOI. *Proc. VLSI Symp. Tech. Dig*, 32-33.

[9] Koo, S. M., Fujiwara, A., Han, J. P., Vogel, E. M., Richter, C. A., & Bonevich, J. E. (2004, Sep). High Inversion Current in Silicon Nanowire Field Effect Transistors. *Nano Lett*, 4(11), 2197-2201.

[10] Young, K. K. (1989). Short-channel effect in fully depleted SOI MOSFETs. IEEE Trans. Electron Dev Feb., 36(2), 399-402.

[11] Yang, F. L., Lee, D. H., Chen, H. Y., Chang, C. Y., Liu, S. D., Huang, C. C., Chung, T. X., Chen, H. W., Huang, C. C., Liu, Y. H., Wu, C. C. , Chen, C. C., Chen, S. C., Chen, Y. T., Chen, Y. H., Chen, C. l., Chan, B. W., Hsu, P. F., Shieh, J. H., Tao, H. J., Yeo, Y. C., Li, Y., Lee, J. W., Chen, P., Liang, M. S., & Hu, C. (2004). 5nm-Gate Nanowire FinFET. Proc. VLSI Symp. Tech. Dig. , 196-197.

[12] Patnaik, V. S., Gheedia, A., & Kumar, M. J. (2008, July, Aug, Sept.,). 3D Simulation of Nanowire FETs using Quantum Models. *SILVACO simulation standard*, 6-12.

[13] SILVACO Inc. (2010). Patrick Henry Drive, Santa Clara, CA 95054 ATLAS User's Manual. , April 20.

[14] Hashemi, P. (2010). Gate-All-Around Silicon Nanowire MOSFETs: Top-down Fabrication and Transport Enhancement Techniques. *PhD Dissertation, Massachusetts Institute of Technology.*

[15] Najmzadeh, M., Bouvet, D., Dobrosz, P., Olsen, S., & Ionescu, A. M. (2009). Investigation of oxidation-induced strain in a top-down Si nanowire platform. *Microelectronic Engineering*, 86(7-9), 1961-1964.

[16] Ramos, J., Severi, S., Augendre, E., Kerner, C., Chiarella, T., Nackaerts, A., Hoffmann, T., Collaert, N., Jurczak, M., & Biesemans, S. (2006). Effective Mobility Extraction Based on a Split RF C-V Method for Short-Channel FinFETs. *Proc. Solid-State Device Research Conference, (ESSDERC)*, 363-366.

[17] Machlup, S. (1954). Noise in semiconductors: spectrum of a two-parameter random signal. *J. Appl. Phys.*, 25, 341-343.

[18] Van der Ziel, A. (1986). Noise in solid state devices and circuits. John Wiley & Sons.

[19] Hooge, F. N., Kleinpenning, T. G. M., & Vandamme, L. K. J. (1981). Experimental studies on 1/f noise. *Rep. Prog. Phys*, 44, 479-531.

[20] Mc Worther, A. L. (1957). Semiconductor surface physics. *University of Pennsylvania Press.*

[21] Vandamme, L. K. J., Li, X., & Rigaud, D. (1994). 1/f noise in MOS devices, mobility or number fluctuations? *IEEE Trans. Electron Dev.*, 41, 1936-1945.

[22] Amarasinghe, N. V., & Celik-Butler, Z. (2000). Complex random telegraph signals in 0.06 μm^2 MDD n-MOSFETs. Solid State Electron Jun., 44(6), 1013-1019.

[23] Kirton, M. J., & Uren, M. J. (1989, Nov). Noise in solid-state microstructures: A new perspective on individual defects, interface states and low-frequency (1/f) noise. *Adv. Phys*, 38(4), 367-468.

[24] Kleinpenning, T. G. M. (1990, Sep.) On 1/f noise and random telegraph noise in very small electronic devices. *Physica B*, 164(3), 331-334.

[25] Hung, K. K., Ko, P. K., Hu, C., & Cheng, Y. C. (1990, Mar). A unified model for the flicker noise in metal-oxide-semiconductor field-effect transistors. *IEEE Trans. Electron Dev.*, 37(3), 654-665.

[26] Daugé, F., Pretet, J., Cristoloveanu, S., Vandooren, A., Mathew, L., Jomaah, J., Nguyen, B., & , Y. (2004, Apr.) Coupling effects and channels separation in FinFETs. *Solid State Electron*, 48(4), 535-542.

[27] Yang, M., Gusev, E. P., Ieong, M., Gluschenkov, O., Boyd, D. C., Chan, K. K., Kozlowski, P. M., D'Emic, C. P., Sicina, R. M., Jamison, P. C., & Chou, A. I. (2003). Performance Dependence of CMOS on Silicon Substrate Orientation for Ultrathin Oxynitride and HfO_2 Gate Dielectrics,. *IEEE Electron Dev. Lett.*, 24(5), 339-341.

[28] Sverdlov, V., Ungersboeck, E., Kosina, H., & Selberherr, S. (2006). Orientation Dependence of the Low Field Mobility in Double- and Single-gate SOI FETs. *Pro. 36th European Solid-State Device Research Conf. (ESSDERC)*, 178-181.

[29] Smirnov, S., Kosina, H., & Selberherr, S. (2003). Substrate Orientation-Dependence of Electron Mobility in Strained SiGe Layers. *Proc. International Conf. on Simulation of Semiconductor Processes and Devices (SISPAD)*, 55-58.

[30] Claeys, C., Mercha, A., & Simoen, E. (2004). Low-frequency noise assessment for deep submicrometer CMOS technology nodes. *J. Electrochem. Soc.*, 151, G307-G318.

[31] Von, Haartman. M., Malm, B. G., & Ostling, M. (2006). Comprehensive study on low-frequency noise and mobility in Si and SiGe pMOSFETs with high-k gate dielectrics and TiN gate. *IEEE Trans. Electron Dev*, 53, 836-843.

[32] Von, Haartman. M., Lindgren, A. C., Hellstrom, P. E., Malm, B. G., Zhang, S. L., & Ostling, M. (2003). 1/f noise in Si and $Si_{0.7}Ge_{0.3}$ pMOSFETs. *IEEE Trans. Electron Dev*, 50, 2513-2519.

[33] Vandamme, L. K. J., & Oosterhoff, S. (1986). Annealing of ion-implanted resistors reduces the 1/f noise. *J. Appl. Phys.*, 59, 3169-3174.

[34] Wei, C., Xiong-Z, Y., Zhou, X., Singh, N., Rustagi, S. C., Lo, G. Q., & Kwong, D. L. (2009, June). Investigation of Low-Frequency Noise in Silicon Nanowire MOSFETs in the Subthreshold Region. *IEEE Electron Dev. Lett.*, 30(6), 668-671.

[35] Petit, C., Salace, G., & Vuillaume, D. (2004). Aluminum, oxide, and silicon phonons by inelastic electron tunneling spectroscopy on metal-oxide-semiconductor tunnel junctions: Accurate determination and effect of electrical stress. *J. Appl. Phys*, 96(9), 5042-5049.

[36] Hall, R. N., Racette, J. H., & Ehrenreich, H. (1960, May). Direct observations of Polarons and Phonons during tunneling in group 3-5 semiconductor junctions. *Phys. Rev. Lett*, 4, 456-458.

[37] Jaklevic, R. C., & Lambe, J. (1966). Molecular vibration spectra by electron tunneling. *Phys. Rev. Lett.*, 17, 1139-1140.

[38] Giaever, I., & Megerle, K. (1961). Study of superconductors by electron tunneling. *Phys. Rev*, 122, 1101-1111.

[39] Chynoweth, A. G., Logan, R. A., & Thomas, D. E. (1962). Phonon- Assisted Tunneling in Silicon and Germanium Esaki Junctions. *Phys. Rev*, 125, 877-881.

[40] Petit, C., & Salace, G. (2003). Inelastic electron tunneling spectrometer to characterize metal-oxide-semiconductor devices with ultrathin oxides. *Rev. Sci. Instrum*, 74, 4462.

[41] He, W., & Ma, T. P. (2003). Inelastic Electron Tunneling Spectroscopy Study of Traps in Ultra-thin High-k Gate Dielectrics. *Appl. Phys. Lett*, 83(26), 5461-5463.

[42] Yanson, I. K., Bogatina, N. I., Verkin, B. J., & Shklyarevskii, O. I. (1972). Asymmetry of Tunnel Spectrum Intensities of Impurity Organic Molecules. *Sov. Phys. JETP*, 35, 540.

[43] Sui, Z., & Herman, I. P. (1993). Effect of strain on phonons in Si, Ge, and Si/Ge heterostructures. *Phys. Rev. B*, 48, 17938-17953.

[44] Tubino, R., Piseri, L., & Zerbi, G. (1972). Lattice Dynamics and Spectroscopic Properties by a Valence Force Potential of Diamondlike Crystals: C, Si, Ge, and Sn. *J. Chem. Phys*, 56(3), 1022.

[45] Kirk, C. T. (1988). Quantitative analysis of the effect of disorder induced mode coupling on infrared absorption in silica. *Phys. Rev. B*, 38, 1255-1273.

Chapter 9

Thermal Transport in Semiconductor Nanowires

J. Anaya, J. Jiménez and T. Rodríguez

Additional information is available at the end of the chapter

1. Introduction

Since the late 1990's semiconductor NWs research is in expansion because of the huge number of potential applications of these one dimensional structures. This research effort is given expression as a vast number of research articles. Semiconductor NWs are the building blocks of advanced nanodevices; e. g. field effect transistors, interconnects, heterostructured devices, as vertically oriented nanoscale laser diodes and HBTs, thermoelectric converters, photovoltaic devices, and sensors [1-5].

The thermal transport in NWs is receiving a great deal of attention, because it constitutes a critical issue for the performance and reliability of NWs based devices. Because of the size reduction a large power density needs to be dissipated, especially in devices operating at high current and power densities; which induces a non negligible Joule heating; therefore, the control of the thermal conductivity is crucial to the thermal extraction from the active parts of these devices [6]. On the other hand, bulk semiconductors are usually good thermal and electrical conductors, which limits their potential performance in thermoelectric devices; however, it was demonstrated that surface engineered semiconductor NWs behave essentially as thermal insulators [2], being at the same time good electrical conductors, this means that semiconductor NWs can have a large themoelectric figure of merit [2,7-9]; which can make them very good thermoelectric materials for waste heat recovery and power generation over a large temperature range.

The experimental measurement of the thermal conductivity at the nanoscale is very challenging [11-13], with very few available contrasted measurements. In this context of lack of experimental results, the development of confident models for describing the thermal conductivity of semiconductor NWs is very important in order to design the thermal management of devices, or the thermoelectric power conversion. All this has created the need for a great modeling effort permitting the design of nanodevices in which the thermal transport plays a capital role.

We present herein a review of the thermal transport in semiconductor NWs. The main experimental achievements are presented, both smooth and surface engineered NWs, also alloyed NWs are considered; these data evidence the relevance of the thermal transport problem in these one dimensional structures. These experimental data, though scarce, permit a certain degree of contrast of the theoretical models. According to this we present a critical overview of the main theoretical approaches to the problem of the thermal conductivity of semiconductor NWs following a historic sequence. Along this way, we review a large number of models, among others the classical approaches based on the Callaway formalism, atomistic models based on molecular dynamics simulations, or the hydrodynamic approaches. One revises the contribution of the different resistive phonon scattering processes. Also, the role of normal scattering processes is discussed, as well as the recent consideration of the scattering by optic phonons. Special attention is paid to the boundary scattering which is the main cause for the thermal conductivity suppression in semiconductor NWs. The role of engineered surfaces is also discussed.

The strengths and weakness of the main models are discussed. Finally, we introduce a predictive approach extended to the full temperature range, based on the Callaway-Holland formalism, which permits an excellent fitting to the experimental data available. Using this approach empiric relations are deduced for the case of Si NWs, which permit to extract the thermal conductivity for a wide range of diameters and temperatures, which presents an unquestionable interest for practical purposes related to the nanodevice design and thermal managing.

2. Experimental approach to the thermal transport in semiconductor nanowires

The measurement of the thermal conductivity at the nanoscale is a challenging experiment. The first measurements of thermal transport of semiconductor NWs were carried out by D. Li et al. [11,13], they measured the thermal conductivity of a few individual Si NWs grown by the vapour liquid solid (VLS) method. The experiments were performed in a microdevice specially designed for measuring the thermal properties of one-dimensional structures [10]. This device, called thermal bridge, consists of a nanowire suspended in between two plates of platinum, one of them acting as the heater, and the other as the thermometer. To guarantee accurate measurements with this device, the junction between the NW and the platinum plates must have an ideally null thermal resistance [10,12].

The results obtained by D. Li et al. [11] showed a strong suppression of the thermal conductivity of the Si NWs as compared with bulk Si (Figure 1-A). Furthermore, a dependence of the thermal conductivity with the NWs diameter was revealed; the smaller the NW diameter the lower the thermal conductivity. On the other hand, the experimental data suggest a change in the temperature dependence of the thermal conductivity of Si NWs for diameters below 25 nm; the experimental data give a linear dependence with T for the thermal conductivity of a NW with 22 nm diameter, in which phonon confinement cannot be neglected [15], with respect to the dependence observed for larger diameter NWs, which follow a behavior similar to the bulk(Figure 1-B).

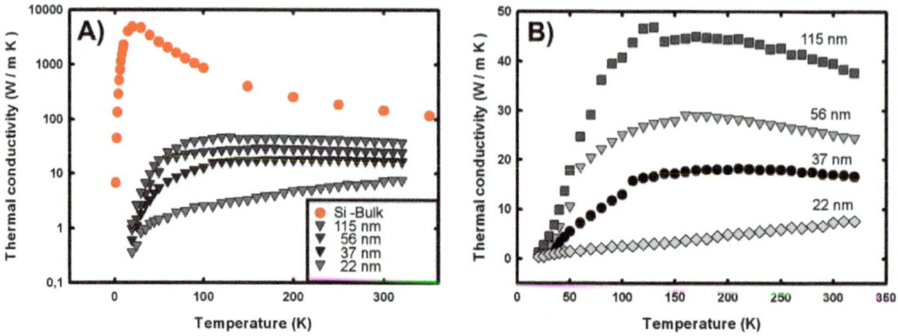

Figure 1. Experimental thermal conductivities as a function of temperature for bulk Si [14], and NWs with different diameters [11] (A), thermal conductivity of NWs in an expanded scale (B) data from [11].

Using the thermal bridge device developed by S. Li et al., the thermal conductivity of other semiconductor NWs has been measured [13,16-20]. All the NWs measured, independently of their nature display a dramatic reduction of the thermal conductivity with the diameter, clearly revealing that the thermal conductivity of NWs is limited by the NWs dimensions. M. C. Wingert et al. carried out thermal conductivity measurements on Ge NWs [20], showing that below a critical diameter of the NW, the thermal conductivity follows a linear dependence with temperature (Figure 2), in agreement with the behavior of the 22 nm Si NW measured in [11,13]. Therefore, this phenomenon can be considered as an intrinsic property of the NWs with reduced dimension, suggesting changes in the phonon transport when phonon confinement effects start to be relevant.

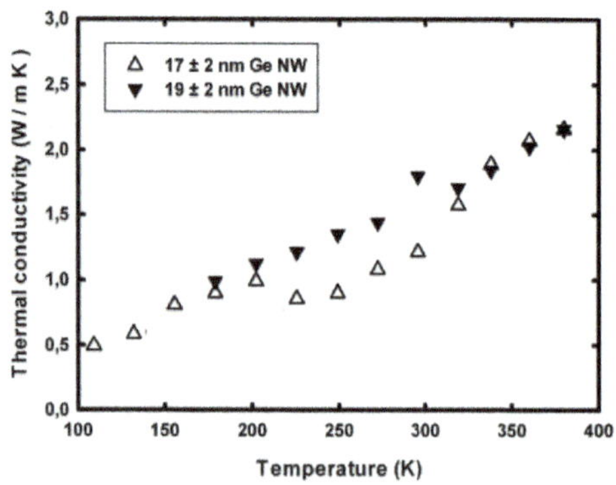

Figure 2. Thermal conductivity of two Ge NWs with diameters below the phonon confinement threshold, showing a linear dependence with temperature [20]

The thermal conductivity of NWs is not only suppressed by the reduced diameter, but, A. Hochbaum et al. showed that the surface corrugation of the NWs is a major issue for the reduction of the thermal conductivity [2]. Using a similar microdevice to the one described in [10], they measured the thermal conductivity of Si NWs fabricated by an aqueous electroless etching method. These NWs present rough surfaces; in particular, the mean roughness height was estimated near 5 nm, with a lateral roughness correlation length of a few nanometers [2]. The experimental thermal conductivities of two NWs of 115 nm diameter and comparable length, but different surface roughness are represented in Figure 3, showing a five times reduction in the thermal conductivity for the rough NW. Furthermore, in NWs with engineered surfaces, one can expect that the thermal conductivity could be reduced down to almost the amorphous limit [2].

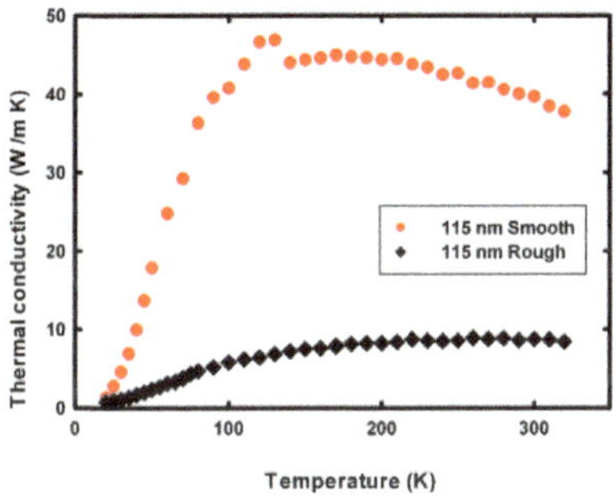

Figure 3. Measured thermal conductivities of two Si NWs of 115 nm diameter, one of them smooth [11] and the other rough [2].

More recently, Y. Park et al. [21] carried out an experiment similar to that performed in [2], but using rough Si NW grown by an VLS method, by which they grew Si NWs with a mean roughness height close to 10 nm, even greater than the NWs used in [2]. However, the thermal conductivity was not so severely suppressed as it did for the electroless etched NWs prepared in [2]. This illustrates that the reduction of the thermal conductivity is not only propitiated by the roughness height, but as J. Lim et al. have recently shown it is the combination of the roughness height and its lateral correlation length, defined as the mean distance between surface hills/depressions, that limits the thermal conductivity in rough NWs [22]. They measured the thermal conductivity of several Si NWs with different surface morphologies, showing that the distance between surface hills/depressions plays a crucial role in the decrease of the thermal conductivity. The SEM images of the NWs measured in [21] compared to the SEM images of the NWs measured in [2], clearly show that the distance between hills/depressions of the NWs of reference [21] are significantly longer than for the NWs of reference [2], which should account for the lower thermal conductivity reduction observed by Lim [22] with respect to the strong reduction reported by Hoschbaum, despite the higher value of the mean roughness height measured in [21].

H. Kim et al. showed the effect of composition on the thermal conductivity of semiconductor NWs. They measured the thermal conductivity of alloyed $Si_{1-x}Ge_x$ NWs with a small Ge fraction [23]. Even for very diluted alloys the thermal conductivity of the NWs showed a dramatic reduction compared to pure Si NWs. Thereafter, H. Kim et al. grew rough $Si_{1-x}Ge_x$ NWs and showed the combined effect of the alloy composition and the surface roughness [24], which permits additional reductions of the thermal conductivity; thus, opening almost unlimited possibilities for designing nanodevices by its thermal conductivity, which is very attractive for thermoelectric devices. These results are summarized in Figure 4, where the experimental values for a smooth Si NW, and a smooth $Si_{1-x}Ge_x$ NW of comparable diameter and a very low Ge fraction (x=0. 004) are shown. The effect of the roughness is summarized in two NWs of the same composition and diameter but differing only by the surface roughness.

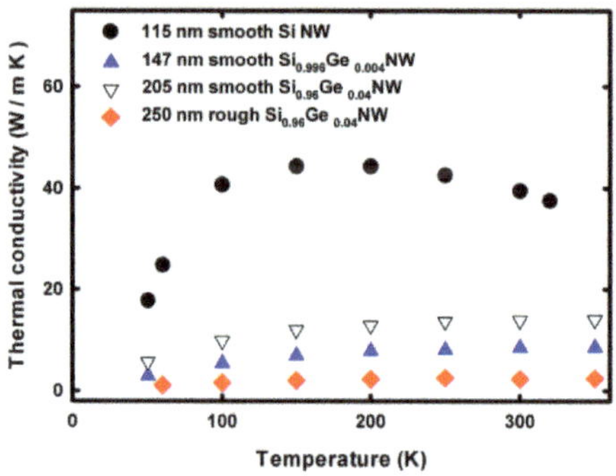

Figure 4. Thermal conductivities of a smooth Si NW from [11] (circles), and alloyed SiGe NWs, smooth NWs data from [23] (triangles) and rough NWs data from [24] (diamonds).

In a recent work, Y. Choongho reported additional thermal conductivity data of $Si_{1-x}Ge_x$ NWs for several compositions [18]; evidencing the relevant role played by the composition, but one cannot be conclusive about the composition bowing, because of the diameter dispersion.

3. Theoretical models

Before the first experimental measurement of the thermal conductivity in a semiconductor NW, a few theoretical approaches already pointed to a thermal conductivity decrease in low dimensional structures. Five years early with respect to the first experimental measurements of NW thermal conductivity, A. Balandin et al. in a pioneering work on the effect of low dimensionality on the thermal transport, presented a theoretical analysis of the lattice thermal conductivity of a semiconductor quantum well [25]. This analysis was built up on the basis that the spatial confinement involves a reduction of the phonon group velocities, therefore, of the phonon relaxation rate. This thermal transport reduction makes evident the increased relevance of the resistive phonon processes in reduced dimension structures. They calculated the thermal conductivity of Si quantum well [25] making use of a reduced Callaway formalism [26], in which only considered the resistive processes given by the Umpklapp scattering, the boundary scattering, and the processes related to the impurity scattering. A detailed description of the Callaway model and the different scattering mechanisms is addressed in paragraph 4. 2. The model proposed in [25] revealed that the thermal conductivity of a Si nanostructure must be substantially reduced with respect to its bulk counterpart. Shortly after, Zou and Balandin applied the same formalism to the study of the thermal conductivity of Si NWs [27]. They modified the simple model described on [25] under the basis of a phonon redistribution by the NW boundaries, which implies a modification of the reduced Callaway expression used in the previous work. On the other hand, the thermal conductivity dependence with the surface roughness of the NWs was also added. The scattering processes considered were a modification of the Umpklapp scattering, the electron-phonon scattering, and a modified boundary scattering taking account of the effect of roughness, expressed in terms of the fraction of specularly to diffusively reflected phonons at the NW boundary [28]. With those modifications, the thermal conductivity of a Si NW with 20 nm diameter was calculated, showing a great dependence of the thermal conductivity reduction with the surface characteristics. Nevertheless, even in the case of the maximum surface contribution, the thermal conductivity calculated was higher, 14 W/mK at 300 K for 20 nm diameter, near twice, than the experimental value reported for a 22 nm diameter Si NW [11,13]. In a different approach, A. Voltz et al. made use of molecular dynamics simulations in thin Si NWs with diameters below 5 nm [29], reaching the same result of a dramatic reduction of the thermal conductivity.

The first experimental data about the thermal conductivity of Si NWs opened the possibility to contrast the theoretical models [11,13], allowing the possibility of developing predictive tools regarding the thermal conductivity of NWs. As mentioned above, the experimental measurements evidenced a strong dependence with the NWs morphology, including diameter and surface topography; as a matter of fact, the construction of a model capable of predicting the thermal conductivity of semiconductor NWS becomes a necessary tool for the design of nanodevices in which the thermal transport becomes a critical issue for performance and reliability. N.Mingo build up a model based on a modification of the ballistic phonon transport in a suspended NW [30]. The model takes account of the scattering events that undergo the phonons in a real semiconductor, and requires the calculation of the complete phonon dispersion relations in the NW. The phonon scattering considered by Mingo, as it was the case for the Balandin models [25,27], are the resistive processes given by Umpklapp, boundary, and impurity scattering mechanisms. At difference of the models reported on references [25,27], where exact expressions for the Umpklapp scattering were used, Mingo introduced two parameters which the values are obtained from the adjustment of their model to the thermal conductivity of bulk Si. [30] These parameters are fixed for the calculation of the thermal conductivity of the NWs; therefore, it can be considered as a free of parameters model. Unfortunately, this model underestimates the thermal conductivity value with respect to the experimental data, requiring of an additional adjustment parameter concerning the boundary scattering; this parameter is defined as an effective diameter, greater than the diameter of the NWS experimentally measured in the SEM [30]. Thereby, using effective diameters up to 30% higher than the experimentally measured ones, Mingo obtained a satisfactory fitting to the experimental values reported for NWs with diameters above the limit for phonon confinement. [30,31] On the other hand, Mingo showed that a reduced Callaway model as the one developed by Balandin et al. [25,27] can give a good fitting to the experimental values if one lowers the cut-off temperature with respect to the bulk [30].

Those models only deal with the resistive scattering processes; this means that the contribution of the normal three-phonon events is neglected. However, it is well known that these processes contribute to the thermal transport resistance through the Herring mechanism [32]. Y. Chen developed a model based on Monte Carlo simulation considering normal and resistive processes [33]. Making use of a complex genetic algorithm and resorting to two free adjusting parameters, they showed a good fitting to the experimental data in the low temperature range ($T<120K$), but overestimated the thermal conductivity for higher temperatures. Once again the behavior exhibited by the 22 nm smooth Si NW was not reproduced by the model, which responded reasonably well above this limit.

Coming back to the first models developed by Zou et al. [27], M. Huang et al. carried out a more detailed treatment of the boundary scattering [34], presenting it as a rational function of the fraction of specularly reflected phonons at the boundary instead of the linear dependence originally used in [27]. Despite of being a model with a more realistic treatment of the boundary scattering the result was not satisfactory. On the other hand, they showed that not only the sound velocities decreased due the reduced diameter of the NWs, but the Debye temperatures had also be lower [34]; which, should account for the good agreement to the experimental data reached by Mingo when using cut-off temperatures lower than the bulk silicon cut-off temperatures in the Callaway model [30].

L. Liang et al. used a simple approach in which they established direct relations between the bulk semiconductor melting point and the nanowire melting point [35]; furthermore, they introduced an exponential expression for describing the boundary scattering due to surface roughness. This simple approach was able of predicting the thermal conductivity at 300 K for all the NWs measured in [11], even for the problematic 22 nm Silicon NW. The anomalous behavior of this thin NW, and thus of those NWs with diameter below the limit for phonon confinement,[11,20] was studied by P. Murphy et al. [36]. In contrast to previous works, they argued that in a strongly disordered surface not all the phonons are equally scattered, being these events dependent on the frequency of the incident phonons. Based on these assumptions, they developed a model splitting the behavior of the phonons in two different contributions, determined by the magnitude of the component of their wave vectors in the direction perpendicular to the NW axis. The behavior is nearly ballistic for phonons with a small transverse wave vector component, while for those with a large transverse wavevector component the phonon scattering was assumed to be purely diffusive, with a mean free path of the order of the NW diameter. Under these assumptions, they showed that for the low temperature range, the calculated thermal conductivity approaches to a linear dependence with the temperature for decreasing NW diameters [36]; further showing that the change from the T^3 law for bulk and thick NWs to the linear dependence for the thinner NWs occurs between 25 nm and 20 nm in diameter, as the experimental measurements demonstrate [11,20].

Z. Tian et al. suggested that the role of the optical phonons must be taken into account for nanostructures [37]. Making use of the phonon-phonon scattering description derived by A. Ward et al. [38], they carried out the calculation of the thermal conductivity contribution of each phonon branch, showing that the optical phonon contribution to the thermal conductivity in NWs can represent near 20% of the total. Under the assumption of a diffusive regime for the boundary scattering, therefore in the Casimir limit [28], they calculated the thermal conductivity of smooth Si NWs in a good concordance with the experimental values. In a previous work, M. Kazan et al. had already considered the optical phonon contribution [39]. They showed how the consideration of the optical phonon decay into acoustic phonons in the frame of the Callaway formalism can cause an improvement in the prediction of the thermal conductivity of bulk Ge. Subsequently M. Kazan et al. applied this method to smooth Si NWs [40]. It should be noted, the Callaway formalism was conceived without the contribution of the optical modes [41-43]; although, it is implicitly taken into account because of the increase of the cut-off temperatures necessary for fitting the experimental data [44]. Therefore, in the modified Callaway model introduced in [40],which explicitly considers the optical contribution, the cut-off temperatures are lower than the cut-off temperatures used when applying the Callaway formalism without considering the optical phonon contribution in a explicit way. Finally, working in the Casimir limit for boundary scattering and making use of a large number of adjustment parameters, they obtained a very good fitting to the thermal conductivity of Si NWs [40].

All those models, which give good fittings to the experimental values of the thermal conductivity for Si NWs work in the diffusive limit for the boundary scattering; this means that one cannot further reduce the thermal conductivity due the boundary scattering when considering rough surfaces. Thereby, when A. Hotchbaum et al. measured experimentally the thermal conductivity of rough Si NWs showing a great reduction in the thermal conductivity of rough NWs compared with the smooth Si NWs [2], none of those models could fit the experimental values.

F. Alvarez et al. showed that the thermal conductivity of semiconductor NWs can be described making use of an analogue hydrodynamic model for the phonon flux [45]. By solving the Guyer-Krumhansl equation for the phonon flux in a NW [46, 47], they reached a very simple expression for the thermal conductivity depending on the bulk thermal conductivity, the dimension of the NWs, and the fraction of specularly reflected phonons at the boundary [48]. In a further study, A.Sellito et al. showed that in a hydrodynamic framework, the roughness of the boundaries may cause a local inversion of the phononic flux, thus extra limiting the thermal conductivity of the NW [49]. This result is consistent with a previous work by A. Moore, who showed how the surface roughness of a NW may cause backscattering events resulting in a more aggressive limitation of the thermal conductivity than the diffusive limit considered in the previous models [50]. On this basis, A. Sellito et al. [51] returned to the hydrodynamic model described in [48], which gave a simple expression for the thermal conductivity, and added the contribution of the backscattering events. With this modification, they attained a good fitting to the thermal conductivity of the smooth Silicon NWs below 150 K [51], besides, obtaining values for the thermal conductivity of the rough NWs close to those reported in reference [2] in the same temperature range.

The effect of the surface roughness in the boundary scattering was studied by P. Martin et al. using perturbation theory [52]; the perturbation due to surface roughness was introduced in the Hamiltonian; also, the existence of a thin layer of native SiO_2 layer surrounding the NW was considered. This model shows values close to the experimental measurements for smooth Si NW, but without a satisfactory fitting to the experimental data of reference [11]. More interesting is the capability shown by the model when one evaluates the thermal conductivity of rough NWs, insomuch as the values are close to those measured by Hochbaum et al. in [2]. On the other hand, they showed that the effect of the boundaries in the thermal conductivity reduction increases the more the smaller the diameter [52]; which suggests a dependence of the thermal conductivity with the diameter and the mean roughness height (η), following the relation $k \propto (D / \eta)^2$.

Other approaches making use of molecular dynamics simulations intended to assess the effect of the roughness in the phonon transport [53, 54], showing that the reduction of the thermal conductivity is strongly dependent on the morphology of the surface. On the other hand, D. Donadio using atomistic simulations [55], suggested that this reduction is due to the flattened phonon dispersion curves causing non-propagating phonon modes. Afterwards, M. Luisier on the basis of the atomistic quantum transport corroborated these results [56]. Moreover, he suggested that there is no need of large mean roughness height to enter in the pure diffusive limit in which the mean free path (MFP) is comparable to the diameter of NW; thus, giving an explanation of why the models that work in this hypothesis for the smooth NWs measured in [11] were getting a good fitting.

J. Shadhu et al developed a model in which they introduced different frequency-dependent regimes for the scattering rates and a complex analysis of the roughness contribution [57]. They found that the correlation length plays a major role in the reduction of the thermal conductivity. Short correlation lengths leads to a larger number of phonons suffering diffusive surface scattering events; even more, a large roughness correlation length implies that only the phonons with long wavelengths, and thus phonons that do not significantly contribute to the thermal transport, are scattered by the surface. The correlation length also determines the incident angles for which the phonons suffer a scattering event. On the other hand, they indicate that the short wavelength modes are strongly attenuated in rough boundaries with short correlation lengths, resulting in a MFP for these phonons shorter than the NW diameter, which can be interpreted as a breakdown of the Casimir limit. With this complex model, they obtained a very good fitting to the experimental thermal conductivities for both rough and smooth Si NWs. Even more, the results achieved in this work are consistent with the recent work of J. Lin that demonstrated experimentally the importance of the correlation length together with the roughness mean height on the thermal conductivity reduction in semiconductor NWs [22].

The idea of a breakdown in the Casimir limit for rough NWs was previously pointed out, but without a physical explanation accounting for it. When C. Guthly et al. measured the thermal conductivity of rough gallium nitride NWs [17], they showed that a very good fitting to the experimental values can be achieved using the Callaway model with a modified boundary scattering. Taking into account that the diffusive limit, characterized by a phonon MFP equal to the diameter of the NW, cannot take account of the roughness effect, they used the MFP as an adjusting parameter. A very good fitting to the experimental values was achieved for values of MFP below the NW diameter, therefore, below the Casimir limit. The result of this paper, together with the confirmation of the violation of the Casimir limit for rough NWs [57], allow the use of simple models based on the Callaway formalism for calculating the thermal conductivity of rough NWs, making use of a reduced effective diameter for the boundary scattering.

When considering alloyed NWs, e. g. $Si_{1-x}Ge_x$ NWs, the thermal conductivity modeling must address the effect of the alloy scattering, besides all the above mentioned scattering terms. N. Yang et al. carried out a first approximation in NWs composed of a mixture of Si isotopes making use of a molecular dynamic simulation [58]. They showed a large reduction in thermal conductivity, due to the isotopic scattering, with a maximum reduction for a composition of 50% between the constituent specimens. Afterwards, J. Chen et al. performed an equivalent molecular dynamics simulation on $Si_{1-x}Ge_x$ NWs [59], showing a strong dependence of the thermal conductivity on the composition of the NW. Furthermore, they found that the thermal conductivity is severely reduced even for small fractions of Ge, what was later on experimentally demonstrated [23]. They estimated a reduction in the thermal conductivity of 50% with respect to pure Si NWs for a Ge fraction of only 5%; however, one experimentally observes a much greater reduction, even for a smaller fraction of Ge in the NW [23]. In another attempt to modeling the thermal conductivity of $Si_{1-x}Ge_x$ NWs, Z. Wang [60] suggested a model on the basis of the previous works made by N. Mingo for Si NWs [30]. They assumed that all the scattering relaxation times, except the impurity relaxation time, can be viewed as independent processes for Si and Ge; therefore, they were added according to the Mathisessen's rule weighted by the composition, while the impurity scattering related to the alloy, is assumed to depend of the composition by a quadratic bowing function. In this way, they obtained a simple model able to achieve a prediction of the thermal conductivity of the order of magnitude of the experimental values obtained in [23]. Unfortunately, although they estimate thermal conductivity values of the order of magnitude of the experimental results, the fitting to the experimental results was not satisfactory, especially in the low temperature range.

Additional work is still needed to supply a good model accounting for the experimental thermal conductivity of $Si_{1-x}Ge_x$ NWs, and in general of alloyed, or doped NWs. If one observes the historical development of the models elaborated for describing the thermal conductivity of pure semiconductor NWs, one can see that the major advances were achieved once reliable experimental data were available. Very recently, new measurements of $Si_{1-x}Ge_x$ NWs have been published by Y. Choongho in a wide range of compositions [18], suggesting that more new models concerning alloyed NWs will arrive in the next future.

4. A Predictive model for the thermal conductivity of NWs

4.1. Preliminary remarks

As discussed in the previous paragraph different approaches have been attempted for the description of the thermal conductivity of semiconductor NWs. The measurement of the thermal conductivity at the nanometer scale is a rather complex experimental exercise; usually, the experimental data are hindered by the incertitude introduced by the NW/ heat sink contact. Therefore, the experimental data providing the thermal conductivity of NWs are scarce, and only a few experiments have supplied reliable thermal conductivity data; these data have been used to check the validity of the thermal transport models allowing a significant advance of the modeling formalisms. Concerning the different approaches used for modeling the thermal conductivity of NWs we have already mentioned that, some of them attained a good fitting to the experimental data using complex calculations, some other required of too many fitting parameters; while some other provided a good fitting in a reduced temperature range. The lack of generalized experimental results is a serious problem for both the design of thermoelectric devices based on NWs, and the thermal management of NW based devices. Therefore, a comprehensive model giving a predictive estimation of the thermal conductivity with the factors that contribute to limit it, e. g. the NW diameter, and the surface roughness, would be of high technological interest. On the other hand, experimental data are restricted to temperatures below room temperature, the data at higher temperatures are not available; however, this information is required for many applications that can operate at higher temperatures. We present herein a predictive model based in the well-known Callaway-Holland formalism that substantially simplifies the estimation of the thermal conductivity, provides an excellent fit to the experimental data, and can be used as a tool for all the technical situations in which the thermal conductivity of NWs is required for the design of NW based devices.

4.2. The Callaway-Holland formalism

The Callaway-Holland formalism has been successfully used for the description of the thermal conductivity of bulk semiconductors [40,42,61]; therefore, it must be conveniently adapted to the nanoscale requirements. It is well known that one of the main consequences of the reduced dimension of the NWs is the appearance of quantum confinement effects; in particular, the phonon confinement is relevant for diameters below 20-25 nm [15]. Many of the applications with NWs are built up with NWs with diameters above this phonon confinement limit; therefore, for these NWs one can reasonably assume a bulk like behavior.

First, we will introduce the main features of the Callaway-Holland formalism, initially developed for the low temperature range [41,43]. Despite the original limitation in temperature, the formalism was demonstrated to be a very useful tool in the calculation of thermal conductivity of semiconductor materials over the full range of temperatures, by the addition of some modifications.

The thermal conductivity is calculated by solving the Boltzmann equation

$$\left(\frac{\partial N}{\partial t}\right)_{\mathbf{c}} - \mathbf{c}\nabla T\frac{dN}{dT} = 0 \tag{1}$$

where N is the phonon density, and *c* the group velocity. In the presence of a small temperature gradient, ∇T , the deviation from equilibrium of the phonon mode k damps out exponentially, and the phonon population decay can be described by a specific relaxation time.

$$\left(\frac{\partial N}{\partial t}\right)_{\mathbf{c}} = \frac{N^0 - N}{\tau(\mathrm{k})} \tag{2}$$

N^0 is the Planck distribution function, and $\tau(\mathrm{k})$ is the relaxation time of the k mode. Callaway split out this expression in order to taking account of the contributions of both normal and resistive phonon processes, giving the following expression for the variation of the phonon density.

$$\left(\frac{\partial N}{\partial t}\right)_{\mathbf{c}} = \frac{N(\boldsymbol{\lambda}) - N}{\tau_N} + \frac{N^0 - N}{\tau_R} \tag{3}$$

Where $N(\lambda)$ is the shifted Planck distribution function towards which the Normal processes approach [62]; it can be expressed in a first approximation as:

$$N(\boldsymbol{\lambda}) \neg\ N(0) + \boldsymbol{\lambda}\left(\frac{\partial N}{\partial \boldsymbol{\lambda}}\right)_{0} = N^0 + \frac{\boldsymbol{\lambda}\mathbf{k}}{K_B T}\frac{e^{\hbar\omega/K_B T}}{\left(e^{\hbar\omega/K_B T} - 1\right)^2} \tag{4}$$

On the other hand, because a small deviation from equilibrium was assumed one can approximate $\frac{dN}{dT} \approxeq \frac{dN^0}{dT}$, therefore, one can simplify the Boltzmann equation as:

$$\left(-\frac{\hbar\omega}{K_B T}\mathbf{c}\nabla T + \frac{1}{\tau_N}\frac{\boldsymbol{\lambda}\mathbf{k}}{K_B T}\right)\frac{e^{\hbar\omega/K_B T}}{\left(e^{\hbar\omega/K_B T} - 1\right)^2} - \left(\frac{1}{\tau_N} + \frac{1}{\tau_R}\right)\left(N - N^0\right) = 0 \tag{5}$$

By introducing a global relaxation time, $\alpha(\mathrm{k})$, the term $N - N^0$ of the equation 5 can be expressed as:

$$N - N^0 = -\alpha(\mathbf{k})\mathbf{c}\nabla T\left(\frac{\hbar\omega}{K_B T^2}\right)\frac{e^{\hbar\omega/K_B T}}{\left(e^{\hbar\omega/K_B T} - 1\right)^2} \tag{6}$$

allowing to transform the Boltzmann equation in the more tractable form:

$$\left(\frac{\alpha(\mathrm{k})}{\tau_C}-1\right)\left(\frac{\hbar\omega}{T}\right)\mathbf{c}\nabla T+\frac{1}{\tau_N}\boldsymbol{\lambda}\mathbf{k}=0 \tag{7}$$

Where $\tau_C=\left(\tau_N^{-1}+\tau_R^{-1}\right)^{-1}$. On the other hand, assuming isotropic media, $\boldsymbol{\lambda}$ must be a constant vector in the direction of the temperature gradient. Thus, one can define a parameter, β, expressed as $\beta=-\frac{\lambda}{\hbar v^2}\frac{T}{\nabla T}$, where v is the speed of sound. Finally, in the acoustic approximation $\mathrm{k}=\frac{c\omega}{v^2}$ and substituting this in the reduced Boltzmann equation, one obtains the following expression for the relaxation time:

$$\alpha=\tau_C\left(1+\left(\frac{\beta}{\tau_N}\right)\right) \tag{8}$$

Using this relaxation time, one can estimate the thermal conductivity through the following equation:

$$\kappa=\frac{1}{(2\pi)^3}\sum_i\int_0^{2\pi}\int_0^{\pi}\left(\int v_i^2\alpha_i(k)C_{ph}(k)_i\cos^2\theta d^3\mathbf{k}\right)d\theta d\phi \tag{9}$$

Where C_{ph} is the phonon specific heat, and θ the angle formed between the phonon wave vector and the temperature gradient. Making use of the branch separation proposed by Holland [42], and considering that the phonon specific heat is expressed as:

$$C_{ph}=\frac{d}{dT}\left(\hbar\omega N^0\right)=\frac{(\hbar\omega)^2}{K_BT^2}\frac{e^{\hbar\omega/K_BT}}{\left(e^{\hbar\omega/K_BT}-1\right)^2} \tag{10}$$

The thermal conductivity transforms into the expression

$$\kappa_i=\sum_i\left(\frac{v_i^2}{8\pi^3}\right)\int_0^{k_D^i}\left(\frac{(\hbar\omega)^2}{K_BT^2}\right)\frac{1}{\tau_C^i}\left(1+\frac{\beta}{\tau_N^i}\right)\frac{e^{\hbar\omega/K_BT}}{\left(e^{\hbar\omega/K_BT}-1\right)^2}\cos^2\theta d^3k \tag{11}$$

where the subindex refers to the different acoustic branches, the longitudinal and the two transversal branches. The explicit form of the β parameter is obtained from the momentum conservation rule in the normal scattering processes; therefore:

$$\int\left(\frac{\partial N}{\partial t}\right)_N \mathbf{k}d^3k=\int\left(\frac{N(\boldsymbol{\lambda})-N}{\tau_N^i}\right)\mathbf{k}d^3k=0 \tag{12}$$

Replacing the term $N(\lambda)-N$ by their explicit form, previously derived in equation 6, one arrives to the equation:

$$\int\frac{e^{\hbar\omega/K_BT}}{(e^{\hbar\omega/K_BT}-1)^2}\left(\frac{\hbar\omega}{K_BT^2}\right)\mathbf{c}\nabla T(\alpha(\kappa)-\beta)\frac{c\omega}{\tau_N^i v^2}d^3\mathrm{k}=0 \tag{13}$$

And making use of $\mathbf{c}^2\nabla T=v_i^2 cos^2\theta$ and replacing α(k) by its value given in equation 8, one obtains:

$$\int\frac{e^{\hbar\omega/K_BT}}{\left(e^{\hbar\omega/K_BT}-1\right)^2}\left(\frac{\hbar\omega}{K_BT^2}\right)\left(\tau_C^i+\left(1-\frac{\tau_C^i}{\tau_N^i}\right)\beta\right)\frac{\omega}{\tau_N v^2}v^2cos^2\theta d^3\mathrm{k}=0 \tag{14}$$

From this expression and taking account of the independence between β and k one obtains

$$\beta=\frac{\int\frac{\tau_C^i}{\tau_N^i}\frac{e^{\hbar\omega/K_BT}}{\left(e^{\hbar\omega/K_BT}-1\right)^2}\left(\frac{\hbar\omega}{K_BT^2}\right)\omega cos^2\theta d^3\mathrm{k}}{\int\left(1-\frac{\tau_C^i}{\tau_N^i}\right)\frac{1}{\tau_N^i}\frac{e^{\hbar\omega K_BT}}{\left(e^{\hbar\omega K_BT}-1\right)^2}\left(\frac{\hbar\omega}{K_BT^2}\right)\omega cos^2\theta d^3\mathrm{k}} \tag{15}$$

Finally, taking into account that $d^3k=\frac{\omega^2}{v^3}d\omega d\Omega_k=\frac{\omega^2}{v^3}sin\theta d\theta d\phi d\omega$, and the change of variable $x=\hbar\omega/K_BT$, the thermal conductivity can be expressed as:

$$\begin{aligned}\kappa=\sum_i\left(\frac{K_B^4T^3}{8\pi^3\hbar^3v_i}\right)&\left\{\int_0^{\pi}\int_0^{2\pi}\int_0^{\frac{\theta_i}{T}}\left\{\tau_C^i\frac{e^x}{\left(e^x-1\right)^2}x^4cos^2\theta sin\theta d\theta d\varphi dx\right\}+\right.\\&+\left.\left(\frac{\int_0^{\pi}\int_0^{2\pi}\int_0^{\frac{\theta_i}{T}}\left\{\frac{\tau_C^i}{\tau_N^i}\frac{e^x}{\left(e^x-1\right)^2}x^4cos^2\theta sin\theta d\theta d\phi dx\right\}}{\int_0^{\pi}\int_0^{2\pi}\int_0^{\frac{\theta_i}{T}}\left\{\frac{1}{\tau_N^i}\left(1-\frac{\tau_C^i}{\tau_N^i}\right)\frac{e^x}{\left(e^x-1\right)^2}x^4cos^2\theta sin\theta d\theta d\phi dx\right\}}\right)\int_0^{\pi}\int_0^{2\pi}\int_0^{\frac{\theta_i}{T}}\left\{\frac{\tau_C^i}{\tau_N^i}\frac{e^x}{\left(e^x-1\right)^2}x^4cos^2\theta sin\theta d\theta d\phi dx\right\}\right\}\end{aligned} \tag{16}$$

Were the integral limits θ_i are the characteristic temperature cut-offs of each branch.

4.3. The phonon scattering terms

The application of the Callaway-Holland formalism to a particular problem in an extended temperature range resumes in the appropriate characterization of the different scattering terms. Therefore, a detailed analysis of the different scattering processes is required.

4.3.1. Three phonon scattering

This kind of process is determined by the exclusive interaction between phonons and follows two ways. In the first one, a phonon of wave vector q_1 combines with a phonon of wave vector q_2 given another phonon of wave vector q_3. In the second process a phonon of wave vector q_1 decays in two phonons of wave vectors q_2 and q_3 respectively.

$$\begin{aligned} q_1 + q_2 &= q_3 \\ q_1 &= q_2 + q_3 \end{aligned} \tag{17}$$

In these processes the energy is conserved; but such conservation does not necessarily hold for the momentum. The processes in which the momentum is not conserved are called Umpkklap, in this process the initial and final moment differs by a non zero reciprocal lattice vector, and the resultant phonons are out of the Brillouin zone; the momentum selection rule for these processes reads as:

$$\begin{aligned} q_1 + q_2 &= q_3' + G \\ q_1 &= q_2 + q_3' + G \end{aligned} \tag{18}$$

Where q'_3 is the "reflected" phonon of wave vector q'_3, and G the reciprocal lattice vector [28]. The phonon wave vector is opposite to q_1 and q_2 phonons, resulting in a process resistive to the heat flow.

Peierls proposed for the Umpklapp processes a scattering rate of the form $\tau_U^{-1} \propto T^{a} e^{\theta/bT}$ [62], where the values of the parameters a and b have been determined in different ways [62-65]. Klemens showed that in the high temperature range the resistive contribution of the three phonon processes obeys to $\tau_U^{-1} = C\omega^2 T$ [64]. Using the Slack's variant [65] of the Peierls form, one can extend the temperature range, the Slack relaxation time for the U processes is given as $\tau_U^{-1} = B\omega^2 T e^{\theta/3T}$. Using this form, Slack established that the thermal conductivities of different pure crystals can be modeled over an extended temperature range giving a good agreement with the experimental values [65]. Furthermore, when the temperature increases, it adopts the same expression as the Klemens relaxation time. Klemens evaluated the constant C as $C = \frac{2\gamma^2 \hbar}{M v^2 \theta}$ [27,64]. Where γ is the Grüneisen parameter, M the average atomic mass in the crystal, and v the velocity of sound. Finally, we can introduce the Umpklapp relaxation time for each phonon branch as:

$$\left(\tau_U^{-1}\right)^i = \frac{2\gamma_i^2\hbar}{Mv_i^2\theta_i}\omega^2 T e^{\theta_i/3T} \tag{19}$$

Regarding the Normal processes, it has been claimed that they do not contribute to the thermal resistance [25, 27, 30, 31, 33]. However, they cannot be neglected because the Callaway-Holland formalism requires of the normal process relaxation time [41-43]. Furthermore, the normal processes can contribute indirectly to the thermal resistance by redistributing the phonon population, which can further participate in resistive scattering processes. These can be explained by the Herring mechanism [32], which redistributes the phonon momentum among the different phonon branches. In this mechanism, the relaxation of transverse phonons consists of its interaction with a longitudinal acoustic phonon, resulting in another longitudinal phonon. On the other hand, the relaxation of the longitudinal phonons occurs via two different processes; they decay in two transverse phonons; or a longitudinal phonon arises from two transverse phonons. Taking account of these processes, the relaxation time suggested by Herring for each normal process can be written as [32]:

$$\left(\tau_N^L\right)^{-1} = C_L\omega^2T^3 \ ; \ \left(\tau_N^T\right)^{-1} = C_T\omega T^4 \tag{20}$$

Where the values of the constants $C_i$can be described in terms of the physical properties of the material, which according to the approximation made by Asen-Palmer et al. [66], take the form:

$$C_L \approx \frac{K_B{}^3\gamma_L^2V}{M\hbar^2v_L^5} \ ; C_T \approx \frac{K_B{}^4\gamma_T^2V}{M\hbar^3v_T^5} \tag{21}$$

where V is the unit cell volume.

4.3.2. Isotope scattering

The impurities are centers for phonon scattering. This mechanism is described as the isotope scattering and is related to the mass difference between the host atoms and the impurity atoms. Even in the case of pure crystals, without impurities, is necessary to consider the existence of isotopes, which efficiently scatter high frequency phonons. The perturbation in the lattice due to a substitutional atom was evaluated by Klemens using perturbation theory [44,64], and the expression for the relaxation time due to isotope mixing can be expressed as

$$\left(\tau_I^i\right)^{-1} = \frac{V\Gamma}{4\pi v_i^3}\omega^4T^4 \tag{22}$$

Where Γ is the mass-fluctuation parameter defined for a single element as $\Gamma = \sum_i f_i \left(1 - \frac{M_i}{M}\right)^2$ where f_i is the fractional atomic abundance of the isotope (impurity) of atomic mass M_i[44].

4.3.3. Boundary scattering

A real crystal has a finite size; therefore, the scattering due to the crystal boundaries cannot be neglected. The boundaries act as highly efficient scattering regions, confining the MFP of the phonons to the order of the dimension of the sample. This problem was discussed by Ziman on the basis of the Casimir theory [28], and for the nanoscaled samples is the dominant scattering mechanism accounting for the reduction of the thermal conductivity. This process can be phenomenologically explained by considering two types of phonon boundary scattering processes; one of them corresponding to the specular phonon scattering, which mainly happens in smooth surfaces; while the other one corresponds to the pure diffusive phonon scattering, which mainly occurs on rough surfaces. Taking account of both processes, which occur with a probability p for the pure specular scattering, and 1-p for purely diffusive scattering, Ziman derived an expression for the MFP of phonons in a finite sample as $\Lambda = \frac{1+p}{1-p}\Lambda_0$ where Λ_0 is the MFP for a perfectly rough boundary [28], where only diffusive processes occur. It depends on the geometry of the sample; in the case of a cylinder, this MFP is equal to the diameter of the cylinder, $\Lambda_0 = D$.

The key issue of this scattering lies in the form of the probability function, p, which has been addressed in different ways in the literature. Commonly, it has been taken equal to zero [25, 30, 33, 37] or as a constant independent of temperature and frequency [27, 34, 35], thereby simplifying the boundary scattering relaxation time. However, a more realistic approximation can be achieved if one eliminates such oversimplifications. Ziman showed that the probability function p depends on the wavelength of the incident phonons [28], being the surface characterized by the mean value of the height deviation over a surface reference level; this parameter is called the asperity parameter,η. In a first approximation, the probability function can be written as $p = exp\left[-{16\pi^3\eta^3}/{\lambda^2}\right]$, which has been used by some authors [50]. Despite being a more accurate way of treating the boundary scattering with respect to the assumption of a constant value of p, the hypotheses assumed in the construction of this probability function were considered by Ziman asnon very realistic. Therefore, Ziman himself developed an alternative treatment, which takes account of the existence of a distribution function for the roughness instead of the average value given by the asperity parameter. In this way, the probability function takes the form $p(\eta, \lambda) = \int P(\eta) Exp\left[-16\pi^3\eta^2/\lambda^2\right] d\eta$ where $P(\eta)$ is the roughness distribution probability. This expression can be simplified if one considers that the reflection probability is a measure of the fraction of area whose average roughness is less than ${\lambda}/{\pi 4}$ [28], reducing the probability function to $p(\eta, \lambda) \approx \int_0^{\lambda/4\pi} P(\eta)d\eta$. This is the form used by Mingo et al. for approaching the thermal conductivity in the low temperature range for rough Silicon NWs with a roughness distribution given as $P(\eta)=\frac{1}{\eta}Exp\left[-\eta/\eta_0\right]$ [67].

Moreover, based on the dependence of the probability function p with frequency and roughness, Soffer addressed a different solution in the presence of a temperature gradient along the sample [68], according to which the probability function takes the form:

$$p(\omega,\eta) = Exp\left[-\left(2\eta\left(\omega/v_i\right)\sin\theta\right)^2\right] \tag{23}$$

where η is the asperity parameter defined by Ziman, and θ the angle of incidence of phonons. This form given by Soffer is the one used in our simulations because we tested that it was the one providing the best results. Finally, adding the limitation in length to the boundary relaxation time, the final expression for the boundary relaxation time can be written as

$$\left(\tau_B^i\right)^{-1} = \left(\frac{1-p}{1+p}\Lambda_0^{-1} + L^{-1}\right)v_i \tag{24}$$

4.4. The thermal conductivity of NWs

Using this relaxation time, the thermal conductivity can be calculated over the full temperature range. The solution of the complex equation 16 was carried out by the Gauss-Kronrod local adaptive method, fixing a tolerance $<10^{-9}$ for enhancing the accuracy. This method provides calculation several times faster than other common numerical methods applied to this kind of problems. In order to test the model, first we carried out a prediction of the thermal conductivity of bulk silicon and germanium, obtaining an excellent fit to the experimental measurements over the full temperature range (Figure 5-A and 5-B).

The parameters used for the calculations are given in table 1. The values for Γ and γ_i were taken from reference [61]; the Grüneisen parameter, γ_i, was approximated to a constant value. Actually, this is crude approach, since the Grüneisen parameter has a dependence with temperature, especially in the low temperature range [44]. The rough simplification of a constant value for this parameter arises from the dependence with the temperature and frequency of the three phonon scattering relaxation times, which is generically expressed as $\tau_j^{-1}=Cts_j \times G_j(\omega)\times F_j(T)$, namely the product of some function of ω, some function of the temperature T and a given constant value for each type of scattering, Cts_j. Therefore, the temperature dependence of each three-phonon scattering process can be completely included in the function $F(T)$; which encloses the temperature dependence of the Grüneisen parameter, while the constant parts of the Grüneisen parameters are encompassed in the constant group Cts_j. The bulk value for the sound velocities of each phonon branch was taken from [44]. Finally, the asperity parameter was chosen assuming a smooth surface, matching the values reported in [40] for smooth silicon NWs.

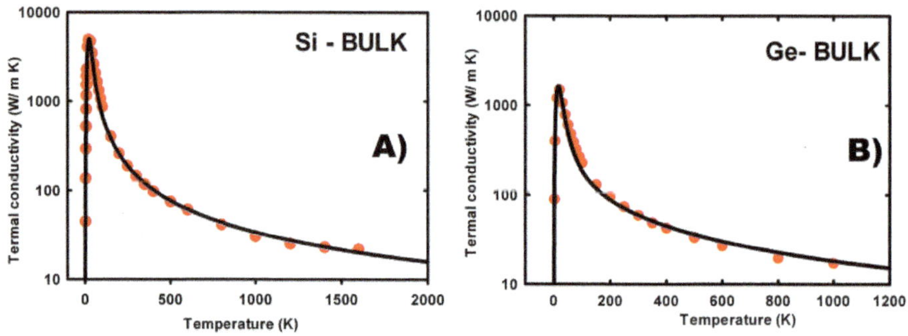

Figure 5. Fitting of the experimental thermal conductivities of bulk Si (A), and bulk Ge(B) by our modified Callaway-Holland model. Experimental data taken from [14].

The values of the cut-off temperatures, θ_i, and the effective size of the sample remains as free parameters in the resolution of the equations. The effective size is characterized by the lengths L and Λ_0, which take the same value as in the bulk for a sake of simplicity.

	L (m)	Λ_0 (m)	η (m)	Γ	θ_L	θ_T	v_L	v_T	γ_L	γ_T
Si	$6x10^{-3}$	$6x10^{-3}$	$1 x10^{-9}$	$2.16 x10^{-4}$	627.7	435.8	8430	5840	1.1	0.66
Ge	$4x10^{-3}$	$4x10^{-3}$	$1x10^{-9}$	$6.08x10^{-4}$	387.1	278.3	4920	3540	1.1	0.66

Table 1. Parameters used in the calculations of the thermal conductivity of bulk Si and bulk Ge.

The equivalent cut-off temperature obtained as $\theta = 3(\theta_L^{-1} + 2\theta_T^{-1})^{-1}$ [69] has a value of 495 K for Si, and 307 K for Ge. In the original Callaway model the cut-off temperature matches the Debye temperature, albeit this parameter has been treated in different ways in further works. The Debye temperature for Si determined from the specific heat takes a value of 645 K [61]. If one uses the sound velocities of the acoustic branches instead of the heat capacity, the cut-off temperature scales up to 710 K [61]; whereas, if one estimates this temperature through the zone-boundary frequencies of the transversal and longitudinal acoustic phonons, the value drops dramatically to 298 K [61]. One can also consider the fact that the Callaway model ignores the contributions of optical phonons, because of the continuum approximation. Fortunately, this limitation can be circumvented by increasing the cut-off temperatures [44]. Thus, the use of the cut-off temperature as a free adjustable parameter implicitly includes the optical phonon contribution.

Finally, the equivalent sample diameter, being of the order of millimeters, does not have a big impact in the thermal conductivity for temperatures above 300 K. This allows to adjust the temperature cut-offs for matching the experimental values in this temperature range. Therefore, the equivalent sample diameter is commonly used to matching the thermal conductivity in the low temperature range [61].

On the other hand, one can assume that the phonon confinement effects only show relevant effects below 25 nm in diameter [15]; this means that one can treat the NWs with diameter larger than 25 nm in a bulk-like approach.

Having established the excellent agreement with the experimental data for bulk Si and Ge, the same procedure was applied to Si NWs, and the results were compared to those measured in [11]. The parameters used for these calculations are given in Table 2. The agreement with the experimental results was excellent over the full temperature range for NWs without phonon confinement (D>25nm). While the results, as expected, show imprecise below the phonon confinement threshold (D< 25nm). The pseudo-linear experimental behavior of the 22 nm NW cannot be predicted by this bulk-like model; which suggests that alternative approaches such as the one conducted in [36] are required for describing the thermal transport in NWs of reduced diameter, exhibiting confinement effects. These results are summarized in Figure 6.

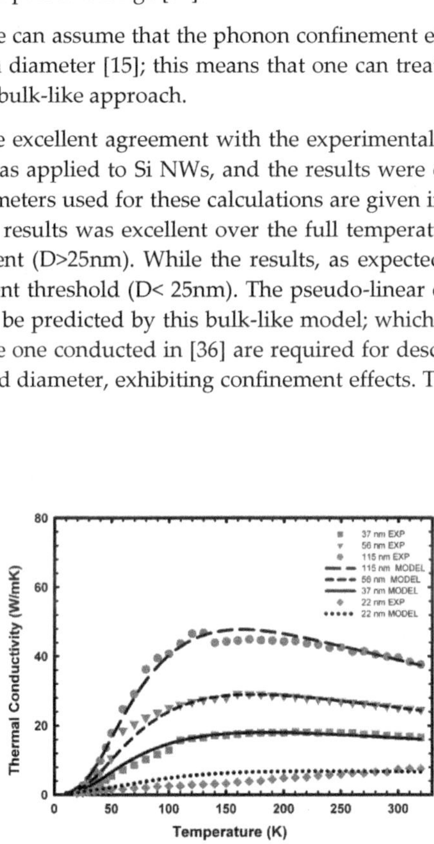

Figure 6. Fitting to the experimental thermal conductivity of Si NWs with different diameters, showing the excellent fit for all of them, except for the 22 nm NW. Experimental data taken from [11]

Several authors have shown that the speed of sound of the different phonon branches is lower in semiconductor NWs than in their corresponding bulk counterparts, even for NWs with diameters larger than the one for phonon confinement [25, 27, 34, 40, 69]. The values of the sound velocities present differences depending upon the method followed to obtaining them; therefore, the sound velocities are left as free parameters when adjusting the experimental thermal conductivity data. The cut-off temperatures for semiconductor NWs must be also lower than their bulk counterparts [30, 34, 40, 69]. This is consistent with the values of the cut-off temperatures and the sound velocities found by fitting the experimental thermal conductivity data, Table 2.

NW	L (m)	Λ_0 (m)	η (m)	r	θ_L (K)	θ_T (K)	v_L (m/s)	v_T(m/s)	γ_L	γ_T
115	5×10^{-9}	115×10^{-9}	1×10^{-9}	2.16×10^{-4}	537.5	302.9	7962	4741	1.1	0.66
56	5×10^{-9}	56×10^{-9}	1×10^{-9}	2.16×10^{-4}	511.1	284.1	7168	4163	1.1	0.66
37	5×10^{-9}	37×10^{-9}	1×10^{-9}	2.16×10^{-4}	479.4	259.3	7066	4106	1.1	0.66
22	5×10^{-9}	22×10^{-9}	1×10^{-9}	2.16×10^{-4}	466.9	141.2	7018	3997	1.1	0.66

Table 2. Parameters used in the calculations of the thermal conductivity of smooth Si NWs of [11].

On the other hand, if one assumes that the crystallinity and the impurity concentration are the same for rough and smooth NWs, the only relevant difference between both types of NWs is the surface roughness; therefore, the thermal conductivity reduction in rough NWs must be exclusively related to the contribution of the roughness to the boundary scattering. The roughness is included in the boundary scattering through the asperity parameter η; choosing a value of η= 1 nm, typically reported for smooth NWs [40], the value of p is close to zero for most of the phonon frequencies, which means that even for such small roughness the boundary scattering is close to the diffusive limit in agreement with the result of reference [56]. Therefore, further increase of η cannot result in a substantial increase of the diffusive scattering contribution, and the dramatic reduction of the thermal conductivity, experimentally observed in rough NWs [2], cannot be accounted for by the alone diffusive boundary scattering. The breakdown of the Casimir limit observed in [57] allows us to introduce in the calculation the concept of an effective Casimir length ($\Lambda_0^{'}<\Lambda_0$) smaller than the classical MFP given by the diameter of the NW consequence of the shrinkage of the acoustic phonon MFP in rough NW. The prediction of the thermal conductivity of rough NWs requires the evaluation of the effective Cassimir length $\Lambda_0^{'}$; for this purpose we used the experimental data of a Si NW with 115 nm diameter and roughness of η=5. 7 nm [2]. Assuming that the only difference with the smooth NW of the same diameter lies on the roughness, all the other parameters showed in Table 2 already used for the calculation of the thermal conductivity of the smooth NW of the same dimension (115 nm) remained unchanged. In Figure 7-A the thermal conductivity calculated for the rough NW with an effective Casimir length $\Lambda_0^{'}$=115 nm gives a value for the thermal conductivity at 300 K only 1 W/mK less than the thermal conductivity of the 115 smooth NW; therefore, the very increase of the asperity parameter cannot account for the abrupt decrease of the thermal conductivity in rough NWs, as long as the phonon MFP is not reduced below the Casimir limit; however, as the MFP is reduced, the thermal conductivity falls down dramatically, reaching the experimental value reported in [2] for an effective Cassimir length of $\Lambda_0^{'}\approx$13 nm, Figure 7-A.

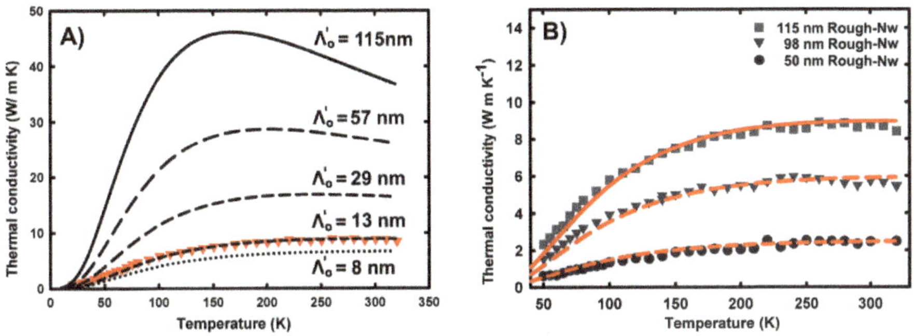

Figure 7. Calculated thermal conductivities of a rough Si NW (115 nm diameter) for different effective Casimir lengths, showing fitting to experimental data (taken from [2]) for Λ_0'=13 nm (A); fitting of the experimental thermal conductivities for rough Si NWs with different diameters, the fitting parameters are summarized in Table. 3 (B), experimental data taken from [2].

This large MFP reduction in rough NWs cannot be explained in terms of the surface scattering mechanism described by Ziman [28], and must be considered in a first instance as an admissible calculation tool for obtaining good experimental data approach as was shown in [17]. The thermal conductivities calculated for NWs with the same geometric characteristics as those measured in [2] are shown in Figure 7-B, evidencing a very good fitting to the experimental data, using the parameters listed in Table 3. The effective Casimir lengths obtained by fitting the experimental values show a significant reduction with respect to the NW diameters, the ideal Casimir lengths; similar reductions have been also reported by other authors [17].

D (nm)	L (m)	Λ_0 (m)	η (m)	θ_L (K)	θ_T (K)	v_L (m/s)	v_T (m/s)	γ_L	γ_T
115	$5\text{x}10^{-9}$	$13\text{x}10^{-9}$	$5.7\text{x}10^{-9}$	537. 5	302. 9	7962	4741	1. 1	0. 66
98	$5\text{x}10^{-9}$	$8.8\text{x}10^{-9}$	$5.7\text{x}10^{-9}$	530. 1	300. 2	7745	4584	1. 1	0. 66
50	$5\text{x}10^{-9}$	$3.3\text{x}10^{-9}$	$5.7\text{x}10^{-9}$	505. 9	274	7159	4160	1. 1	0. 66

Table 3. Parameters used in the calculations of the thermal conductivity of rough NW of [2].

4.5. A predictive expression for the thermal conductivity of Si NWs

A big constraint for the applications of NWs in which the thermal conductivity plays a major role is the lack of experimental data over a large range of diameters and temperatures, or in its defect the absence of a model supplying such information. In view of the difficulties inherent to thermal conductivity measurements at the nanoscale, and the variety of diameters and surface roughness, the development of a model capable of predicting the thermal conductivity in a wide range of diameters and surface roughness acquires a capital interest.

We have shown that one can achieve an excellent fit of the experimental thermal conductivity of semiconductor NWs using a modified Callaway-Holland formalism. This agreement should permit us the prediction of the thermal conductivity for NW diametersand the temperature ranges for which experimental data are not available. Particularly relevant is the case of the high temperature range, above 400 K, for which there are not experimental measurements available. Since the model for bulk Si provides a very good result in the high temperature range, it is reasonable to assume that the same model for NWs without phonon confinement works well in the high temperature range.

In our model, the fitting to the thermal conductivity is achieved using the speeds of sound and the cut-off temperatures for transverse and longitudinal modes, as free parameters. By relating those parameters to the NW diameter, one could predict the thermal conductivity for any NW diameter of a given roughness. In the case of the smooth NWs, a very good result was obtained for NWs with a diameter above the phonon confinement limit. Unfortunately, there are only three diameters for which reliable experimental data are available. However, additional data exist when one considers rough NWs; assuming that the reduction in the thermal conductivity of these NWs with respect to smooth NWs is exclusively related to the contribution of the roughness to the boundary scattering, which does not modify the sound velocities, nor the cut-off temperatures, one can extend the range of available experimental data. This hypothesis was confirmed with rough NWs of 115 nm diameter, for which an excellent fitting was achieved by acting only on the boundary relaxation time, keeping all the other parameters the same as the ones used for the smooth 115 nm NW (see section 4. 4.). Therefore, one can use the values of the adjustable parameters, speed of sound, and cut off-temperature, of both smooth and rough NWs to build up the predictive expression for the thermal conductivity.

The values of the speed of sound and cut-off temperature estimated for each NW diameter for which experimental data are available are plotted in Figure 8 (dots). In this figure one can observe that the behavior of the both velocities, the speed of sound of the longitudinal acoustic branch and the transverse acoustic branch respectively, have the same smooth form. Similar behavior is observed for the cut-off temperatures for each branch. This allows us to assume that those free parameters can be phenomenologically expressed as a function of the NW diameter. In the same figure, we show the nonlinear regression for each free parameter, showing in all the cases an adjustment of $r^2>0.99$ to a generic function expressed as

$$f(D)=\frac{1+aD}{b+cD} \tag{25}$$

Therefore, with these phenomenological relations one is allowed to interpolate the values of the free parameters used in the calculation of the thermal conductivity of Si NWs in a range of diameters between the phonon confinement limit to slightly above the hundred of nms in diameter, using for that the adjustment parameters given in Table 4. Note that these parameters only represent numbers suitable for describing the sound velocities and the temperature cut-offs as a function of the NW diameter.

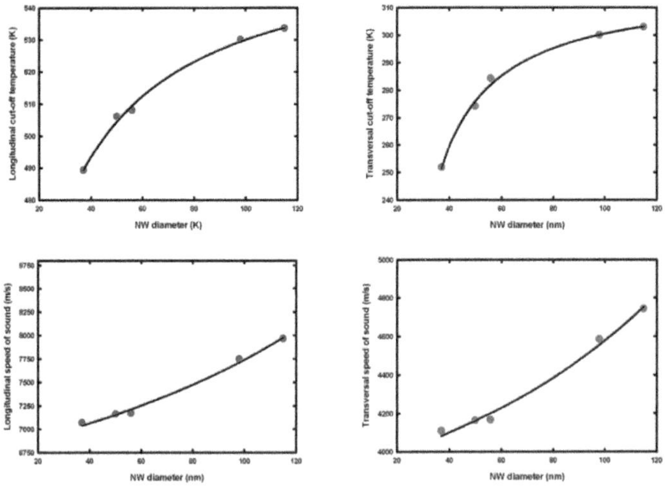

Figure 8. Plots of the cut-off temperatures, and sound speeds for the longitudinal and acoustic phonon branches vs the NW diameter. The symbols represent the values obtained by fitting the available experimental data. The lines are the fits using the phenomenological three parameter equation 25 in the text body.

	a	b	c
v_L	$-2,362\text{x}10^{-3}$	$1,478\text{x}10^{-4}$	$-4,91310^{-7}$
v_T	$-2,563\text{x}10^{-3}$	$2,566\text{x}10^{-4}$	$-9,41410^{-7}$
θ_L	$1,668\ \text{x}10^{-1}$	$3,671\text{x}10^{-3}$	$2,969\text{x}10^{-4}$
θ_T	$-5,153\ \text{x}10^{-2}$	$2,395\text{x}10^{-3}$	$-1,621\text{x}10^{-4}$

Table 4. Parameter values from the non linear regressions to the data of Figure 8.

Using these parameters one can construct the thermal conductivity vs T plots for any NW diameter. These plots are shown in Figure 9-A. The thermal conductivity vs the NW diameter for two fixed temperatures, 300K and 600K, are represented in Figure 9-B; one observes that at high temperature the influence of the diameter on the thermal conductivity is flattened out with respect to what happens at lower temperatures; which roughly can be interpreted in terms of the balance between boundary scattering processes and three phonon processes. In the low temperature range (T<300K) the boundary scattering is dominant, while in the high temperature range the three phonon scattering processes increase its weight.

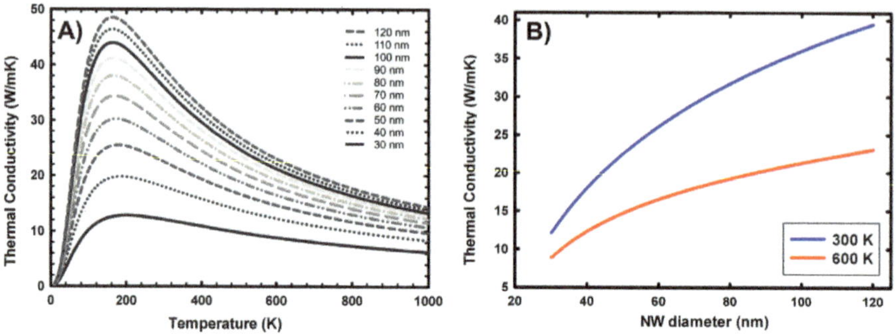

Figure 9. Calculated thermal conductivity for arbitrary diameters using the parameters given in Table 3 (A). Thermal conductivities at 300K and 600K as a function of the NW diameter (B)

5. Summary and perspectives

An overview of the thermal transport in semiconductor NWs has been presented. Both experimental and theoretical aspects have been discussed, evidencing that the theoretical models need of reliable thermal conductivity measurements to contrast the models, in order to advance in the understanding of the thermal transport in semiconductor NWs. Several clues need to be addressed in the close future; in particular, the role of optical phonons scattering, the scattering mechanisms in low diameter NWs, in connection with the phonon confinement phenomena; and a comprehensive view of the boundary scattering in rough surfaces, for which a breakdown of the Casimir limit occurs; also, the modeling of the thermal transport in alloyed NWs has to be addressed because of the importance of alloys for engineering the thermal conductivity. We have developed a model based on a modified Callaway-Holland formalism, which agglomerates the contributed ideas provided by different authors. This model permits an excellent fitting of the experimental results over an extended temperature range. Using the fitting of this model to the available experimental results one established empiric relations that permit to build up a predictive tool, which opens the access to the thermal conductivity of any Si NW diameter above the phonon confinement limit. This formalism can be extended to any other semiconductor without loss of generality, provided that reliable experimental thermal conductivity data for a few NWs diameters are available.

Acknowledgement

This work was funded by the Spanish Government (MAT-2007-66181-C03 and MAT-2010-20441-C02)

Author details

J. Anaya[1], J. Jiménez[1] and T. Rodríguez[2]

1 GdS Optronlab, Ed I+D, University of Valladolid, Valladolid, Spain

2 Electronic Technology, ETSIT, Polytechnic University of Madrid, Madrid, Spain

References

[1] Cui Y, Lieber C.M. Functional Nanoscale Electronic Devices Assembled using Silicon Nanowire Building Blocks. Science 2001;291, 851.

[2] Hochbaum A.I, Chen R, Delgado R.D, Liang W, Garnett E.C, Najarian M, Majumdar A, Yang P. Rough Silicon Nanowires as High Performance Thermoelectric Materials. Nature 2008;451, 163.

[3] Kayes B.M, Atwater H.A, Lewis N.S. Comparison of the device physics principles of planar and radial p-n junction nanorod solar cells. J. Appl. Phys. 2005;97, 114302.

[4] Xie X.Q, Liu W.F, Oh J.I, Shen W.Z. Optical absorption in c-Si/a-Si:H core/shell nanowire arrays for photovoltaic applications. Appl. Phys. Lett. 2011;99, 033107.

[5] Singh N, Buddharaju K.D, Manhas S.K, Agarwal A, Rustagi S.C, Lo G.Q, Balasubramanian N, Kwong D.L. Si, SiGe Nanowire Devices by Top–Down Technology and Their Applications. IEEE Trans. Electron Devices 2008;55, 3107.

[6] Westover T, Jones R, Huang J.Y, Wang G, Lai E, Talin A.A. Photoluminescence, Thermal Transport, and Breakdown in Joule-Heated GaN Nanowires. Nano Lett. 2009;9, 257.

[7] Shi L, Yao D, Zhang G, Li B. Large thermoelectric figure of merit in Si1−xGex nanowires. Appl. Phys. Lett. 2010;96, 173108.

[8] Mingo N. Thermoelectric figure of merit of II–VI semiconductor nanowires. Appl. Phys. Lett. 2004;85, 5986.

[9] Mingo N. Thermoelectric figure of merit and maximum power factor in III–V semiconductor nanowires. Appl. Phys. Lett. 2004;84, 2652.

[10] Shi L, Li D, You C, Jang W, Kim D, Yao Z, Kim P, Majumdar A. Measuring Thermal and Thermoelectric Properties of One-Dimensional Nanostructures Using a Microfabricated Device. J. Heat Transfer 2003;125, 881.

[11] Li D, Wu Y, Kim P, Shi L, Yang P, Majumdar A. Thermal conductivity of individual silicon nanowires. Appl. Phys. Lett. 2003;83, 2934.

[12] Hippalgaonkar K, Huang B, Chen R, Sawyer K, Ercius P, Majumdar A. Fabrication of Microdevices with Integrated Nanowires for Investigating Low-Dimensional Phonon Transport. Nano Lett. 2010;10 4341.

[13] Li D.Y, Wu Y, Fan R, Yang P.D, Majumdar A. Thermal conductivity of Si/SiGe superlattice nanowires. Appl. Phys. Lett. 2003;83, 3186.

[14] Glassbrenner J.C, Slack G.A. Thermal Conductivity of Silicon and Germanium from 3°K to the Melting Point. Phys. Rev. 1964;134, 4A 1058.

[15] Campbell I. H, Fauchet P.M. The effects of microcrystal size and shape on the one phonon Raman spectra of crystalline semiconductors. Solid St. Commun. 1986;58, 739.

[16] Zhou J, Jin C, Seol J.H, Li X, Shi L. Thermoelectric properties of individual electrodeposited bismuth telluride nanowires. Appl. Phys. Lett 2005;87, 133109.

[17] Guthy C, Nam C, Fischer J.E. Unusually low thermal conductivity of gallium nitride nanowires. J. Appl. Phys. 2008;103, 064319.

[18] You C. Large Thermoelectric Figure-of-Merits from SiGe Nanowires by Simultaneously Measuring Electrical and Thermal Transport Properties. Nano Lett. 2012;12, 2918.

[19] Shi L, Hao Q, Choongho Y, Mingo N, Xiangyang K. Thermal conductivities of individual tin dioxide nanobelts. Appl. Phys. Lett. 2004;84, 2638.

[20] Wingert M.C, Chen Z.C.Y, Dechaumphai E, Moon J, Kim J, Xiang J, Chen R. Thermal Conductivity of Ge and Ge–Si Core–Shell Nanowires in the Phonon Confinement Regime. Nano Lett 2011;11, 5507.

[21] Park Y, Kim J, Kim H, Kim I, Lee K, Seo D, Choi H, Kim W. Thermal conductivity of VLS-grown rough Si nanowires with various surface roughnesses and diameters. Appl. Phys. A 2011;104,7.

[22] Lim J, Hippalgaonkar K, Andrews S.C, Majumdar A, Yang P. Quantifying Surface Roughness Effects on Phonon Transport in Silicon Nanowires. Nano Lett. 2012;12, 2475.

[23] Kim H, Kim I, Choi H, Kim W. Thermal conductivities of Si1−xGex nanowires with different germanium concentrations and diameters. Appl. Phys. Lett. 2010;96, 233106.

[24] Kim H, Park Y, Kim I, Kim J, Choi H, Kim W. Effect of surface roughness on thermal conductivity of VLS-grown rough Si1−xGex nanowires. Appl. Phys. A 2011;104, 23.

[25] Balandin A, Wang K.L. Significant decrease of the lattice thermal conductivity due to phonon confinement in a free-standing semiconductor quantum well. Phys. Rev B 1998;58, 1544.

[26] Parrott J. E., Stuckes A. D. Thermal Conductivity of Solids, New York: Methuen; 1975.

[27] Zou J, Balandin A. Phonon heat conduction in a semiconductor nanowire. J. Appl. Phys. 2001;89, 2932.

[28] Ziman J.M. Electrons and Phonons New York: Oxford University Press; 1967.

[29] Volz S.G, Chen G. Molecular dynamics simulation of thermal conductivity of silicon nanowires. Appl. Phys. Lett. 1999;75 2056.

[30] Mingo N. Calculation of Si nanowire thermal conductivity using complete phonon dispersion relations. Phys. Rev. B 2003;68, 113308.

[31] Mingo N, Yang L. Predicting the Thermal Conductivity of Si and Ge Nanowires. Nano Lett. 2003;3, 1713.

[32] Herring C. Role of Low-Energy Phonons in Thermal Conduction. Phys. Rev 1954;95, 954.

[33] Chung J.D, McGaughey A.J.H, Kaviany M. Role of Phonon Dispersion in Lattice Thermal Conductivity Modeling. Transactions of the ASME 2004;126, 376.

[34] Huang M, Chong W, Chang T. The lattice thermal conductivity of a semiconductor nanowire. J. Appl. Phys. 2006;99 114318.

[35] Liang L.H, Li B. Size-dependent thermal conductivity of nanoscale semiconducting systems. Phys. Rev. B 2006;73, 153303.

[36] Murphy P.G, Moore J.E. Coherent phonon scattering effects on thermal transport in thin semiconductor nanowires. Phys. Rev. B 2007;76, 155313.

[37] Tian Z, Esfarjani K, Shiomi J, Henry A.S, Chen G. On the importance of optical phonons to thermal conductivity in nanostructures. Appl. Phys. Lett 2011;99, 053122.

[38] Ward A, Broido D.A. Intrinsic phonon relaxation times from first-principles studies of the thermal conductivities of Si and Ge. Phys. Rev B 2010;81, 085205.

[39] Kazan M, Pereira S, Coutinho J, Correia M.R, Masri P. Role of optical phonon in Ge thermal conductivity. Appl. Phys. Lett 2008;92, 211903.

[40] Kazan M, Guisbiers G, Pereira S, Correia M, Masri P, Bruyant A, Volz S, Royer P. Thermal conductivity of silicon bulk and nanowires: Effects of isotopic composition, phonon confinement, and surface roughness. J. Appl. Phys 2010;107, 083503.

[41] Callaway J. Model for Lattice Thermal Conductivity at Low Temperatures. Phys. Rev. 1959;113, 1046.

[42] Holland M.G. Analysis of Lattice Thermal Conductivity. Phys. Rev 1963;132, 2461.

[43] Carruthers P. Theory of Thermal Conductivity of Solids at Low Temperatures. Rev. Mod. Phys 1961;33, 92.

[44] Srivastava G.P. The Physics of Phonons, Bristol:Adam Hilger; 1990.

[45] Alvarez F.X, Jou1 D. Size and frequency dependence of effective thermal conductivity in nanosystems. J. Appl. Phys, 2008;103, 094321.

[46] Guyer R.A, Krumhansl J.A. Solution of the Linearized Phonon Boltzmann Equation. Phys. Rev 1966;148, 766.

[47] Guyer R.A, Krumhansl J.A. Thermal Conductivity, Second Sound, and Phonon Hydrodynamic Phenomena in Nonmetallic Crystals. Phys. Rev 1966;148, 778.

[48] Alvarez F.X., Jou D, Sellitto A. Phonon hydrodynamics and phonon-boundary scattering in nanosystems. J. Appl. Phys, 2009;105, 014317.

[49] Sellitto A. Alvarez F.X, Jou D. Second law of thermodynamics and phonon-boundary conditions in nanowires. J. Appl. Phys. 2010;107, 064302.

[50] Moore A.L, Saha S.K, Prasher R.S, Shi L. Phonon backscattering and thermal conductivity suppression in sawtooth nanowires. Appl. Phys. Lett 2008;93, 083112.

[51] Sellito A, Alvarez F.X, Jou D., Temperature dependence of boundary conditions in phonon hydrodynamics of smooth and rough nanowires. J. Appl. Phys 2010;107, 114312.

[52] Martin P, Aksamija Z, Pop E, Ravaioli U. Impact of Phonon-Surface Roughness Scattering on Thermal Conductivity of Thin Si Nanowires. Phys. Rev. Lett. 2009;102, 125503.

[53] Kosevich Y.A, Savin A.V. Reduction of phonon thermal conductivity in nanowires and nanoribbons with dynamically rough surfaces and edges, EPL 2009;88, 14002.

[54] Liu L,Chen X. Effect of surface roughness on thermal conductivity of silicon nanowires. J. Appl. Phys 2010;107, 033501.

[55] Donadio D, Galli G. Atomistic Simulations of Heat Transport in Silicon Nanowires. Phys. Rev. Lett. 2009;102, 195901.

[56] Luisier M. Investigation of thermal transport degradation in rough Si nanowires. J. Appl. Phys 2011;110, 074510.

[57] Sadhu J, Sinha S. Room-temperature phonon boundary scattering below the Casimir limit. Phys. Rev. B 2011;84, 115450.

[58] Yang N, Zhang G, Li B. Ultralow Thermal Conductivity of Isotope-Doped Silicon Nanowires. Nano Lett. 2008;8, 276.

[59] Chen J, Zhang G, Li B. Tunable thermal conductivity of Si_{1-x} Ge_x nanowires. Appl.Phys.Lett 2009;95, 073117.

[60] Wang Z, Mingo N. Diameter dependence of SiGe nanowire thermal conductivity. Appl. Phys. Lett 2010;97, 101903.

[61] Morelli D.T., Heremans J.P, Slack G.A. Estimation of the isotope effect on the lattice thermal conductivity of group IV and group III-V semiconductors. Phys. Rev B 202;66, 195304.

[62] Peierls R. E. Quantum Theory of Solids New York: Oxford University Press; 1955.

[63] Sharma P.C, Frank Rose M. Three-phonon scattering processes and their role in phonon thermal conductivity of silicon. Journal Solid State Chemistry 1998;73, 92.

[64] Klemens P.G. Solid State Physics, Vol 7, p.1 New York: Academic; 1958.

[65] Slack G. A. Thermal Conductivity of CaF2, MnF2, CoF2, and ZnF2 Crystals. Phys. Rev. 1961;122, 1451.

[66] Asen-Palmer M. Bartkowski K, Gmelin E, Carona M, Zhernov A.P, Inyushkin V.A, Taldenkov A, Ozhogin V.I, Itoh K.M, Haller E.E. Thermal conductivity of germanium crystals with different isotopic compositions. Phys. Rev. B 1997;56, 9431.

[67] Heron J. S, Fournier T, Mingo N, Bourgeois O. Mesoscopic Size Effects on the Thermal Conductance of Silicon Nanowire. Nano Lett. 2009;9, 1861.

[68] Soffer S.B. Statistical Model for the Size Effect in Electrical Conduction. J. Appl. Phys. 1967;38, 1710.

[69] Omar M.S, Taha H.T. Lattice dislocation in Si nanowires. Physica B 2009;404, 5203.

Chapter 10

Hybrid Silicon Nanowires: From Basic Research to Applied Nanotechnology

Muhammad Y. Bashouti, Matthias Pietsch,
Kasra Sardashti, Gerald Brönstrup,
Sebastian W. Schmitt, Sanjay K. Srivastava,
Jürgen Ristein, Jordi Arbiol, Hossam Haick and
Silke Christiansen

Additional information is available at the end of the chapter

1. Introduction

Silicon nanowires (SiNWs) have attracted particular attention in nanotechnology applications due to their unique advantages with respect to electrical, optical and thermoelectric properties compared to the planar thin films [1-3]. SiNWs show strong absorption of visible light in thin layers, which makes them attractive for the use in novel thin film concepts as in photovoltaics [4]. Like all the one-dimensional (1D) nanostructures, SiNWs have a large surface to volume ratio and thus surface-dominated properties that can be tuned, e.g. by suitable surface functionalization to be applied in sensitive sensors [5, 6]. Surface functionalization affects the overall electrical properties of SiNWs as dopants surface concentration, surface recombination rate, density of surface states, etc.[2, 7-10]. Atop Silicon atoms on SiNW surfaces can be terminated with a wide variety of molecules bearing covalent interfacial bonds i.e. Si-C [2, 7-10], Si-N [11] and Si-O [12]. Up to now we essentially exploited the Si-C bonds that permit a thorough control of the SiNW surface yielding a strongly reduced tendency to the undesirable oxidation of the SiNWs [2]. With such surface functionalization methods at hand, control and prediction of the electrical transport properties have become so viable that functionalized SiNWs can be considered as auspicious nanoscale building blocks in future high-performance nano-devices in the areas of electronics, opto-electronics, photovoltaics and sensing [13-18]. In this chapter we present a chemically-based surface functionalization method based on Si-C bond surface termination that permits to fine-tune the electrical properties of VLS-grown SiNWs [2, 7-10, 19].

To date, electrical properties of SiNWs were controlled by doping with either boron (p-type) or phosphorus (n-type) [20]. Electrical transport measurements indicate a reduced resistivity according to higher doping levels and a tunable conductivity of doped SiNWs depending on the gate potential (V_g) applied. Although changes of V_g provide temporary control of SiNW conductivity, a more permanent and robust method is desirable for SiNWs to be useful in nanoelectronic device applications.

A promising approach to control the electronic properties of SiNWs is to use hybrid monolayers (either polar or nonpolar molecules) that affects the surface state density and displays an electrical potential on the surface/interface which modify the work function, electron affinity, surface Fermi level, and band offset/band bending at an interface [21-23]. This functionalization effect is a general one and can be obtained with non-molecular treatments as well [24]. Nevertheless, the use of molecules, especially organic ones, allows a systematic tuning of the desired dipole moment by an appropriate choice of their functional groups [25] or the intermolecular interactions between self-assembled molecules [10]. A surface functionalization of SiNWs with organic functionalities can be as important as the effect of the diameter and the wire orientation. In fact, due to the very high surface-to-volume ratio, the functionalization of the surface and the immediate surroundings of the wire is expected to have a dominant impact on its properties. Some computational studies have shown the importance of passiveness in SiNWs [26, 27]. Similarly, the large surface-to-volume ratio in SiNWs can be exploited in SiNW functionalization and used to modify transport properties. SiNWs tend to form a layer of SiO_2 with thickness 1–2nm under ambient conditions. Subsequent HF treatment can be used to remove this oxide layer. Different Si surface treatments can be performed to passivate SiNW surfaces with organic functionalities and obtain desired electrical surface properties [2, 7-9].

2. Definition: What hybrid material means?

Although the term "hybrid material" is used to describe many diverse systems straddling a widespread area of different materials, we restrict it by defining the concept as a composition and structure with/without interactions between the inorganic and organic units. In general we can distinguish between two possible classes. *Class I* hybrid materials are those that show weak interactions between the two phases, such as van der Waals, hydrogen bonding or weak electrostatic interactions. *Class II* hybrid materials are those that show strong chemical interactions between the components as in is our case.

2.1. Artificial strategies towards hybrid materials

There are two main approaches which are used to form hybrid materials:

1. Building block approach: A well-defined preformed matrix is applied to react with building blocks to form the final hybrid material. Regularly, the matrix consists of at least one

functional group that allows an interaction with the building blocks. A representative example of this approach is the bonding of organic molecules to silicon surfaces.

2. In situ formation of the components: One or more structural units are formed from the precursors that are transformed into a novel network structure. In these cases well-defined discrete molecules are transformed to multidimensional structures, which often show totally different properties from the original precursors. Typical examples of this approach are polymer reactions and Sol-Gel techniques.

We use the building block approach since it has a main advantage associated with the *in situ* formation in which the structural unit will remain well-defined and usually does not suffer from significant structural changes during the hybrid-matrix formation. Additionally, the building blocks or the matrix can be designed independently in such a way to give the best performance and to be designed to our desire. For example polar or non-polar molecules can be attached to the silicon matrix for sensing application or for FET's applications.

2.2. Chlorination/Alkylation process

The organic building blocks were connected to SiNW surfaces following the building-block approach. We report on the functionalization of SiNWs with alkyl chains using a versatile two step chlorination/alkylation process [8]. The two step process found to be gentle in the sense that it did not break the matrix (i.e.) SiNWs or change their diameters.

The SiNWs were terminated with alkyl chains using a two-step chlorination/alkylation route (see Figure 1). Before any chemical treatment, each sample was cleaned by N_2 flow. After being cleaned, H-terminated SiNWs were prepared by etching the amorphous SiO_x coating by exposing the SiNWs to buffered HF solution (pH=5) for 60 s and then NH_4F for 30 s. The samples were then rinsed in water for <10 s to limit oxidation, dried in flowing N_2(g) for 10 s, and immersed into a saturated solution of PCl_5 in C_6H_5Cl (0.65 M) that contained a few grains of $C_6H_5OOC_6H_5$ that act as a radical initiator. The reaction solution was heated to 90-100 °C for 5-7 min. Then the sample was removed from the reaction solution and rinsed in tetrahydrofuran (THF) followed by rinsing in methanol and then dried under a stream of N_2(g). The Chlorine (Cl) terminated SiNWs were alkylated by immersion in 0.5 M alkyl Grignard, R-MgCl, in THF, with R= methyl (CH_3; hereinafter C1), ethyl (CH_3CH_2; hereinafter C2), propyl ($CH_3(CH_2)_2$; hereinafter C3), butyl ($CH_3(CH_2)_3$; hereinafter C4), pentyl ($CH_3(CH_2)_4$; hereinafter C5), or hexyl ($CH_3(CH_2)_5$; hereinafter C6), or Octyl($CH_3(CH_2)_7$; hereinafter C8), or Nonyl($CH_3(CH_2)_8$; hereinafter C9), or Decyl($CH_3(CH_2)_7$; hereinafter C10).

The reaction was performed for 5 to 4320 min (72 h) at 80 °C. At the end of the reaction time, the samples were removed from the reaction solution, rinsed in THF and methanol, and dried under a stream of N_2(g). 2D Si (100) surfaces were alkylated following the procedure mentioned above, but with two minor differences:

i. the samples were chlorinated (i.e., immersed in the hot solution of PCl_5) for 60 min

ii. the alkylation time for all samples was 24 h.

These conditions were found to give the maximal coverage of alkyl chain on 2D Si(100). Alkylation of 2D Si(100) for 48 and 72 h gave similar coverage to those processed for 24 h.

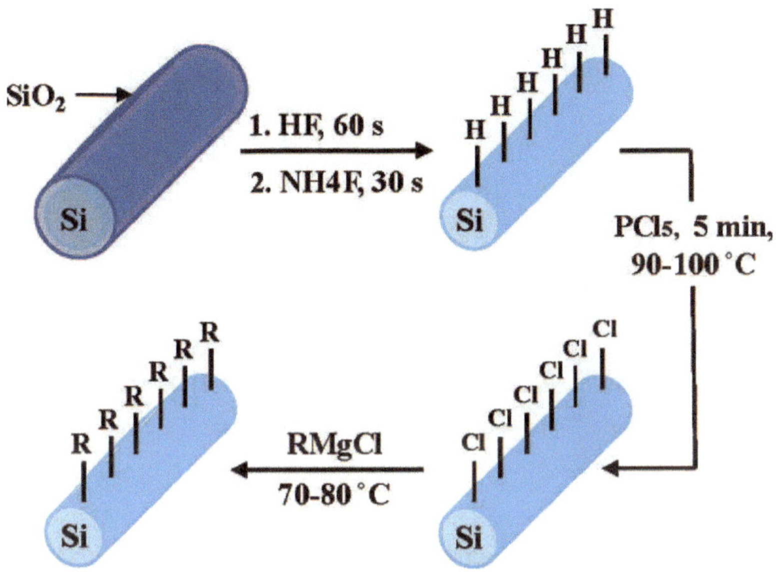

Figure 1. Scheme for illustrating the functionalization of SiNWs through the two step chlorination/alkylation process.

3. Fabrication of silicon nanowires

The SiNWs discussed in this chapter are grown by the well-known vapor-liquid-solid (VLS) mechanisms and a short description of the SiNWs growth will be given [28]. Si and Au form an eutectic that melts at 363°C while they individually melt at temperatures well above 1000°C. To grow SiNWs small gold droplets are dispensed on a Si wafer and inserted in a chemical vapor deposition (CVD) chamber. Then the samples heated up to 500°C i.e. above the eutectic temperature and below the melting points of each element. Si is supplied in vapor phase using silane (SiH_4) as the precursor. Si atoms enter the Au droplets causing their supersaturation of Si and the subsequent deposition of excess Si at the droplet-substrate interface. The VLS growth mechanism has its name due to the different states of the Si during the process. Different doping of the SiNWs can be obtained by introducing doping gases during the growth. Additionally, by the right choice of nanoparticle size, SiNWs diameters ranging from a few hundred nanometers down to a few nanometers can be obtained. The parameters of the CVD processes that are used for the SiNWs discussed in this paper are: growth pressure of 0.5 to 2.0 mbar, heater temperature of between 600°C and 700°C (since the heater was not in direct contact with the sample later temperature was considerably lower) and silane flow rate of 5 sccm using 5 to 10 sccm Ar as carrier gas. The grown SiNWs exhibit random orientation and usually gold nanoparticles on their tips (Figure 2).

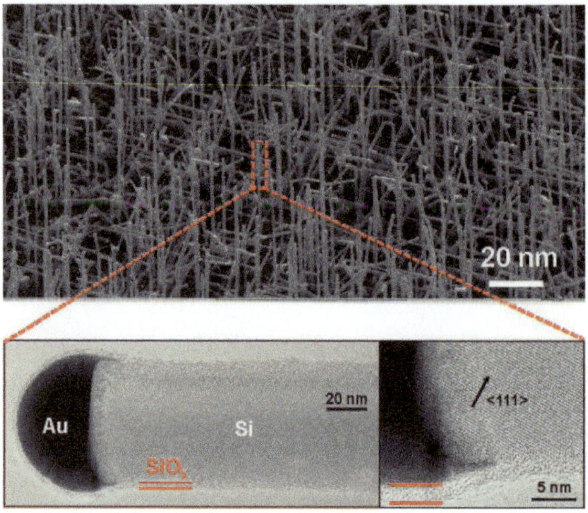

Figure 2. SEM image of SiNWs prepared by CVD on Si(111) substrates (upper part) and TEM images of an individual SiNW (lower part). The images indicate that the studied SiNWs consists of a 50±10 nm diameter Si core coated with 1-2nm of native SiO_x shell.

The Au nanoparticle from the catalyst is very mobile and diffuses over the SiNW surface (and inside the bulk regime).

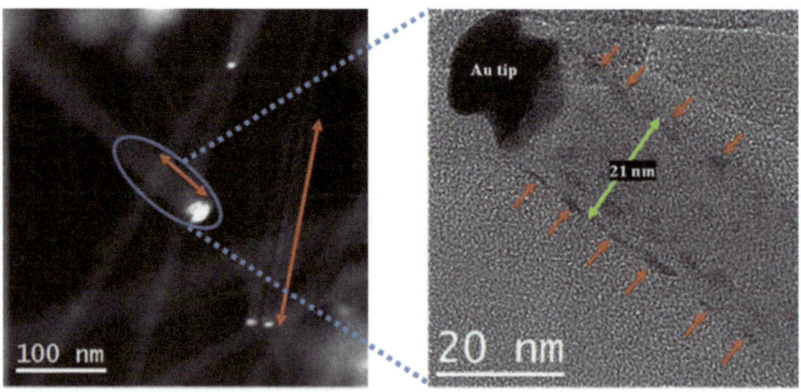

Figure 3. (a) HAADF-STEM general view of SiNWs, the red lines show the Au nanocluster extension. (b) HRTEM details of the selected area in Figure 3a.

This effect is enhanced in thin NWs and it was found from our previous report that Au nanoparticles can accelerate the oxidation on SiNWs [29]. 25-nm SiNWs show Au in the NW (in brighter contrast in Fig 3). Notice that Au is not only localized on the catalyst tip, but extends up to 300 nm apart from the tip. Figure 3a correspond to the HAADF-STEM general view of the NWs, the red lines show the Au nanocluster extension. Fig 3b corresponds to HRTEM details of the marked area in Fig 3a. In this case we have clearly observed that the presence of Au clusters (darker contrast), enhances oxidation rate in the thinner NWs relative to the thicker ones.

4. Molecular functionalization of CVD-SiNWs

Before any chemical treatment, SiNWs show a native oxide as illustrated previously in Figure 2. The oxide layer can be detected by X-ray photoemission spectroscopy (XPS) of the oxide core level spectrum. The Si2p of the oxide peak appears in the 101-104 eV binding energy range. In order to attach molecules to the SiNW surface, the native oxide has to be removed. This can be done by shortly immersing the SiNWs in HF and NH_4F solution. The XPS survey of the as-prepared H-SiNWs proved the thorough removal of oxides by the absence of any peak in the 101-104 eV as can be clearly seen in Figure 4.

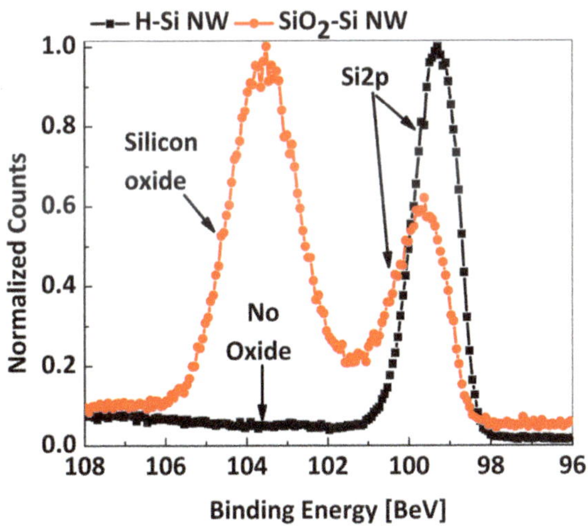

Figure 4. Si_2p emission of SiO_2-SiNW and H-SiNW.

It should be noted that the Si2p emission of SiO_2-SiNW is shifted towards higher binding energy as will be explained in the solar cell section. The emission of the C1s regime before and after termination is also compared in order to follow the covalent attachment of the molecules on the surfaces. Comparing the two emissions, an additional peak has to be introduced for molecule-terminated SiNW surfaces to obtain an appropriate deconvolution (figure 5). Figure 5 depicts the XPS spectrum of C1s emission region for the hydrogen- and alkyl-terminated SiNWs. H-SiNWs' spectrum was fitted to two C-C at 285.20±0.02 eV, and C-O at 286.69±0.02 eV while the alkyl-SiNW was fitted to three peaks: C-Si at 284.11±0.02 eV, C-C at 285.20±0.02 eV, and C-O at 286.69±0.02 eV. The new peak at C-Si at 284.11±0.02 eV confirm the chemical bonding of organic molecules onto the SiNW surface. Peaks were typically adjusted to produce fits that minimized the difference between the full widths at half-maximum (fwhm).The center-to-center distances were fixed at 1.10±0.10 eV between the C-Si and the C-C emissions and at 2.60±0.10 eV between the C-O and the C-Si emissions. To attain the molecular coverage of surface atomic sites, the area under the C-Si peak is divided by the one under the Si2p peak (sum of $Si2p_{1/2}$ and $Si2p_{3/2}$) and normalized by the scan time. The coverage of the alkyl functionalities was compared with the highest value obtained for methyl (C1) functionality; in this instance "highest" means nearly full coverage of the Si atop sites, achieved after alkylation times of >24 h. This comparison is expressed throughout the text as "$(C\text{-}Si/Si2p)_{alkyl}/(C\text{-}Si/Si2p)_{max.methyl}$". Occasionally, a small signal due to oxygen was observed at 532.02±0.02 eV (O1s) and assigned to adventitious adsorbed hydrocarbons having oxygen bonded to carbon (286.69±0.02 eV) as shown before and after alkyl treatment. It is reasonable to assume that these adventitious hydrocarbons could stem from the wet chemical processing with THF solvent, THF/methanol rinse after functionalization, and/or carbonaceous materials present in the laboratory environment.

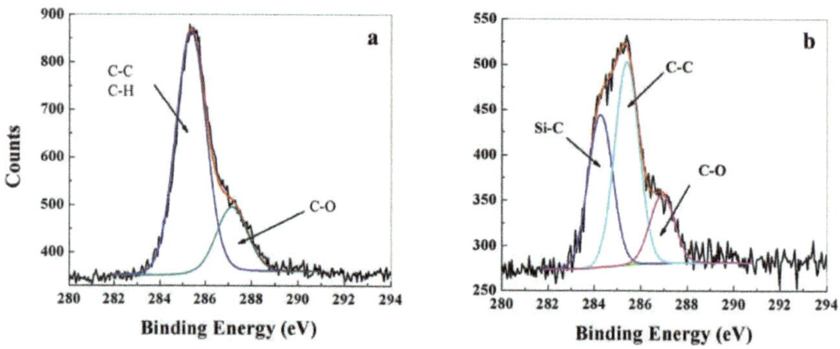

Figure 5. XPS data from the C1s emission region of the hydrogen and alkyl-terminated SiNWs. (a) H-SiNWs show two peaks: C-C at 285.20±0.02 eV, and C-O at 286.69±0.02 eV. (b) Alky-SiNWs, show three peaks: C-Si at 284.11±0.02 eV, C-C at 285.20±0.02 eV, and C-O at 286.69±0.02 eV.

The absence of SiO_x signal in the Si2p spectral surveys supports the argument that the O1s signal arises from physical bonding of hydrocarbons. Moreover, lack of F1s signal (which appears at 686.01±0.02 eV binding energy), confirmed that the $NH_4F_{(aq)}$-etched Si surface was not terminated by Si-F species.

Figure 6 shows the "$(C\text{-}Si/Si2p)_{alkyl}/(C\text{-}Si/Si2p)_{max.methyl}$" ratio as a function of alkylation time. As shown in the figure, the longer molecular chain requires a longer time for alkylation. For example, the time required to achieve 92±3 % of the saturation level in the adsorption curve,

Γ_{sat}, for C1 functionality, stands at 20±2 min. Increasing the chain length to C6 increases Γ_{sat} to 120±10 min, while the maximum coverage that could be achieved is merely 56±10 % of C1-SiNWs. Longer alkylation times, up to 24 h, changed the total coverage by 5±3 %. Alkylation times of 24 to 72 hours did not further change the coverage on SiNWs and planar Si(100).

Table 1 summarizes the Γ_{sat} values for many of the molecules presented in our former studies, along with the related $(C\text{-}Si/Si2p)_{alkyl}/(C\text{-}Si/Si2p)_{max.methyl}$ ratios. As observed in the table, C2-SiNWs show a "$(C\text{-}Si/Si2p)_{ethyl}/(C\text{-}Si/Si2p)_{max.methyl}$" value of 68±3 %, indicating that C2 groups can be packed at a relatively high density without major steric effects. C3, C4, C5, C6, C8 and C10 produced 56±5 %, 49±5 %, 50±10 %, 56±10 %, 56 ±10 %, and 80±10 % coverage, respectively, much higher than those of the same molecules on 2D Si (100) substrates (30±20 %,). Furthermore, the time required to achieve maximum coverage of the molecules on SiNWs is 4-30 times shorter than that of the 2D substrates. Using simple geometrical considerations, the gain in the steric hindrance on surfaces of 50±10 nm (in diameter) SiNWs is marginal. This cannot explain the higher molecular coverage on SiNWs, compared to 2D Si(100). Indeed, gain in steric hindrance between the adsorbed organic molecules starts to be important in NWs with <10-20 nm in diameter. This conclusion claims that the observed differences in coverage are related to different kinetics and higher available reactive sites on SiNWs, compared to 2D Si(100). The observation of 0.11±0.02 eV higher binding energy for a specific alkyl-terminated SiNWs, compared with equivalent 2D Si(100) surfaces, could further support the higher reactivity of SiNW atop sites. The broadening of Si2p fwhm in the SiNWs case suggests, compared to 2D Si(100), that wider distribution of bond strengths exist on the surface, thus leading to a wider distribution of reactive sites [30].

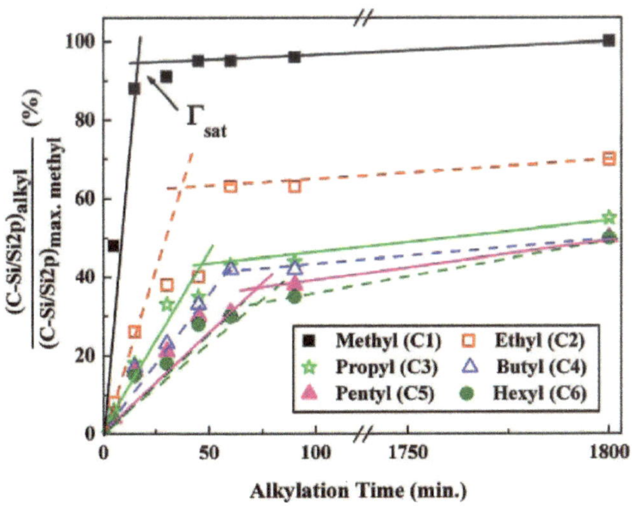

Figure 6. Coverage of alkyl molecules of SiNW atop sites as a function of alkylation time. Γ_{sat} stands for the time required to achieve92± 3 % of the saturation level of the adsorption curve.

		Silicon Nanowire (SiNW)		2D Si(100)
Alkyl molecule	Γ_{sat}(min)	C-Si/Si2p	Max. Coverage	Max. Coverage
C1	20±2	0.135±0.001	100±5%	100±5%
C2	50±10	0.093±0.002	68±5%	60±20%
C3	60±10	0.075±0.003	56±5%	30±20%
C4	65±10	0.066±0.004	49±5%	30±20%
C5	90±10	0.068±0.005	50±5%	30±20%
C6	120±10	0.076±0.006	56±5%	30±20%
C8	"/>120	0.075±0.007	56±5%	30±20%
C10	"/>120	0.108±0.009	80±5%	30±20%

Table 1. Summary of the XPS results for C1-C6 alkyl chains bonded to SiNWs and 2D Si(100) surfaces via Si-C bond

4.1. Early stages of oxide growth in H-SiNWs

To examine the thermal stability of H-SiNWs, the native oxide shell on SiNWs was etched away by hydrofluoric acid and ammonium fluoride. Subsequently, freshly etched SiNWs were annealed in ambient conditions (with average humidity of 30 %) in seven distinct temperatures of 50 °C, 75 °C, 150 °C, 200 °C, 300 °C, 400 °C and 500 °C, each for five different time-spans: 5 min, 10 min, 20 min, 30 min and 60 min. After each heating stage, the specimens were examined by XPS. To identify various sub-oxide states in the Si2p spectrum, spectral decomposition was performed by a curve-fitting procedure, using Gaussian-Lorentzian functions after a Shirley background subtraction as shown as example in Figure 7. The sub-oxides states and their respective chemical shifts (Δ) relative to the $Si2p_{3/2}$ (at 99.60± 0.02 eV) are as follows: $Si2p_{1/2}$ (Δ=0.60 eV), Si_2O (Δ=0.97 eV), SiO (Δ=1.77 eV), Si_2O_3 (Δ=2.50 eV), and SiO_2 (Δ=3.87 eV). The shift in binding energy is dependent on temperature and time (dashed line in Figure 7a). Generally speaking, upon the formation of sub-oxides and SiO_2, the Si2p peak moves to higher binding energies to compensate the band bending and surface Fermi level shift as the oxide layer grows (not shown). The intensity of a sub-oxide (I_{ox}) is expressed by the ratio of the integrated area under its peak to the Si2p peaks overall area ($Si2p_{3/2}$+$Si2p_{1/2}$) i.e. the total intensity of unoxidized silicon atoms. Sub-oxide intensity can be expressed in terms of the number of monolayers formed corresponding to the intensity of 0.21 per monolayer [31].

Figure 8 depicts the variation of the average number of monolayers over time for individual sub-oxides (I_{ox}) and the overall oxide intensity ($I_{tot}=I_{Si2O}+ I_{SiO}+I_{Si2O3}+I_{SiO2}$) for low-temperature (T < 200 °C, Figure 8a) and high-temperature (T ≥200 °C, Figure 8b) oxidation of H-SiNWs. Comparison of the sub-oxides growth in Figure 8 illustrates their temperature and time dependence. In both temperature ranges, Si_2O is present even in the etched surfaces with intensity of I_{ox}=0.02±0.01 i.e. ~0.2 monolayer and its intensity plateaus at the value of I_{ox}=0.06±0.01 (~0.25 to 0.3 monolayer) at the very early stages exhibiting negligible growth over the whole time range. SiO and Si_2O_3 exhibit approximately identical behavior and values in both ranges despite the plateau region occurs 15 minutes earlier in the high temperature range. They exhibit no peak at t=0 and an intensity of 30-60 % of the maximum intensity after 5 min and 20 min for high and low temperature regime, respectively. However, at T ≥ 200 °C, SiO and Si_2O_3 have 0.2 monolayer coverage that is twice as large as coverage in T < 200 °C (~0.1 monolayer). The behavior of the transient sub-oxides versus oxidation time is similar to that of SiNWs exposed to molecular oxygen wherein the intensities of the transient sub-oxide almost remain constantly below one monolayer. Conversely, SiO_2 intensity was found to strongly rely on the temperature and time and two regimes of fast and slow growth were clearly recognized.

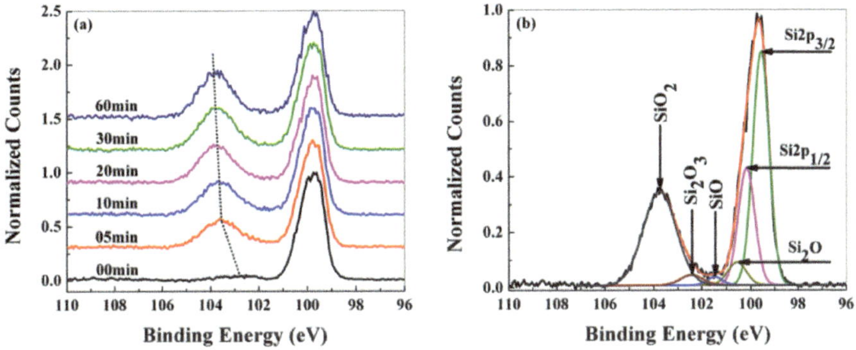

Figure 7. (a) An example of Si2p XPS spectra of the H-SiNW surfaces annealed at 500 °C for different times; the inclined dashed line represents the blue shift upon formation of the oxide layer. (b) Decomposed Si2p sample spectrum of oxidized SiNWs illustrating four different suboxides including Si_2O, SiO, Si_2O_3 plus the stoichiomertic SiO_2 at 100.6 eV, 101.7 eV 102.5 eV and 103.4 eV, respectively.

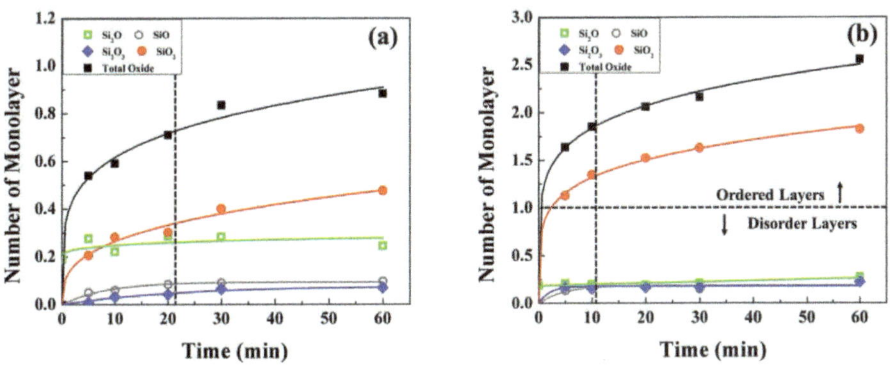

Figure 8. Number of monolayers over time for the sub-oxides Si_2O (green), SiO (gray), Si_2O_3 (blue), SiO_2 (red) and their sum as total oxide (black) in the two low-temperature (a) and high-temperature (b) regimes. The vertical dashed lines represent Γ_{sat}, while the horizontal one in (b) defines the onset for ordered oxide layer formation.

At higher temperatures ($T \geq 200$ °C), the time required to achieve 70±10 % of the final intensity (defined as Γ_{sat}) for SiO_2 is 10±2min, while at lower temperatures ($T<200$ °C), Γ_{sat} shifts to 18±2min. On the other hand, Γ_{sat} for the total oxide amount lies approximately on the same times as SiO_2 demonstrating its sizable weight in the total oxide amount. Therefore, temperature dependence of oxide growth can be well observed in Figure 9a that depicts the total oxide vs. time for all tested temperatures. According to the TEM measurements the extension of the annealing period at 500 °C from 1 hour to 24 hours results in an increase of oxide layer thickness from 4.5nm to 6nm. Thus, 23 hours of further annealing leads to less than 35 % increase in oxide thickness that is noticeably smaller than the growth observed at the starting times and the graph after Γ_{sat} can be assumed as a plateau like sub-oxides plots.

Early stages of high temperature oxidation in H-SiNWs can be modeled by the modified Deal-Grove model taking into account partly or completely sub-oxides as growth sites with varying concentrations over time and temperature [32, 33]. Nonetheless, above Γ_{sat} a linear Deal-Grove regime cannot be accommodated with the approximately flat tail of the SiO_2 plot and a limiting mechanism is required to interpret such a behavior [34]. Between Γ_{sat} and maximum oxidation (1 hour) at high temperatures, a significant fraction of the superficial Si atoms are surrounded by fourfold occupied oxygen sites with high local strains that act as a barrier for further oxidation. Under this conditions, oxidation hindrance can be caused by two phenomena: self-limiting oxide film formation (seen in 2D Si) [32] and self-retarding oxidation known for SiNWs and their particular geometry [35]. However, self-retarded oxidation becomes more significant on curved surfaces like SiNWs at diameters below 35 to 44 nm due to the extremely small amount of viscous flow in the oxide layer that could relax the applied stresses [36]. On the other hand, repetition of the same measurements for Si2D resulted in the same flat end for the I_{ox}-t plot at $T \geq 200$ °C (Figure 9a and 9b). Therefore, the inhibited oxide growth in the high temperature region is better understood as a result of self-limited oxidation rather than geometrical compressive that can be neglected for the current NWs. Non-retarded growth of oxide at 500 ˚C (Figure 9a) can be associated with partial melting at NW surfaces causing accelerated oxygen diffusion and NWs fragmentation (6 times length decrease) which led to deeper oxygen penetration and larger interfacial areas (Figure 9b and c).

At low temperatures Γ_{sat} of SiO_2 is shifted to longer times (18±2min) and transient sub-oxides are the main sources for the total oxide intensity at times between 0 and Γ_{sat} (Figure 8b). Based on our earlier work, heating above 450 ºC can remove all the Si-H bonds on flat Si wafers [37]. At lower temperatures and initial times (i.e. before $<\Gamma_{sat}$), the oxidation occurs in the presence of functionalizing H atoms and OH groups (resulted from water vapor interaction with the surfaces). Therefore, oxidation proceeds thorough Si-Si backbond oxidation forming sequentially peroxidic and disiloxanic bonds in the first few atomic layers. Nevertheless, heating in the atmosphere of 30 % average humidity, water assistance to early oxide growth and corresponding oxidation-hydroxylation cannot be neglected [38, 39]. To understand the relation between the rate-determining step and the surface bonds strength, we studied methyl-termination of

SiNWs, which naturally provides the same coverage as hydrogen functionalities but with Si-C bonds, instead [19, 30]. Interestingly, when exposed to air at room temperature for 44 days, SiO_2 formation was relatively suppressed, while Si_2O intensity was equal to the one for H-SiNW [8, 40]. After the 44-day period, methyl-terminated SiNWs showed I_{SiO2} intensity of 0.11; it almost equals that of H-SiNWs after 1 hour. This can be explained by the lower susceptibility of Si-Si underlying bonds in Si-C covered surfaces compared to Si-H covered ones [30]. For further oxidation at lower temperatures however, the electron or hole tunneling mechanism proposed for Si(100) room temperature oxidation encounters difficulties to be proven considering the fact that neither individual sub-oxides nor total oxide amount reaches the level of a single monolayer [41]. Alternatively, rupture of the surface bonds may act as the second oxidation source after local development of the backbond Si-O. In this respect, in methyl-terminated SiNWs monotonic growth of sub-oxides and SiO_2 has been observed after a dramatic decline in the amount of Si-C concentration. Similarly, bond propagation can be conceived in H-SiNWs with a relatively lower resistance to the oxidizing species [30].

Figure 9d shows the temperature dependence of the total oxide formation rate in the 2D Si and H-SiNWs specimens before Γ_{sat} for all the temperatures investigated in this study. High and low temperature data points are fitted in two different lines for each specimen. Near-linear behavior of oxidation rate logarithm vs. reciprocal of temperature at the early stages of oxidation allows us to calculate respective activation energies (E_A^{ox}) based on the Arrhenius equation. In the high temperature region 2D Si and SiNWs exhibit E_A^{ox} of 46.35 m eV and 48.22 m eV, respectively. Such a minor difference in the activation energy magnitudes denotes again that geometrically-induced stresses affect negligibly the rate-determining step. Large divergence from values of activation energies for the linear Deal-Grove model (1.83 to 2.0 eV) approves that the rate is rather determined by the growth sites development than by surface oxidation reaction. In addition to sub-oxide islands, near-interfacial remnant silicon atoms may contribute to the oxidation process as growth sites [33, 42]. Moreover, higher absolute intensities and oxidation rate values for SiNWs at 500 °C beyond the first 10 minutes compared to 2D Si necessitate the existence of a growth-site-enhancing mechanism as partial melting of SiNWs. This mechanism improves the oxygen permeability to deeper layers and increases the interfaces due to NW fragment formation (Figure 9c).

For the low temperature regime E_A^{ox} values of 35.20 m eV for 2D Si and 23.31 m eV for SiNWs were also calculated. Lower activation energies for SiNWs may correspond to the strain induced by dangling bonds rebonding at the (110) and (111) adjoining surfaces as well as higher ratio of ambient exposure area to the oxide/silicon interface area [43]. Stretching weakens the very first layers of silicon backbonds and results in their higher susceptibility to bond rupture and suboxide formation. Conversely, bending stresses are able to increase robustness of hydride terminations (di- and trihydrides) on the surface and lower the SiNW oxidation rate at higher temperatures where surface bond propagation determines the overall rate (for 75 °C and 150 °C where rates are lower than 2D Si).

All the above descriptions suggest the following mechanism (schematically shown in Figure 10): (i) At low temperatures, Si-Si backbond oxidation and later Si-H bond propagation are the rate-determining steps. (ii) At higher temperatures, the oxygen diffusion is considered as the rate-determining step initially through controlling the growth sites concentration (conserving some Si-H interfacial bonds) where disorder monolayer still dominates the oxide intensity. Beyond Γ_{sat}, the reaction shifts to almost zero-order which denotes a self-limiting process because initially formed oxide shells prevent a further oxidation. A fraction of the Si-H bonds converted to Si-OH according to H_2O molecules present in the surrounding atmosphere.

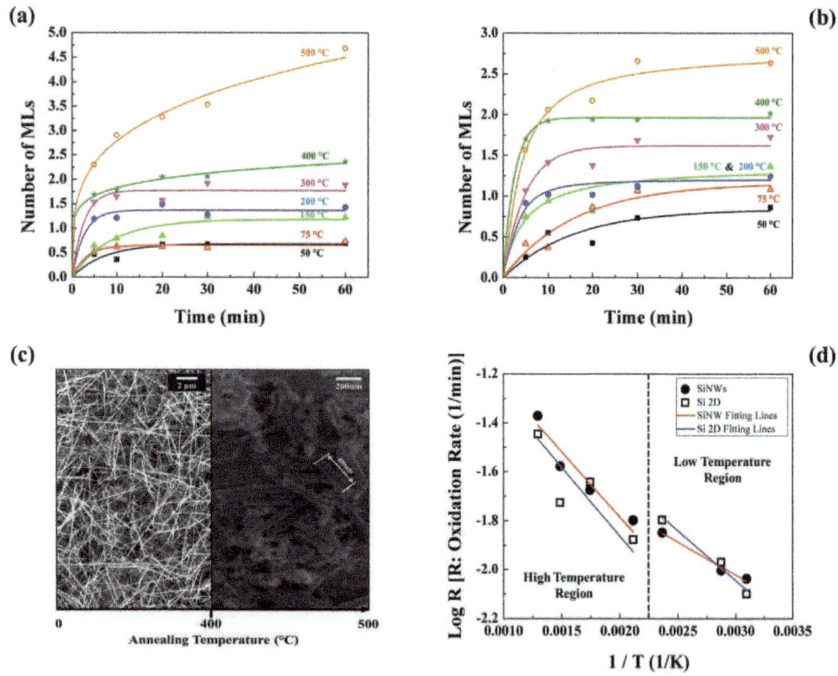

Figure 9. Oxide monolayer growth over time for H-SiNW surfaces (a) and 2D H-Si (b) oxidized at different temperatures for time spans ranging from 5 to 60 min. (c) Comparison between pristine (right) and melted (left) SiNWs. (d) Temperature dependence of oxide formation in H-SiNWs and 2D Si in air between 50°C and 500°C. For all temperatures, the annealing time is below Γ_{sat}.

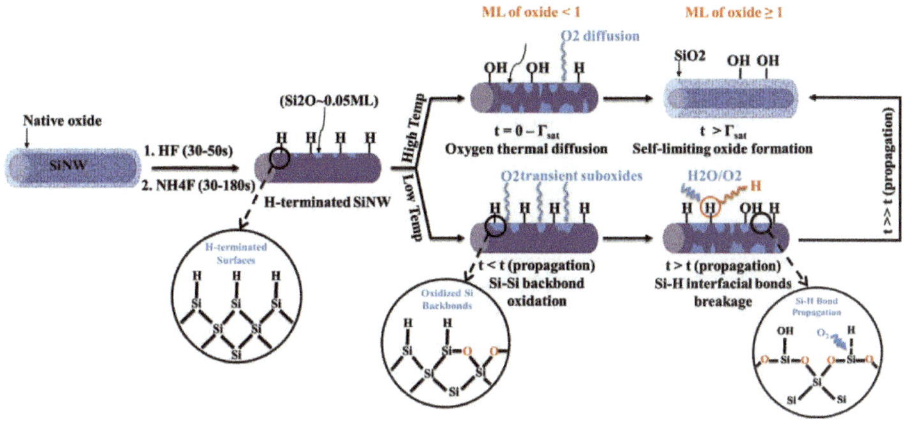

Figure 10. Schematic graph summarizing the inferred oxidation mechanism for H-SiNWs. The blue spots represent the sub-oxide islands. Interfaces at different stages of low-temperature oxidation as well as initial H-terminated one sare depicted in more detail in circles.

4.2. Stability of molecules on SiNWs in long time

To investigate the stability of the alkylated SiNWs, freshly prepared SiNWs having maximum coverage of alkyl functionalities (mostly obtained after alkylation time of 24 h) were exposed to ambient air for several hundreds of hours (Table 2). Oxide intensity in SiNWs functionalized by methyl and ethyl equals zero up to 48 hours. After 336 h of exposure, oxide intensity reached 0.03 (~0.14 monolayer) for C1-SiNWs and 0.05 (~0.23monolayer of oxide) for C2-SiNWs. In contrary, after 24 hours of exposure, oxide intensity in SiNWs terminated with C3 to C6 rose to 0.01 (~0.05 monolayer of oxide), 0.02 (~0.10 monolayer of oxide), 0.02 (~0.10 monolayer of oxide), and 0.01 (~0.05monolayer of oxide), respectively. This implies that SiNWs terminated with C3-C6 alkyl molecules exhibit less stability against oxidation compared to C1- and C2-SiNWs. Nevertheless, after 336 hours, oxide intensity of C3- to C6-terminated SiNWs have approximately the same magnitudes (~ 0.13). From the data in Tables 1 and 2, with a wider perspective, it can be deduced that the shorter the alkyl chain, the lower the SiO_x/Si2p ratio and the less liable to oxidation the terminated SiNWs.

These observations can be explained by the fact that the shorter the molecular chains, the lower the coverage of the alkyl molecules and the higher the probability of interaction between oxidizing agents (O_2 and H_2O) and molecule-free sites (or pinholes).The oxidation degree of the SiNWs was compared to planar Si(100) surfaces alkylated with the same series of molecules. C1-SiNWs exposed to air over a period of 336 h illustrated ca.4-fold lower oxidation intensity than the equivalent Si(100) surfaces, despite alike initial coverage levels. Similarly, C2-SiNWs exhibited ca. 3-folds lower oxide intensity than C2-Si(100). In contrast, (C3-C6)-SiNWs oxide amount was comparable to those of planar (C3-C6)-Si(100) after 336 h. These observations could be attributed to stronger Si-C bond on SiNWs surfaces. This can be supported by the shift in Si-C energy bond of C1 terminations from SiNW (284.22±0.02 eV) to planar Si(100) (284.11±0.02 eV). Stronger bonding of surface functionalities has also been observed in H-terminated SiNWs. Respective studies suggest that the robustness of the SiNW's surface hydride (or alkyls) is a consequence of bending stresses being developed at the edge of two adjoining facets of a SiNW.

	exposure time to air			
Sample	0 hr	24 hr	48 hr	336 h
C1-SiNW	0	0	0	0.03
2D C1-Si(100)	0			0.11
C2-SiNW	0	0	0	0.05
2D C2-Si(100)	0	0.03	0.08	0.13
C3-SiNW	0	0.01	0.07	0.13
2D C3-Si(100)	0			0.15
C4-SiNW	0	0.02	0.07	0.13
2D C4-Si(100)	0			0.15
C5-SiNW	0	0.02	0.06	0.14
2D C5-Si(100)	0			0.16
C6-SiNW	0	0.01	0.06	0.12
2D C6-Si(100)	0	0.04	0.08	0.18

Table 2. Summary of the Oxidation intensity (SiOx/Si2P Ratio) for Alkylated SiNWs and 2D Si(100) Surfaces, Having Maximum Coverage (Mostly, Obtained after 24 hrs of Alkylation Time), at Different Exposure Times to Ambient Air.

4.3. Stability of Si-C bond on SiNWs

SiNW surfaces were terminated by CH_3 and compared with equivalent planar CH_3–Si surfaces having same surface coverage with Si–C bonds. Figure 11 shows the coverage of methyl functionalities as a function of alkylation time. Comparing the alkylation kinetics on SiNWs with 2D Si(100) showed two main notable differences: (I) 92±3 % of the maximum coverage on 2D Si(100) was achieved after 1000±50 min alkylation; (II) the reaction kinetics on 2D Si(100) at short and long alkylation times are rather different than on SiNWs. These differences in the kinetic behavior could be attributed mainly to differences in the surface energy. The cohesive energy becomes higher upon decreasing the dimensions of a given structure. Based on earlier work it can be seen that 50±10 nm (in diameter) SiNW surfaces has higher cohesive energy (~4.4 eV) than 2D Si(100) surfaces (~3.4 eV)[30, 44]. Furthermore, the broadening of Si2p fwhms in the case of SiNWs (data not shown), as compared to 2D Si(100) and 2D Si(111), suggests that wider distribution of Si–C bond strengths and/or ligands angle variations exists on the surface. Fig. 12 depicts variation of the SiO_2 intensity and molecular coverage over exposure time for both NWs and planar specimens of silicon. Since determining surface coverage requires accounting for escape length of photoelectrons, an "impartial" indicator was required to examine the robustness of the data presented in Fig. 12. With this in mind, ToF-SIMS experiments, by which the methyl coverage is proportional to the $SiCH_3^+$

(or $CH_3^{\cdot}$) fragments, can be carried out. ToF-SIMS results showed similar concentrations (within 5 % experimental error) of characteristic fragments in both CH_3–SiNWs and 2D CH_3–Si(100) substrates, marking the reliability of the XPS data presented in Fig. 11 and 12.

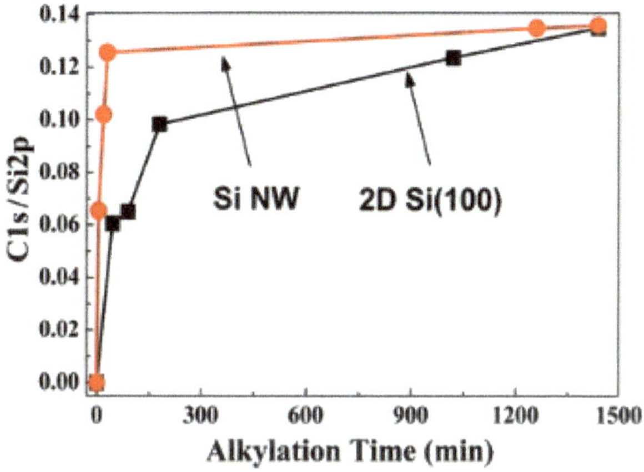

Figure 11. Coverage of methyl functionalities on 2D and SiNW as a function of alkylation time.

As shown in Fig. 12a, CH_3–SiNWs exposed to air over one month showed ca. 3-fold less oxidation than the equivalent 2D CH_3–Si(100) surfaces, albeit similar molecular coverage's in both structures. To attain a wider perspective, we made a comparison between 2D CH_3–Si(111) surfaces (naturally of 15–20 % coverage)and 2D CH_3–Si(100).

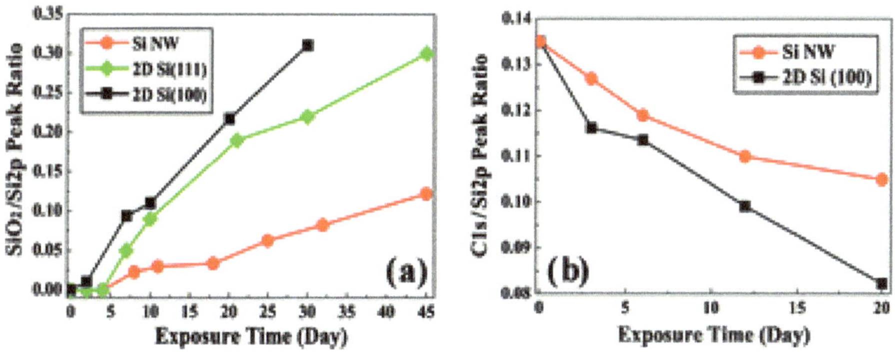

Figure 12. (a) Ratio of the oxidized Si2p peak area to the bulk Si2p peak area for the methyl modification of SiNWs, 2D Si(100), and 2D Si(111), exposed to air over extended time periods. (b) Si–C decay as a function of exposure time for SiNW and 2D Si(100). The measurements were done three times for each sample and averages have been taken.

Interestingly, in spite of notably lower coverage, 2D CH_3–Si(111) surfaces exhibit higher stability against oxidation than the equivalent 2D Si(100) surfaces, while they showed ca. 2-fold higher oxide intensity than the CH_3–SiNWs. This can also be referred to surface bond energy. The observation of ~0.11 ±0.02 eV higher binding energy for a CH_3–SiNW, compared to equivalent 2D CH_3–Si(100) surfaces, could further be ascribed to the higher reactivity of atop sites. Indeed, the CH_3–SiNW's C1s spectra showed 22 % decrease in the Si–C bond signal throughout 20 days, at the time which 2D CH_3–Si(100) specimen showed ~34 % decrease of the same signal (Fig. 12b). Surface bond strength dependency on structural geometry can be backed by the reports of Lee and co-workers on stronger H-termination of SiNWs relative to 2D Si surfaces suggested that the robustness of the SiNWs hydride is a consequence of bending stresses, where re-bonding of dangling bonds at the edge of two adjoining facets of a SiNW takes place [45].

	Oxidation time CH_3-SiNWs		Oxidation time CH_3-SiNWs	
Segment	0 days	4 days	0 days	4 days
Si-O	0.037±0.002	0.063±0.003	0.021±0.001	0.039±0.001
Si-C	0.142±0.006	0.091±0.009	0.074±0.001	0.039±0.001
Si-CH3	0.239±0.011	0.202±0.013	0.382±0.041	0.301±0.010
Si_2OC2H_6	0.004±0.001	0.008±0.001	0.004±0.0002	0.003±0.0002
SiO_2	0.071±0.002	0.139±0.002	0.030±0.0005	0.066±0.0003

Table 3. $SiC_xH_y^+$ peak intensities of ToF-SIMS spectra of SiNWs before and after 4 days exposure to an oxidizing atmosphere

To understand the relation between strength of the Si–C bond and the oxidation resistance, ToF-SIMS fragments of CH_3–SiNWs and 2D CH_3–Si(100) surfaces were collected after different exposure periods to the oxidizing agents. Table 3presents the selected ToF-SIMS spectrum peak intensities of CH_3–SiNW before and after 4-day exposure to pure O_2(20 %)–N_2(80 %) environment of 10–15 % relative humidity (RH). After 4 days of oxidation, Si_xO signal (where x = 1,2) of SiNW samples increased by 70–95±10 % while the concentration of Si–C and Si–CH_3 bonds decreased by 36±2 % and 15± 2 %, respectively. Higher changes in the Si_xO signal suggest higher molecule-free Si–Si backbonds susceptibility to bond rupture upon interaction with oxidant molecules, than the Si–C bonds. And higher changes in the Si–C (36±2 %) signal than Si–CH_3(15±2 %) indicates that Si-Si bonds adjacent to the CH_3 groups are less prone to oxidation than Si-C surface bonds. Moreover, the results point out the early breakage of Si–Si back-bonds to form Si–O–Si back-bonds. After oxidation of the entire Si–Si back-bonds, oxidizing species start attacking the Si–C bonds. For 2D CH_3–Si(100) surfaces, the Si_xO signal increased by 86–120±10 % while the concentration of Si–C and Si–CH_3 bonds decreased by 47±2 % and 21±2 %, respectively, after 4 days of exposure to the oxidizing atmosphere. Evaluating the relative changes obtained for 2D CH_3–Si(100) surfaces and CH_3–SiNWs calls for two main conclusions: (1) CH_3–SiNWs are more stable than 2D CH_3–Si(100) surfaces, in consistency with the XPS observation, (2) The oxidation mechanism of 50±10 nm (in diameter) CH_3–SiNWs is similar to that of 2D CH_3–Si(100) surfaces.

4.4. Effect of chain length of alkyl molecules

According to theoretical simulations, steric effects (van der Waals diameter) can hinder formation of dense packing of alkyls longer than C1. Therefore, increasing length of the alkyl chains increases the van der Waals diameter from 2.5 Å (in the case of C1) to more than 4.5-5.0 Å for longer alkyl chains. The latter van der Waals diameters are much larger than the internulecular distance between adjacent Si atoms (3.8 Å), a matter that decreases their adsorption rate and limits their coverage to maximum 50-55 % of a monolayer of Si surface sites. With these findings in mind, we found the chain length dependence in Figure 13 to be consistent with the similar decay of the coverage reported by others on flat Si(111) surfaces for longer chains than in the present work, with, no increase of coverage above 7 carbon atoms [46]. Based on these inconsistencies it can be inferred that the higher coverage obtained for >C6 chain lengths is artificial and/or not significant owing to the problems in fitting of the XPS data. In contrast, few experimental observations suggest that the coverage behavior of chains longer than C6 on SiNWs might not be artificial, as justified by the following explanations: (1) ToF SIMS experiments (which eliminate potential artificial observations) have shown higher absolute coverage long chain lengths, in good consistency with XPS observations. (2) In the case of 2D Si(100) surface, the results of ToF-SIMS have shown (more or less) the expected coverage versus chain length behavior; at the time, similar behavior was not observed in the case of SiNWs. (3) The experimental error and/or accuracy of the peak fitting are relatively small (4-11 %).

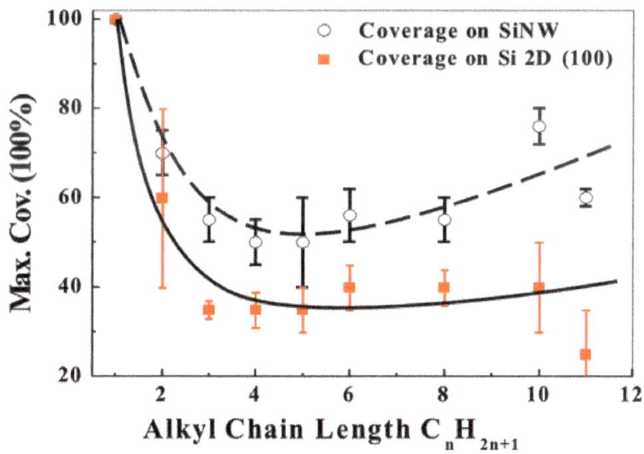

Figure 13. Max-alkyl versus alkyl chain length on SiNWs and, for comparison, on 2D Si(100) surfaces.

It is reasonable to consider passivation of SiNWs by alkyl molecules as a function of two main factors: (I) molecule-molecule lateral interaction and (II) molecule-substrate vertical interaction. For short alkyl chains (C1-C5), which exhibit liquid-like behavior and thermal fluctuations [47] the determining factor is the vertical interaction. Increasing the chain length to C6-C10 forms a solid-like phase, where the lateral interactions become more dominant in the consequent coverage of the SiNW surfaces. It is worth pointing out that the lateral interactions between long alkyl chains might be formed during the physisorption stage, before the covalent bindings between the carbon and silicon atoms (or chemisorption) arise [48].

5. Integration of hybrid SiNWs into technological devices

Hybrid SiNWs modified by covalent Si-C bonds showed excellent atmospheric stability, high tansconductance values, low surface-defect densities, and allowing formation of technically-feasible air-stable SiNW field-effect transistors (FETs) and SiNW-photovoltaics [9, 19]. Here we demonstrate the electrical characteristics of such SiNW-based devices molecularly modified by the two-step chlorination/alkylation method. Analyses suggest that extrinsic effects (i.e. surface states and surface charge) accompanied by the adsorption of molecular layers significantly affect the FET and solar cell characteristics.

5.1. Fabrication of the FET

SiNWs were manufactured through a top-down approach based on electron beam lithography rather than the bottom-up VLS growth. It initiates from a bonded SOI wafer, with a thin (100 nm) top Si (100) layer (resistivity of 10–20 Ωcm) insulated from the (100) silicon substrate (resistivity of 0.8–1.2 Ωcm) by a silicon dioxide layer (380 nm). Surface was covered by a bilayer PMMA and mask definition was then performed by high-resolution e-beam lithography. The resist was then developed in a solution of MiBK:IPA (1:3). The pattern was transferred from the PMMA to the top SiO_2 layer by buffered HF etching. The central region, where the silicon was defined, was linked through small connections to the device leads. Subsequently, a 35 wt % KOH solution, saturated with isopropyl alcohol (IPA), was used to etch a single NW in the central region. Due to the anisotropy in etching, NW exhibits a trapezoidal cross-section. After this step, the aluminum contacts were defined by means of a lift-off process. The SiNWs, after the silicon etching, had a length (L) of 1mm, a top width (w) of 70 nm, a height (t) of 100 nm, and trapezoid base (W) (of 200 nm, as illustrated in figure 14. The diagonal is based on (111) planes, whereas atop side is made up of a (100) plane. The available starting material had a resistivity of 10–20 Ωcm. At the end of the fabrication process, the Al electrodes were coated with PMMA to protect them from etching agents (e.g., buffer HF, NH_4F, etc.) during the alkylation process. To assess the electrical characteristics of the various SiNW FETs, voltage-dependent back-gate measurements were performed in a probe station (Cascade Microtech Femtoguard, Summit/S300-11741B-6, USA) under the ambient conditions. In these measurements, back-gate voltages (V_g) of -20 V to +20 V, in steps of 1 V, were applied to the silicon substrate, of which the doping level is high enough such that the band bending in the Si substrate is negligible under operating conditions. In the present device geometry (figure 14), over 80 % of the applied gate potential is estimated to fall across the gate oxide, justifying the use of a large gate-voltage increment. For each gate voltage, the I–V characteristics were measured between the drain (d) and source (s) electrodes, contacted by a 0.5μm radius coax probe (73Ct-CMIA/05), at a bias range of between -6 V and +6 V, in 100-mV steps and/or from -0.2 to +0.2V in 4-mV steps, under ambient conditions.

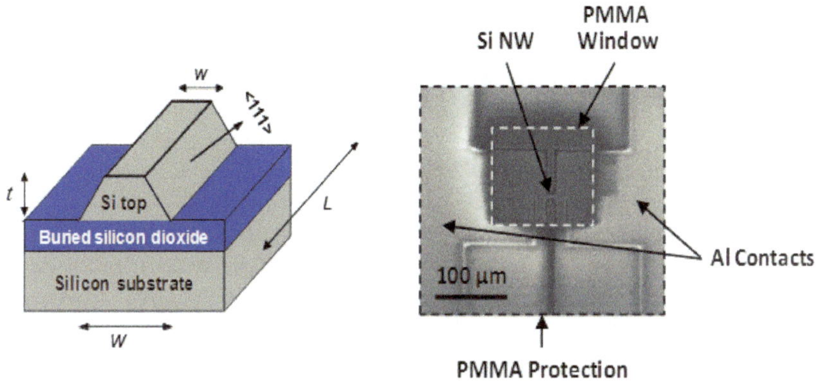

Figure 14. In the right schematic the trapezoidal SiNW insulated from the silicon substrate by a buried SiO_2 layer is shown. The SiNWs used in this study have a length (L) of 1μm, a top width (w) of 70 nm, and a height (t) of 100 nm, The diagonal is based on (111) planes, whereas the small top is based on (100) plane. W (trapezoid base) =200 nm. The left image is a SEM image of the SiNW device, which is coated with PMMA protection layer.

5.1.1. Electrical effects of organic functionalities

Figure 15 shows the current–voltage characteristics of SiO_2-covered SiNW FETs and butyl-terminated SiNW FETs (as a representative example of the molecule-terminated SiNW FETs) at different gate voltages. The source–drain current (I_{sd}) of both samples increased with increasing gate voltage, indicating that the transport through the semiconducting SiNW is dominated by negative carriers (electrons), as it applies to the n-type SiNW. Figure 15 depicts tansconductance (g) versus back-gate voltage (V_g) in the SiO_2-SiNW-FET, butyl-SiNW-FET, 1,3-dioxan-2-ethyl-SiNW-FET, and propyl-alcohol-SiNW-FET. The data presented in figure 15 contain three distinctive features that seem to be at odds with one another and need to be explained. The first distinctive feature is an explicit dependence of the SiNW FET characteristics on the adsorbed molecules.

The zero-gate voltage (V_g=0), small-bias (V_{sd}=0–0.2V) tansconductance depends on the type of absorbed molecule. Particularly, one notices the following trend in the work function observed for these molecules (Table 4).: propyl-alcohol-SiNWs (0.5±0.1 μS)< SiO_2-SiNWs (2.8±0.4 μS) < allyl-SiNWs (7.3±0.8 μS)< butyl-SiNWs (7.6±0.9 μS) < 1,3-dioxan-2-ethyl-SiNWs (10.1±0.9 μS). Furthermore, the threshold voltage (V_{th}), which is roughly defined as the minimum gate voltage for the low-bias conductance to reach the saturation level (figure16) correlates well with the work function in freshly prepared 1,3-dioxan-2-ethyl, butyl and propyl-alcohol functionalities altering the V_{th} values to -4V,-2V, and+2V,respectively. These correlations seem to suggest that the influence of the molecules on the operation of the SiNW FETs plays its most crucial role through a shift in the threshold gate voltage, as the effective work-function of the gate metal, or, equivalently, the electron affinity of the SiNW has been modified by introduction of the molecular layer. However, such a suggestion would be difficult to rationalize because the present molecular layer does not reside between the gate and the SiNW, but is located on the surface of the SiNW. Thus the molecular layer is not expected to directly shift the effective gate voltage [9]. Additionally, sign of the observed conductance shift is opposite to that of the expected work-function influence on the SiNW channel expected. For example, allyl, butyl, and 1,3-dioxan-2-ethyl functionalities are expected to impede the conduction through the Si channel because of the reduced Si electron affinity (see Table 4) but indeed an enhanced small-bias conductance was observed. In contrary to the distinctive feature for small gate-voltage characteristics, there are two other features concerning changes brought about by the adsorption of all the molecules. The first distinctive feature is the sizeable enhancement of the source–drain conductance at large, positive, gate voltages (V_g > +3V) from the one for SiO_2-SiNWs upon introduction of the molecules. However, this effect is almost independent of either the magnitude or the sign of work-function, essentially leading to the same curve as in figure 16. The second feature is that the tansconductance of the molecularly-terminated SiNW-FET devices are significantly reduced from that of the SiO_2-SiNWs at large negative gate bias (V_g < -5V). For SiO_2-SiNW FETs, the tansconductance decreases from 4.54 μS at Vg=+20 V only slightly to 1.47 μS at Vg=-20 V, indicating absence of on/off characteristics in this V_g region [9].

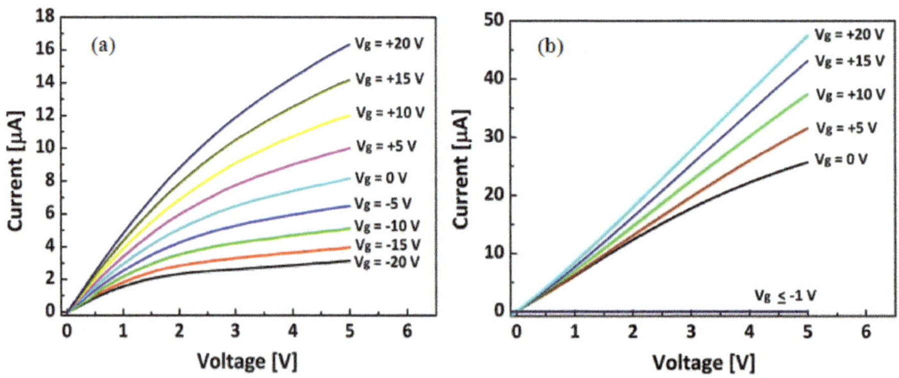

Figure 15. I–V_{ds} curves of (a) SiO_2-SiNW FET (b) butyl-SiNW FET at different back-gate voltages (V_g)

With adsorption of molecules, the channel current is more effectively (by orders of magnitude) turned off by the negative gate voltage, resulting in a significant improvement in the on/off ratio of the FET device. Particularly, on/off ratios of $9x10^3$, $4x10^3$, and $1x10^2$ are observed for 1,3-dioxan-2-ethyl-, butyl-, and propyl-alcohol-terminated SiNW-FETs, respectively. Nevertheless, it is worth to notice that models based on shifts in the threshold voltage from molecular dipolar layers are unable to explain the second and third characteristic observations.

Sample	work-function [meV]
Propyl-alcohol Si(111)	+500±50
SiO2-Si(111)	0±50
Allyl-Si(111)	-150±50
Butyl-Si(111)	-300±50
1,3-dioxan-2-ethyl-Si(111)	-400±50

Table 4. Change in work function due to molecule adsorption on Si(111) samples, as compared with SiO_2-Si(111). The measured work function of SiO_2-Si(111) samples was 4050 meV. The band banding was measured and found to be constant (80±5 meV) in all studied samples.

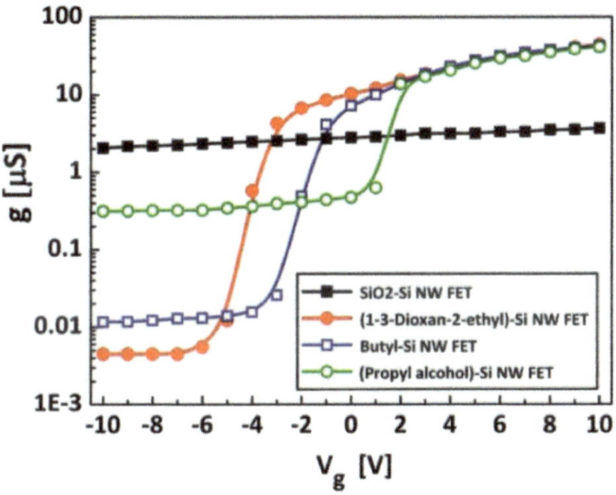

Figure 16. Tansconductance (g) versus back-gate voltage (V_g) in SiO_2-SiNW FET, butyl-SiNW FET, 1,3-dioxan-2-ethyl-SiNW FET, and propyl alcohol- SiNW FET. The tansconductance values were obtained at V_{sd}=0.2 V.

While the transistor characteristics for SiO_2-SiNW FETs have relative stability in air, properties of molecule-covered SiNW-FETs were found to evolve upon air exposure. The time dependence of the average tansconductance at V_g=0 V of various SiNWs versus time in air is illustrated in Fig.17, by a normalized parameter, namely γ (=g_t /$g_{t=0}$). For SiO_2-SiNWs, g is constant over the entire period of investigation. For both 1,3-dioxan-2-ethyl-SiNWs and butyl-SiNWs, g decreased with time to 30–40 % of its initial values over a period of one month (see figure 17). For SiNWs covered with propyl-alcohol, the observed decrease in g was more gradual.

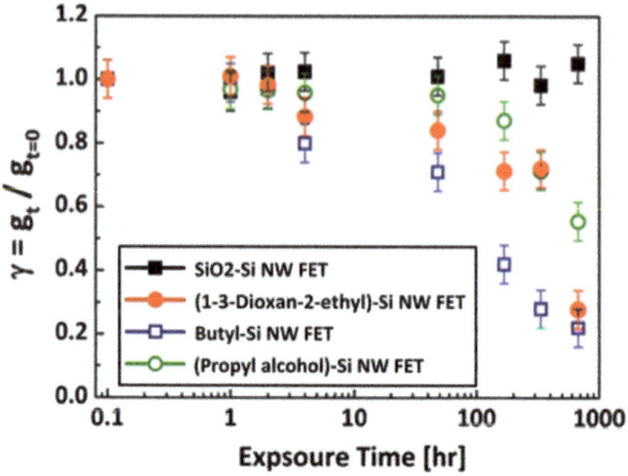

Figure 17. Normalized average conductance at V_g=0 of SiO_2-SiNW FET, butyl-SiNW FET, dioxan -propyl -SiNW FET, and propyl alcohol SiNW FET, versus air exposure time.

5.2. Photovoltaic cell based hybrid SiNWs

We compare three different surface terminations of SiNWs: (i) samples for which the oxide shell was removed by a standard HF dip, named H-SiNWs, (ii) samples with single monolayer of oxide grown after HF dip (14 days), shown by SiO_2-SiNWs; (iii) samples further processed by methyl functionalization immediately after the HF dip, marked with CH_3-SiNWs.

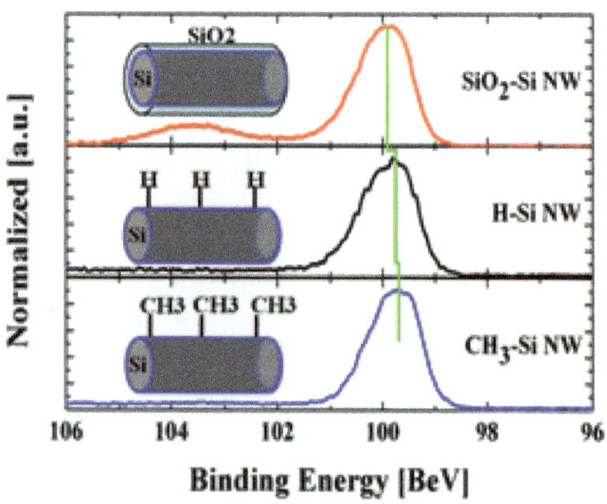

Figure 18. XP spectra of the Si2p region of SiO_2-SiNW, H-SiNW and CH_3-SiNW. The vertical solid line indicates the position of the chemically unshifted $Si2p_{3/2}$ component in each spectrum.

The observed binding energy of the $Si2p_{3/2}$ bulk signal allows determining the difference between the Fermi level and the energy of the valance-band maximum at the surface. We derive the binding energy of the Si2p signal with respect to the valance band maximum from the comparison between the Si 2p core level and valance band spectra of the three samples and obtain the value of E_v-$E_{Si2p3/2}$ = 98.72±0.02 eV. This is similar to previous studies done by Himpsel et al [49] and Hunger et al. i.e. 98.74 eV [50]. Correspondingly, the data for E_F-E_V yields: 0.83 eV, 0.98 eV and 1.05 eV for CH_3-SiNW, H-SiNW and SiO_2-SiNW, respectively. The values yield slight downward surface band bending (B.B.) for the H-SiNW and SiO_2-SiNW and upward B.B for the CH_3-SiNW as will be discussed later.

5.2.1. Work function and band bending of the heterojunction

Work function of all the samples measured independently by the Kelvin probe method as $\Phi_{SiO2\text{-}SiNW}$= 4.32 eV, $\Phi_{H\text{-}SiNW}$=4.26 eV, $\Phi_{CH3\text{-}SiNW}$=4.22 eV. Combining these data with the Fermi level position relative to the band edges (E_F-E_V), yields electron affinity values (χ) of 4.29 eV, 4.12 eV and 3.93 eV for SiO_2-SiNW, H-SiNW and CH_3-SiNW, respectively, according to χ= Φ-E_g +(E_F-E_V). Summarizing these data, the band diagram depicted in figure 19 can be obtained. The bulk Fermi level position was obtained from the specific resistance (1-5 Ωcm) of the n-type samples and gave E_F-E_V=0.88±0.02 eV. We get a slight upward band bending of 0.05 eV which is consistent with flat band conditions for the CH_3-SiNW sample within our error margin. Identical results were found in photoemission studies on two dimensional methyl terminated Si(111) by Hunger et al. We get downward band bending corresponding to electron accumulation at the surface of the H-SiNW and SiO_2-SiNW. This uncommon aspect clearly indicates a yet unidentified surface doping process supporting the n-type character. As a tentative explanation, theory of Dittrich et al. can be referred that holds specific water based functional groups responsible for this effect [51]. For the SiO_2-SiNW, we add the band edges of the SiO_2 top layer and obtain almost a symmetric heterojunction with identical valance and conduction band discontinuities. With E_g=1.12 eV, estimating the electron affinity of the bulk silicon as 4.05 eV, the surface dipole, δ_{ss}, caused by the different monolayer surfaces were calculated to be +0.24 eV, +0.07 eV and -0.12 eV for SiO_2-SiNW, H-SiNW and CH_3-SiNW respectively. The surface dipoles are attributed to different bond charge distribution between the three samples.

5.2.2. Photoelectron Yield Spectroscopy (PYS)

The degree of surface band bending provides a measure for the quality of the electronic passivation on a semiconductor surface. Flat band conditions can usually be achieved only when surface defect states like dangling bonds are essentially passivated making them electronically inactive.

The improved electronic properties of the interface also influence directly the photoemission of electrons from the valance band edge and from surface states inside the band gap. These photoemission processes are studied by photoelectron yield spectroscopy (PYS).

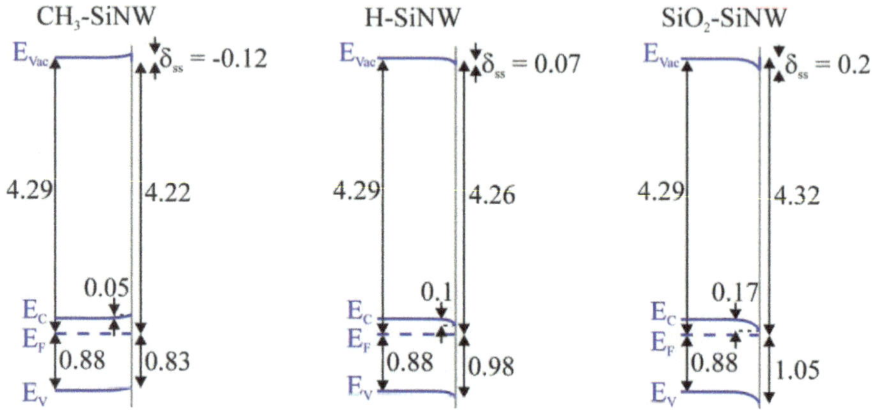

Figure 19. Surface energy band diagram of CH_3-SiNW, H-SiNW and SiO_2-SiNW surfaces. All the numbers are in eV units.

The photoelectron yield spectra of the three different NW samples investigated in this work are shown in figure 20. Each PY spectrum shows two thresholds near 5.0±0.2 eV and 4.2±0.2 eV. The higher energy band corresponds to the valence band density of states of the silicon while the lower energy band associates with defects in the band gap. These spectra essentially represent the occupied density of states. Since the reference energy for yield spectroscopy is the vacuum level (E_{Vac}), we independently used the work function data from the Kelvin measurements (E_{Vac}-E_F) to mark the spectral position of the Fermi level by the dotted vertical lines in figure 20.

The photoelectron intensity at lower photon energies compared to this level is due to thermal occupation of states beyond the Fermi level. The band edges of the two emissions were plotted by dashed lines as can be seen in figure 20. To this end, we can distinguish between the two emissions, for example, the CH_3 show higher emission of 10^2 at 4.7 eV (note the logarithmic scale). Obviously the ratio between defect states and valence band states is significantly different for the three samples. For a systematic comparison we have redrawn the yield spectra in figure 21. The top energy axes in those plots are now related to the valence band maximum instead of the vacuum level by using the electron affinities of the three samples (see figure 21) and the band gap energy of silicon.

The spectra are further normalized at the energy of 0.76 eV below the valence band maximum where they should be strongly dominated by the valence band emission only and thus, the PYS should be identical for all the three samples at this energy point (See figure 21). By this procedure, the relative defect contributions are scaled for all three samples to the Si valence band density of states and are better comparable. The highest defect density is seen for the SiO_2-SiNWs and in this case the defect and the valence band emissions are hardly resolved as two different bands (see also figure 21a).

In the case of the H-SiNW, the defect band is already lower by a factor of about 2, and it can be well distinguished from the valence band emission. Finally, the lowest defect density is observed for the CH_3-SiNWs. For this sample, however, a third band between 4.7 and 5.3 eV is clearly visible and can be distinguished by a significantly smaller slope of the spectrum in this energy region compared to the spectra of the other two samples.

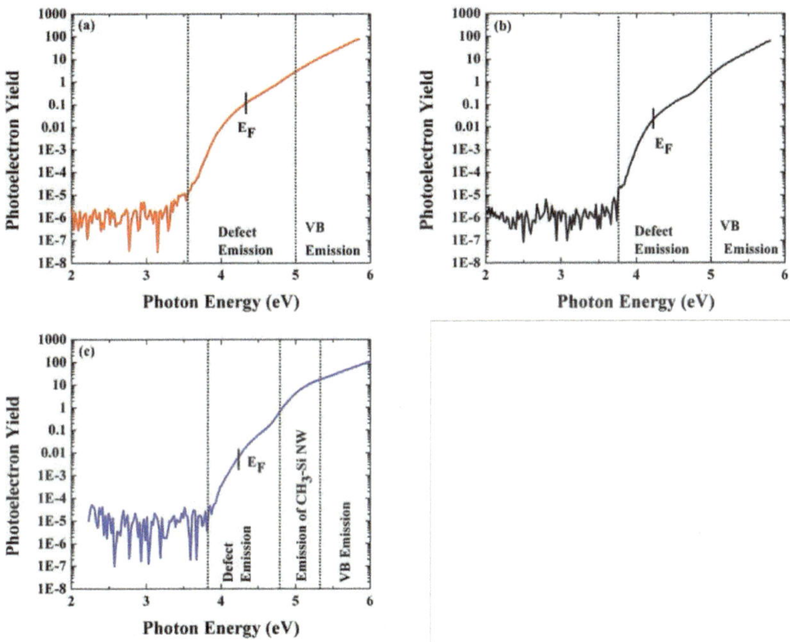

Figure 20. Photoemission yield spectrum of (a) SiO_2-SiNW, (b) H-SiNW, (c) CH_3-SiNW.

This additional band is even more obvious in the spectroscopic density of states that is a derivative of the yield spectra, as shown in figure 21b. Microscopic origin of this additional photoemission band which is presumably related to the methyl functionalization is not yet clear and is subject of further investigations. It is noteworthy to mention that the bandgap energies for SiO_2-SiNWs and CH_3-SiNWs are ~8.9 eV and 7.3 eV, respectively [52]. These values notably differ and thus elucidate the importance of the surface termination with molecules for any electronic application of NWs.

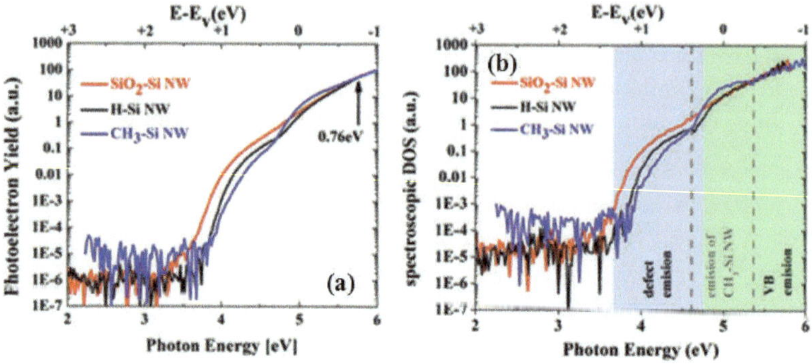

Figure 21. Photoelectron yield Y(*hν*) spectra and spectroscopic density of states of SiO_2-SiNW, H-SiNW and CH_3-SiNW

5.2.3. I-V curves of solar cells processed with functionalized SiNWs

Here we demonstrate a hybrid organic / inorganic solar cell based on a combination of a polymer (PEDOT:PSS) constituting the hole conductor and SiNWs play the role of light absorber and electron conductor. NWs employed to assemble the photovoltaic cells are synthesized on the basis of metal-assisted etching rather than VLS growth. Solar cell performance differs between cells of SiNWs with various functionalities [53].

Surfaces of those SiNWs were terminated by H-, SiO_2- and CH_3- and integrated into a solar cell through identical processes. In all the cells the photogenerated electron-hole pairs are separated at a heterojunction which forms at the SiNW/polymer interface (see figure 22a). The H-SiNWs can just be excluded from further solar cell studies due to the fact that such surfaces indicate hydrophobic behavior and the spun polymer does not wet the SiNW properly.

This hybrid solar cell concept follows the principle proposed by Lewis et al. [54] who made use of the advantageous separation in orthogonal directions of light absorption in the SiNWs and charge carrier separation at the wrap around heterojunction. This SiNW-based cell concept has the following advantages: (i) efficient absorption of light in SiNWs [55] (ii) short diffusion distance for carriers to the wrapped around heterojunction; (iii) use of the air-stable and robust polymer, PEDOT:PSS, as an efficient hole conductor; (iv) the SiNW/ polymer solar cell design utilizes only 1 % of Si used in thin film cells of identical thicknesses. However, for photovoltaic application full coverage of SiNWs by PEDOT:PSS is desirable but not yet achieved (approximately 320nm was covered on top SiNW edge as can be seen in figure 22a.

SiNWs serve mainly as light trapping structures; reduce diffusion distance to the nanowires length while uncovered surfaces represent recombination centers for charge carriers. Further studies on wetting agents added to PEDOT:PSS may result in full coverage and higher efficiencies. Figure 22b shows the current density-voltage (*J-V*) characteristics of CH_3-SiNW/ PEDOT:PSS and SiO_2-SiNW/PEDOT:PSS solar cells under AM1.5 illumination. SiO_2-SiNW/ PEDOT:PSS samples bear a short circuit current (J_{sc}) of 1.6mA/cm^2, an open circuit voltage

(V_{oc}) of 320 mV, a fill factor (FF) of 0.53 and a conversion efficiency (η) of 0.28 %. CH_3-SiNW/PEDOT:PSS devices exhibit improved performance with J_{sc}, V_{oc}, FF and μ magnitudes of 7.0 mA/cm^2, 399 mV, 0.44 and 1.2 %, respectively. The low values of FF and J_{sc} in both types of samples are attributed to high contact resistances (Rs 300Ω).

It should be notéd that by decreasing the resistance we can improve the efficiency of the solar cell as shown by different groups [56-58]. In spite of notably low efficiencies relative to the values nowadays reported for efficient solar cells, their comparative increase (of about a factor of 4) upon methylation that proves a very promising prospective for this kind of surface functionalization.

The improved performance of the CH_3-SiNWs is attributed to the removal of the tunnelling oxide and the reduction of defect states related PY which is caused by a reduced defect density at the heterojunction interface. The advantage of methyl functionalization for the solar cell application has more aspects. The negative surface dipole polarity (-0.12 eV) of the CH_3-SiNW leads to a favorable barrier formation between Si and PEDOT:PSS. Lewis et al. report that charge transfer rates for CH_3-Si do not differ from those of H-Si so that CH_3-monolayers represent no appreciable tunneling barrier for charge transfer at room temperature [59].

Moreover, using one monolayer of CH_3 at the interface, allow a more efficient charge coupling between silicon and polymer. By reducing the defect density at CH_3-SiNWs (see PYS section), the surface recombination is decreased and cause an improvement in V_{oc}. According to the Shockley diode equation $V_{oc} = k_B T/q \cdot \ln(J_{sc}/J_0)$, where J_0 is the saturation current and the other variables have their usual meanings. Since increase of J_{sc} influences V_{oc} as well, it has to be mentioned that the V_{oc} gain observed cannot rely on this effect alone.

Assuming a similar J_0, which correspond to the removal of the tunneling oxide without changing the interface properties, the increase of J_{sc} would lead to a V_{oc} gain of $\Delta V_{oc} = kT/q \cdot \ln(J_{sc,methyl}/J_{sc,oxide}) = k_B T/q \cdot \ln(7.0/1.6) = 0.037$ V. Regarding the observed gain of $\Delta V_{oc} = 0.079$ V, only a reduced surface recombination (as measured by PY) and/or a favorable barrier formation (surface dipole) can establish consistency with the experimental data.

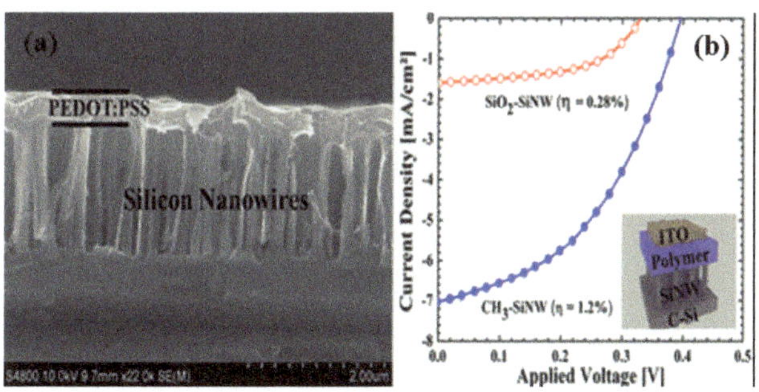

Figure 22. (a) Tilted view of the heterojunctionSiNW/PEDOT:PSS. (b) J-V characteristic under AM1.5 illuminations of the radial heterojunction solar cells from CH_3-SiNW and SiO_2-SiNW. Inset: schematic view of solar cell device structure.

A potential explanation of the increase in the PY of CH_3-SiNW and an increased efficiency of processed solar cells could be that CH_3-SiNWs simply absorb the solar light better than the other SiNW. This can experimentally falsified by the equal total absorbance (A), reflection (R) and transmission (T) of all three types of samples (see figure 23.

All the three spectra (SiO_2-SiNW, CH_3-SiNW, H-SiNW) show the same basic structure and the same A, R, T in short wavelengths ($h\nu > E_g$). In wavelengths smaller than ~1000 nm, the three samples behave the same: reflection is around 10 %, the transmission is roughly 0 % and thus the absorption around 90 % for the whole visible spectrum. This tendency is observed even in multi monolayers of oxide. However, for wavelengths larger than ~1200 nm (attribute to surface state regime) the transmission is around 50 % for all three samples and the reflection is also nearly 50 %, except for the SiO_2-SiNW sample which shows 40 %. The lower reflection in SiO_2-SiNW is attributed to light adsorption of the surface state.

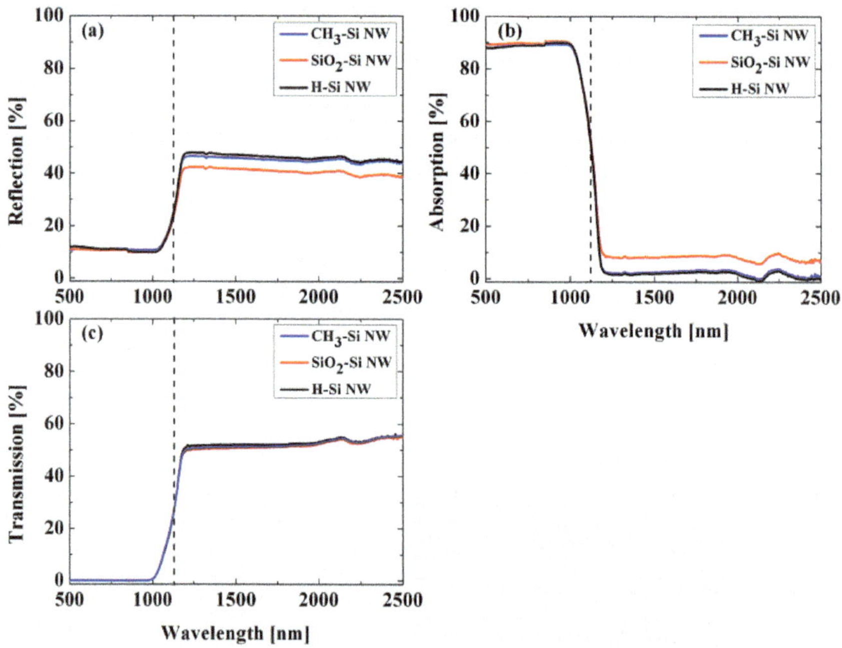

Figure 23. Spectra of SiO_2-SiNW, H-SiNW and CH_3-SiNW: (A) Reflectance (B) Transmission (C) absorbance. The dash line indicates the wavelength that corresponds to the SiNW band gap energy.

In summary, we terminated the SiNWs with organic molecules. The stability of the organic found to be depended on their coverage and surface. CH_3 shows the highest oxide resistivity, since it gives 100 % of coverage. To utilize the advantage of these molecules, we integrate them in FETs and solar cells. We showed that we can tune the properties of the FETs by attaching molecules on the surface of the channel i.e. SiNW. The solar cells of hybrid SiNW shows higher efficiency than native oxide. This can be attributed to the low defects in the band gap as had been shown by photoelectron yield spectra.

Author details

Muhammad Y. Bashouti[1], Matthias Pietsch[1], Kasra Sardashti[1], Gerald Brönstrup[1], Sebastian W. Schmitt[1], Sanjay K. Srivastava[1], Jürgen Ristein[3], Jordi Arbiol[4], Hossam Haick[2] and Silke Christiansen[1,5]

1 Max-Planck-Institute for the Science of Light, Erlangen, Germany

2 The Department of Chemical Engineering and Russell Berrie Nanotechnology Institute, Technion – Israel Institute of Technology, Haifa, Israel

3 Institue for Technical Physics, University of Erlangen-Nuremberg, Erlangen, Germany

4 Institució Catalana de Recerca i EstudisAvançats (ICREA) and Institut de Ciència de Materials de Barcelona,ICMAB-CSIC, Bellaterra, Spain

5 Institute of Photonic Technology, Jena, Germany

References

[1] Bandaru, P.R. and P. Pichanusakorn, *An outline of the synthesis and properties of silicon nanowires.* Semiconductor Science and Technology, 2010. 25(2).

[2] Assad, O., et al., *Stable Scaffolds for Reacting Si Nanowires with Further Organic Functionalities while Preserving Si-C Passivation of Surface Sites.* Journal of the American Chemical Society, 2008. 130(52): p. 17670-+.

[3] Barth, S., et al., *Synthesis and applications of one-dimensional semiconductors.* Progress in Materials Science, 2010. 55(6): p. 563-627.

[4] Bronstrup, G., et al., *Statistical model on the optical properties of silicon nanowire mats.* Physical Review B, 2011. 84(12).

[5] Paska, Y., et al., *Enhanced Sensing of Nonpolar Volatile Organic Compounds by Silicon Nanowire Field Effect Transistors.* Acs Nano, 2011. 5(7): p. 5620-5626.

[6] Paska, Y. and H. Haick, *Interactive Effect of Hysteresis and Surface Chemistry on Gated Silicon Nanowire Gas Sensors.* Acs Applied Materials & Interfaces, 2012. 4(5): p. 2604-2617.

[7] Bashouti, M.Y., et al., *Chemical Passivation of Silicon Nanowires with C(1)-C(6) Alkyl Chains through Covalent Si-C Bonds.* Journal of Physical Chemistry C, 2008. 112(49): p. 19168-19172.

[8] Bashouti, M.Y., et al., *Covalent Attachment of Alkyl Functionality to 50 nm Silicon Nanowires through a Chlorination/Alkylation Process.* Journal of Physical Chemistry C, 2009. 113(33): p. 14823-14828.

[9] Bashouti, M.Y., R.T. Tung, and H. Haick, *Tuning the Electrical Properties of Si Nanowire Field-Effect Transistors by Molecular Engineering.* Small, 2009. 5(23): p. 2761-2769.

[10] Paska, Y. and H. Haica, *Controlling properties of field effect transistors by intermolecular cross-linking of molecular dipoles.* Applied Physics Letters, 2009. 95(23).

[11] Qiu, T. and P.K. Chu, *Self-selective electroless plating: An approach for fabrication of functional 1D nanomaterials.* Materials Science & Engineering R-Reports, 2008. 61(1-6): p. 59-77.

[12] Chen, W.W., et al., *Etching Behavior of silicon nanowires with HF and NH4F and surface characterization by attenuated total reflection Fourier transform infrared spectroscopy: Similarities and differences between one-dimensional and two-dimensional silicon surfaces.* Journal of Physical Chemistry B, 2005. 109(21): p. 10871-10879.

[13] Chan, C.K., et al., *High-performance lithium battery anodes using silicon nanowires.* Nature Nanotechnology, 2008. 3(1): p. 31-35.

[14] Chen, L.J., *Silicon nanowires: the key building block for future electronic devices.* Journal of Materials Chemistry, 2007. 17(44): p. 4639-4643.

[15] Cui, Y. and C.M. Lieber, *Functional nanoscale electronic devices assembled using silicon nanowire building blocks.* Science, 2001. 291(5505): p. 851-853.

[16] Cui, Y., et al., *High performance silicon nanowire field effect transistors.* Nano Letters, 2003. 3(2): p. 149-152.

[17] Goldberger, J., et al., *Silicon vertically integrated nanowire field effect transistors.* Nano Letters, 2006. 6(5): p. 973-977.

[18] Peng, K.Q., et al., *Silicon nanowires for rechargeable lithium-ion battery anodes.* Applied Physics Letters, 2008. 93(3).

[19] Haick, H., et al., *Electrical characteristics and chemical stability of non-oxidized, methyl-terminated silicon nanowires.* Journal of the American Chemical Society, 2006. 128(28): p. 8990-8991.

[20] Morales, A.M. and C.M. Lieber, *A laser ablation method for the synthesis of crystalline semiconductor nanowires.* Science, 1998. 279(5348): p. 208-211.

[21] Ashkenasy, G., et al., *Molecular engineering of semiconductor surfaces and devices.* Accounts of Chemical Research, 2002. 35(2): p. 121-128.

[22] Cohen, R., et al., *Molecular electronic tuning of Si surfaces.* Chemical Physics Letters, 1997. 279(5-6): p. 270-274.

[23] Ishii, H., et al., *Energy level alignment and interfacial electronic structures at organic/metal and organic/organic interfaces (vol 11, pg 605, 1999).* Advanced Materials, 1999. 11(12): p. 972-972.

[24] Anagaw, A.Y., R.A. Wolkow, and G.A. DiLabio, *Theoretical study of work function modification by organic molecule-derived linear nanostructure on H-silicon(100)-2 x 1.* Journal of Physical Chemistry C, 2008. 112(10): p. 3780-3784.

[25] Hiremath, R.K., et al., *Molecularly controlled metal-semiconductor junctions on silicon surface: A dipole effect.* Langmuir, 2008. 24(19): p. 11300-11306.

[26] Aradi, B., et al., *Theoretical study of the chemical gap tuning in silicon nanowires.* Physical Review B, 2007. 76(3).

[27] Leu, P.W., B. Shan, and K.J. Cho, *Surface chemical control of the electronic structure of silicon nanowires: Density functional calculations.* Physical Review B, 2006. 73(19).

[28] Wagner, R.S., et al., *Study of Filamentary Growth of Silicon Crystals from Vapor.* Journal of Applied Physics, 1964. 35(10): p. 2993-&.

[29] Sivakov, V.A., et al., *Silicon nanowire oxidation: the influence of sidewall structure and gold distribution.* Nanotechnology, 2009. 20(40).

[30] Bashouti, M.Y., et al., *Silicon nanowires terminated with methyl functionalities exhibit stronger Si-C bonds than equivalent 2D surfaces.* Physical Chemistry Chemical Physics, 2009. 11(20): p. 3845-3848.

[31] Haber, J.A. and N.S. Lewis, *Infrared and X-ray photoelectron spectroscopic studies of the reactions of hydrogen-terminated crystalline Si(111) and Si(100) surfaces with Br-2, I-2, and ferrocenium in alcohol solvents.* Journal of Physical Chemistry B, 2002. 106(14): p. 3639-3656.

[32] Thanikasalam, P., T.K. Whidden, and D.K. Ferry, *Oxidation of silicon (100): Experimental data versus a unified chemical model.* Journal of Vacuum Science & Technology B, 1996. 14(4): p. 2840-2844.

[33] Dimitrijev, S. and H.B. Harrison, *Modeling the growth of thin silicon oxide films on silicon.* Journal of Applied Physics, 1996. 80(4): p. 2467-2470.

[34] Deal, B.E. and A.S. Grove, *General Relationship for Thermal Oxidation of Silicon.* Journal of Applied Physics, 1965. 36(12): p. 3770-&.

[35] Liu, B.Z., et al., *Oxidation of silicon nanowires for top-gated field effect transistors.* Journal of Vacuum Science & Technology A, 2008. 26(3): p. 370-374.

[36] Buttner, C.C. and M. Zacharias, *Retarded oxidation of Si nanowires.* Applied Physics Letters, 2006. 89(26).

[37] Miyazaki, S., et al., *Surface Fermi level position of hydrogen passivated Si(111) surfaces.* Applied Physics Letters, 1996. 68(9): p. 1247-1249.

[38] Cerofolini, G.F., G. LaBruna, and L. Meda, *Gas-phase room-temperature oxidation of (100) silicon.* Applied Surface Science, 1996. 93(3): p. 255-266.

[39] Cerofolini, G.F. and L. Meda, *Mechanisms and kinetics of room-temperature silicon oxidation.* Journal of Non-Crystalline Solids, 1997. 216: p. 140-147.

[40] Reddy, S., P.O. Assad, and H. Haick, *Highly stable organic modification of Si(111) surfaces: Towards reacting Si with further functionalities while preserving the desirable chemical properties of full Si-C atop site terminations.* Journal of the American Chemical Society, 2008. 130(29): p. 9184-+.

[41] Cerofolini, G.F., D. Mascolo, and M.O. Vlad, *A model for oxidation kinetics in air at room temperature of hydrogen-terminated (100) Si.* Journal of Applied Physics, 2006. 100(5).

[42] Uematsu, M., H. Kageshima, and K. Shiraishi, *Microscopic mechanism of thermal silicon oxide growth.* Computational Materials Science, 2002. 24(1-2): p. 229-234.

[43] Ma, D.D.D., et al., *Small-diameter silicon nanowire surfaces.* Science, 2003. 299(5614): p. 1874-1877.

[44] Zhang, R.Q., et al., *Structures and energetics of hydrogen-terminated silicon nanowire surfaces.* Journal of Chemical Physics, 2005. 123(14).

[45] Sun, X.H., et al., *FTIR spectroscopic studies of the stabilities and reactivities of hydrogen-terminated surfaces of silicon nanowires.* Inorganic Chemistry, 2003. 42(7): p. 2398-2404.

[46] Wallart, X., C.H. de Villeneuve, and P. Allongue, *Truly quantitative XPS characterization of organic monolayers on silicon: Study of alkyl and alkoxy monolayers on H-Si(111).* Journal of the American Chemical Society, 2005. 127(21): p. 7871-7878.

[47] Schwartz, D.K., *Mechanisms and kinetics of self-assembled monolayer formation.* Annual Review of Physical Chemistry, 2001. 52: p. 107-137.

[48] Nuzzo, R.G., B.R. Zegarski, and L.H. Dubois, *Fundamental-Studies of the Chemisorption of Organosulfur Compounds on Au(111) - Implications for Molecular Self-Assembly on Gold Surfaces.* Journal of the American Chemical Society, 1987. 109(3): p. 733-740.

[49] Himpsel, F.J., G. Hollinger, and R.A. Pollak, *Determination of the Fermi-Level Pinning Position at Si(111) Surfaces.* Physical Review B, 1983. 28(12): p. 7014-7018.

[50] Hunger, R., et al., *Chemical and electronic characterization of methyl-terminated Si(111) surfaces by high-resolution synchrotron photoelectron spectroscopy.* Physical Review B, 2005. 72(4).

[51] Dittrich, T., et al., *On the origin of the positive charge on hydrogenated Si surfaces and its dependence on the surface morphology.* Surface Science, 1999. 437(1-2): p. 154-162.

[52] Salomon, A., et al., *What is the barrier for tunneling through alkyl monolayers? Results from n- and p-Si-Alkyl/Hg junctions.* Advanced Materials, 2007. 19(3): p. 445-+.

[53] Sivakov, V., et al., *Silicon Nanowire-Based Solar Cells on Glass: Synthesis, Optical Properties, and Cell Parameters.* Nano Letters, 2009. 9(4): p. 1549-1554.

[54] Boettcher, S.W., et al., *Energy-Conversion Properties of Vapor-Liquid-Solid-Grown Silicon Wire-Array Photocathodes.* Science, 2010. 327(5962): p. 185-187.

[55] Bronstrup, G., et al., *A precise optical determination of nanoscale diameters of semiconductor nanowires.* Nanotechnology, 2011. 22(38).

[56] Ozdemir, B., et al., *Silicon nanowire - poly(3,4-ethylenedioxythiophene)-poly(styrenesulfonate) heterojunction solar cells.* Applied Physics Letters, 2011. 99(11).

[57] Yuan, C.Z., et al., *Hierarchically structured carbon-based composites: Design, synthesis and their application in electrochemical capacitors.* Nanoscale, 2011. 3(2): p. 529-545.

[58] Shiu, S.C., et al., *Morphology Dependence of Silicon Nanowire/Poly(3,4-ethylenedioxythiophene):Poly(styrenesulfonate) Heterojunction Solar Cells.* Chemistry of Materials, 2010. 22(10): p. 3108-3113.

[59] Maldonado, S. and N.S. Lewis, *Behavior of Electrodeposited Cd and Pb Schottky Junctions on CH3-Terminated n-Si(111) Surfaces.* Journal of the Electrochemical Society, 2009. 156(2): p. H123-H128.

Chapter 11

Wet – Chemically Etched Silicon Nanowire Solar Cells: Fabrication and Advanced Characterization

Björn Hoffmann, Vladimir Sivakov,
Sebastian W. Schmitt, Muhammad Y. Bashouti,
Michael Latzel, Jiří Dluhoš, Jaroslav Jiruse and
Silke Christiansen

Additional information is available at the end of the chapter

1. Introduction

In the past decade the usage of solar energy production has increased rapidly and has become a reasonable alternative to fossil energy sources. Solar energy conversion is a challenge of the millennium and solar cells are one eagerly thought way of converting solar energy into electrical energy. Solar cells are divided into three generations according to Martin Green [1]: The 1st generation (1st gen) is till to date dominating production and installation and makes use of bulk, essentially silicon wafers with planar cell geometry. Today, a typical silicon based cell uses 200μm thick Si wafers as substrates. The 2nd generation (2nd gen) uses thin-films rather that bulk wafers and materials can be wide spread. The idea is to use only very thin layers of photovoltaically active material on a passive mechanically supporting material e.g. glass. Such cells offer the option to use cheap substrates and even flexible substrates [2] so that cost saving potentials as well as new applications can be envisaged. Until today, thin-film cells (2nd gen) do not live up to their full potential due to the fact that they are 50%-60% less efficient than wafer cells (1st gen). Solar cell concepts termed 3rd gen aim at the combination of low production costs and high efficiencies. In order to reach this goal, unique optical and physical properties of complex nanostructures as well as novel materials and compounds are exploited. Silicon nanowires (SiNWs) are a very promising candidate for next generation PV. The basic material is abundant and cheap and the semiconductor industry gained intense experience with its handling. The nanostructuring and cell preparation can also be performed in cheap and large scale processes. Thus it could easily be integrated into existing PV lines.

1.1. Reasons for nanowires in PV concepts

The conversion efficiency of solar cells is affected by many factors. The internal quantum efficiency is already quite good for a long time and thus the industry concentrated more on optimizing the contacts for efficiency enhancement or reducing the overall costs. Whereas one efficiency record follows another in research [3], industry scale solar cells, however, are only improving slowly [4]. While optimizing factors like contact resistance or quantum efficiency only permit a moderate overall efficiency enhancement, the optical absorption is a property that can encourage large improvements of the SC efficiency. By using silicon nanostructures, the absorption of sunlight can be improved from typically 60% to over 90% in the wavelength range between 400 and 1100 nm [5–9]. This increased absorption is very encouraging and makes the nanostructures so interesting for SC research as well as for future industrial production.

One of many ways in order to reach this increased absorption of sunlight is the use of nanowires constituting the SC absorbers. The nanowires act as light trapping specimens due to its strong scattering capacities [10] and for the same reason as antireflective coatings [11–13]. A lot of different methods exist that produce such highly absorbing nanostructures which are divided into bottom-up growth mechanisms [14–16] and top-down etching processes [17–20]. This work will concentrate on a metal assisted wet-chemical etching (MA-WCE) process that does not need any vacuum equipment and can be performed even on large surfaces in a few minutes.

The application of micro- and nanowires in SCs is quite a new field of PV research. Having used the "Web of Knowledge" search engine, graphs were generated which are presented in Figure 1[1]. From the first publication in 2002 [21], the topic has rapidly increased in popularity and it has reached nearly 250 publications in 2011. The number of citations also shows a growing increase of nearly 5000 citations in 2011.

Figure 1. Publications and citations on nanowire-based solar cells in the last 10 years.

1 http://www.webofknowledge.com/ Search properties: title includes (nanowire OR nanorod OR microwire OR microrod) AND (solar)

2. Experimental part

2.1. Top-down fabrication of silicon

SiNWs can be produced by many different methods, normally divided into bottom-up growth mechanisms and top-down etching processes. While growing of SiNWs normally is conducted by using the vapor-liquid-solid (VLS) mechanism [14], which needs high temperatures, toxic gases and expensive vacuum equipment, top-down etching can be performed in easier ways. One differentiates between physical (dry) and chemical (wet) etching. Physical or dry etching is usually conducted in reactive gases offered in a reactive ion etching (RIE) setup [22], while chemical etching is usually based on wet chemistries, preferably on the basis of hydrofluoric acid for the case of silicon. Both methods have assets and drawbacks, but when it comes to the question of fast and cost-effective fabrication with the perspective of mass production, wet-chemical (WC) etching is intriguing.

Since details of WC etching processes were already discussed in a previous Intech book chapter [20], this work will only give a short summary and point out the latest optimization processes. The SiNW substrates that constitute the SC absorber are fabricated in a two-step MA-WCE procedure. Therefore, the substrate material, which is typically a lowly doped (1-5 Ohmcm) n-type Si wafer, passes four different etching solutions. In the beginning, the native oxide is removed with 2% hydrofluoric acid (HF, solution I). In the following step (solution II), silver (Ag) nanoparticles are formed on the surface of the wafer in a mixture of 0.02 molar (M) $AgNO_3$ and 5M HF in a ratio of 1:10. The duration of this step influences the density of particles and thus also the forming SiNW geometry. This procedure needs to be less than 30 seconds. Otherwise, the Si is completely coated by a continuous Ag layer and no access of the etching solutions to the Si surface can be achieved. Controlled SiNW properties can be achieved when using the following etching steps. Solution III consists of a mixture of 5M HF and 30% H_2O_2 in a ratio of 1:1. In this step the Si is oxidized by the interplay of H_2O_2 and the nanoscale Ag particles that form the thermodynamically unstable intermediate compound AgO(OH) which is a very efficient oxidant that oxidizes the silicon beneath the Ag nanoparticle. The resulting SiO_2 is dissolved by the HF in the solution. Thus, the Si is removed below the Ag particles which dig into the Si wafer and thus move away from the wafer surface inducing the SiNW formation. The direction of this Ag nanoparticle movement is anisotropic following certain crystal orientations and depends on the temperature as well [20]. Under controlled conditions, the SiNWs form perpendicularly to the surface of a Si(100) wafer and are oblique to the surface of a Si(111) wafer following <100>-directions [18]. In the last step (solution IV), concentrated nitric acid is used to remove the remaining Ag nanoparticles. Figure 2 illustrates the etching steps process.

During the SiNW formation, an increasing absorption can be observed after a few seconds. The shiny Si wafer surface turns matt finished and changes to a brown color which becomes deep black after one minute of etching. After three minutes the SiNWs reach a length of 1μm and elongate with a rate of 1 μm/min after that.

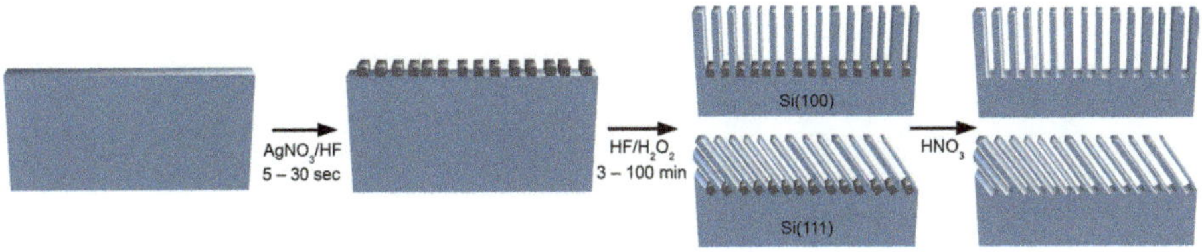

Figure 2. The WCE process is based on a sequence of four solutions starting with oxide removal (solution I: dilute HF, not shown), followed by solution based Ag nanoparticle deposition (solution II), Ag nanoparticle assisted catalytic etching of Si (solution III) and Ag nanoparticle removal (solution IV).

2.2. Fabrication of Semiconductor-Insulator-Semiconductor (SIS) solar cells

SiNWs enable several concepts of efficient solar cell geometries. The charge carrier dividing junction (p-n or Schottky type) can be applied in a radial [23,24] or axial [25,26] concept inside the NWs and a possible solution is also the wrapping of NWs with a second material [27]. In this work, we used a semiconductor-insulator-semiconductor (SIS) heterojunction solar cell concept that has already been developed in the 1970s [28–30]. Therefore, the complex SiNW structure is uniformly coated with a very thin tunneling barrier and a thicker layer of aluminum (Al) doped zinc oxide (ZnO) as a second, transparent wide band gap, degenerately doped semiconductor. In order to achieve this conformal coating, atomic layer deposition (ALD, Oxford Opal) is used. This technique can deposit many materials in a self-limiting process with a thickness control at the Angstrom level even on highly complex surfaces [31,32]. Two different classes of precursors are needed for the ALD deposition of metal oxides like Al_2O_3, TiO_2 or ZnO: metal containing molecules and an oxidizing agent. The two precursors are pulsed into the ALD vacuum chamber one after the other by using very short pulse durations of typically <100ms, while longer inert gas purging steps are used between the precursor pulses to remove residual gas from the chamber. The special advantage of an ALD process is that the precursors are forming a self-limiting monolayer that does not allow further deposition of that precursor. The oxidizing agent can be water or an oxygen plasma which removes the functional groups of the previously deposited metal precursor molecules and forms the metal oxide [33,34]. This step is also self-limiting so that a completed cycle forms a stoichiometric monolayer of the desired material even on 3D surfaces. In the case of Al_2O_3, trimethylaluminum and oxygen plasma were used for its formation and for ZnO deposition, diethylzinc and water pulses are applied. The doping of ZnO with aluminum was realized by substituting every 20th cycle of ZnO by one cycle of Al_2O_3 [35,36].

Two different insulating layer materials were tested to form the barrier layer in the SIS cell: TiO_2 and Al_2O_3. Depending on the material, 10 to 30 Å of the barrier layer were deposited onto the SiNWs. Subsequently, the second semiconductor, Al doped ZnO (AZO), which acts as transparent front contact at the same time, is deposited in the same ALD reactor. The thickness of this layer was typically around 450 nm which resulted in a complete filling of the gaps in between the SiNWs and also created a continuous AZO layer on top of the SiNWs for efficient charge carrier conduction. Figure 3 shows the three steps of the ALD process.

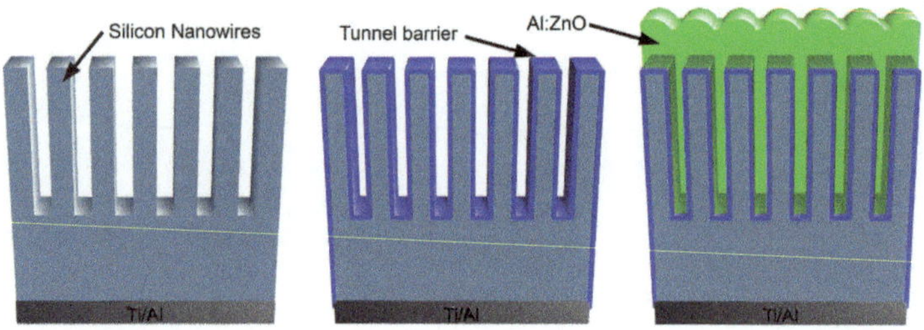

Figure 3. Schematic of the layout of an SIS solar cell concept based on SiNWs: etched SiNWs (left) are coated with a thin tunnel barrier (middle) and subsequently with a transparent front contact – Al doped ZnO (right).

The structural analysis of the completed SIS solar cells was performed by scanning electron microscopy (SEM) techniques including 3D analysis supported by a sequence of focused ion beam (FIB) etching steps. Thereby, the gap filling between SiNWs and SiNW topologies were investigated. Figure 4 shows two colored cross section micrographs of as etched SiNWs and of a completed SIS solar cell that contains wrapping layer stacks on the SiNWs of barrier layer and AZO.

Additionally, the complete SCs were analyzed by glow discharge optical emission spectrometry (GD-OES, Horiba Jobin Yvon JY 5000 RF) to obtain a depth profile of the compositions of the material [37,38]. In a GD plasma source, cathodic sputtering is used to remove material layer-by-layer from a sample surface with an accuracy of a few nanometers [39]. The sample atoms are transported into a plasma plume, in which they are excited and ionized through collisions with electrons, ions and the metastable carrier gas (here: Ar).

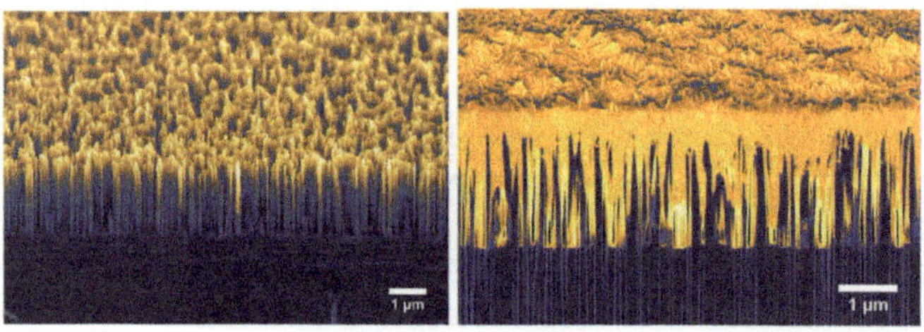

Figure 4. Left: colored SEM micrograph of as etched SiNWs. Right: colored SEM micrograph of a FIB polished cross section of a completed SIS solar cell. The SiNWs (blue) are embedded in barrier layer and Al:ZnO (AZO) (orange).

The analyte isotopes of interest can then be detected via downstream spectroscopic techniques. In the presented study, OES is performed by analyzing the intensities of characteristic emission lines of excited analytes with a spectrometer (typically a Paschen–Runge monochromator or Czerny–Turner polychromator) with A GD-OES Such an analysis was performed on a SiNW SC with a TiO_2 tunnel barrier in the optimization of the solar cell production process. In the SEM profile in Figure 5, the SC can be divided into three regions:

region 1 shows the front contact which solely consists of aluminum doped zinc oxide (AZO). The second region is composed of the SiNW array with a wrapping TiO_2 layer which is filled with AZO. In the third region, the bulk Si wafer is reached. From the measured spectra several interesting information can be extracted [40]. In region 1 the Al content is found to be 2.5 mass% (m%) leading to an ideal conductivity of the AZO layers. In region 2, the silicon content increases only slowly in depth which leads to the assumption that the wires are tapered. This can be interesting for further theoretical investigations regarding optical simulations of light trapping effects. After having checked the residual silver, it is revealed that the cleaning process of the nitric acid did not completely remove the metal. According to these findings, the cleaning process was improved to eliminate residual silver contaminations.

The GD-OES measurements enable another interesting analysis of the silicon nanostructure: Knowing the total mass and thickness of the tunnel barrier layer, the surface of the nanostructures can be calculated. Thereby, the surface enhancement of the etched nanostructures as compared to a flat wafer can be derived. In the case of the TiO_2 barrier SiNW SC (Figure 5), an enhancement factor of 6.2 was calculated. This factor is strongly dependent on the etching conditions and can easily reach values of 40 and more. For a solar cell, a higher surface amount is a clear disadvantage because higher junction areas result in higher surface recombination of charge carriers. On the other hand a higher surface increases the absorption of light and leads to an overall higher charge carrier collection at the interface. Thus, an optimum has to be found between the counteracting mechanisms.

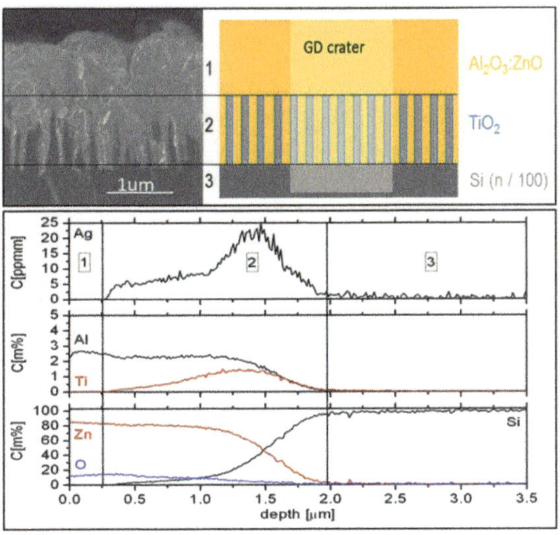

Figure 5. Top: SiNW SC cross section and scheme, showing three different regions inside the cell. Bottom: GD-OES spectra illustrating the depth dependent mass% of the elements Si, O, Zn, Al, Ti and Ag. Adapted from [40].

During the research state, typical SCs had only a size of a few mm² and thus did not need metal front contacts. In the later course of cell development, first demonstrators that had the size to power electrical devices were needed. Therefore, two different prototypes with gold front contact grids and sizes of 36cm² and 1.4cm² were fabricated (see Figure 6). To demonstrate the functionality of the SCs, the small demonstrator was contacted to a motor with a rotating wheel and was illuminated by a battery powered flashlight. Even though the flashlight delivered only a fraction of the sunlight's power, the SC produced enough energy to drive the motor.

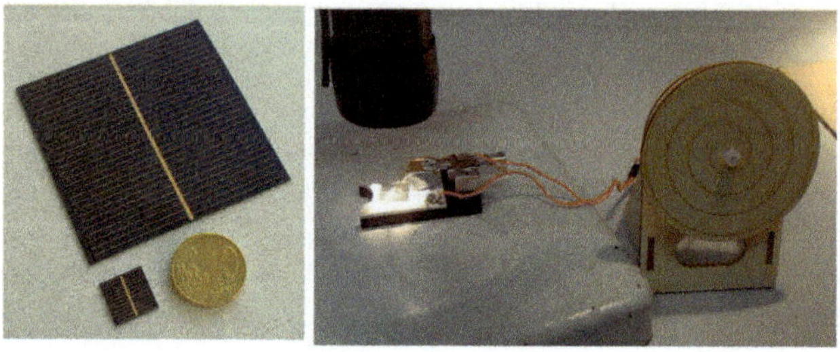

Figure 6. Left: SiNW SC prototypes with gold front contacts. The large cell has a size of 36 cm², the small cell measures 1.4 cm². For comparison: a 10 €-cent coin. Right: The 1.4 cm² cell under a battery-driven flashlight produces enough power to rotate a small toy device.

3. Characterization

The characterization of our fabricated SCs can be divided into investigation of structural and electro-optical properties. Understandably, the conversion efficiency is the most important property. It is determined in a current-voltage (I-V) measurement under a solar simulator that reproduces the certified AM 1.5 solar spectrum. An optical analysis was performed with an integrating sphere in order to investigate the connection between light absorption and efficiency. Figure 7 shows the measured photocurrent under illumination at -2V of 12 different solar cells and the respective optical absorption (mean value between 400 – 1100 nm). In most cases a connection between these two values is clearly visible. Therefore, it is crucial to optimize the absorption by tailoring the nanostructured material composite. Furthermore it can be seen that even a small reduction in absorption from 90% to 88% (wafer 1 and 4) can cause a serious drop in photocurrent from 35 mA to 15 mA. Although other effects may play a role, this effect cannot be neglected.

After these findings the best solar cells were analyzed in detail regarding energy conversion and quantum efficiency as well as local conversion properties in terms of light and electron beam induced current mappings (LBIC/EBIC).

Over 120 solar cells with three different types of Si wafers and 7 different etching conditions were fabricated. For the record cell a n-type Si(111) wafer with a doping of 1-9 Ohm*cm was used that was etched for 30s/6min in the two etching solutions. The Al_2O_3 tunneling barrier had a thickness of 12 Å. From this cell a small piece of 8mm² size was analyzed in the solar simulator. The SC shows a clearly visible diode-like behavior in the dark with a rectification factor of 6600 at +/-2V. Under AM1.5 illumination the record cell reaches an open circuit voltage V_{OC} of 453 mV, a short circuit current density J_{SC} of 35 mA/cm² and a fill factor (FF) of 57%. This results in a conversion efficiency of 9.1% which is an impressive result for a nanowire based solar cell with as little optimization as we did, e.g. in terms of surface passivation [41,42] or size and shape of the SiNWs [22,43,44]. The corresponding dark and illuminated J-V curves are shown in Figure 8.

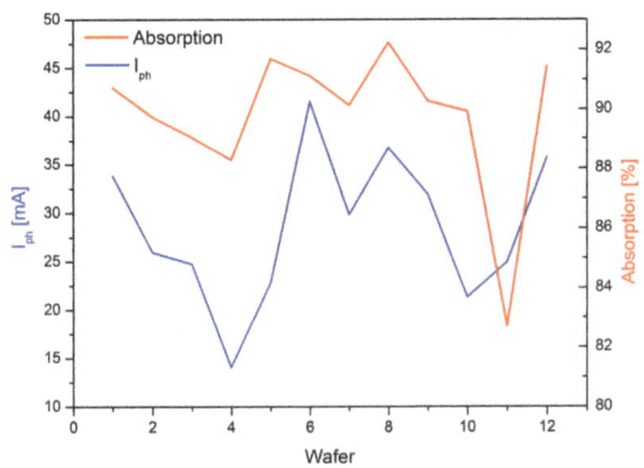

Figure 7. Measured photocurrent (at -2V) and absorption of 12 different solar cells. A correlation is clearly visible.

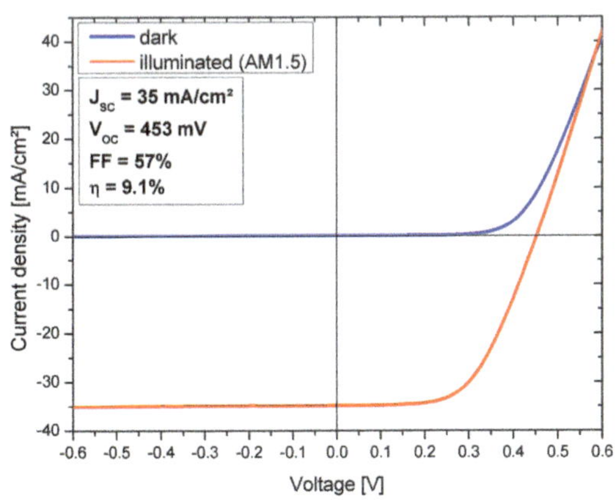

Figure 8. Dark and illuminated current density – voltage (J-V) curves of our record SiNW-SC out of 120 test samples. Very high J_{SC} combined with average V_{OC} and fill factor yield in a conversion efficiency of 9.1%.

3.1. Local electro-optical properties

To determine the influence of structural inhomogeneity on the electrical properties, it is necessary to generate localized charge carriers inside the solar cell and perform a mapping across the cell while monitoring the generated external current between front and back contacts. Two different methods enable such an analysis: light and electron beam induced current (LBIC [45] & EBIC [25, 46]) measurements. In LBIC, a laser is scanned over the solar cell surface and the generated current is measured for each spot. In this way an efficiency mapping can be obtained. By using lasers with different wavelengths the response of the solar cell to distinct excitation energies can be investigated. Furthermore, an external quantum efficiency (EQE) mapping can be calculated from such a measurement. The disadvantage of LBIC is its large laser spot focus which inhibits the analysis of μm-sized defects. Here EBIC becomes suitable. The electron beam of an SEM can be focused on a few nm and with a low acceleration voltage the expansion of excited charge carriers can be kept at a minimum of <100nm. Every incident electron generates thousands of charge carrier pairs that can be separated by the nearby SIS junction. This enables the visualization of the internal charge carrier separation at the SiNW - front contact interface.

The LBIC and EBIC analyses were performed on the 1.4 cm^2 prototypes with metal front grid shown in Figure 6. For the LBIC measurements, a custom built system with four different lasers with wavelengths of 405, 544, 633 and 1064 nm was applied. Figure 9 shows the mapping results and an EBIC mapping of the same cell as well in order to be able to draw a comparison. The LBIC mappings are normalized to show the dependence of the power output on the light wavelength. Nearly no current is generated at 405 nm. At 544 nm, several areas are visible that have better or worse efficiency. These areas arise from an inhomogeneous etching process which generates areas with slightly higher or lower absorption. At 633 nm two small areas reach a generated current of 25 μA. Visible for the naked eye, these areas appear darker than the remaining surface. At 1064 nm, the light is not absorbed in the SiNWs themselves anymore but in the bulk wafer below. Therefore the signal is very homogeneous as the nanowires do not influence the current generation anymore. The EBIC mapping can be compared best with the 633 nm LBIC mapping. Due to the higher resolution of the EBIC, this measurement shows much more details.

In order to explain the different current amounts in the LBIC measurements in a better way, the EQE of the SC was measured. The results are shown in Figure 10. The wavelength-dependent behavior is in accordance with the determined LBIC mappings. At 405 nm, the EQE is at nearly zero and 633 nm exhibits the highest EQE from all four laser wavelengths. Although the EQE curve is similar in shape to typical commercial solar cell EQE results, only lower values are reached. The reasons are still under investigation. Furthermore, Figure 10 shows the optical absorption of the SC which was calculated by 1 minus the measured total reflectance. In Figure 11, a high resolution large scale EBIC scan of a SC demonstrator is shown. The simultaneously gathered SEM and EBIC signals are coded in brown and green color schemes. The EBIC signal clearly shows regions of lower or higher current due to etching variations and also many circular defects which seem to be the result of pinholes in the tunnel barrier layers. In several circular defects, a dirt particle in the middle seems to be the cause of the short circuit.

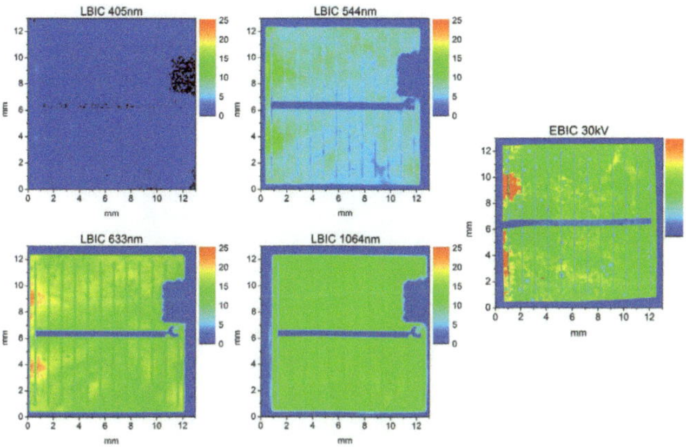

Figure 9. LBIC mapping at four different laser wavelengths and the corresponding EBIC measurement at 30 kV accelerating voltage. The inactive region in the top-right of each LBIC measurement is caused by shading from the mounting.

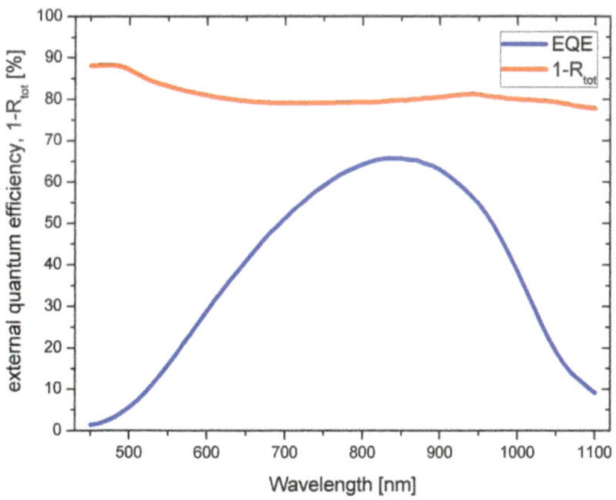

Figure 10. Absorption (1-R_{tot}) and EQE of a SiNW-based solar cell prototype.

The EBIC measurement is a powerful tool not only for observing the surface of a solar cell but also for looking inside the cell when combining sequences of EBIC analyses and FIB sectioning and polishing. Combined with a focused ion beam (FIB), information on sub-surface structures can be obtained. In a tomographic approach, a section into the SiNW based SC is carried out and a sub-surface slice of a few μm in depth and length is polished. The prepared area can subsequently be investigated by EBIC mapping and typical SEM SE imaging. Further slicing and polishing and repeated EBIC analysis permit a 3D reconstruction of EBIC as well as SE data resulting in tomographic information. Figure 12 represents the concept of a cross sectional EBIC measurement: The electron beam scans over the ion beam polished cross section of a SiNW based SIS SC and the current that flows between the front and back contact is amplified and mapped by using the electron beam position. A resulting image is also shown.

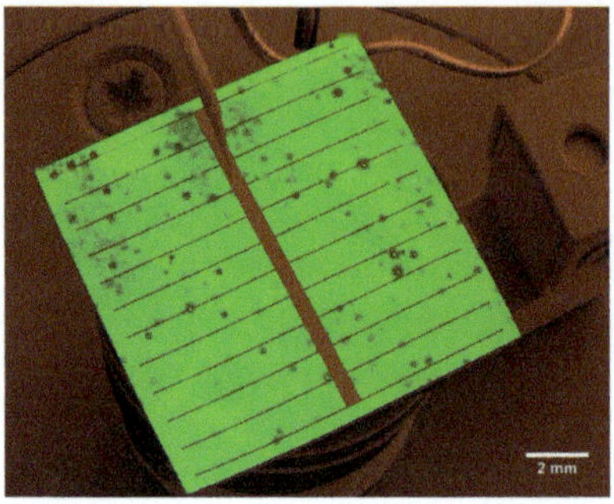

Figure 11. A large scale combined SEM/EBIC image of a solar cell surface. Point defects appear as darker dots and larger sized contrast differences indicate structural variations due to inhomogeneous etching.

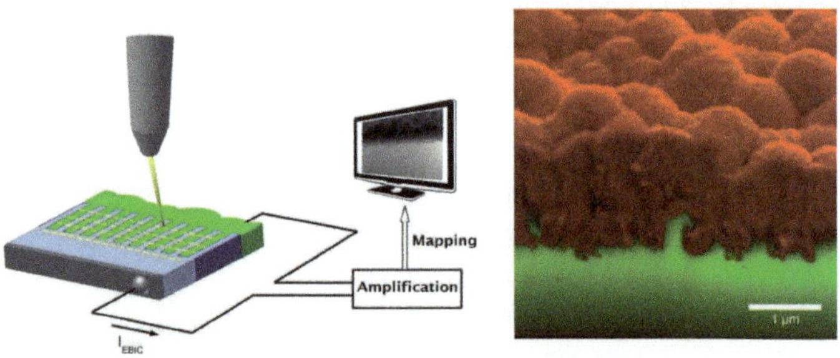

Figure 12. Left: Scheme of a cross sectional EBIC measurement of a SiNW based SIS SC. Right: colored dual-signal micrograph that shows the SEM image in red and the EBIC in green. A pronounced EBIC current of an individual SiNW is clearly visible in green.

The 3D-EBIC is devoted to solve an important question in the field of SiNW solar cells: Are the nanowires active photovoltaic components or do they only act as passive anti-reflecting building blocks? It is generally accepted that SiNWs absorb light by generating charge carrier pairs [10]. However, whether it is possible to separate electrons and holes prior to their recombination is a question to be answered for each and every material and cell concept again. To answer these interesting and important questions a dual beam SEM/FIB system (Tescan Lyra 3) with an integrated quantitative EBIC detector (Point electronic and Kammrath & Weiss) is used. The gallium ion beam of the FIB cuts a volume of 10x10x10 μm^3 of the solar cell perpendicular to the surface slice by slice. The electron beam observes each slice from an angle of 55° to the perpendicular and creates a secondary electron image as well as an EBIC image. Afterwards, these images can be combined in order to create a 3D reconstruction of the solar cell. In Figure 13, three slices of a 3D reconstruction are displayed. The assembled dataset offers a multitude of visualizations which can be shown best in a video. Therefore, the microscopy manufacturer Tescan provides a video of the analysis of our SiNW-SIS-SC which can be viewed via this internet link: http://www.youtube.com/watch?v=X9O6pwcN4Tg. It shows the possibilities in signal combination as well as the visualization of complex 3D surfaces.

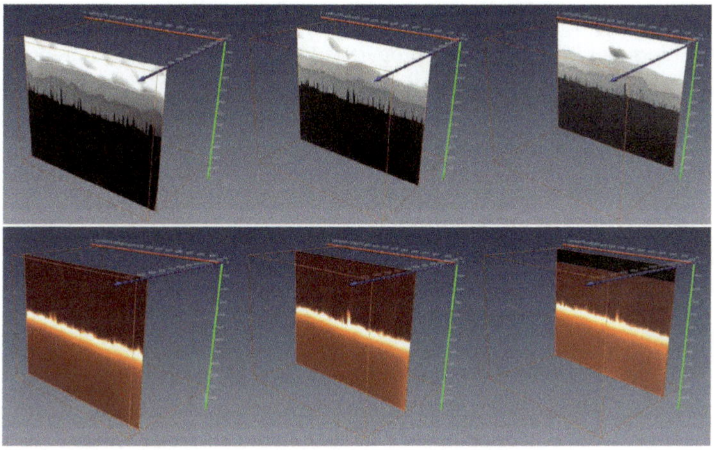

Figure 13. Three single slices out of 400 that show the SE-signal (top) as well as the EBIC signal (bottom) at the same position.

The 3D determination of electrical properties becomes very important in times of emerging nanostructured solar cells. Our results which are shown in Figure 12 and 13, prove that nanowires can be active photovoltaic components. This can be seen in the second slice where a single nanowire is visible in the EBIC image. Otherwise, many NWs are visible in the SEM image that cannot be seen in the EBIC signal. The difference between active and passive PV component seems to lie in the diameter of the wire. NWs, which are smaller than 50nm, exhibit no EBIC signal. That can be explained by the high surface to volume ratio, which leads to higher recombination, or a lesser absorption and thus no charge carrier generation occurs. These information are of great value when it comes to further optimization of the initial etching process. The nanowire dimensions should be in an area where high absorption is still achieved and charge carrier separation can take place inside the wires.

4. Outlook on nanowire research

The field of nanowire based solar cell research develops very fast and thus new methods and materials are coming up continuously. Our group is driving this type of research forward very actively and we are even constantly improving our own records. This outlook will give the reader some ideas about the progress of NW cell research in our lab. By further improving nanowire based cell concepts, we are aiming to achieve the goal of seeing these cells on top of a roof one day.

4.1. Silicon nanowire arrays formed by reactive ion etching

Reactive ion etching (RIE) in combination with nanosphere lithography (NSL) represents a new way of forming hexagonal arrays of SiNWs with controllable geometries in wafers [22] or even multicrystalline silicon (mc-Si) thin films on glass. [44] These structures offer several advantages regarding further systematic research on SiNW based optoelectronic and solar cell devices. Since device constituting properties like spectral absorption, reflection and the area and distribution of charge separating surfaces (p/n-junctions, heterojunctions etc.) are dependent on SiNW array geometries, a control of these geometries creates the opportunity for systematic device development. Figure 14 shows an example of the potential of this highly flexible and scalable process which results in large areas of nanostructured surfaces. The nanospheres arrange in self-aligned hexagonal colloidal 2D crystals on the wafer surface. The initial sphere diameter determines the lattice constant of the structure, whereas oxygen-plasma reduced spheres pre-determine the diameter of the SiNWs.

Figure 14. Left: RIE of a NSL patterned wafer surface. The nanospheres act as a shadow mask for the RIE process which results in aligned arrays of SiNWs. Inset: The lattice constant of the hexagonal arrays is pre-determined by the initial sphere diameter d_i, the diameter of the structures d_r can be controlled by the reduction of the initial spheres. Right: With the described method, the structuring of large surfaces is possible as it can be seen in the SEM micrograph.

RIE etched SiNWs may not offer the high throughput and cheap production of WCE NWs, but the predefined non-random structure offers many advantages in the control of optical and electrical properties in NW SCs compared to the random structures of chemically etched NWs and thus permit tailoring and optimizing SC efficiencies. Moreover, the etched SiNWs usually have a smaller surface to volume ratio, thus permitting smaller surfaces that need passivation [41].

4.2. Novel electrodes based on silver nanowires or graphene

One-dimensional (1D) metal nanostructures, especially single crystalline nanowires, have attracted intensive interest because of their great potential for practical applications in nanoscale electronics, optoelectronics, and molecular sensing devices. However, one of the most significant challenges still facing device fabrication and processing is the large scale reproducibility of the nanowires into highly controlled arrays with high uniformity. Herein, we demonstrate an extremely simple method for large-scale and highly ordered silver nanowire (AgNW) monolayer synthesis via a thermal assembling process. These AgNWs offer new possibilities in the formation of transparent and highly conductive contacts.

The AgNW were synthesized by wet chemistry. For a typical synthesis, 5 ml of ethylene-glycol (EG) was heated to 175±10 °C for 1 h under magnetic stirring. Subsequently, Cu-additive solution (40 μL, 4 mM $CuCl_2$) was injected together with the polymer PVP (1.5 ml with a concentration of 0.147 M) into the heated EG. The volume, systematically influences can design the final shape and size of the AgNW. Finally, 1.5 ml of a 0.094 M $AgNO_3$ was injected slowly (for 10min). After 1 to 1.5 h, the reaction was stopped and cooled to room temperature by a water bath. Products were then washed with ethanol three times.

SEM images with low and high magnification of Ag nanowires (AgNW) are presented in Figure 15. The AgNW had a typical length of 30±10 μm, with diameter of 50±10 nm.

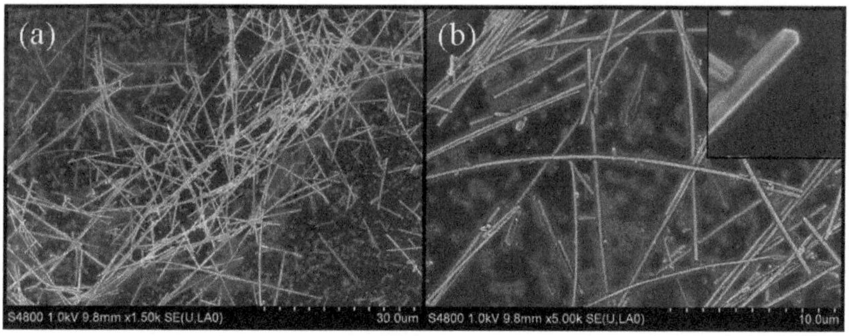

Figure 15. SEM micrographs of AgNW prepared by thermal assembly in the presence of PVP polymer. (a) A low-magnification SEM image reveals the highly uniform shape of AgNWs, (b) A higher magnification SEM image shows the high aspect ratio of AgNW. Inset is a SEM image that illustrates the pentagonal cross-section of the AgNWs.

Graphene with its outstanding optical and electrical properties offers many advantages when applied as contact material to SiNW SCs. The high in-plane conductivity as well as the high optical transparency recommends graphene as a new type of TCMs (transparent conductive materials) and it might substitute or support the typical TCO in our SiNW SC concept. It can be easily transferred to desired surfaces and its flexibility offers advantages for flexible SCs (e.g. on plastic substrates or metal foils). Figure 16 shows a transferred layer of graphene on top of RIE fabricated SiNWs. Preliminary investigations strongly promise excellent contact properties and thus research on this material will be forced with a lot of effort.

Figure 16. A graphene monolayer was deposited on top of SiNWs that are created by RIE to act as front contact. In the upper part the contact needle of the EBIC module can be seen.

5. Conclusions and outlook

In the last years not only have we developed and realized a novel nanowire based solar cell concept, but we have also developed advanced analysis methods for such kind of nanostructured devices. By systematically using cheap and abundant materials and avoiding expensive vacuum equipment for the nanostructuring process, we generated solar cells with over 9% conversion efficiency. As these values are the results of early stage research, further improvement up to an efficiency of 12-15% are expected. Furthermore, the concept will be transferred to thin films like multicrystalline silicon on glass or flexible substrates.

The development of 3D-EBIC will be most likely of great influence for the analysis of 3-dimensional nanostructured solar cells and thus it is also a great advance for the PV research field.

Acknowledgements

The authors gratefully acknowledge Florian Talkenberg for the ALD processes, Arne Bochmann for the LBIC measurements and Maria Hadjipanayi for the EQE investigations. Many thanks go to Gerald Brönstrup and Felix Voigt for fruitful discussions.

The authors gratefully acknowledge financial support by the European Commission projects within the framework of FP7, "ROD-SOL" and "Fiblys".

Author details

Björn Hoffmann[1,2], Vladimir Sivakov[2], Sebastian W. Schmitt[1], Muhammad Y. Bashouti[1], Michael Latzel[1], Jiří Dluhoš[3], Jaroslav Jiruse[3] and Silke Christiansen[1,2]

1 Max Planck Institute for the Science of Light, TDSU Photonic Nanostructures, Erlangen, Germany

2 Institute of Photonic Technology, Department Semiconductor Nanostructures, Jena, Germany

3 TESCAN, a.s., Brno, Czech Republic

References

[1] Green MA. (2001) Third generation photovoltaics: Ultra-high conversion efficiency at low cost. Prog Photovolt Res Appl. 9(2),123-35. http://doi.wiley.com/10.1002/pip.360

[2] Kaufmann CA, Neisser A, Klenk R, Scheer R. (2005) Transfer of Cu(In,Ga)Se2 thin film solar cells to flexible substrates using an in situ process control. Thin Solid Films. 480-481,515-9. http://dx.doi.org/10.1016/j.tsf.2004.11.067

[3] Green MA, Emery K, Hishikawa Y, Warta W, Dunlop ED. (2012) Solar cell efficiency tables (version 39). Prog Photovolt Res Appl. 20(1),12–20. http://onlinelibrary.wiley.com/doi/10.1002/pip.2163/full

[4] Bagnall DM, Boreland M. (2008) Photovoltaic technologies. Energy Policy. 36(12), 4390–6. http://linkinghub.elsevier.com/retrieve/pii/S0301421508004552

[5] Du QG, Kam CH, Demir HV, Yu HY, Sun XW. (2011) Broadband absorption enhancement in randomly positioned silicon nanowire arrays for solar cell applications. Opt Lett. 36(10),1884–6. http://www.ncbi.nlm.nih.gov/pubmed/21593923

[6] Tsakalakos L, Balch J, Fronheiser J, Shih M-Y, LeBoeuf SF, Pietrzykowski M, et al. (2007) Strong broadband optical absorption in silicon nanowire films. J Nanophotonics. 1(1),013552. http://link.aip.org/link/JNOACQ/v1/i1/p013552/s1&Agg=doi

[7] Srivastava SK, Kumar D, Singh PK, Kar M, Kumar V, Husain M. (2010) Excellent antireflection properties of vertical silicon nanowire arrays. Sol. Energy Mater. Sol. Cells. 94(9),1506–11. http://linkinghub.elsevier.com/retrieve/pii/S0927024810000863

[8] Srivastava SKSK, Kumar D, Singh PKP, Kar M, Kumar V, Husain M. (2011) Fabrication of silicon nanowire arrays based solar cell with improved performance. Sol. Energy Mater. Sol. Cells. 95(1),215–8. http://linkinghub.elsevier.com/retrieve/pii/S092702481000214X

[9] Sivakov V, Andrä G, Gawlik A, Berger A, Plentz J, Falk F, et al. (2009) Silicon nanowire-based solar cells on glass: synthesis, optical properties, and cell parameters. Nano Lett. 9(4),1549–54. http://dx.doi.org/10.1021/nl803641f

[10] Brönstrup G, Garwe F, Csáki A, Fritzsche W, Steinbrück A, Christiansen S. (2011) Statistical model on the optical properties of silicon nanowire mats. Phys. Rev., B. 84(12),1–10. http://link.aps.org/doi/10.1103/PhysRevB.84.125432

[11] Cao L, White JS, Park J-S, Schuller J a, Clemens BM, Brongersma ML. (2009) Engineering light absorption in semiconductor nanowire devices. Nat Mater. 8(8),643–7. http://www.ncbi.nlm.nih.gov/pubmed/19578337

[12] Xie WQ, Oh JI, Shen WZ. (2011) Realization of effective light trapping and omnidirectional antireflection in smooth surface silicon nanowire arrays. Nanotechnology. 22(6),065704. http://www.ncbi.nlm.nih.gov/pubmed/21212474

[13] Muskens OL, Rivas JG, Algra RE, Bakkers EP a M, Lagendijk A. (2008) Design of light scattering in nanowire materials for photovoltaic applications. Nano Lett. 8(9), 2638–42. http://www.ncbi.nlm.nih.gov/pubmed/18700806

[14] Wagner R, Ellis W. (1964) Vapor-Liquid-Solid Mechanism of Single Crystal Growth. Appl Phys Lett. 4(5),89–90. http://ieeexplore.ieee.org/xpls/abs_all.jsp?arnumber=4850930

[15] Hofmann S, Ducati C, Neill RJ, Piscanec S, Ferrari a. C, Geng J, et al. (2003) Gold catalyzed growth of silicon nanowires by plasma enhanced chemical vapor deposition. J Appl Phys. 94(9),6005. http://link.aip.org/link/JAPIAU/v94/i9/p6005/s1&Agg=doi

[16] Sunkara MK, Sharma S, Miranda R, Lian G, Dickey EC. (2001) Bulk synthesis of silicon nanowires using a low-temperature vapor–liquid–solid method. Appl Phys Lett. 79(10),1546. http://link.aip.org/link/APPLAB/v79/i10/p1546/s1&Agg=doi

[17] Peng K, Fang H, Hu J, Wu Y, Zhu J, Yan Y, et al. (2006) Metal-particle-induced, highly localized site-specific etching of Si and formation of single-crystalline Si nanowires in aqueous fluoride solution. Chemistry. 12(30),7942–7. http://www.ncbi.nlm.nih.gov/pubmed/16871502

[18] Sivakov V, Bro◎nstrup G, Pecz B, Berger a., Radnoczi GZ, Krause M, et al. (2010) Realization of Vertical and Zigzag Single Crystalline Silicon Nanowire Architectures. J Phys Chem C. 114(9),3798–803. http://pubs.acs.org/doi/abs/10.1021/jp909946x

[19] Huang Z, Geyer N, Werner P, de Boor J, Gösele U. (2011) Metal-assisted chemical etching of silicon: a review. Advanced Materials. 23(2),285–308. http://doi.wiley.com/10.1002/adma.201001784

[20] Sivakov V, Voigt F, Hoffmann B, Gerliz V, Christiansen S. (2011) Wet - Chemically Etched Silicon Nanowire Architectures◎: Formation and Properties. In: Hashim A, editor. Nanowires - Fundamental Research. Intech; 2011. p. 45 - 80.http://www.intechopen.com/articles/show/title/wet-chemically-etched-silicon-nanowire-architectures-formation-and-properties

[21] Huynh WU, Dittmer JJ, Alivisatos AP. (2002) Hybrid nanorod-polymer solar cells. Science (80-). 295(5564),2425–7. http://www.sciencemag.org/content/295/5564/2425.abstract

[22] Garnett E, Yang P. (2010) Light trapping in silicon nanowire solar cells. Nano Lett. 10(3),1082–7. http://www.ncbi.nlm.nih.gov/pubmed/20108969

[23] Garnett EC, Yang P. (2008) Silicon nanowire radial p-n junction solar cells. J. Am. Chem. Soc. 130(29),9224–5. http://www.ncbi.nlm.nih.gov/pubmed/18576622

[24] Ke Y, Wang X, Weng XJ, Kendrick CE, Yu YA, Eichfeld SM, et al. (2011) Single wire radial junction photovoltaic devices fabricated using aluminum catalyzed silicon nanowires. Nanotechnology. 22(44),445401. http://www.ncbi.nlm.nih.gov/pubmed/21983364

[25] Hoffmann S, Bauer J, Ronning C, Stelzner T, Michler J, Ballif C, et al. (2009) Axial p-n junctions realized in silicon nanowires by ion implantation. Nano Lett. 9(4),1341–4. http://www.ncbi.nlm.nih.gov/pubmed/19256535

[26] Wen C-Y, Reuter MC, Bruley J, Tersoff J, Kodambaka S, Stach EA, et al. (2009) Formation of compositionally abrupt axial heterojunctions in silicon-germanium nanowires. Science (80-). 326(5957),1247–50. http://www.ncbi.nlm.nih.gov/pubmed/19965471

[27] Tsai S-H, Chang H-C, Wang H-H, Chen S-Y, Lin C-A, Chen S-A, et al. (2011) Significant Efficiency Enhancement of Hybrid Solar Cells Using Core-Shell Nanowire Geometry for Energy Harvesting. ACS Nano., 5(12), 9501-10. http://pubs.acs.org/doi/abs/10.1021/nn202485m

[28] Shewchun J, Dubow J, Myszkowski A, Singh R. (1977) The operation of the semiconductor-insulator-semiconductor (SIS) solar cell: Theory. J Appl Phys. 49(2),855–64. http://ieeexplore.ieee.org/xpls/abs_all.jsp?arnumber=5104464

[29] Shewchun J, Dubow J, Wilmsen C, Singh R, Burk D, Wager J. (1978) The operation of the semiconductor-insulator-semiconductor solar cell: Experiment. J Appl Phys. 50(4),2832–9. http://ieeexplore.ieee.org/xpls/abs_all.jsp?arnumber=5105997

[30] Genis AP, Smith PA, Emery K, Singh R, DuBow JB. (1980) Efficient indium tin oxide/polycrystalline silicon semiconductor-insulator-semiconductor solar cells. Appl Phys Lett. 37(1),77. http://link.aip.org/link/APPLAB/v37/i1/p77/s1&Agg=doi

[31] Ritala M, Leskela M. (2001) Atomic layer deposition. Handbook of thin film materials. 1,103–59. http://cambridgenano.com/papers/atomic layer deposition ritala leskela.pdf

[32] Leskelä M, Ritala M. (2002) Atomic layer deposition (ALD): from precursors to thin film structures. Thin Solid Films. 409(1),138–46. http://linkinghub.elsevier.com/retrieve/pii/S0040609002001177

[33] Dingemans G, Seguin R, Engelhart P, Sanden MCMVD, Kessels WMM. (2010) Silicon surface passivation by ultrathin Al 2 O 3 films synthesized by thermal and plasma

atomic layer deposition. Phys. Status Solidi RRL. 4(1-2),10–2. http://doi.wiley.com/10.1002/pssr.200903334

[34] Profijt HB, Potts SE, van de Sanden MCM, Kessels WMM. (2011) Plasma-Assisted Atomic Layer Deposition: Basics, Opportunities, and Challenges. J Vac Sci Technol A. 29(5),050801. http://link.aip.org/link/JVTAD6/v29/i5/p050801/s1&Agg=doi

[35] Saarenpää H, Niemi T, Tukiainen A, Lemmetyinen H, Tkachenko N. (2010) Aluminum doped zinc oxide films grown by atomic layer deposition for organic photovoltaic devices. Sol. Energy Mater. Sol. Cells. 94(8),1379–83. http://linkinghub.elsevier.com/retrieve/pii/S0927024810001716

[36] Yamada A, Sang B, Konagai M. (1997) Atomic layer deposition of ZnO transparent conducting oxides. Appl Surf Sci. 112,216–22. http://linkinghub.elsevier.com/retrieve/pii/S0169433296010227

[37] Angeli J, Bengtson A, Bogaerts A, Hoffmann V, Hodoroaba V-D, Steers E. (2003) Glow discharge optical emission spectrometry: moving towards reliable thin film analysis: a short review. J Anal At Spectrom. 18(6),670. http://pubs.rsc.org/en/content/articlehtml/2003/ja/b301293j

[38] Schmitt SW, Gamez G, Sivakov V, Schubert M, Christiansen SH, Michler J. (2011) Chemical and optical characterisation of atomic layer deposition aluminium doped ZnO films for photovoltaics by glow discharge optical emission spectrometry. J Anal At Spectrom. 26(4),822. http://xlink.rsc.org/?DOI=c0ja00158a

[39] Pisonero J, Fernández B, Pereiro R, Bordel N, Sanz-Medel A. (2006) Glow-discharge spectrometry for direct analysis of thin and ultra-thin solid films. Trends Analyt Chem. 25(1),11–8. http://dx.doi.org/10.1016/j.trac.2005.04.019

[40] Schmitt SW, Venzago C, Hoffmann B, Sivakov V, Hofmann T, Michler, J., Christiansen, S. Gamez G. (2012) Glow discharge techniques in the chemical analysis of photovoltaic materials. Prog Photovolt Res Appl. http://doi.wiley.com/1021002/pip.2264.

[41] Bashouti MY, Stelzner T, Berger A, Christiansen S, Haick H. (2008) Chemical Passivation of Silicon Nanowires with C 1 –C 6 Alkyl Chains through Covalent Si–C Bonds. J Phys Chem C. 112(49),19168–72. http://dx.doi.org/10.1021/jp8077437

[42] Dan Y, Seo K, Takei K, Meza JH, Javey A. (2011) Dramatic reduction of surface recombination by in situ surface passivation of silicon nanowires. Nano Lett. 11(6) 2527 http://pubs.acs.org/doi/abs/10.1021/nl201179n

[43] Kim DR, Lee CH, Rao PM, Cho IS, Zheng X. (2011) Hybrid Si Microwire and Planar Solar Cells: Passivation and Characterization. Nano Lett. 11(7), 2704 http://www.ncbi.nlm.nih.gov/pubmed/21609002

[44] Schmitt SW, Schechtel F, Amkreutz D, Bashouti M, Srivastava, S. K., Hoffmann B, Rech B, et al. (2012) Nanowire arrays in multi-crystalline silicon thin films on glass: a promising material for photovoltaics, opto-electronics and fundamental research in nano-technology. Nano Lett. 12(8), 4050.

[45] Bajaj J, Tennant WE. (1990) Remote contact LBIC imaging of defects in semiconductors. J Cryst Growth. 103(1-4),170–8. http://www.sciencedirect.com/science/article/pii/002202489090186O

[46] Donolato C. (1983) Theory of beam induced current characterization of grain boundaries in polycrystalline solar cells. J Appl Phys. 54(3),1314. http://link.aip.org/link/JAPIAU/v54/i3/p1314/s1&Agg=doi

Section 3

Carbon Based Nanowires

Chapter 12

Carbon Nanotubes Filled With Ternary Chalcohalides

Marian Nowak and Marcin Jesionek

Additional information is available at the end of the chapter

1. Introduction

Since the discovery by S. Iijima in 1991 [1] various potential applications have been proposed for carbon nanotubes (CNTs): sensors, field emission displays, nanometer-sized semiconductor devices and hydrogen storage media. There is a huge literature stream related to nanotube research. On a fundamental level, there are still challenges to mass-produce controlled nanostructures at reasonable cost and new features. There is also a great demand for control versatile electronic characteristics of CNTs. One strategy is to use the CNTs themselves, controlling useful properties *via* their radii and morphologies. An alternative approach leading to new features of CNTs, i.e. directional action on their versatile electronic characteristics, is based on filling them with condensed substances from a wide range of materials [2-8]. CNTs are sp^2 graphene carbon cylinders capable of hosting a variety of species, including 1D crystals of metals, metal salts and oxides; semiconductors; superconductors; and chains of fullerene or endofullerene molecules (see Refs. cited in [7,8]). Such objects are distinguished in their unique physical and chemical properties from both hollow nanotubes and the encapsulated substances, which permits one to purpose-tailor "nanowires" and "nanotubes" with unique physical and chemical properties.

Among materials encapsulated within CNTs are both elementary substances and the binary compounds formed from the group 15–16–17 elements. Figure 1 presents a typical high resolution transmission electron microscopy (HRTEM) image of single-walled carbon nanotube (SWNT) filled with one of the metal halogenides, i.e., BaI_2@SWNT. However, one should note the possible filling of SWNTs with the alone helical iodine chains (Figure 2). Of course, the multi-walled carbon nanotubes (MWNTs) can be filled with the elemental forms (e.g., Bi, Sb, S, Se, Te, I_2), halogenides (e.g., consistent of Cl, Br, I_2), oxides (e.g., Sb_2O_3), and chalcogenides (e.g., Sb_2S_3, CdS, SnSe, SnTe, HgTe), too. A comprehensive list of such carbon nanostructures has been reviewed by A. Eliseev at al. [7] and P. Lukanow et al. [8].

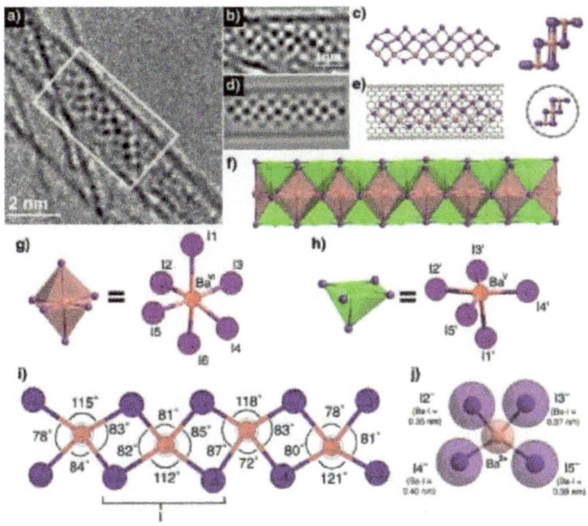

Figure 1. a) HRTEM image of BaI_2/SWNT composite. b) Detail from boxed region in (a), for which image noise was reduced with a deconvolution filter. c) Structure plot based on a 2D peak- mapping analysis of (b); end-on view on the right. d) Simulated HRTEM image, e) complete structure model of BaI_2/SWNT composite. f) Coordination model of 1D BaI_2 chain. g) 5- coordinate, h) 6-coordinate BaI_6 polyhedra and ball-and-stick models derived from (c). i) and j) Models showing derived equatorial bond angles and distances for BaI_4 units along the center of the 1D BaI_2 chain (ionic radii in (j) indicated by diffuse spheres) (Reproduced from [9] with permission. Copyright Wiley-VCH Verlag GmbH).

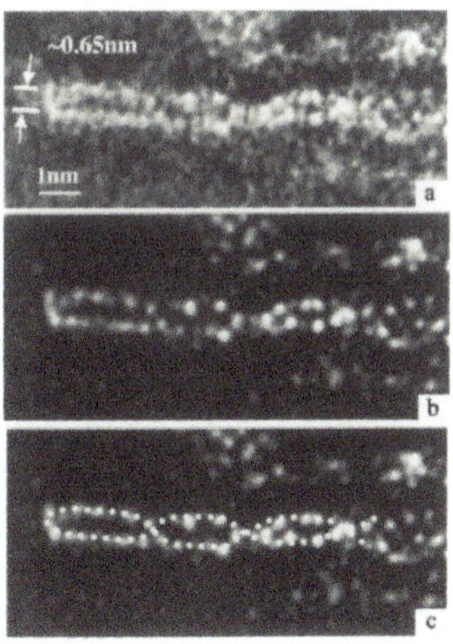

Figure 2. a) Higher resolution Z-contrast image of a SWNT containing two strands of iodine. (b) Maximum entropy processed image to reduce the noise. Two constrictions can be seen, consistent with the projection of a double helix configuration with the two chains spiraling along the inside walls of the nanotube as shown in the overlay in (c). The period of the helix is about 5 nm, and the maximum separation of the two strands is ~0.65 nm, much less than the 1.3 -1.4 nm diameter of the SWNT (Reproduced from [10] with permission. Copyright the American Physical Society).

This review summarizes results of recent investigations [11-14] of carbon nanotubes (CNTs) filled with ternary chalcohalides formed from the group 15–16–17 elements, i.e. antimony sulfoiodide (SbSI) and antimony selenoiodide (SbSeI) known as SbSI-type materials. Some recent original data are presented, too. It should be underlined that the bulk SbSI being a semiconducting ferroelectric has an unusually large number of very attractive and suitable properties (see Refs. cited in [15-17]). Among them there are the photoferroelectricity, pyroelectric, pyrooptic, piezoelectric, electromechanical, electrooptic, photorefractive and nonlinear optical effects. Therefore SbSI is taken into consideration as a valuable material for many applications (see the literature cited in [15-17]).

SbSI type materials have been synthesized in different ways (see Refs. in [15-17]. The most often used method consists in fusion of stoichiometric amounts of antimony, sulfur and iodine or antimony iodide, and antimony sulphide with antimony iodide. The procedure requires high temperature (723 - 873 K) and long reaction time (1-3 days). Recently [15,17], a novel sonochemical method for direct preparation of semiconducting and ferroelectric SbSI nanowires has been established. The determined [15,17] value of the indirect forbidden energy band gap of SbSI gel E_{gIf}=1.829(27) eV is well compared to the bulk value of band gap of SbSI reported in the literature (see Refs. in [15,17]). The maximum of dielectric constant $\varepsilon=1.6\cdot10^4$ of SbSI nanowires was observed at Curie temperature T_c=292(1) K [17,18] that well corresponds with the phase transition in bulk SbSI crystals. It should be underlined that SbSeI can be sonochemically prepared in the form of crystalline nanowires [17,19], too.

2. Preparation of SbSI@CNTs and SbSeI@CNTs

There are known a few methods for filling CNTs with different substances [4,6-8]: catalytic synthesis of nanotubes using the metals as catalysts, capillary drawing-in of molten materials or materials dissolved in solvents having a low surface tension, saturation with metal vapor as well as electrochemical methods based on passing the electrical current through an electrolyte containing dissolved metal atoms. In this paper we present another method for inserting materials into the inner cavity of a nanotube. Our method is based on sonochemistry. It is well known [20-25] that ultrasound can induce new reactivities leading to the formation of unexpected chemical species. What makes sonochemistry unique is the remarkable phenomenon of cavitation [25,26]. Comparing sonochemical method of preparing materials with the traditional ones, it can be seen that ultrasound irradiation can be used at room temperature and ambient pressure to promote heterogeneous reactions that normally occur only under extreme conditions of hundreds of atmospheres and degrees (e.g., [20-22]).

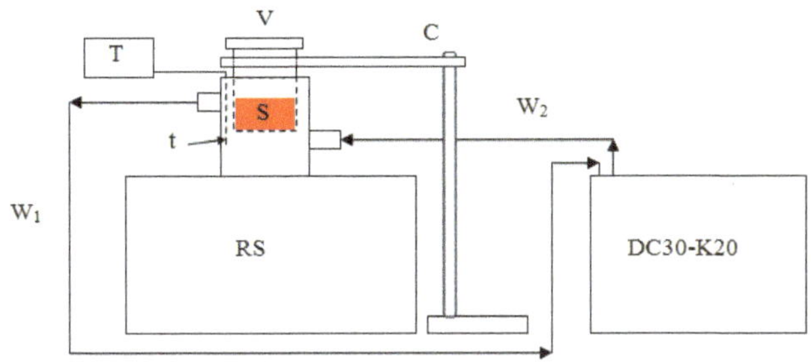

Figure 3. Schematic illustration of the set-up for ultrasonic filling of CNTs with SbSI and SbSeI (RS– ultrasonic reactor; DC30-K20 – refrigerated circulator bath; W_1, W_2 – thermal cycle; C – stand; S –sample; V – Pyrex glass cylinder; T – temperature controller; t – temperature sensor) [13].

Both SbSI and SbSeI were prepared in CNTs ultrasonically from the constituents (the elements Sb, I, and S or Se) [11-14]. Methanol and ethanol served as the solvents for these reactions, respectively. All the reagents used in experiments were of analytical purity and were used without further purification. Antimony (99.95 %), selenium (99.5 %) and multi–walled CNTs (659258–10G, 90+%) were purchased from Sigma–Aldrich. Sublimated sulfur (pure p.a.), iodine (pure p.a.), absolute methanol (pure p.a.), and absolute ethanol (pure p.a.) were purchased from POCH S.A. (Gliwice, Poland). In a typical procedure of fabrication SbSI@CNTs [11], the elemental mixture with stoichiometric ratio of e.g. 0.380 g Sb, 0.099 g S and 0.394 g I, was immersed with 0.282 g of CNTs in 40 ml absolute methanol, which was contained in a 54 ml Pyrex glass cylinder of 20 mm inside diameter. In the first procedure of fabrication SbSeI@CNTs [14], the elemental mixture with stoichiometric ratio of e.g. 0.156 g Sb, 0.101 g Se and 0.165 g I_2, was immersed with 0.102 g of CNTs in 10 ml absolute ethanol, which was contained in a 54 ml Pyrex glass cylinder of 20 mm inside diameter. Furthermore, the syntheses of SbSeI in CNTs have been done in [14] for twice and four times greater contents of Se in the mixture (the obtained products are discussed in the next section). Vessels with the substrates were closed during experiments [11, 14] to prevent volatilization of the precipitants in long time tests. Bottoms of the vessels were planar or semispherical and 1 mm in thickness. Figure 3 presents scheme of the technological set-up. The cylinders were partly submerged in water in an ultrasonic reactor (InterSonic IS–UZP–2, frequency 35 kHz, with 75 W electrical power and 2.6 W/cm^2 power density guaranteed by the manufacturer). Temperature of the bath was established using the Haake DC–K20 refrigerated circulator bath (Thermo Scientific). The sonochemical processes were continued for 3 h at 323 K temperature of the water in the ultrasonic reactor. During the sonications sols were formed. It was observed that the color of the suspension changed gradually indicating the growth process of the SbSI and SbSeI. To control this process, measurements of optical diffusive reflectance $R_d(\lambda)$ of the sample were performed (Figure 4). It was assumed that the sonochemical process is finished when the spectral characteristics of $R_d(\lambda)$ do not change with time.

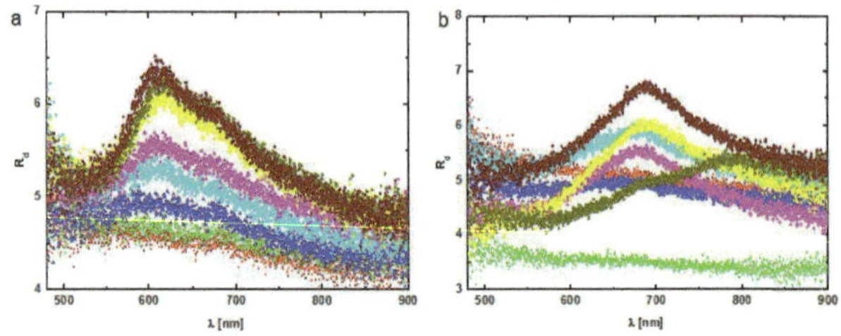

Figure 4. Change of diffusive reflectance during the sonication of Sb, S and I in methanol (a) and of Sb, Se and I in ethanol (b) (■ – after 5 min. of sonication, T=323 K; ● – 20 min. ▲ – 40 min. ▼ – 60 min.; ♦– 80 min.;, ◄ – 120 min.; ► – 180 min.; ● – after washing) [13].

When the sonication processes were finished, a dark sols were obtained. They were centrifuged using the MPW–223e centrifuge, MPW Med. Instruments (Poland), to extract the products. Then the liquids above the sediments were replaced with pure methanol or ethanol to wash the precipitates in the cases of SbSI@CNTs and SbSeI@CNTs, respectively. The centrifugation and washing were performed 5 times in both cases. At the end methanol as well as ethanol were evaporated from the samples during the drying in air at room temperature, so a brown-purple (Figure 5a) and dark brown–purple (Figure 5b) substances were obtained in the cases of SbSI@CNTs and SbSeI@CNTs, respectively.

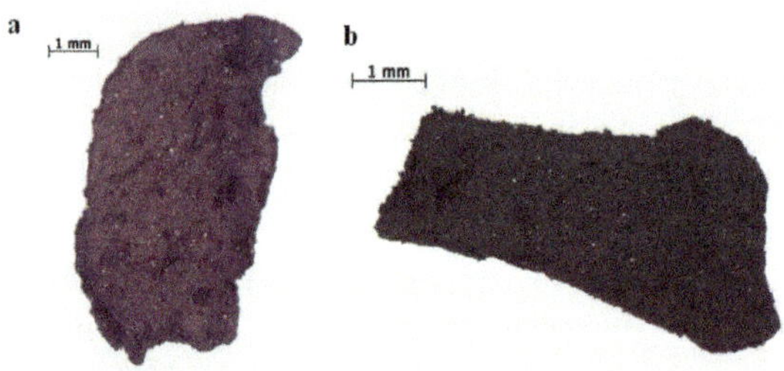

Figure 5. Dried MWCNTs filled ultrasonically with SbSI in methanol and SbSeI in ethanol [13].

3. Morphology, composition and crystal structure of SbSI@CNTs and SbSeI@CNTs

Characterization of the multi–walled CNTs filled with SbSI or SbSeI [11-14] was accomplished using different techniques, such as powder X–ray diffraction (XRD) performed on a JEOL JDX-7S X-ray diffractometer with graphite monochromatized Cu K_α radiation (λ = 0.154056 nm), scanning electron microscopy (SEM) and energy dispersive X–ray spectroscopy (EDS) performed on a Hitachi S-4200 microscope with EDS Thermo Scientific spectrometer, high–

resolution transmission electron microscopy (HRTEM) and selected area electron diffraction (SAED) performed on a JEOL-JEM 3010 microscope.

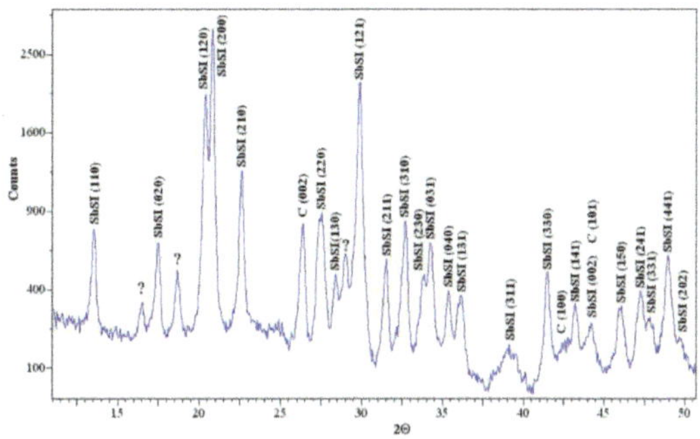

Figure 6. The powder XRD pattern of dried CNTs filled with SbSI ultrasonically in methanol [11].

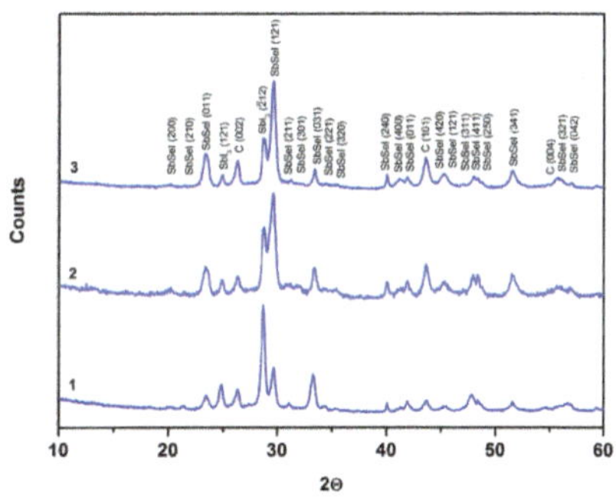

Figure 7. The powder XRD patterns of dried multi–walled CNTs filled with SbSeI ultrasonically in ethanol with different compositions of the reactants [14] (1- 0.101 g Se, 2- 0.202 g Se, 3- 0.404 g Se; 0.156 g Sb, 0.165 g I_2, and 0.102 g of CNTs in 10 ml absolute ethanol; described peaks correspond to the crystalline SbSeI [27,28], C [29,30] and SbI_3 [31,32]).

The powder XRD pattern of the MWNTs filled with SbSI is shown in Figure 6. The well-defined, sharp diffraction lines suggest the well-crystallized substance. It was found that the diffraction lines can be divided into three groups. In the first group, containing most of the lines, the peaks can be indexed to be a pure orthorhombic phase for SbSI with the cell constants a=0.858 nm, b=1.017 nm, and c=0.414 nm. The identification was done using the PCSI-WIN computer program and the data from JCPDS-International Centre for Diffraction Data 2000. The intensities and positions of the peaks are in good agreement with literature values for SbSI [33]. The second group of diffraction lines can be indexed to be a carbon phase $P6_3mc$ with the cell constants a=0.2470 nm and c=0.6790 nm [30]. The third group of a few additional X-ray diffraction lines for 2θ equal 16.51°, 18.66° and 29.00° has given the following values of d_{hkl}: 0.536 nm, 0.475 nm, and 0.308 nm, respectively. These lines may be explained as follows. The XRD investigations [34] of SbSI sonochemically produced in methanol show the coexistence of phases with $Pna2_1$ and *Pnam* crystal symmetry that are characteristic for ferroelectric and paraelectric domains, respectively. Hence, such data obtained at 298 K represent a structure of just below or very near the transition temperature of SbSI. Another explanation for the additional X-ray diffraction lines can be given taking into account existence of unknown phase in the investigated material. This problem must be studied in the future.

XRD			SAED		HRTM		Literature data					
							C [30]		SbSeI [28]		SbI_3 [32]	
2θ	I_{rel}	d_{hkl} [pm]	d_{hkl} [pm]	hkl	sign	d_{hkl} [pm]	d_{hkl} [pm]	hkl	d_{hkl} [pm]	hkl	d_{hkl} [pm]	hkl
20.166	8	440.0			2	432(13)			434.9	200		
21.468	8	413.6							401.3	210		
23.324	35	381.1							383.66	011		
24.961	17	356.4									351.41	121
26.245	29	339.3			1	332(5)	339.5	002				
28.797	49	309.8									311.63	$\bar{2}12$
29.602	100	301.5							303.13	121		
31.183	13	286.6							287.7	211		
31.768	10	281.4							279.3	301		
33.422	22	267.9							265.62	031		
34.488	9	259.8							259.51	221		
35.457	8	253.0							253.3	320		
40.013	17	225.1	223.35	$\bar{2}40$					223.35	240		
41.154	13	219.2							217.45	400		
41.869	15	215.6					213.9	100	215.88	321		
43.578	32	207.5	206.35	002			204.02	101	206.35	002		
45.117	17	200.8							200.65	420		
45.448	17	199.4							197.14	112		
47.029	8	193.1							195.85	331		
47.966	15	189.5							189.17	411		
48.459	13	187.7							187.81	250		
51.546	21	177.2							175.3	341	177.12	143
54.88	8	167.2					169.75	004				
55.806	14	164.6							165.9	321		
56.981	10	161.5							161.7	042		
			151.56	$\bar{2}42$					151.56	242		

Table 1. Comparison of interplanar spacings determined by XRD (Figure 7), SAED (Figure 14) and HRTEM (Figure 12) of multi-walled CNTs filled with SbSeI sonochemically in ethanol with literature data for CNTs, SbSeI orthorhombic crystals, and SbI_3 monoclinic crystals.

The powder XRD patterns of the CNTs filled ultrasonically with SbSeI are shown in Figure 7. One can see that the best results (the lowest intensity of SbI_3 peaks) have been obtained for four times greater content of Se in the mixture (compare curves in Figure 7). These results indicate that the CNTs should be filled with SbSeI with the excess of Se. The well–defined, sharp diffraction lines suggest the well–crystallized substance (curve 3 in Figure 7). It was found that the diffraction lines can be divided into three groups (Table 1). In the first group, containing most of the lines, the peaks can be indexed to be a pure orthorhombic phase for SbSeI with the cell constants a=0.8698 nm, b=1.0412 nm, and c=0.4127 nm [27,28]. The intensities and positions of the peaks are in good agreement with literature values for SbSeI [27,28]. The second group of diffraction lines can be indexed to be a carbon phase $P6_3mc$ with the cell constants a=0.2470 nm and c=0.6790 nm [29,30]. The third group of a few additional X–ray diffraction lines can be attributed to the monoclinic antimony iodide (SbI_3) with the cell constants a=0.7281 nm, b=1.0902 nm, c=0.8946 nm, and β=109.930 ° [31,32]. It can be explained as follows. In ethanol iodine reacts with antimony forming antimony iodide (1). This substance is soluble in alcohols [35]. When the alcohol is evaporated, the SbI_3 crystallizes. It is probable that (despite the washing of the CNTs sonochemically filled with SbSeI) some portion of SbI_3 still exists in the product. This problem should be solved in the future.

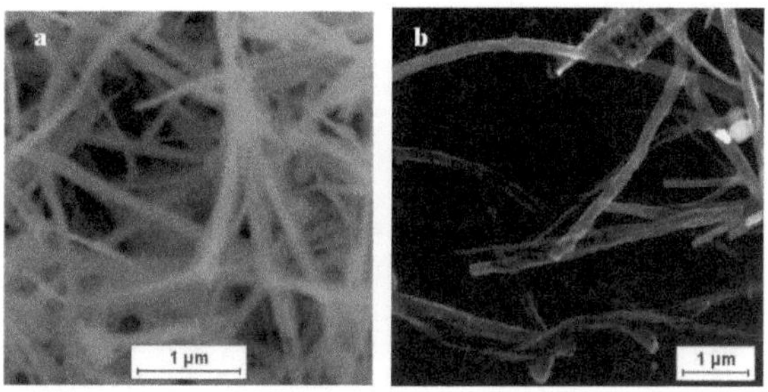

Figure 8. The typical SEM micrographs of dried gels of SbSI@CNTs [11] and SbSeI@CNTs [14].

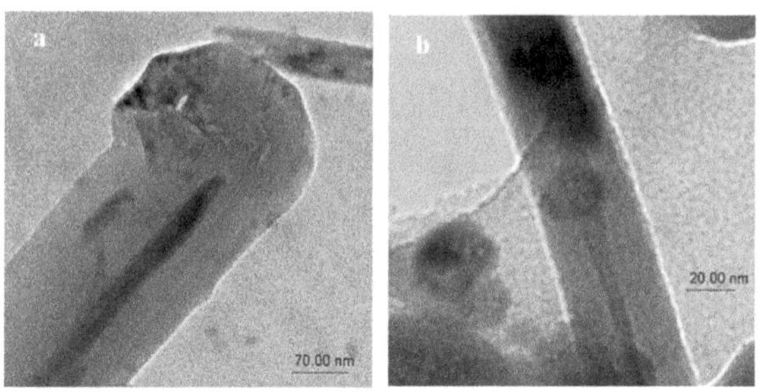

Figure 9. Typical TEM images of an individual SbSI@CNT [11] and SbSeI@CNT [14].

The typical SEM micrographs of the dried SbSI@CNTs and SbSeI@CNTs prepared sonochemically in methanol and ethanol, respectively, are shown in Figures 8 and 9. The samples used for SEM and EDS observations were prepared by dispersing a small quantity of the CNTs filled with SbSI or SbSeI in ethanol followed by ultrasonic vibration for 10 min. One or two drops of the nanoparticle solutions were deposited on naturally oxidized silicon single crystal plate and dried in 120 min at room temperature in vacuum. One can see SbI_3 nanoparticles around some of the SbSeI@CNTs (Figure 9). Probably, these nanoparticles have been grown from the solution (SbI_3 dissolved in ethanol) during drying of the product. The SbI_3 by-product is the rest of intermediate compound obtained during sonochemical synthesis of SbSeI [36]. The dimension of SbI_3 nanoparticle has been about 30 nm. The chemical composition and crystal structure of these nanoparticles have been proved [14].

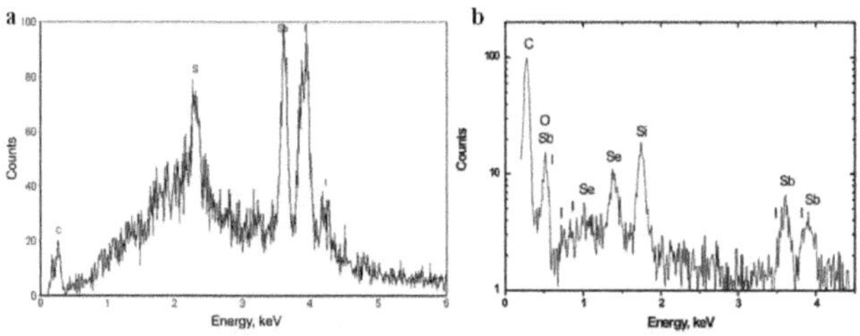

Figure 10. The EDS spectra of dried multi–walled CNTs filled ultrasonically with SbSI [11] and SbSeI [14] (the investigated material was deposited on naturally oxidized Si single crystal).

Element	Results of EDS investigations			
	SbSI@CNTs [11]		SbSeI@CNTs [14]	
	Concentration of all detected elements (at. %)	Concentration of components without C (at. %)	Concentration of components without Si and O (at. %)	Concentration of components without Si, O and C (at. %)
Sb	5.1(3)	46(2)	2.1	44
S	1.7(3)	21(3)	–	–
Se	–	–	1.3	28
I	3.8(3)	33(2)	1.3	28
C	89.4(13.4)	–	95.3	–

Table 2. Atomic concentration of components determined by EDS of the multi–walled CNTs filled sonochemically with SbSI in methanol and with SbSeI in ethanol.

The EDS spectra of dried multi–walled CNTs filled ultrasonically with SbSI and SbSeI are presented in Figure 10. The measured atomic concentrations of Sb, S, Se, I and C (without Si and O from the supporting wafer) are presented in Table 2. Taking into consideration the large excess of Sb and the deficit of S or Se as well as I detected by EDS, it should be underlined that the EDS analysis performed on sonochemically produced alone SbSeI gel confirmed an elemental atomic ratio of 0.41:0.26:0.33 for Sb, Se, and I [19]. Therefore, as it was suggested in [19], probably a separate phase of Sb and I is present, and presumably it is adsorbed on the surface of the SbSeI. A similar case was recently discovered in the XPS investigations of sonochemically prepared SbSI nanowires [37]. It is improbable that the excess Sb is present within the ordered SbSI and SbSeI nanowires inside the CNTs, since the interplanar spacings corresponds to stoichiometric materials (Figures 11 and 12). In [37] the antimony subiodide was suggested as the hypothetical material of the surface layer on SbSI nanowires. Such Sb_3I subiodide was obtained sonochemically [38]. Probably, there is the same compound on the surface of sonochemically prepared SbSeI nanowires. The existence of this substance will be investigated in the near future.

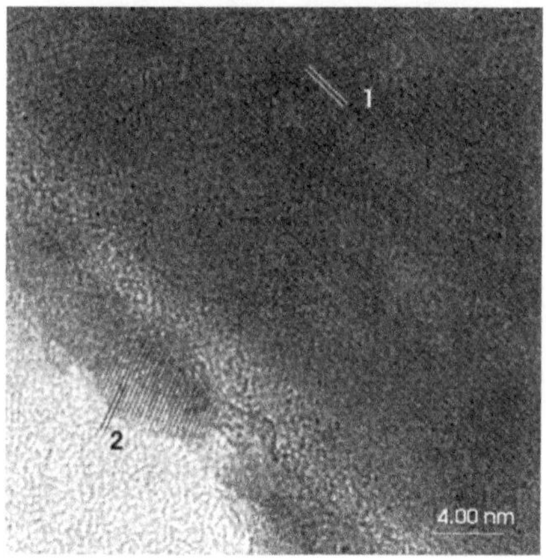

Figure 11. HRTEM image of an individual MWCNT filled with SbSI sonochemically in methanol. The fringe spacings of 0.319(2) nm (1) and 0.209(2) nm (2) correspond to the interplanar distances between the (220) planes of SbSI crystal and (101) planes of carbon nanotube, respectively [11].

The TEM (Figure 9) and HRTEM images (Figures 11 and 12) of an individual CNTs sonochemically filled with SbSI and SbSeI reveal that the products consist of coaxial nanocables. The lateral dimensions of the SbSI@CNTs and SbSeI@CNTs have been in the ranges from 30 to 200 nm and from 20 to 170 nm, respectively [11,14]. Lengths of these nanocables reach up to several micrometers in both cases. The HRTEM images of individual CNTs sonochemically filled with SbSI (Figure 11) exhibit good crystallinity of the SbSI and its clear (220) lattice fringes parallel to the nanocable axis. It indicates the growth of SbSI inside the CNT in [001] direction. The interplanar spacing is about 0.319(2) nm, which coincide with the interplanar spacing 0.32494 nm of (220) planes of the orthorhombic structure of conventional SbSI [33]. Figure 11 shows also the lattice fringes of the CNT walls. The fringe spacings of 0.209(2) nm match with the 0.21390 nm interplanar distances between the (101) planes of carbon nanotubes [30]. All these results correspond well with the XRD patterns (Figure 6) of the CNTs sonochemically filled with SbSI. The SAED pattern (Figure 13) recorded on the end of multi–walled CNT filled with SbSI (presented in Figure 9a) indicates the interplanar spacings appropriate for CNTs as well as SbSI crystals (see Table 3).

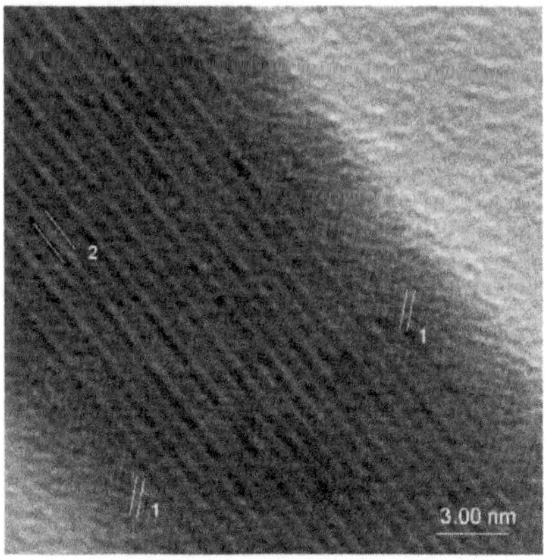

Figure 12. HRTEM image of an individual multi–walled CNT filled with SbSeI sonochemically in ethanol. The fringe spacings of 0.332(5) nm (sign 1) and 0.432(13) nm (sign 2) correspond to the interplanar distances between the (002) planes of carbon nanotube and (200) planes of SbSeI crystal, respectively [14].

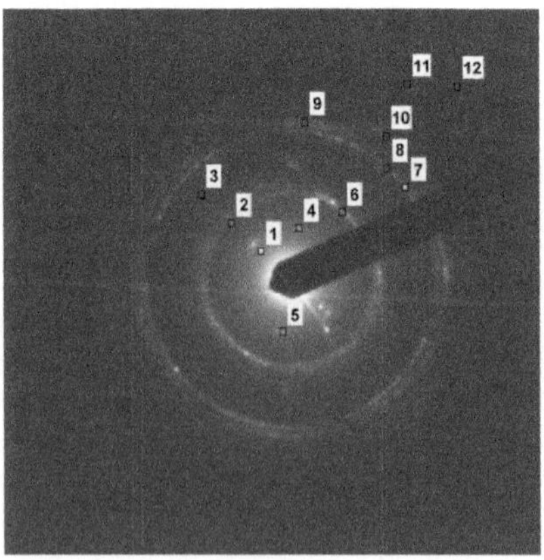

Figure 13. SAED pattern of MWCNT filled with SbSI ultrasonically in methanol [11] (shown in Figure 9). The diffraction patterns correspond to the interplanar distances presented in Table 3.

Sign	Results of the SAED d_{hkl} (nm)	Literature data			
		for C [29, 30]		for SbSI [27, 33]	
		d_{hkl} (nm)	(hkl)	d_{hkl} (nm)	(hkl)
1 reflex	0.4360	–	–	0.43402	(120)
		–	–	0.42450	(200)
4 reflex	0.3732	–	–	0.38465	(011)
5 reflex	0.3470	0.33950	(002)	0.35036	(111)
2 reflex	0.2189	0.21390	(100)	0.21663	(330)
6	0.2089	0.20402	(101)	0.20800	(002)
3 reflex	0.1446	–	–	0.14244	(422)
		–	–	0.14244	(531)
7 reflex	0.1190	–	–	0.12031	(143)
		–	–	0.11960	(181)
8 circle	0.1210	0.12350	(110)	–	–
9 circle	0.1155	0.11606	(112)	–	–
		0.11464	(105)	–	–
		0.11316	(006)	–	–
10 circle	0.1036	0.10425	(202)	–	–
11 circle	0.07874	0.07954	(122)	–	–
12 circle	0.07038	0.07720	(206)	–	–

Table 3. Comparison of interplanar spacings determined by SAED (Figure 13) of multi–walled CNT filled with SbSI ultrasonically in methanol with literature data for CNTs and SbSI bulk crystals.

The HRTEM images of an individual CNTs sonochemically filled with SbSeI (Figure 12) exhibit good crystallinity of the SbSeI and its clear (200) lattice fringes parallel to the nanocable axis. It indicates the growth of SbSeI inside the CNT in [001] direction. The interplanar spacing is about 0.432(13) nm, which coincide with the interplanar spacing 0.4349 nm of (200) planes of the orthorhombic structure of conventional SbSeI [28]. Figure 12 shows also the lattice fringes of the CNT walls. The fringe spacings of 0.332(5) nm match with the 0.3395 nm interplanar distances between the (002) planes of carbon nanotubes [30]. All these results correspond well with the XRD patterns (Figure 7, Table 1) of the CNTs sonochemically filled with SbSeI. The SAED pattern (Figure 14) recorded on the multi–walled CNT filled with SbSeI (presented in Figure 12) indicates the interplanar spacings appropriate for SbSeI crystals (see Table 1). Surprisingly, the presence of SbSeI nanowires or SbSeI material on the outer walls of the MWCNT has not been observed in the HRTEM and TEM images. Probably, the process of capillary suction of molecules into carbon nanotubes facilitates the process of SbSeI nanocrystal growth.

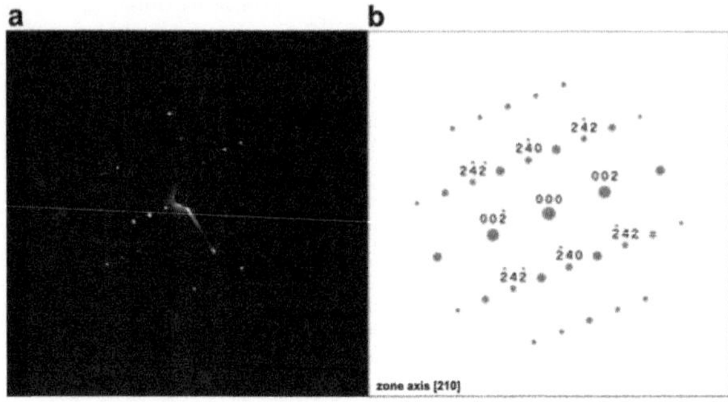

Figure 14. Electron diffraction pattern of multi–walled CNT filled with SbSeI ultrasonically in ethanol [14] in the orientation close to the [210] zone axis (a) and its simulated diagram (b). The diffraction patterns correspond to the interplanar distances presented in Table 1.

4. Mechanisms of sonochemical preparation of SbSI@CNTs and SbSeI@CNTs

It is known that the mode of insertion dictates the nature and morphology of the obtained filling of CNTs [39]. When the filling is induced via solution–deposition, small discrete encapsulates are obtained, whereas when it is obtained via capillarity, continuously filled CNTs are observed. Probably the last happens when CNTs are filled sonochemically by SbSI [11] as well as by SbSeI [14]. As in the case of sonochemically produced alone SbSI [15] and SbSeI [19], the transient high–temperature and high–pressure field produced during ultrasound irradiation provide a favorable environment for the 1D growth of the SbSI and SbSeI nanocrystals from elements inside multi–walled CNTs in ethanol, though the bulk solution surrounding the collapsing bubbles is at relatively low temperature (T=323 K) and atmospheric pressure.

The probable reaction routes of synthesis of SbSI and SbSeI in CNTs, and the mechanisms of nanowires formation using elemental Sb, S or Se and I in the presence of methanol and ethanol under ultrasonic irradiation can be summarized as follows:

a. iodine, I_2, dissolved in methanol or ethanol reacts with antimony and forms the antimony triiodide, SbI_3, also dissolved in ethanol [40]

$$2Sb + 3I_2 \rightarrow 2SbI_3 \tag{1}$$

b. dehydrogenation, dehydration as well as decomposition of ethanol (2) and methanol (3) in or close to the cavitation bubbles leads to the formation of hydrogen and water [41,42]

$$
\begin{aligned}
&C_2H_5OH \leftrightarrow C_2H_4O + H_2 \ , \\
&C_2H_5OH \leftrightarrow C_2H_4 + H_2O \ , \\
&C_2H_5OH \leftrightarrow C_2H_5O + H
\end{aligned}
\tag{2}
$$

$$
\begin{aligned}
&CH_3OH \leftrightarrow \ CH_3^{\cdot} + OH^{\cdot}, \\
&OH^{\cdot} + CH_3OH \leftrightarrow H_2O + CH_2OH^{\cdot}, \\
&CH_3^{\cdot} + CH_3OH \leftrightarrow CH_4 + CH_2OH^{\cdot}, \\
&CH_2OH^{\cdot} + M \leftrightarrow CH_2O + H^{\cdot} + M \ , \\
&CH_2O + M \leftrightarrow \ H^{\cdot} + CHO^{\cdot} + M \ , \\
&CHO^{\cdot} + M \leftrightarrow CO + H^{\cdot} + M \ , \\
&H^{\cdot} + CH_3OH \leftrightarrow H_2 + CH_2OH^{\cdot}
\end{aligned}
\tag{3}
$$

c. the sonolysis of water yields the $H^{\bullet}$ and $OH^{\bullet}$ radicals

$$H_2O)))H^{\bullet} + OH^{\bullet} \tag{4}$$

d. the ultrasonic irradiation facilitates the reduction of chalcogens to the active forms of S^{-2} and Se^{-2} (see Refs. in [24,43]) that react with the in-situ generated $H^{\bullet}$ radicals forming H_2S and H_2Se [44]

$$S \ + \ 2H^{\bullet} \rightarrow \ H_2S \tag{5}$$

$$Se \ + \ 2H^{\bullet} \rightarrow \ H_2Se \tag{6}$$

e. opening of the CNT ends due to the acids (H_2S, H_2Se) [7] or the action of ultrasounds [45,46].

f. CNT's suck the released H_2S or H_2Se molecules and SbI_3 in ethanol needed to build SbSI and SbSeI. This is allowed by the capillary effect [47]. It is well known [48,49], that open CNTs are impregnated with excess of the precursor solutions under ultrasonic conditions to ensure that most of the tubes can be filled by the impregnating solution. From the thermodynamic point of view, the dissolution of the impregnating solution sticking to external walls of CNTs into the washing medium necessitates the condition of the solvation energy gain smaller than zero [48,49]. On the other hand, the energy gain for the capillary filling of a CNT must be smaller than the solvation energy to ensure that the solution is stable in the internal cavity of CNTs [48,49].

g. H_2S or H_2Se reacts with SbI_3 to yield SbSI and SbSeI molecules

$$SbI_3 + H_2S \rightarrow SbSI + H_2 + I_2 \tag{7}$$

$$SbI_3 + H_2Se \rightarrow SbSeI + H_2 + I_2 \tag{8}$$

h. SbSI or SbSeI molecules, under the microjets and shockwaves formed during collapsing of the bubbles, are pushed towards each other in CNTs and are held by chemical forces. Therefore, the nuclei of SbSI or SbSeI are formed as a result of the interparticle collisions (see [24]).

i. the freshly formed nuclei in the solution are unstable and have the tendency to grow into double chains $[(SbSI)_{\infty}]_2$ or $[(SbSeI)_{\infty}]_2$ consisting of two chains related by a twofold screw axis and linked together by a short and strong Sb-S or Sb-Se bonds [50]. Local turbulent flow associated with cavitation and acoustic streaming greatly accelerates mass transport in the liquid phase;

j. the SbSI and SbSeI chains can be readily crystallized into a 3D lattice in the CNTs through van der Waals interactions. Induced by this structure, crystallization tends to occur along the c-axis, favoring the stronger covalent bonds over the relatively weak, inter-chain van der Waals forces [51]. Thus, this solid material has a tendency to form highly anisotropic, 1D structures also inside the CNTs.

The (7) and (8) were the basic reactions used in different methods of preparation of bulk SbSI-type crystals [52]. Despite the tubes filling, the observed nanowire type morphology of the product (Figures 11 and 12) is possibly due to the inherent chain type structure and growth habit of the SbSI as well as SbSeI crystals [50].

5. Optical properties of SbSI@CNTs and SbSeI@CNTs

Optical absorption spectra have been extensively used as one of the most important tools for probing the energy gaps (E_g) and band structures of semiconductors [53]. There are several methods for measuring them. These include diffuse reflectance spectroscopy (DRS). When a material, consisting of many particles or nanoparticles, is illuminated some of the impinging radiation penetrates the sample and some is reflected from its surface. The portion that penetrates the sample is scattered at a large number of points in its path as well as it is transmitted through the particles a number of times. Only the part of this radiation that is returned to the surface of the sample and comes back out is considered to be diffuse reflection. This component has been investigated extensively by many authors [54-56]. DRS is a suitable, not destructive, and simple method of investigation, especially important in examinations of porous, nanocrystaline materials and gels [57]. It is impossible to examine such materials applying specular reflection and it is also extremely difficult to determine the pathlength in an optical transmittance of them. The following factors are related to high spectral quality of diffuse reflectance:

1. dilution of the sample with a non-absorbing matrix ensures a deeper penetration of the incident beam into the sample which increases the contribution of the scattered component in the spectrum and minimizes the specular reflection component;

2. smaller particles improve the quality of DRS spectra because the contribution of reflection from the surface is reduced.

The optical diffuse reflection spectroscopy (DRS) of SbSI@CNTs and SbSeI@CNTs was carried out [11,14] on a spectrophotometer SP-2000 (Ocean Optics Inc.) equipped with an integrating sphere ISP-REF (Ocean Optics Inc.). Spectra were recorded at room temperature, from 350 to 1000 nm. The standard WS-1 (Ocean Optics Inc.) was used as a reference. The diffuse reflectance values were converted to the Kubelka–Munk function (known to be proportional to the absorption coefficient) shown by [54-56]

$$F_{K\text{-}M}(R_d) = \frac{(1-R_d)^2}{2R_d} \sim \alpha \tag{9}$$

where R_d describes the coefficient of diffuse reflectance and α is the absorption coefficient of light in the investigated material.

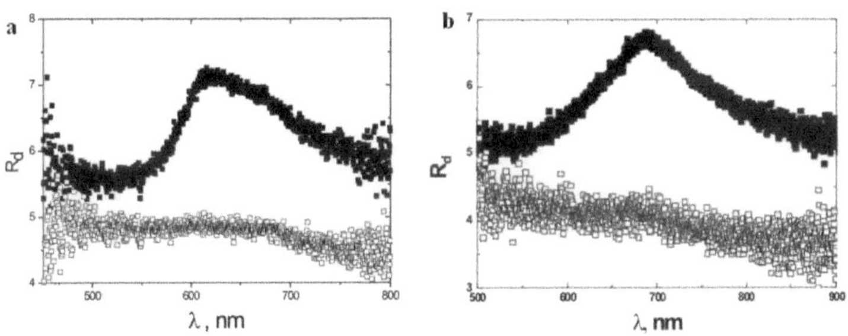

Figure 15. Comparison of the diffuse reflectance spectra of the multi–walled CNTs filled with SbSI (a) [11] and SbSeI (b) [14] (■) ultrasonically (in methanol and ethanol, respectively), and of the empty multi–walled CNTs (□) in methanol and ethanol.

In Figure 15 the diffuse reflectance spectra of CNTs filled with SbSI and SbSeI are compared with the spectrum registered for hollow CNTs. In the first cases one can see the characteristic for semiconducting materials edges of fundamental absorption around 615 nm and 687 nm. However, the diffuse reflectance decreases also with increasing wavelengths (Figure 15), probable due to the large amount of free carriers absorbing light. Figure 16 presents the spectra of Kubelka–Munk functions derived from the diffuse reflectance data presented in Figure 15.

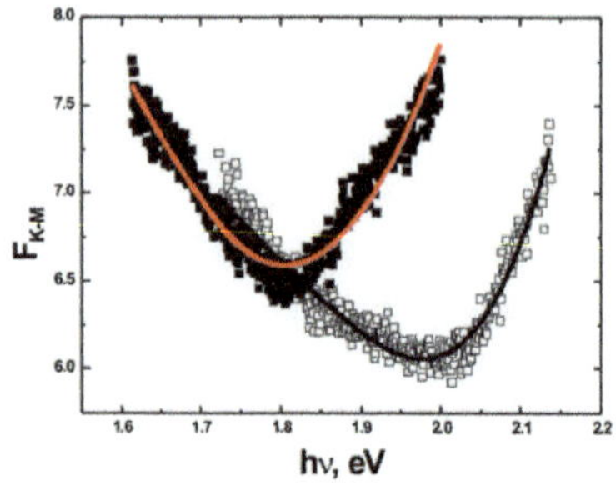

Figure 16. Comparison of fitted spectra of Kubelka–Munk functions calculated for diffuse reflectances of multi–walled CNTs filled with SbSeI (■) with the data reported for multi–walled CNTs filled with SbSI (□). Solid curves represent the least square fitted theoretical dependences (description in the text; values of the fitted parameters are given in Table 4) [14].

Fitted parameters	**CNTs filled with SbSeI [14]**			**SbSeI nanowires [19]**	**CNTs filled with SbSI [11]**	**SbSI nanowires [57]**
	Values determined assuming					
	Indirect allowed absorption	Indirect allowed absorption without phonon statistics – normalized for hν	Indirect forbidden absorption			
χ^2	0.1223	5.592	78.14	–	–	–
E_{glf} [eV]	–	–	1.6(1)	1.63	1.871	1.854
E_{gla} [eV]	1.61(6)	1.72(4)	–	–	–	–
E_{ph} [eV]	0.099(6)	–	–	–	–	–
A_{60} [1/eV^{3}m]	–	–	41(2)	94.8	95.6	157
A_{123} [10^{12}m^{-3}]	11.14(1)	11(2)	11.11(1)	–	8.24	–
A_{50} [1/eV^{2}m]	27.7(6)	–	–	–	–	–
A41 [1/eV^{2}m]	–	60(1)	–	–	–	–
A_0 [1/m]	1.019(1)	1(1)	1.02(2)	0.01071	2.702	0.0213
E_U [eV]	–	–	–	0.0810	–	0.147
A_U [1/m]	–	–	–	$1.09 \cdot 10^{-10}$	–	$84.4 \cdot 10^{-9}$

Table 4. Comparison of parameters of multi–walled CNTs filled with SbSI and SbSeI with the literature data reported for SbSeI and SbSI nanowires produced sonochemically (E_{gla} – indirect allowed energy gap; E_{glf} – indirect forbidden energy gap; E_{ph} – phonon energy; E_U – Urbach energy; A_0 – constant absorption; A_{50} A_{60}, A_{120} – proportionality factors determined from the fitting of the spectra of F_{K-M} functions evaluated from the DRS data).

The Kubelka–Munk (F_{K-M}) spectra were least square fitted with theoretical dependences appropriate for different mechanisms of absorption [57]. The best fitting was obtained for the sum of indirect allowed absorption with absorption/emission of phonons (α_1), free carrier absorption (α_2) and constant absorption term (α_3) (see the literature in [53,57]):

$$\alpha = \alpha_1 + \alpha_2 + \alpha_3 \tag{10}$$

where

$$\alpha_1 = A_{50} \frac{\left(h\nu - E_{gIa} + E_{ph}\right)^2}{\exp\left(\frac{E_{ph}}{k_B T}\right) - 1} \qquad for \qquad E_{gIa} - E_{ph} < h\nu \leq E_{gIa} + E_{ph} \tag{11}$$

$$\alpha_1 = A_{50} \left\{ \frac{\left(h\nu - E_{gIa} + E_{ph}\right)^2}{\exp\left(\frac{E_{ph}}{k_B T}\right) - 1} + \frac{\left(h\nu - E_{gIa} - E_{ph}\right)^2}{1 - \exp\left(-\frac{E_{ph}}{k_B T}\right)} \right\} \qquad for \qquad h\nu > E_{gIa} + E_{ph} \tag{12}$$

$$\alpha_2 = A_{123} \cdot \lambda^2 \tag{13}$$

$$\alpha_3 = A_0 \tag{14}$$

E_{gIa} represents the indirect allowed energy gap, E_{ph} is the phonon energy, A_0, A_{50}, A_{125} are constant parameters, hν is the photon energy, T is the temperature, k_B is the Boltzmann constant. The determined values of these parameters are given in Table 4. In the same table they are compared with the literature data reported for SbSeI [19] and SbSI [57] nanowires produced sonochemically.

It should be noted that absorption spectra [58-62] of bulk SbSeI single crystals have been fitted as indirect allowed absorption without phonon statistics – normalized for hν (α_4)

$$\alpha_4 = \frac{A_{41}}{h\nu}(h\nu - E_{gIa})^2 \qquad for\ h\nu > E_{gIa} \tag{15}$$

where E_{gIa} represents the indirect allowed energy gap, A_{41} is the proportionality factor. It should be noted that the indirect allowed optical energy gap described by formulae (17) was discovered also in doped single crystals SbSeI:Co [61], SbSeI:V [60], SbSeI:Cr [59], and

SbSeI:Ni [58]. Therefore, we fitted the same spectrum of $F_{K\text{-}M}$ of the CNTs filled with SbSeI (Figure 16) with theoretical dependence appropriate for the sum of indirect allowed absorption without phonon statistics – normalized for hν (α_4), free carrier absorption (α_2) and constant absorption term (α_3). Values of the fitted parameters are given in Table 4, too. However, the value of sum of squared residuals (χ^2) in the last fitting is over six hundred times larger then in the case of fitting with formula (10) (Table 4).

One should remember that the $F_{K\text{-}M}$ spectrum of sonochemically produced alone SbSeI nanowires [19] was best fitted for indirect forbidden absorption (α_5), Urbach ruled absorption (α_6) and constant absorption term (α_3), where

$$\alpha_5 = A_{60}\left(h\nu - E_{gIf}\right)^3 \qquad for\ h\nu > E_{gIf} \tag{16}$$

$$\alpha_6 = A_U \exp\left[\frac{h\nu}{E_U}\right] \tag{17}$$

Therefore, we fitted the same spectrum (Figure 16) of $F_{K\text{-}M}$ of the CNTs filled with SbSeI with theoretical dependence appropriate for the sum of α_5, α_2 and α_3 (the fitting did not take into account the α_6 because it is in contradiction with the experimental data described by α_2). One can see that the value of sum of squared residuals (χ^2) in the last fitting is over forty five times larger than in the case of fitting with the indirect allowed absorption (Table 4).

The observed free carrier absorption of light in the case of CNTs filled with SbSI and SbSeI is evoked by the CNTs material because it is absent in the case of alone SbSI and SbSeI nanowires [19,57]. Instead of it the Urbach absorption was reported in the latter cases. May be the latter mechanism of absorption exists also in CNTs filled with SbSI and SbSeI but it is hidden by the free carrier absorption. It is known (see [47]) that a metal atom intercalated inside the internal cavity of a CNT displays a tendency towards the transfer of some part of the valence electrons to the outer surface of the nanotube, where unoccupied electronic states exist. As a result of such a transfer there arises an additional mechanism of electrical conduction, related to the travel of an electron about those states.

The nanocrystalline SbSeI filling the CNTs is a semiconductor with little smaller energy gap of 1.61 eV than the alone SbSeI nanowires (Table 4). The established type of indirect allowed energy gap of SbSeI in CNTs is different from the indirect forbidden energy gap of sonochemically produced alone SbSeI nanowires [19]. However, one should remember that absorption spectra [58-62] of bulk SbSeI single crystals have been fitted with indirect allowed absorption without phonon statistics – normalized for hν. Therefore, the indirect allowed energy gap of SbSeI in CNTs is very probable and the light absorption in the alone SbSeI nanowires should be studied again.

6. Electrical and photoelectrical properties of SbSI@CNTs and SbSeI@CNTs

In [13] for the first time electrical investigations have been performed on circular samples of SbSI@CNT gel prepared sonochemically. The diameter and thickness of the sample were 4 mm and 1.5 mm, respectively. The largest opposite surfaces of the sample were covered with a silver paste (SPI Supplies) and electrical connection from these electrodes was made by copper wire. The measurements were performed applying impedance analyzer Agilent 4294A in the frequency (f) range 200 Hz – 20 MHz with amplitude of 0.01 V and zero bias. The impedance investigations were made in vacuum in the temperature region from 273 K to 353 K using a LN2 cryostat VPF-700 (Janis Research Company, Inc.). The metallic chamber served as a Faraday cage in order to eliminate undesired external interference. The temperature was controlled using 331 Temperature Monitor, (Lake Shore). Measurements were performed in darkness to avoid the excess carriers that can be photogenerated in SbSI@CNT. LabView program has been developed for computer-controlling the experiment, data acquisition and analysis.

Figures 17a and 17b present influence of temperature on Bode plots for SbSI@CNT gel prepared ultrasonically in methanol. The impedance amplitude ($|Z|$) decreases while the phase (ϕ) increases with increasing frequency. Using formulae: $ReZ = |Z| \cos\phi$ and $ImZ = |Z| \sin\phi$, the Nyquist plots (Figure 17c) have been constructed. The two arcs in Nyquist plots, in the temperature range from 273 K to 293 K, suggest the need of using non Debye model to interpret the data. Hence, the equivalent circuit to interpret the impedance of SbSI@CNT gel should contain the constant phase element (CPE) with impedance given by

$$Z_{CPE} = \left[A(j\omega)^n \right]^{-1} \tag{18}$$

where: parameter A is a constant, $j^2 = -1$, $\omega = 2\pi f$, the exponent n equals 1 for ideal capacitor, 0.5 in the case of diffusion processes, 0 for ideal resistor, and -1 for ideal inductor [63]. The porous materials have values of n from the range $0.9 \div 1$ [63].

For temperatures higher than T=293 K only one arc in Nyquist plots is observed (Figure 17c). Probably, it is evoked by the ferroelectric-paraelectric phase transition of SbSI near Curie temperature T_C=292(1) K [17,18].

The experimental data have been best fitted (Figure 17) employing complex nonlinear curve fitting using ZView 2 program [64]. Fitting of the Nyquist plots has been done using equivalent circuits reported for different models (e.g. Voigt circuit, Maxwell circuit, and their combinations). The best fitting was obtained using equivalent circuit shown in Figure 18. It is consistent of serial connection of two parallel circuits (resistors: R1, R2 and CPE elements: CPE1, CPE2) with inductance L. Probably, the first parallel circuit (R_1 – CPE_1) represents properties of SbSI component while the second parallel circuit (R_2 – CPE_2) describes properties of CNTs and the electrodes. The inductance L is dedicated to the presence of CNTs [65,66].

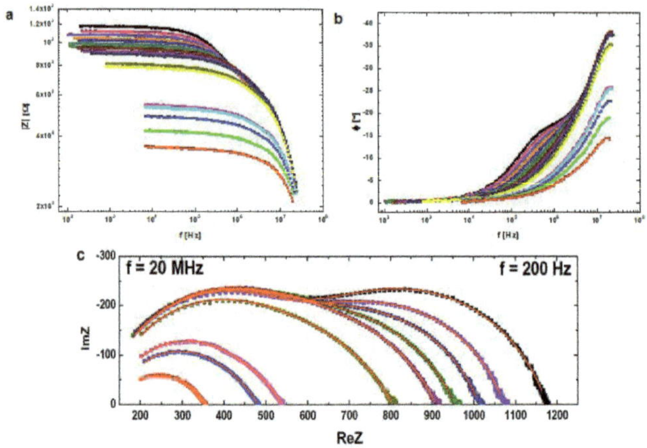

Figure 17. Influence of temperature on Bode (a, b) and Nyquist (c) plots for SbSI@CNT gel prepared ultrasonically in methanol [13] (■ – 273 K, ■ – 283 K, ■ – 293 K, ■ – 303 K, ■ – 313 K, ■ – 323 K, ■ – 333 K, ■ – 343 K, ■ – 353 K). Solid curves represent the fitted theoretical dependences calculated for an equivalent circuit presented in Figure 18. The fitted parameters are given in Figures 19-21.

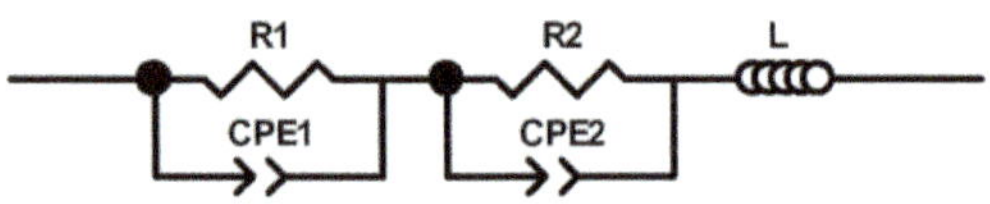

Figure 18. Equivalent circuit for SbSI@CNT gel prepared ultrasonically in methanol [13].

Every semiconducting ferroelectric SbSI nanowire (in CNT) can be represented by parallel connection of resistance and capacitance. There appear a few groups of possible explanation of physical origin of CPEs:

1. CPE can be related to macroscopic heterogeneities (porosity, grain sizes, surface roughness, etc.) occurring at the sample electrodes [67];

2. CPE can be associated with ion (anion) adsorption at the roughened electrodes [67];

3. CPE can be related to random mixtures of a conductor and an insulator that can be described by the effective medium approximation [68];

4. CPE can be due to distribution of activation energy of the reduction process [69];

5. CPE can also include contributions from dynamic disorders such as diffusion [68].

In the present case, we are able to offer (after [70]) some additional suggestions, as follows. The CPE represents time-dependent phenomena which, in the case of polarization fluctuations, may be an indicator of the cooperative nature of the fluctuations of individual dipoles. Specifically, fluctuating dipoles interact with their surroundings.

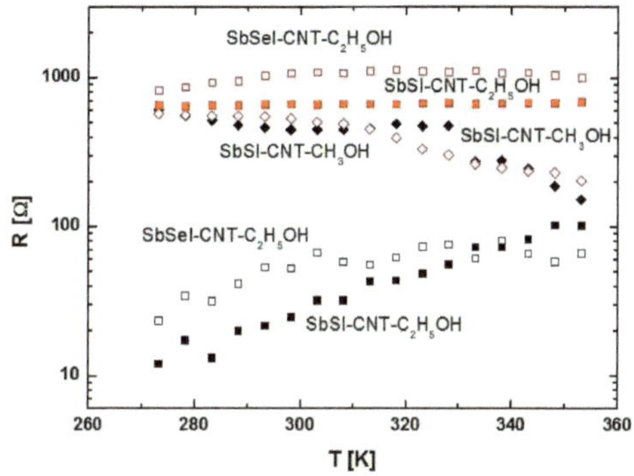

Figure 19. Comparison of the temperature dependences of resistance parameters of the equivalent circuits describing Nyquist plots registered in the cases of SbSI@CNT ultrasonically prepared in methanol [13] (◇ – R_1, ◆ – R_2) (Figures 17 and 18), and SbSI@CNT (■ – R_1, ■ – R_2) and SbSeI@CNT (□– R_1, □ – R_2) ultrasonically prepared in ethanol.

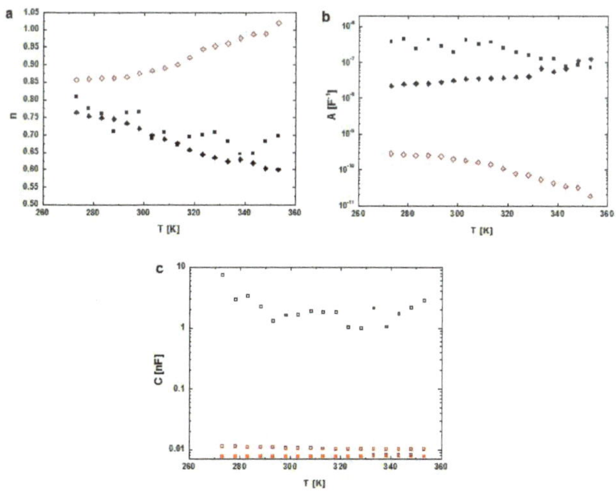

Figure 20. Comparison of the temperature dependences of capacitance parameters of the equivalent circuits describing Nyquist plots registered in the cases of SbSI@CNT ultrasonically prepared in methanol [13] (◆– n_1, A_1; ◇ – n_2, A_2) (Figures 17 and 18), and SbSI@CNT (▪ – n, A; ■ – C) and SbSeI@CNT (■ – C_1, □ – C_2) ultrasonically prepared in ethanol.

In order to interpret experimental data, it is essential to have a model equivalent circuit that provides a realistic representation of the sample. The investigated SbSI@CNT and SbSeI@CNT gel is a complicated object that contains carbon nanotubes (filled with SbSI or SbSeI) connected with each other in parallel and in series, as well as interfaces (i.e. nano-wires - nanowires or nanowires - electrodes). Of course the models should not possess too many components, because the fitting of many variables will be unreliable. Therefore the interpretation of the components of the equivalent electric circuits presented in Figure 18 is very difficult. To obtain simpler image of the observed phenomena experiments with single SbSI@CNT and SbSeI@CNT are needed.

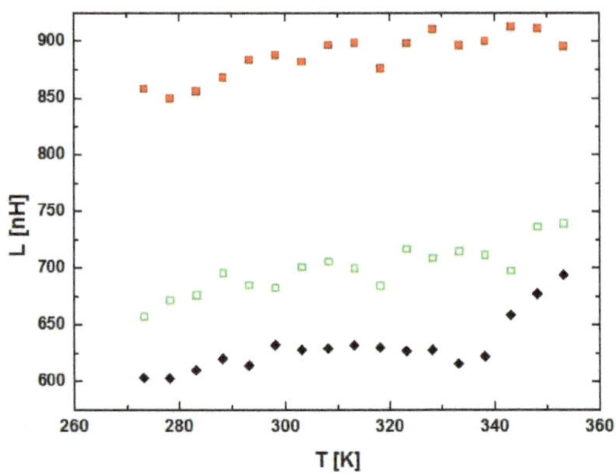

Figure 21. Comparison of the temperature dependences of inductance parameters of the equivalent circuits describing Nyquist plots registered in the cases of SbSI@CNT ultrasonically prepared in methanol [13] (◆) (Figures 17 and 18), and SbSI@CNT (■) and SbSeI@CNT (□) ultrasonically prepared in ethanol.

Figure 22 presents comparison of the temperature dependences of relaxation time of SbSI@CNT ultrasonically prepared in methanol with the relaxation time of SbSI@CNT and SbSeI@CNT ultrasonically prepared in ethanol. One can see the influence of the liquid used during sonochemical filling of CNTs on properties of the product. In the case of SbSI@CNT ultrasonically prepared in methanol there are two relaxation times in temperatures lower than T=293 K. Above this temperature only one relaxation time is needed to describe the properties of SbSI@CNT. Such behavior can be connected with the ferroelectric-paraelectric phase transition of SbSI near Curie temperature T_C=292(1) K [17,18]. This phenomenon needs additional investigations.

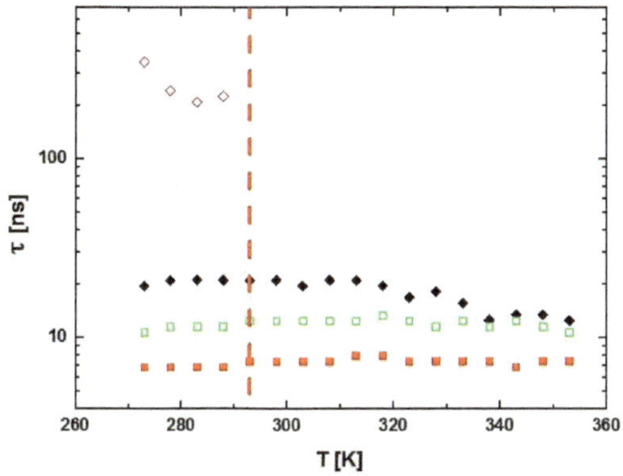

Figure 22. Comparison of the temperature dependences of relaxation time of SbSI@CNT ultrasonically prepared in methanol (♦ – τ_1, ◇ – τ_2) (Figure 17), and SbSI@CNT (■) and SbSeI@CNT (□) ultrasonically prepared in ethanol [13].

It should be noted, that similar equivalent circuits have been well fitted to the Nyquist plots registered also in the cases of SbSI@CNT and SbSeI@CNT ultrasonically prepared in ethanol. Comparison of the temperature dependences of the fitted parameters are given in Figures 19-21.

To overcome the mentioned above problems with interfaces (i.e., nanotubes - nanotubes and nanotubes - electrodes), an essential prerequisite is to build reliable interconnections between the CNTs and the external electrical circuits. To address this need, various chemical and physical processes have been explored to build such interconnections [71]. Changxin Chen et al. [72,73] have presented ultrasonic nanowelding process, with which one can fabricate reliable bonding between SWCNTs and metal electrodes. Ultrasonic nanowelding was carried out in a FB-128 ultrasonic wire bonder [72,73]. An Al_2O_3 single crystal with a 2,500 μm^2 circular pressing surface and a RMS roughness of 0.2 nm was mounted onto the bonder to act as the welding head. A clamping force of 78.4 mN was applied to press the welding head against the nanotube and electrodes. At the same time an ultrasonic vibration with a frequency of 60 kHz and an ultrasonic power of about 10 mW was applied to the welding head through an ultrasonic transducer. The ultrasonic energy was transferred to the bonding interface through the ultrasonic welding head. Thus the ends of SWCNTs and electrodes were welded together under the combined action of the ultrasonic energy and a clamping force. The welding process was carried out at room temperature for a duration of 0.2 s.

Figure 23 presents SEM and AFM images of carbon nanotubes on Si/SiO_2 substrate with Au electrodes after ultrasonic welding. In our ultrasonic nanowelding set-up the generator ADG70-100P-230-NO 70 kHz (Rinco Ultrasonics) with converter C 70-2 (Rinco Ultrasonics) has been used. The sonotrodes have been equipped with SiC single crystals. The preliminary electrical tests of SbSI@CNTs ultrasonic nanowelded with different microelectrodes on different substrates are successful. The obtained results will be published in a near future.

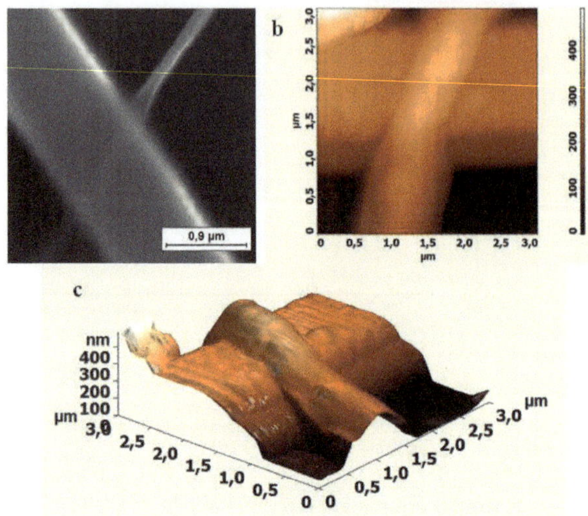

Figure 23. SEM and AFM images of carbon nanotubes on Si/SiO_2 substrate with Au electrodes after ultrasonic welding.

5. Conclusions

The presented very simple, sonochemical synthesis of nanophase SbSI or SbSeI in CNTs at low temperature is a convenient, fast, mild, efficient and environmentally friendly route for producing novel type of hybrid nanomaterials. The resulting SbSI@CNTs and SbSeI@CNTs composites are highly anisotropic 1D structures whose electronic and optical properties are considerably modified with respect to the encapsulating nanotube. The fabricated SbSI and SbSeI fillings of CNTs are single crystalline in nature and in the form of nanowires.

Since the sonochemical process was carried out at ambient temperature and pressure, it may be predicted that upscaling of this method will lead to large quantities of SbSI@CNTs and SbSeI@CNTs. Further studies on the properties of the sonochemically prepared SbSI and SbSeI in CNTs are underway. Taking into account the recently reported sonochemical preparation of pure $SbS_{1-x}Se_xI$ nanowires [36], one should expect the sonochemical preparation of quaternary $SbS_{1-x}Se_xI$ semiconductors within CNTs. The possibility of change the energy gap (and other parameters) with changing molar composition (x) of the quaternary compounds [36] gives opportunity to tail the functional properties of CNTs filled with $SbS_{1-x}Se_xI$. It should be remember that the physical properties of such quaternary compounds, formed as solid solutions, can be tailored with stoichiometric composition. It should be an advantage of using such material in different devices. It seems that this approach can be extended to the preparation of some other ternary and quaternary nanomaterials formed from the group 15-16-17 elements within CNTs. It should provide a "bottom–up" approach for the manufacture of future nanoscale devices.

This review summarizes all so far published results of investigations on SbSI@CNTs and SbSeI@CNTs. One can see that the properties of these 1D hybrid nanomaterials still remain little known. Meantime, these nanostructures should provide promising materials for fundamental investigations on nanoscale ferroelectricity and piezoelectricity as well as materials for some applications.

Obviously, the presented new materials as the other one-dimensional semiconductor nanostructures should receive considerable attention from the scientific and engineering communities due to their potentially useful novel electronic properties.
The ultrasonic nanowelding of SbSI@CNTs and SbSeI@CNTs with metal microelectrodes meets the need of future large-scale, simply, convenient, fast, and efficient technology to build reliable interconnections between the CNTs filled with ternary chalcohalides formed from the group 15–16–17 elements and the external electrical circuits.

Acknowledgements

The authors are indebted to Dr. P. Szperlich, Prof. J. Szala, Prof. S. Kochowski, Dr. A. Starczewska, Dr. M. Kępińska, and Dr. T. Rzychoń of the Silesian University of Technology, Katowice (Poland), and Prof. D. Stróż and Prof. R. Wrzalik of the University of Silesia, Katowice (Poland) for collaboration in investigations of carbon nanotubes filled with ternary chalcohalides. This review was partially supported by the NCN (Poland) under contract No. UMO-2011/01/B/ST5/06273.

Author details

Marian Nowak* and Marcin Jesionek

*Address all correspondence to: Marian.Nowak@polsl.pl

Institute of Physics, Silesian University of Technology, Poland

References

[1] Iijima, S. (1991). Helical microtubules of graphitic carbon. *Nature*, 354, 56-58.

[2] Ajayan, P. M., & Iijima, S. (1993). Capillarity-induced filling of carbon nanotubes. *Nature*, 361, 333-334.

[3] Ugarte, D., Stöckli, T., Bonard, J. M., Châtelain, A., & de Heer, W. A. (1998). Filling carbon nanotubes. *Appl. Phys. A*, 67, 101-105.

[4] Monthioux, M. (2002). Filling single-wall carbon nanotubes. *Carbon*, 40, 1809-1823.

[5] Tasis, D., Tagmatarchis, N., Bianco, A., & Prato, M. (2006). Chemistry of Carbon Nanotubes. *Chem. Rev.*, 106, 1105-1136.

[6] Rakov , Eduard G. (2006). Chemistry of Carbon Nanotubes. *Chapter 5, in Nanomaterials Handbook, Taylor & Francis Group, LLC.*

[7] Eliseev, A., Yashina, L., Kharlamova, M., & Kiselev, N. (2011). One-Dimensional Crystals inside Single-Walled Carbon Nanotubes: Growth, Structure and Electronic Properties. *In: J. M. Marulanda (Ed.) Electronic Properties of Carbon Nanotubes, Rijeka: InTech,* 127-156, http://www.intechopen.com/books/electronic-properties-of-carbon-nanotubes/one-dimensional-crystals-inside-single-walled-carbon-nanotubes-growth-structure-and-electronic-prope, accessed 21 June 2012).

[8] Lukanov, P., Tîlmaciu, C. - M., Galibert, A. M., Soula, B., & Flahaut, E. (2011). Filling of Carbon Nanotubes with Compounds in Solution or Melted Phase. *in: R. Klingeler, R. B. Sim (eds.) Carbon Nanotubes for Biomedical Applications, Carbon Nanostructures, Berlin: Springer-Verlag.*

[9] Sloan, J., Grosvenor, S. J., Friedrichs, S., Kirkland, A. I., Hutchison, J. L., & Green, M. L. H. (2002). A One-Dimensional BaI_2 Chain with Five- and Six-Coordination, Formed within a Single- Walled Carbon Nanotube, Angew. *Chem. Int. Ed*, 41, 1156-1159.

[10] Fan, X., Dickey, E. C., Eklund, P. C., Williams, K. A., Grigorian, L., Buczko, R., Pantelides, S. T., & Pennycook, S. J. (2000). Atomic Arrangement of Iodine Atoms inside Single-Walled Carbon Nanotubes. *Phys. Rev. Lett*, 84, 4621-4624.

[11] Nowak, M., Jesionek, M., Szperlich, P., Szala, J., Rzychoń, T., & Stróż, D. (2009). Sonochemical growth of antimony sulfoiodide in multiwalled carbon nanotube, Ultrason. *Sonochem*, 16, 800-804.

[12] Stróż, D., Nowak, M., Jesionek, M., & Bałdys, K. (2010). Structure of antimony sulfoiodide ultrasonically prepared in carbon nanotubes. *Solid State Phenomena,* 163, 88-92.

[13] Jesionek, M. (2011). Preparation and properties of hybrid materials containing nanocrystaline SbSI and SbSeI in carbon nanotubes (in Polish). *PhD thesis. Instytute of Physics, University of Silesia, Katowice.*

[14] Jesionek, M., Nowak, M., Szperlich, P., Stróż, D., Szala, J., Jesionek, K., & Rzychoń, T. (2012). Sonochemical growth of antimony selenoiodide in multiwalled carbon nanotube, Ultrason. *Sonochem*, 19, 179-185.

[15] Nowak, M., Szperlich, P., Bober, Ł., Szala, J., Moskal, G., & Stróż, D. (2008). Sonochemical Preparation of SbSI Gel, Ultrason. *Sonochem*, 15, 709-716.

[16] Nowak, M., Mroczek, P., Duka, P., Kidawa, A., Szperlich, P., Grabowski, A., Szala, J., & Moskal, G. (2009). Using of textured policrystalline SbSI in actuators. *Sensors and Actuators A*, 150, 251-256.

[17] Nowak, M. (2010). Photoferroelectric nanowires. *in: N. Lupu (Ed.) Nanowires Science and Technology. Rijeka: InTech*, 269-308, http://www.intechopen.com/articles/show/title/photoferroelectric-nanowires, accessed 21 June 2012).

[18] Szperlich, P., Nowak, M., Bober, Ł., Szala, J., & Stróż, D. (2009). Ferroelectric properties of ultrasonochemically prepared SbSI ethanogel. *Ultrason. Sonochem*, 16, 398-401.

[19] Nowak, M., Kauch, B., Szperlich, P., Jesionek, M., Kępińska, M., Bober, Ł., Szala, J., Moskal, G., Rzychoń, T., & Stróż, D. (2009). Sonochemical Preparation of SbSeI Gel. *Ultrason. Sonochem*, 16, 546-551.

[20] Suslick, K. S., & Price, G. J. (1999). Applications of ultrasound to materials chemistry. *Annu. Rev. Mater. Sci*, 29, 295-326.

[21] Jin, Ho. Bang, & Suslick, K. S. (2010). Applications of Ultrasound to the Synthesis of Nanostructured Materials. *Adv. Mater*, 22, 1039-1059.

[22] Abedini, R., & Mousavi, S. M. (2010). Preparation and enhancing of materials using ultrasound technique:polymers, catalysts and nano-structure particles. *Petroleum & Coal*, 52, 81-98.

[23] Mason, T. J. (2003). Sonochemistry and sonoprocessing: the link, the trends and (probably) the future,. *Ultrason. Sonochem*, 10, 175-179.

[24] Gedanken, A. (2004). Using sonochemistry for the fabrication of nanomaterials,. *Ultrason. Sonochem*, 11, 47-55.

[25] Cravotto, G., & Cintas, P. (2006). Power ultrasound in organic synthesis: moving cavitational chemistry from academia to innovative and large-scale applications. *Chem. Soc. Rev*, 35, 180-196.

[26] Suslick, K. S., Didenko, Y., Fang, M. M., Hyeon, T., Kolbeck, K. J., Mc Namara, W. B., Mdleleni, M. M., & Wong, M. (1999). Acoustic cavitation and its chemical consequences. *Phil. Trans. Roy. Soc. A*, 357, 335-353.

[27] Ibanez, A., Jumas, J. C., Olivier-Fourcade, J., Philippot, E., & Maurin, M. (1983). On chalcogenide-iodates of antimony SbXI (X=S, Se, Te): structure and Mossbauer spectra of ^{121}Sb. *J. Solid State Chem*, 48, 272-283.

[28] Antimony Selenide Iodide (2000). JCPDS-International Centre for Diffraction Data. PCPDFWIN Card File , 2(76-1354)

[29] Hassel, O., & Mark, H. (1924). Über die Kristallstruktur des Graphits. *Z. Phys*, 25, 317-337.

[30] Carbon (2000). JCPDS-International Centre for Diffraction Data. PCPDFWIN Card File , 2(75-1621)

[31] Pohl, S., & Saak, W. (1984). Zur Polymorphie von Antimontriiodid. Die Kristallstruktur von monoklinem SbI_3 . *Z. Kristallogr*, 169, 177-184.

[32] Antimony Iodide (2000). JCPDS-International Centre for Diffraction Data. PCPDFWIN Card File , 2(75-1417)

[33] Antimony Sulfide Iodide (2000). JCPDS-International Centre for Diffraction Data. PCPDFWIN Card File , 2(74-0149)

[34] Starczewska, A., Wrzalik, R., Nowak, M., Szperlich, P., Jesionek, M., Moskal, G., Rzychoń, T., Szala, J., Stróż, D., & Maślanka, P. (2009). Influence of the solvent on ultrasonically produced SbSI nanowires. *Ultrason. Sonochem*, 16, 537-545.

[35] Properties of inorganic compounds. (2012). National Physics Laboratory. http://www.kayelaby.npl.co.uk/chemistry/3_2/3_2.html, accessed 21 June).

[36] Nowak, M., Kauch, B., Szperlich, P., Stróż, D., Szala, J., Rzychoń, T., Bober, Ł., Toroń, B., & Nowrot, A. (2010). Sonochemical preparation of $SbS_{1-x}Se_xI$ nanowires. *Ultrason. Sonochem*, 17, 487-493.

[37] Nowak, M., Talik, E., Szperlich, P., & Stróż, D. (2009). XPS analysis of sonochemically prepared SbSI ethanogel. *Appl. Surf. Sci*, 255, 7689-7694.

[38] Nowak, M., Szperlich, P., Talik, E., Szala, J., Rzychoń, T., Stróż, D., Nowrot, A., & Solecka, B. (2010). *Ultrason. Sonochem*, 17, 219-227.

[39] Sloan, J., Cook, J., Chu, A., Zwiefka-Sibley, M., Green, M. L. H., & Hutchison, J. L. (1998). Selective Deposition of UCl_4 and $(KCl)_x(UCl_4)_y$ inside Carbon Nanotubes Using Eutectic and Noneutectic Mixtures of UCl_4 with KCl. *J. Solid State Chem*, 140, 83-90.

[40] Rolsten, R. F. (1961). Iodide Metals and Metal Iodides. *New York: Wiley*.

[41] Mizukoshi, Y., Nakamura, H., Bandow, H., Maeda, Y., & Nagata, Y. (1999). Sonolysis of organic liquid: effect of vapour pressure and evaporation rate. *Ultrason. Sonochem*, 6, 203-209.

[42] Gutiĕrrez, M., & Henglein, A. (1988). Sonolytic decomposition of poly(vinylpyrrolidone), ethanol, and tetranitromethane in aqueous solution. *J. Phys. Chem*, 92, 2978-2981.

[43] Li, B., Xie, Y., Huang, J., & Qian, Y. (1999). Sonochemical synthesis of silver, copper and lead selenides. *Ultrason. Sonochem*, 6, 217-220.

[44] Li, H.- L., Zhu, Y.- C., Chen, S.- G., Palchik, O., Xiong, J.-P., Koltypin, Yu., Gofer, Y., & Gedanken, A. (2003). A novel ultrasound-assisted approach to the synthesis of CdSe and CdS nanoparticles. *J. Sol. State Chem*, 172, 102-110.

[45] Lu, K. L., Lago, R. M., Chen, Y. K., Green, M. L. H., Harris, P. J. F., & Tsang, S. C. (1996). Mechanical Damage of Carbon Nanotubes by Ultrasound. *Carbon*, 34, 814-816.

[46] Chunwei, Yang., Xinguo, Hu., Dianlong, Wang., Changsong, Dai., Liang, Zhang., Haibo, Jin., & Simeon, Agathopoulos. (2006). Ultrasonically treated multi-walled carbon nanotubes (MWCNTs) as PtRu catalyst supports for methanol electrooxidation. *J. Power Sources*, 160, 187-193.

[47] Eletskii, A. V. (2004). Sorption properties of carbon nanostructures. *Physics-Uspekhi*, 47, 1119-1154.

[48] Qiang, F. U., Weinberg, G., & Dang-sheng, S. U. (2008). Selective filling of carbon nanotubes with metals by selective washing. *New Carbon Materials*, 23, 17-20.

[49] De Gennes, P. G., Brochard-Wyart, F., & Quere, D. (2003). Capillarity and Wetting Phenomena: Drops, Bubbles, Pearls, Waves. *Berlin: Springer*, 51-52.

[50] Voutsas, G. P., & Rentzeperis, P. J. (1986). The crystal structure of the quaternary compound $SbSe_{0.75}S_{0.25}I$. *Z. Kristallogr*, 178, 289-295.

[51] Molnar, B., Johannes, R., & Haas, W. (1965). Properties of single-crystal SbSI. *Bull. Am. Phys. Soc*, 10, 109.

[52] Gerzanich, E. I., Lyakhovitskaya, V. A., Fridkin, V. M., & Popovkin, B. A. (1982). *in: E. Kaldis (Ed.), Current Topics in Materials Science, Amsterdam: North-Holland*, 10, 55-190.

[53] Pankove, J. I. (1971). Optical processes in semiconductors. *New Jersey: Prentice-Hall Inc.*

[54] Kortum, G. (1969). Reflectance Spectroscopy. *Berlin: Springer-Verlag.*

[55] Philips-Invernizzi, B., Dupont, D., & Caze, C. (2001). Bibliographical review for reflectance of diffusing media. *Opt. Eng*, 40, 1082-1092.

[56] Dahm, D. J., & Dahm, K. D. (2007). Interpreting Diffuse Reflectance and Transmittance: A Theoretical Introduction to Absorption Spectroscopy of Scattering Materials. *Norfolk: IM Publications LLP.*

[57] Nowak, M., Kauch, B., & Szperlich, P. (2009). Determination of energy band gap of nanocrystalline SbSI using diffuse reflectance spectroscopy,. *Rev. Sci. Instrum*, 80, 046107 .

[58] Dong-Woon, Shin., Seung-Cheol, Hyun., Sang-An, Park., Yoeng-Geun, Kim., Chang-Dae, Kim., & Wha-Tek, Kim. (1994). Optical properties of undoped and Ni-doped V-VI-VII single crystals. *J. Phys. Chem. Solids*, 55, 825-830.

[59] Seung-Cheol, Hyun., Young-Geun, Kim., Mi-Yang, Kim., Jeong-Dae, Koh., Byong-Seo, Park., & Wha-Tek, Kim. (1995). Optical properties of undoped and chromium-doped V^A-VI^A-VII^A single crystals. *J. Mater. Sci*, 30, 6113-6117.

[60] Sang-An, Park., Mi-Yang, Kim., Ji-Young, Lim., Byong-Seo, Park., Jeong-Dae, Koh., & Wha-Tek, Kim. (1995). Optical Properties of Undoped and V-Doped VA-VIA-VIIA Single Crystals. *Phys. Status Solidi B*, 187, 253-260.

[61] Soonie, Jeon., Gijun, Cho., Wha-tek, Kim., & Sook-Il, Kwun. (1988). *Solid State Commun*, 68, 1043-1046.

[62] Yeong-Koo, Baik., Jae-Keun, Kim., Wha-Tek, Kim., & Sook-Il, Kwun. (1988). Optical Properties of SbSI:Co and SbSeI : Co Single Crystals. *New Phys. (Korean Physical Society)*, 28, 291-295.

[63] Barsoukov, E., & Macdonald, J. R. (2005). Impedance Spectroscopy Theory, Experiment, and Applications. *2nd edition, Wiley, New Jersey.*

[64] ZView Version 3.1c, Copyright. (2007). Scribner Associates. *Inc. written by Derek Johnson.*

[65] Tselev, A., Woodson, M. , Ch, Qian. , & Liu, J. (2008). Microwave Impedance Spectroscopy of Dense Carbon Nanotube Bundles. *Nano Lett.*, 8, 152-156.

[66] Plombon, J. J., O'Brien, K. P., Gstrein, F., & Dubin, V. M. (2007). High-frequency electrical properties of individual and bundled carbon nanotubes. *Applied Physics Lett*, 90, 063106.

[67] Barsoukov, E., & Macdonald, J. R. (2005). Impedance Spectroscopy, Theory, Experiment and Applications. *2nd edition, Wiley-Interscience, New York.*

[68] Metikos-Hukovic, M., Omanovic, S., & Jukic, A. (1999). Impedance spectroscopy of semiconducting films on tin electrodes. *Electrochimica Acta*, 45, 977-986.

[69] Ricciardi, S., Ruiz-Morales, J. C., & Nuñez, P. (2009). Origin and quantitative analysis of the constant phase element of a platinum SOFC cathode using the state-space model. *Solid State Ionics*, 180, 1083-1090.

[70] Osman, R. A. M., & West, A. R. (2011). Electrical characterization and equivalent circuit analysis of $(Bi_{1.5}Zn_{0.5})(Nb_{0.5}Ti_{1.5})O_7$ Pyrochlore, a relaxor ceramic. *Journal of Applied Physics*, 109, 074106-8.

[71] Changxin, Chen., & Yafei, Zhang. (2009). Nanowelded Carbon Nanotubes. From Field-Effect Transistors to Solar Microcells,. *Series: NanoScience and Technology, Springer, Chapter 4: Ultrasonic Nanowelding Technology Between Carbon Nanotubes and Metal Electrodes*, 978-3-642-01498-7, 47-62.

[72] Changxin, Chen., Liyue, Liu., Yang, Lu., Eric-Wai, Siu., Kong, Yafei., Zhang, Xinjun., & Sheng, Han. Ding. (2007). A method for creating reliable and low-resistance contacts between carbon nanotubes and microelectrodes. *Carbon*, 45, 436-442.

[73] Changxin, Chen., Liyue, Liu., Yang, Lu., Eric-Wai, Siu., Kong, Yafei., Zhang, Xinjun., Sheng, Han., & Ding, . (2007). Corrigendum to "A method for creating reliable and low-resistance contacts between carbon nanotubes and microelectrodes". *Carbon*, 45, e1, 436-442.

Chapter 13

Synthesis and Characterization of Advanced Carbon-Based Nanowires – Study of Composites Actuation Capabilities Containing These Nanowires as Fillers

V. Salles, L. Seveyrat, T. Fiorido, L. Hu, J. Galineau,
C. Eid, B. Guiffard, A. Brioude and D. Guyomar

Additional information is available at the end of the chapter

1. Introduction

Electromechanical properties in polymers can be employed to create a large number of sensors and actuators [1-2]. For example they could replace the piezoelectric materials commonly used in Micro Electromechanical Systems (MEMS). Even if the electromechanical coupling is relatively weak for polymers, they can generate high strains due to electrostrictive and Maxwell effects which are a quadratic function of the applied electric field as opposed to a linear function for piezoelectric materials. Other advantages of the polymers are their ease of processability, flexibility and cheapness.

Actuation capabilities of polymers depend on their intrinsic electrical, dielectric and mechanical properties. In order to optimize these properties, one way consists on the incorporation of nano-objects in the polymer matrix. Due to the nanometer-scale of the fillers dimensions, an important change of the properties can be obtained with relatively low ratios. In addition, the effect is more pronounced for conductive nano-objets since higher permittivities are expected to be obtained with a reduced amount of fillers [3]. Carbon-based nanocomposites are good candidates for electroactive nanocomposites: some studies were performed particularly on carbon black, carbon nanotubes, graphite or graphene fillers [4-9]. This part of the chapter is devoted to the influence of carbon-based nanowires (SiC@C, $SiC@SiO_2$, $Fe_3C@C$) as fillers on polyurethanes. The microstructural, dielectric, mechanical, electromechanical and electromagnetic properties are presented and discussed in the following sections.

Among the available electroactive polymers, polyurethanes are of great interest for a wide range of actuator applications due to their significant electric field-induced strains [10-11].

The total electrical field induced strain S can be expressed from electrostricition and Maxwell stress [12-13].

$$S = S_{electrostriction} + S_{Maxwell} = M_{33}^{*} E^2 \quad (1)$$

With M^*_{33} is the apparent electromechanical coefficient and E the applied electrical field.

The Mawwell stress effect comes from the interaction force between the positive and negative charges present on the electrodes

$$S_{Maxwell} = -\frac{\varepsilon_r^{'} \varepsilon_0}{Y} E^2 \quad (2)$$

With ε'_r the real part of the polymer relative permittivity, ε_0 the vacuum permittivity, and Y the Young's modulus.

The electrostriction represents the coupling between the electrical polarization and mechanical response in the material and is related to the electrostrictive coefficient Q and the relative permittiy ε'_r.

$$S_{electrostriction} = Q \varepsilon_0^2 (\varepsilon_r^{'} - 1)^2 E^2 \quad (3)$$

It was shown in previous studies [4, 8, 14] that in such polymers as PU, the Maxwell stress can be neglected.

Based on experimental studies [12, 14-15], the Q coefficient is found to be inversely proportional to the product of the Young's Modulus and permittivity according to:

$$Q \propto \frac{1}{\varepsilon_0 \varepsilon'_r Y} \quad (4)$$

As a consequence, an empirical proportionality relationship is also obtained for the M coefficient :

$$M \propto \frac{\varepsilon_0 (\varepsilon'_r - 1)^2}{\varepsilon'_r Y} \quad (5)$$

This chapter is divided into two main parts. The fabrication procedures and the characterization of the as-prepared nanowires are detailed in the first part whereas the second part is dedicated to show how these nanomaterials can act on the electromechanical properties of EAPs when they are introduced as fillers, demonstrating their high potentiality to fabricate innovative actuators.

2. Fabrication and characterization of carbon-based nanowires

Among the different fabrication processes able to lead to one-dimensional nanostructures, one can distinguish two main approaches: one based on vapour deposition and one other based on polymer pyrolysis, corresponding to two well-known methods presented in this paper: Chemical Vapor Deposition (CVD) and electrospinning (ES) respectively. The interest in studying such two processes lies on fabrication of nanofibers which tend to be mono-crystalline in the first case and more polycrystalline in the second one, with nano-sized crystals.

2.1. Elaboration and characterization of silicon carbide nanowires by CVD

We report here a simple process to fabricate SiC-based nanowires coated with either silica, SiC@SiO_2, or carbon, SiC@C. In both cases, SiC nanocables (NC) were prepared from solid precursors of Si and C, *i.e.* SiO(s) and polypropylene (PP) respectively [16]. These reactants are much cheaper compared to high-purity gaseous reactants used in conventional CVD syntheses. In a typical experiment, the experiments were carried out in a furnace equipped with an alumina tube. PP was placed in an alumina boat, followed by a second alumina boat containing an equimolar mixture of Si(s) and SiO_2(s) or simply a SiO(s) powder, which was partially covered by a graphite condensation plate (Figure 1). All of the experiments were conducted under argon at a very low flow rate (10 mL.min^{-1}), at 1400 °C during 10 h.

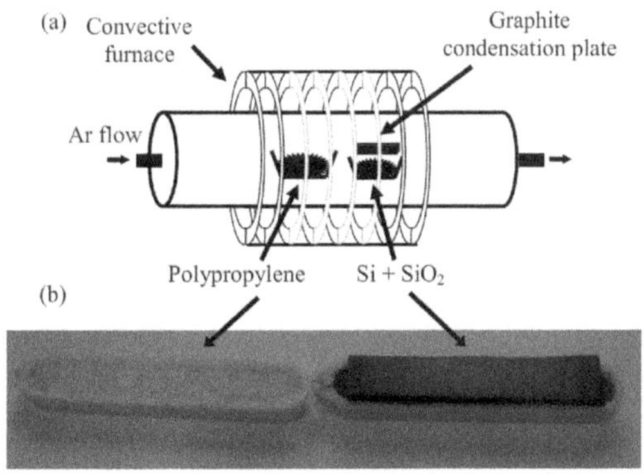

Figure 1. a) Experimental setup for the growth of SiC-based nanowires; b) Picture of the two alumina boats containing the reagents and the graphite condensation plate [16]

At high temperature, a reaction between Si and SiO_2 involves the formation of gaseous SiO which is able to react with carbonated species to initiate the nanowires growth. While using a PP/SiO weight ratio of 1 (maximum), SiC@SiO_2 NCs are formed whereas a ratio of 2 leads to the growth of SiC@C NWs (Figures 2 and 3). It is thus possible to control the silica thickness from 2 to 10 nm by adjusting the PP/SiO weight ratio from 1 to 0.25 for instance.

The as-obtained NWs have a diameter ranging from 20 to 40 nm. The NWs core consists of 3C-SiC, the cubic polytype of SiC, and the shell is either composed of amorphous silica (Figure 2) or turbostratic carbon (Figure 3). The high-purity of such structures was confirmed by EELS analyses and Raman investigations allowed to determine the P-type doping of the SiC-based NWs.

Figure 2. SiC@SiO2 NCs prepared from 1 g PP and 4 g SiO(s). a) Slightly out-of-focus TEM image of the crude product, where the amorphous silica layer can be easily distinguished from the SiC core (the corresponding well-focused image is included as an insert); b) HRTEM image showing a 10 nm silica layer around the SiC core [16]

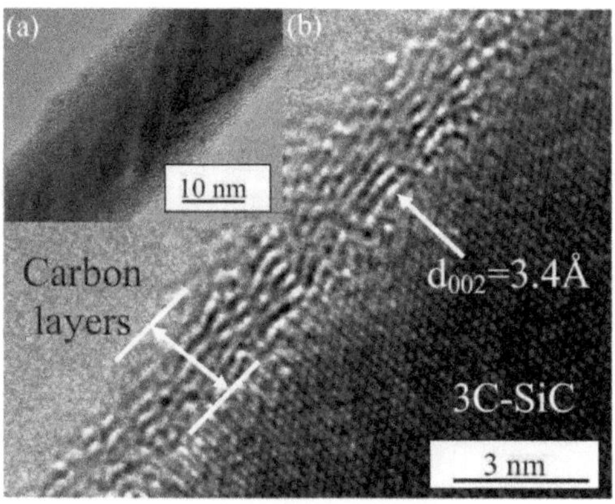

Figure 3. a,b) HRTEM images of a 30 nm SiC@C NC prepared by increasing the amount of PP to 8 g for every 4 g SiO(s) [16]

2.2. Elaboration and characterization of iron carbide nanowires by electrospinning

Electrospinning is a process which has been mainly studied and used during the last decade to prepare nanowires/nanofibers. We also have noticed an increasing interest of the scientific community during this period (Figure 4).

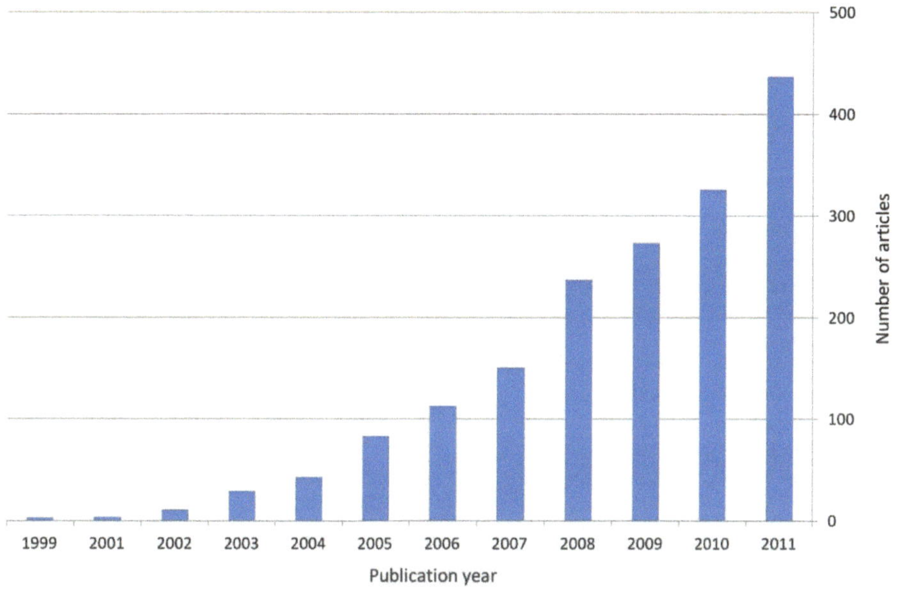

Figure 4. Histogram representing the number of scientific articles published per year during the last decade (Research performed on 4th June 2012 with "Web of Science", with "electrospinning" AND "nanofiber")

A lot of effort has been focused on the electrospinning process due to its versatility, which permits the formation of polymeric fibers as well as, after treatment, oxide and non-oxide inorganic fibers at submicro- and nanoscales [17-18]. This process is easy to scale-up to large amounts and has been recently employed to develop non-oxide ceramic nanowires. In this field, the synthesis of SiC [19-20], B_4C [21], B_4C/SiC [22], GaN [23] and BN [24] fibers has been reported for diameters smaller than 1μm. This production technique allows to precisely control the chemical composition of sub-microscale and nanoscale fibers which are serious candidate for applications ranging from smart textile to biomedical, including automotive and environmental areas.

Electrospinning consists in electrostatic stretching of a molten polymer or a solution containing a polymer solubilized in a solvent (the most common procedure). A high-voltage (HV) is applied between a spinneret (needle) and a metallic target (Figure 5).

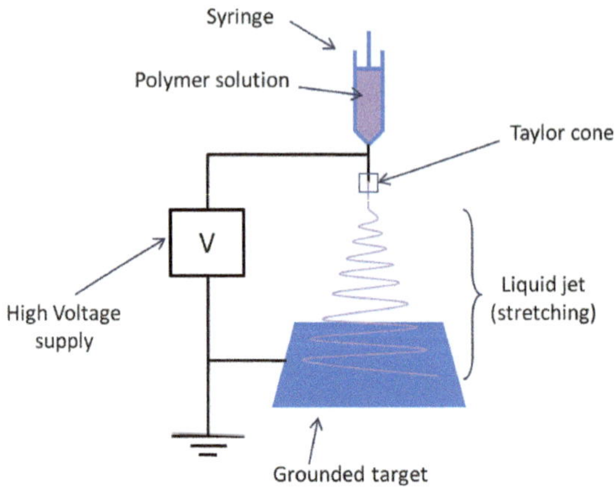

Figure 5. Schema of a typical electrospinning setup

With a low and continuous flow rate of solution through the needle, the solution drops without application of voltage but when the electric field is high enough to overcome the surface tension, a Taylor cone is formed at the needle tip and the solution is then accelerated toward the target in the form of a polymer jet. With stable working conditions, a continuous filament can be produced and collected as a solid 1D-material since the main part of the solvent evolves during the jet acceleration, in case of wet spinning. solvent evolves during the jet acceleration. Its final diameter, conventionally in the submicrometer scale, is about 1000 times lower than the inner spinneret diameter. Numerous parameters directly influence the morphology of the as-spun fibers/filaments. It is possible to fabricate dense, porous, beaded, flattened or perfectly cylindrical fibers simply by adjusting the solution properties :

- surface tension,
- electrical conductivity,
- viscoelastcity,
- polymer concentration,
- solvent(s) volatility,

and/or the spinning conditions :

- high-voltage,
- working distance,
- polymer/solution flow rate,
- inner needle diameter,
- type of target,

• moisture degree.

In case of fabrication of inorganic fibers by electrospinning, a specific precursor is employed [17-24]. Iron-based materials (iron oxide and iron carbide) can be fabricated from a mixture of polyvinyl pyrrolidone (PVP), bringing the spinnability property, and iron (II) acetate ($FeAc_2$) in acetic acid and ethanol (Figure 6) [25].

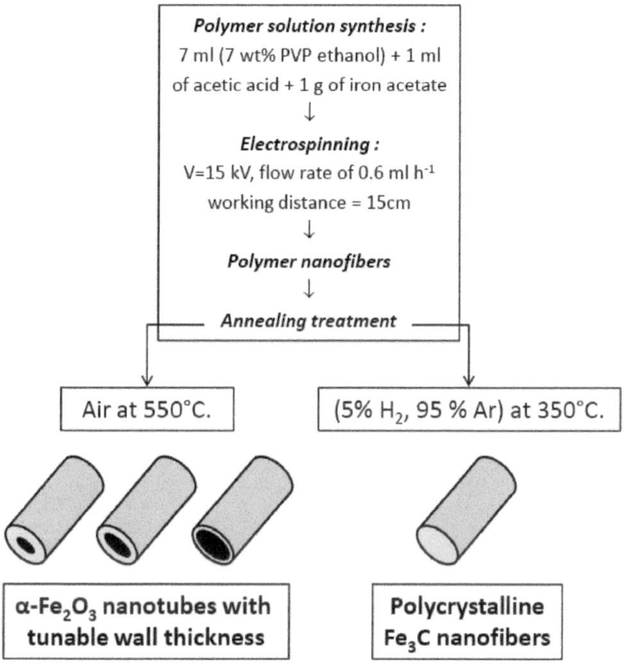

Figure 6. Experimental process with the different nanostructures obtained [25]

After stirring, to obtain a homogeneous and brown colored solution, the polymer solution is loaded into a syringe (5 mL) before being spun on a metallic target. After optimization of all spinning parameters, homogeneous filaments characterized by a smooth surface and an average diameter ranging from 200 to 500 nm can be obtained according to the precursor content (*i.e.* $FeAc_2$/PVP ratio) in the initial solution (Figure 7). An average value was calculated on 100 fibers of each sample giving a linear evolution of the fiber diameter versus the $FeAc_2$/PVP ratio: 440 nm, 395 nm, 260 nm and 220 nm for ratios of 2.63, 1.75, 0.87 and 0.43 repectively. This phenomenon is mainly attributed to a decrease of solid content when the ratio is decreased since the PVP content is the same in all the solutions (corresponding to 7 wt%). We can notice that iron acetate is stored and sampled under argon in order to prevent it from degradation over time.

A pyrolysis step was then carried out under argon and hydrogen (95% Ar and 5% H_2) with a heating rate of 3°C/min and an holding time of 4h at 450°C before cooling down. This thermal treatment allowed the organic part decomposition as well as the precursor conversion into an inorganic material. The thermal behaviour of the filaments during the pyrolysis step is still under investigation but we know that the last weight loss is due to PVP gaseous evolution up to 400-450 °C. Above this temperature, there is no more weight variation of the filaments. Temperatures above 800 °C were not studied.

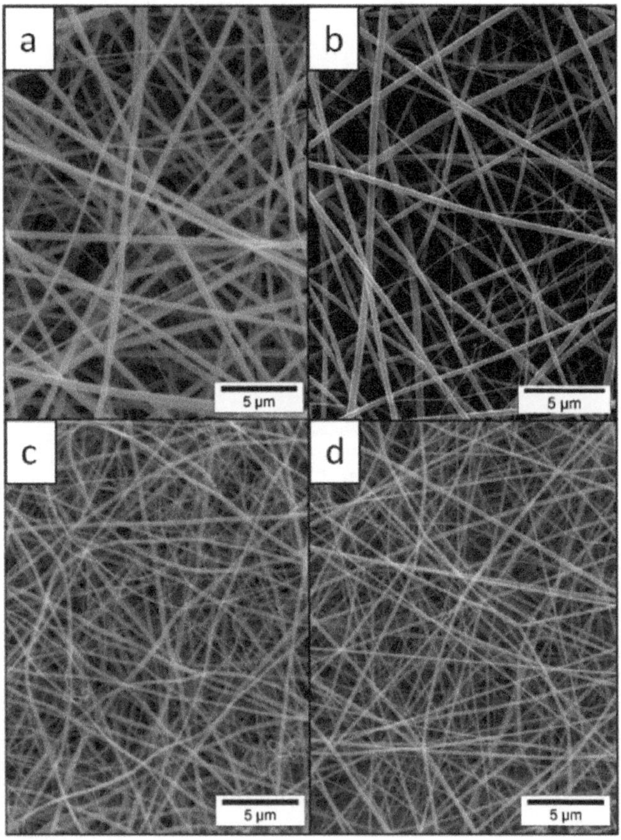

Figure 7. SEM images of the various PVP/$FeAc_2$ nanofibers with different wt ratios (FeAc2/PVP). (a) ratio = 2.63; (b) ratio = 1.75; (c) ratio = 0.87; (d) ratio = 0.43 [25]

The weight loss occuring during treatment is logically accompanied by a shrinkage in the macroscopic sample dimensions (already high-lighted and described in a previous work [24]), as well as in the filament diameter (Figure 8).

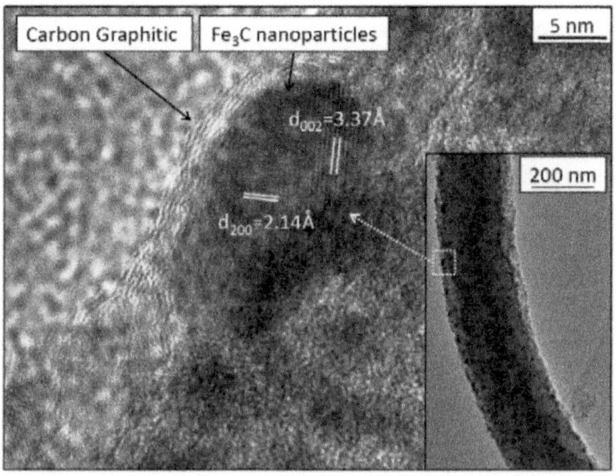

Figure 8. HRTEM of Fe_3C nanoparticles surrounded with graphitic planes. Inset: TEM images of Fe_3C nanofibers prepared with $FeAc_2$/PVP = 0.87 [25]

As XRD analysis was inadequate (no signal) to characterize the type of material composing constituting the inorganic filaments, transmission electronic microscopy (TEM) was performed on several filaments (Figure 8). The first main information is that the filaments are made of nanoparticles homogeneously dispersed in a matrix made of a different material. High-resolution TEM (HR-TEM) was then used to observe that the small nanoparticles have a spherical shape, and interplanar distances of 3.37 and 2.14 Å both measured on the same nanoparticle indicating the presence of pure iron carbide, corresponding to crystalline planes (002) and (200), repectively, of the Fe_3C orthorhombic phase. Moreover, these cementite domains are surrounded by well-known concentric graphitic planes forming a coating layer of 2-3 nm on the particle surface. The reduction in a hydrogen atmosphere of the as-electrospun PVP/$FeAc_2$ fibers is responsible for the formation of polycrystalline Fe_3C nanofibers embedded in carbon graphitic planes, ensuring stability and cohesion of the Fe_3C nanoparticles. The presence of graphiic carbon was confirmed by Raman spectroscopy with D and G bands at 1360 and 1594 cm^{-1} respectively [25].

The functionalization of organic matrices for electroactuation application has been studied and is presented below, using either SiC-based or Fe_3C-based nanowires as inorganic fillers. Carbon-based fibers, SiC@C and Fe_3C, were expected to be interesting because of their carbon coating which should enhance the surface electrical conductivity of the fillers, thus promoting the charge displacement at a nano- or a submicro-scale inside the polymer matrix. This should have a positive effect on the electrical polarization of the polymer host, directly linked to its ability to be deformed when it is subjected to an electric field.

3. Integration of the nanowires into polymer and study of composites actuation capabilities

3.1. Elaboration of nanocomposites

The two polymers were polyether-based aromatic thermoplastic polyurethane (PU), purchased from Lubrizol Corporation: Estane 58888 NAT 0.21 (PU88) and Estane X-4977 NAT039 (PU75). They differ by the ratio of hard to soft segments. Hard segments (HS) comprise the 4.4′ methylene bis(phenyl isocyanate) (MDI) and 1,4-butanediol (BDO) and soft segments are poly(tetramethlene oxide) (PTMO). The PU88 has a density of 1.13 g/cm^3, a hardness of 88 shore A and the HS content is about 46%; the PU75 has a density of 1.07 g/cm^3, a hardness of 75 shore A and the HS content is less, near 26% [26]. HS amount is an important factor because the physical properties depend directly on the degree of phase separation between the hard and the soft phases.

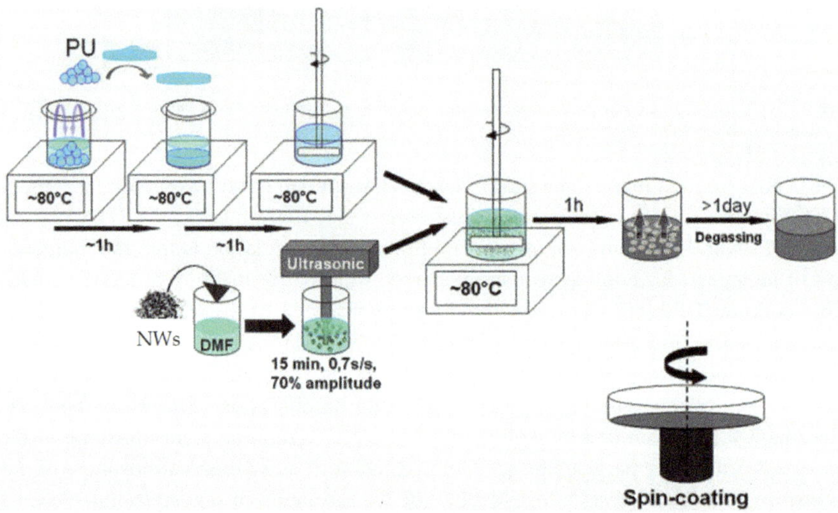

Figure 9. Elaboration of the PU88-SiC nanocomposite solution by solution casting method and spin coating [26]

PU88-SiC@C and PU88-SiC@SiO$_2$ nano-composite films were prepared following a solution casting method (Figure 9). Before use, the PU granules were heated at 80°C for 3 h to remove any traces of water. They were then put in N,N-dimethylformamide (DMF, Sigma-Aldrich D158550, 99%). The solution was maintained at 80°C for 2 h. The SiC nanowires were dispersed into DMF using an ultrasonic processor (Hielsher UP400S, 400 W, 24 kHz, 7 mm diameter sonotrode) under the following experimental conditions depending on the type of the nanowire: 10 to 20 min duration, 80% amplitude, 80% pulse. This solution was added to that of the polymer and the resultant mixture was heated at 80 °C for 1 h under mechanical stirring, until a homogeneous and viscous solution was obtained. This solution was degassed for 24 hours at room temperature before spin coating on glass plates using a Laurell WS400 BX spin coater. This was a two steps process of 10 and 15 seconds with different rotational speed depending on the desired thickness: 200/300 rpm for 80 μm and 500/800 rpm for 35 μm. The films were then placed in an oven at 60°C for 24 hours and then removed from the glass substrate. A second heating treatment at 125°C (below the melting temperature) for 3 hours was performed in order to eliminate any residual traces of solvent.

The elaboration method was improved for the PU75-Fe_3C@C nanocomposites by using closed recipients during the dissolution and mixing steps so that the evaporation of the DMF solvent was avoided. In addition, the solution was applied on glass plates with an Elcometer 3700 doctor blade film applicator (Figure 10). This technique produced large surface films with homogeneous thickness. The following thermal treatments were the same as described above.

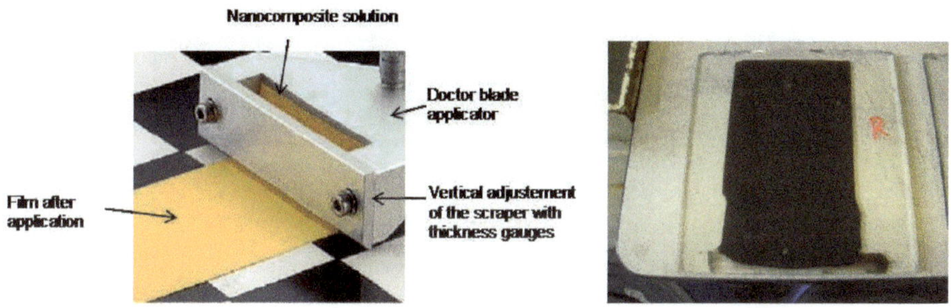

Figure 10. Deposition of the PU75-Fe_3C nanocomposite solution with Elcometer 3700 doctor blade applicator

For both methods (Figures 9 and 10), the weight fractions of fillers varied between 0 and 4 %.

It is important to note that spin coating is widely used as a shaping method for EAPs that are introduced inside actuators devices, however, the method so-called "Doctor Blade" allows to prepare polymer or composite films with larger dimensions which is interesting for studies necessitating numerous samples for analyses (electrostriction, mechanical test, DSC, SEM,...), with verification of reproducible results as well.

3.2. Characterization of the nanocomposites

3.2.1. Structural characterization of the nanocomposites

Microstructural observations (Figure 11), carried out by SEM using a cryo-fracture preparation, showed that the nanowires seem to be aligned preferentially in planes parallel to the surface of the film. Whatever the shaping method used, taking into account the external forces applied on the long chains of PU, it is assumed that there is no difference of the nanowires organization in the polymer host. The homogeneity of this distribution was checked over the entire width, and from the bottom to the top surface.

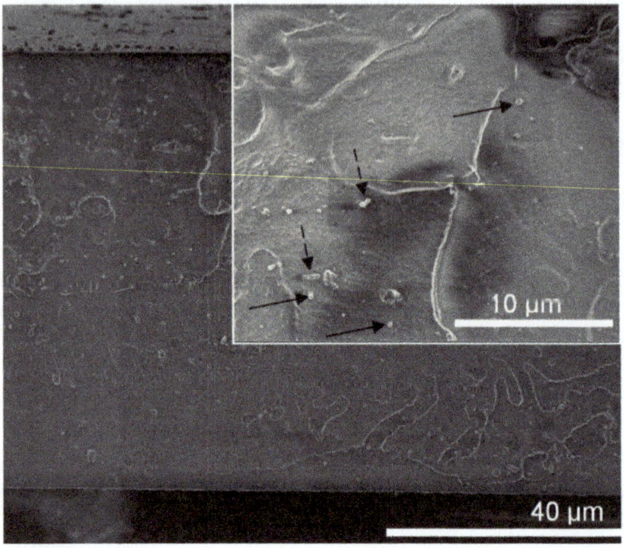

Figure 11. SEM observations performed on samples PU75-Fe_3C (2.5 wt% of Fe_3C); Arrows indicate the presence nanofillers with perpendicular (solid arrows) and non perpendicular (dashed arrows) orientations compared to the fracture plane

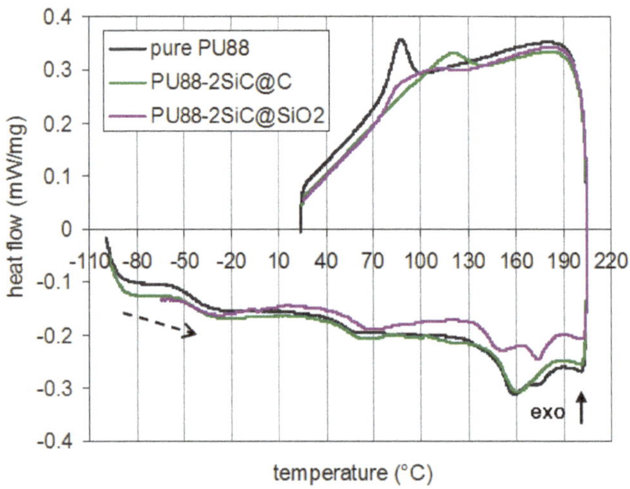

Figure 12. DSC thermograms on heating and cooling of pure PU88, PU88-2SiC@C and PU88-2SiC@SiO_2 nanocomposites films

The Differential Scanning Calorimetry (DSC) technique was used to study the effect of the nanowires on the structure of the host polymer. The thermal analysis was performed with DSC 131 evo from Setaram, under nitrogen atmosphere. The samples (about 20 mg) were placed in closed 100-μl aluminum crucibles and cooled from ambient temperature down to -100°C, then heated to 210°C and finally cooled to RT. The heating and cooling ramps were performed at 10°C.min^{-1}.

The thermograms for the pure PU88, and the PU88-2SiC@C and PU88-2SiC@SiO_2 compositions are presented below (Figure 12).

The glass transition temperature for the soft segments (Tg), melting temperatures (T_{II} and T_{III}), heat of fusion (ΔHm), crystallization temperature (T_c) and heat of crystallization (ΔHc) are summarized in Table 1.

	Tg (°C)	T_{II} (°C)	T_{III} (°C)	ΔHm (J/g)	T_C (°C)	ΔHc (J/g)
Pure PU88	-42.5	157	173	9.6	88	-8.6
PU-0.5SiC@C	-40.8	159	173	8.9	111	-9.9
PU-1SiC@C	-40.0	157	174	10.1	115	-8.8
PU-2SiC@C	-42.0	160	172	8.3	118	-8.4
PU-3SiC@C	-43.1	160	174	8.8	119	-7.0
PU-1SiC@SiO_2	-40.9	150	173	8.9	88-110	-10.7
PU-2SiC@SiO_2	-47.6	158	173	10.0	88	-10.4
PU-3SiC@SiO_2	-48.3	153	171	10.4	88-110	-11.5
PU-4SiC@SiO_2	-39.6	150	171	10.3	110	-11.8

Table 1. DSC results obtained for the PU88-SiC nanocomposite films

The first thermal phenomenon near -45°C was related to the glass transition (Tg_{SS}) of the soft segments (SS). This value can give an estimation of the amount of hard segments (HS) dissolved in the soft domains and as a consequence an indication of the purity of the soft phase [27-28]. When the fraction of nanowires was increased, no significant change of Tg was observed. It was thus concluded that the degree of HS-SS mixing was not modified by the incorporation of nanowires.

Some effects of relaxation of the polymer chain, a modification at the SS/HS interface or thermal endotherm seemed to explain the behavior of heat flow near 60°C. At temperatures in the range 150-170°C, it was observed a bimodal endotherm that can be related to the micromixing of non-crysytalline or semi-crystalline hard and soft phases followed by the melting of crystalline HS [29]. Another interpretation of the bimodal peak could be the melting of crystalline hard segments: the two peaks represent the two characteristic lengths of HS [27]. In both cases, the increase of the samples' crystallinity would be accompanied by an increase of these temperatures and/or an increase of the total enthalpy of the 2 phenomena.

With the incorporation of SiC@C nanowires, the endothermal peaks remained unmodified, temperature and corresponding enthalpy did not change. With $SiC@SiO_2$, one can observe only a slight decrease of the temperature of the first endotherm related to the micromixing of hard and soft phases.

During the cooling experiments, a more important influence of the nanowires on the crystallization behavior was observed. When the amount of nanowires was increased, the crystallization temperature was raised but without a significant change in enthalpy. The nanowires seemed to act as nucleating agents improving the crystallization. There is a more important effect on the crystallization of the polyurethane with the incorporation of SiC@C compared to $SiC@SiO_2$ (Figure 12).

The same study has been performed on nanocomposite films based on PU75 and $Fe_3C@C$ nanowires and the values of Tg and melting temperatures and heat of fusion are summarized in Table 2. The PU75 is less rigid than the PU88, as indicated by the value of the glass transition temperature which is 25°C under the PU88 one. With the incorporation of $Fe_3C@C$, there is no modification of the thermal properties during heating and cooling, and so no morphological change of the PU75 polymer host.

	Tg (°C)	T_{II} (°C)	T_{III} (°C)	ΔHm (J/g)
Pure PU75	-68.6	153	174	2.3
PU75-2.5$Fe_3C@C$	-69.0	152	175	3.0
PU75-5$Fe_3C@C$	-71.7	152	174	2.4

Table 2. DSC results obtained for the PU75-$Fe_3C@C$ nanocomposite films

It can be concluded from these DSC results that, whatever the PU, no major microstructural modification of the matrix was observed with the incorporation of nanowires.

3.2.2. Dielectric properties of the nanocomposites

Prior to the electrical measurements, the samples were coated with 20 nm thick gold by sputtering. The relative permittivity ε', the dielectric loss factor tan δ and the real part of the electrical conductivity σ' were obtained with a Solartron 1255 impedance / gain-phase analyzer and a 1296 dielectric interface. The voltage was set at 1 V_{RMS} and a frequency range was performed in the range 0.1-10^5 Hz.

Figure 13 presents the electrical conductivity versus frequency of some PU88/SiC nanowires composites. In all samples, two frequency regions can be observed: one for which the conductivity was almost constant (low frequencies < 10Hz) and which is attributed to charge displacements and another one (high frequencies > 10Hz) where the conductivity increased with the frequency and corresponds to the dielectric losses [30].

Compared to the $SiC@SiO_2$, the SiC@C nanowires lead to a slightly more important increase of the polyurethane conductivity. This is in goog agreement with the conductive carbon layer which covers the SiC wires, instead of the SiO_2 one.

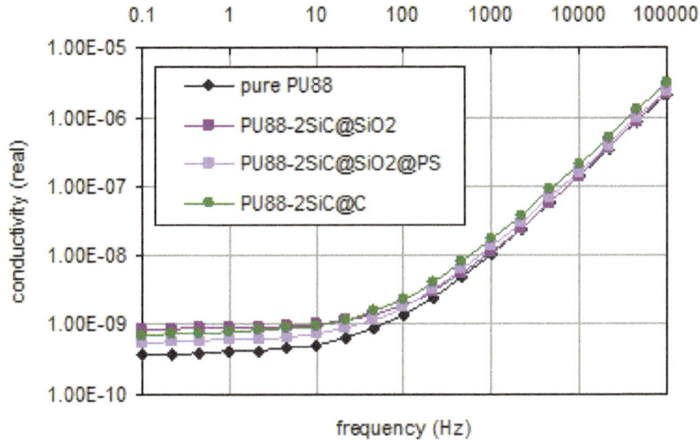

Figure 13. The electrical conductivity versus frequency for pure PU and PU88-2SiC composites

The frequency/conductivity curves show that even if the conductivity is slightly enhanced with the incorporation of nanowires, the percolation threshold is not reached even with higher concentrations of SiC (3% tested) or SiO_2 (4% tested). This threshold could be estimated as the content located between that for which the conductivity still depended on the frequency and that for which it became completely independent of the frequency.

Figure 14 exhibits the permittivity versus frequency of some PU88/SiC nanowires composites

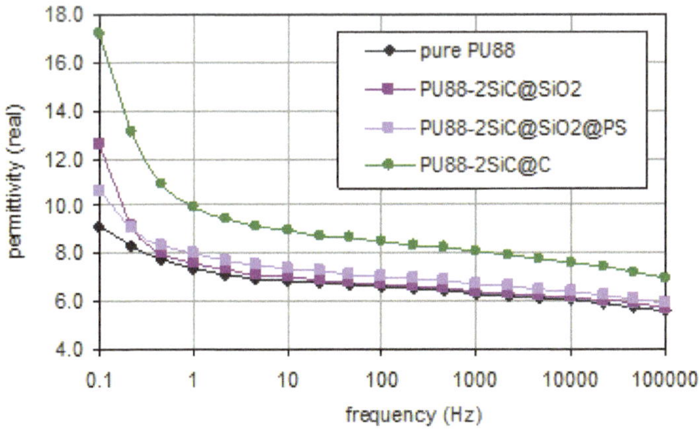

Figure 14. The relative permittivity versus frequency for pure PU and PU88-2SiC composites.

Whatever the sample, the permittivity increases as the frequency is lowered, which is in good agreement with Maxwell Wagner mechanism [31]. The composites present an higher permittivity than the pure polyurethane one: for example, at 0.1 Hz, it increases from 9 for pure PU to 12 for PU88-2SiC@SiO_2 and to 18 for PU88-2SiC@C.

The effect is more pronounced for the SiC@C compared to SiC@SiO_2 certainly because the first type of filler has a conductive carbon layer on the surface instead of the insulating SiO_2 layer.

For the PU75-Fe_3C system, the same dielectric characterization was made (Figure 15).

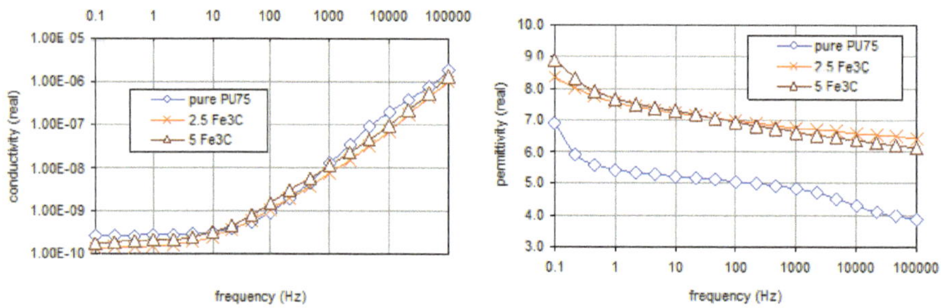

Figure 15. The conductivity (a) and relative permittivity (b) versus frequency for composites with varying SiC contents (a) PU88-SiC@C and (b) PU88-SiC@SiO_2

The conductivity curves for the PU75 system shows like for the PU88 two regions depending on the value of the frequency. Whatever the frequency, the conductivity is not really modified with or without Fe_3C.

Compared to the SiC@C filler, the Fe_3C@C has a very slight influence on the permittivity of the samples at low frequency. But at a frequency near 1 kHz, the gain was of the same order of magnitude for the two fillers i.e. 1.35 fold.

3.2.3. Mechanical properties of the nanocomposites

The mechanical behaviour of the composite films was evaluated on 25 x 10 mm^2 samples, using a house made tensile test (Figure 16). On one end, the sample was clamped onto a Newport platin and the other end onto a force sensor. The stage was moved using a function generator at a frequency of 50 Hz and the dynamic force was acquired on an oscilloscope. Measurements were performed over a deformation range from 0 to 100%. The Young's modulus was determined at the beginning of the curve strain / stress for elongations less than 4%.

Figure 17 exhibits the evolution of the Young's modulus of the nanocomposites films as a function of the content of nanowires.

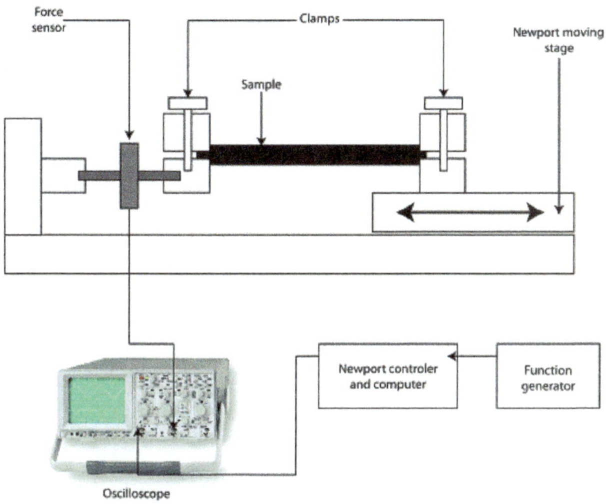

Figure 16. Experimental set-up for measuring the mechanical properties of the nanocomposite films

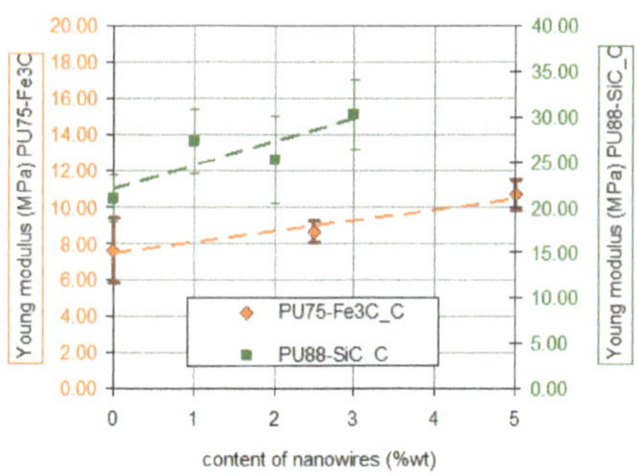

Figure 17. The Young's modulus versus the content of nanowires for PU88-Si@C and PU75-Fe_3C@C

For the two polyurethane systems, it is observed an expected little increase of the Young's modulus with the incorporation of nanowires. For the same content of nanowires, for example 2.5%, there is a 1.3 gain factor for PU88-SiC@C and 1.2 gain factor for PU75-Fe_3C@C.

3.2.4. Electromechanical capabilities

a. Electric field-induced thickness strain measured by a double-beam interferometer at low frequency (0.1Hz).

The electric field-induced thickness strain S was measured with a double-beam laser interferometer (Agilent 10889B), with a precision in the order of 10 nm. The film samples of 25 mm diameter were placed on a horizontal stainless steel disc (20 mm in diameter), and a second brass disc placed on the upper side of the film allowing the application of a bipolar electric field at 0.1 Hz. The voltage was supplied by a function generator (Agilent 33250A) amplified 1000-fold through a high-voltage lock-in amplifier (Trek 10/10B). A guard ring was added to the sample surface and grounded so as to eliminate any surface leakage currents. Figure 18 gives photographs of the experimental set-up for strain measurements.

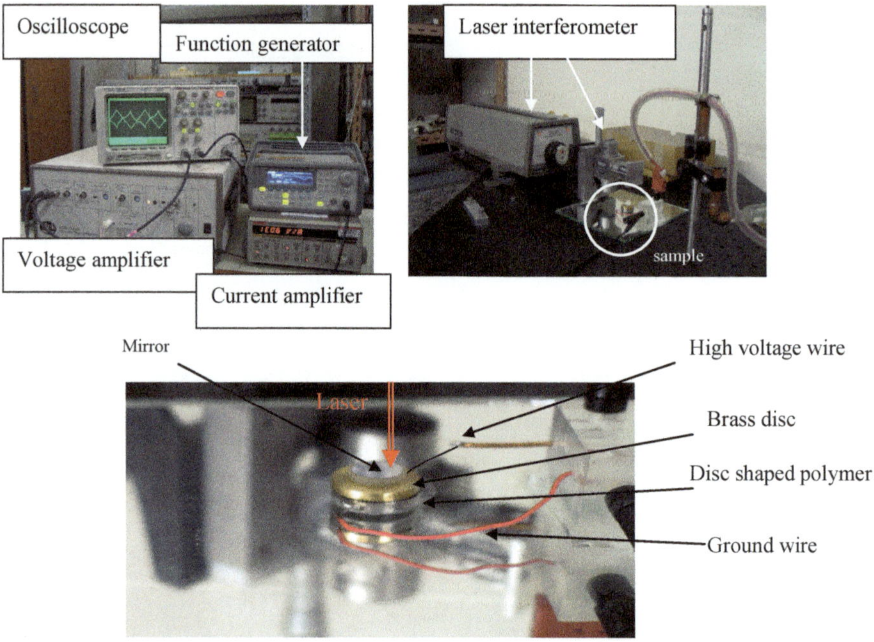

Figure 18. Experimental set-up for strain measurements with laser interferometer at 0.1 Hz.

Figure 19-a) presents the thickness strain amplitude – S_3 (S_3 is negative since the film is compressed) versus the electric field amplitude of pure and filled with 0.5% SiC@C PU88 films of different thicknesses.

All studied films exhibited the same type of strain variations (except the 20 μm – thick nanocomposite because electric breakdown occurred above 18V/μm): a first region where the strain increases regularly with the electric field amplitude indeed and a second region at higher fields where an asymptotic value is reached. This is the result of the hyperelastic behaviour of such PU materials.

The deformation depends on the thickness of the film: lower electric fields were required to obtain a given strain level before saturation as the film's thickness increased. The 20μm-thick nanocomposite film presented the maximal thickness strain of about 18% at 18 V/μm. At moderate electric values, the greatest effect was observed for the 100 μm-thick PU-SiC@C film since the strain was equal to 10% at a field as low as 6 V/μm.

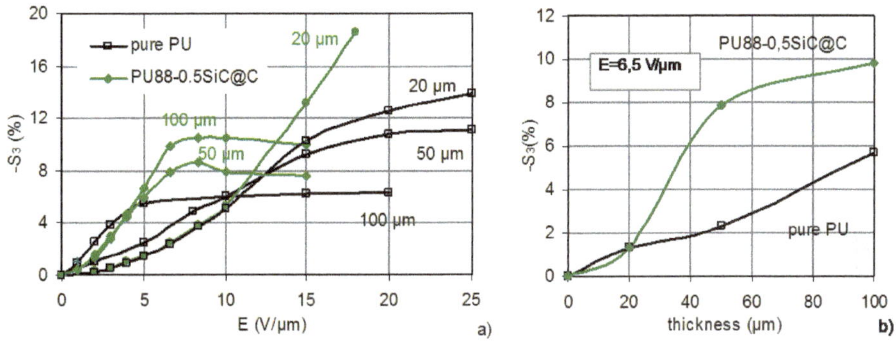

Figure 19. Evolution of the strain amplitude versus the electric field amplitude for various thicknesses of pure PU88 and PU88-0.5SiC@C films at 0.1 Hz. (b) The evolution of the strain amplitude versus the thickness of pure PU88 and PU88-0.5SiC@C films at E= 6.5 V/μm [11]

As the relative permittivity was found to be higher in the SiC@C composites, the results showed an increase of the interfacial polarization induced by a pronounced space charges effect. Space charges accumulate in PU and more effectively in PU-SiC@C composites and it was still evidenced by the large dependence of the strain on the film thickness, particularly for low and medium electric field values (E < 10 V/μm). These charges are generally electrons injected in the bulk from the electrodes and/or impurities (ions) in the sample which are trapped to the different structural interfaces of the material. The consequence of the presence of space charges is the appearance of high electric field local densities, located at the various trapping sites (defects, interfaces), thus yielding a non-uniform field distribution in the film thickness and consequently a high deformation [32].

However, the film must be thick enough to present such a strain improvement by space charges effect, as it is shown in Figure 19-b which gives the evolution of strain at 6.5 V/μm electric field versus thickness. If the film was too thin, the strain was not increased because the interfaces that were able to trap the space charges started to overlap and as a consequence charge homogenization occured [14, 33]. When the film thickness increased, the strain also increased, with a more pronounced effect for the nanocomposite. It should be emphasized that the polyurethane exhibited a considerable ability to collect space charges certainly due to the two-phase structure composed of hard and soft segments inducing many interfaces. The presence of space charges were experimentally evidenced by studying the steady-state current – electric field characteristic of the films [11].

DSC measurements presented in previous section were performed in order to determine whether the SiC@C nanowires themselves or polymer chains/nanowires' interfaces behaved as additionnal space charge traps and/or if they increased the level of phase mixing between soft and hard segments and consequently the structural interface density. The DSC results show that there is no modification of the phase mixing so it can be concluded that space charge effects in the composites were not enhanced by an increase of the structural interface density, but rather by the SiC@C nanowires themselves, acting as charge traps.

Another study was performed with SiC@SiO_2 nanowires as fillers with the laser interferometer method at low frequency [34]. The electric field induced thickness strain response has been investigated for the polystyrene-grafted silica coating of SiC NWs (PU88-2SiC@SiO_2@PS) nanocomposites and compared to the pure polyurethane film and the PU88-2SiC@SiO_2 nanocomposite without polystyrene grafting.

Polyurethane polymer filling with both types of nanowires leads to a higher electromechanical response than for the pure PU88 film. At a moderate electric field of 10 V/μm, 2SiC@SiO_2@PS loading increased the strain level of pure PU88 by a factor of 2.2. The improvement came partially from polystyrene grafting since the PU-2SiC@SiO_2 films showed only a 1.7 times increase. It was shown in the article that since a higher strain was observed for the PU88-2SiC@SiO_2@PS films, the polystyrene chains grafting seemed to act as a dispersing agent, thus yielding a homogeneous distribution of NWs in the composite.

b. Deflection strain under electric field of a diaphragm type dielectric actuator working at high frequency (near 1kHz).

Another way to investigate the electromechanical properties of polymer films is to measure the expension in the 31 direction (which is directly proportional to the 33 contraction). It was chosen to investigate it using a diaphragm, which is also a good candidates for integration of these polymer films into Microsystems [35].

The diaphragm device (Figure 20) comprised a sample holder and the electroded sample.

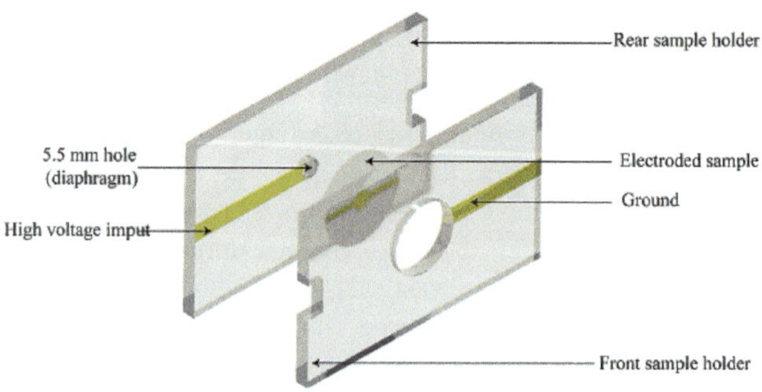

Figure 20. Diaphragm device used for the measurement of the deflection strain under electrical field at high frequency [35]

The sample holder was fabricated in thick PMMA sheets. Two diameter holes of 5.5 mm and 20 mm were drilled into the two parts of the device. The 5.5 mm hole constituted the diaphragm. Gold electrode was sputtered onto each part for electrical connections. Samples of 25 mm in diameter were electroded using gold sputtering as shown in the figure 19 and then aligned with the 5.5 mm hole and the electrode on the rear part of sample holder. The sample adhered onto the sample holder by electrostatic forces.

The polymer film was excited in a pseudo piezoelectric mode by applying a DC electrical field in addition to the AC excitation. The sinusoidal voltage was applied using a function generator (Agilent 33200A) and amplified by the Trek Model 10/10 high voltage power amplifier The DC component induces a constant polarization of the membrane, and thus, a piezoelectric behavior. In this study, all data were recorded in pseudo-piezoelectric mode and at the first resonant frequency. In addition, the DC bias voltage was chosen equal to the AC voltage magnitude. The membrane deformation was recorded using a Polytech OFV-5000 vibrometer equiped with a OFV-505 sensor head. The reflectivity of the gold electrode was used to monitor the deflection. The deflection strain is defined as S_d=S/d where d is the deflection and h the sample thickness. The deflection is caused by the compressive strain undergone by the film rigidily clamped at the edge.

The deflection strain magnitude for PU75 loaded with 0, 2.5 and 5wt% of Fe_3C@C has been studied as a function of the square applied electrical field magnitude [36]. The results are presented in figure 21.

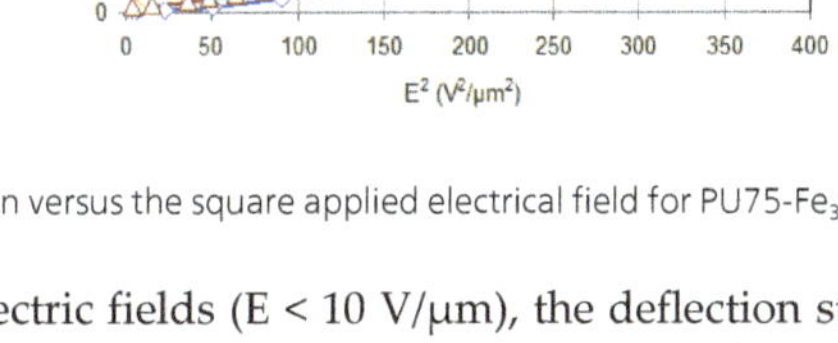

Figure 21. The deflection strain versus the square applied electrical field for PU75-Fe_3C@C nanocomposites films

At low and medium electric fields (E < 10 V/μm), the deflection strain increased quasi-linearly with E^2, thus confirming the electrostrictive nature of the electromechanical response. For higher electric field, a saturation effect corresponding to an asymptotic constant value of strain was observed and can be attributed to the saturation of the electrical polarization [37].

The deflection is increased by 2 to 7 fold from pure PU75 to 2.5% load PU depending on the electrical field magnitude, and is higher than that measured on the same PU filled with carbon black [35]. A considerable deflection strain of more than 130 % was found for the PU75-2.5Fe_3C@C nanocomposite for a 18 V/μm electrical field.

With higher amount of Fe_3C nanowires, the deflection decreases. This decrease could be related to an agglomeration of the nanowires. Some complementary structural charaterizations are under progress by SEM and Focused Ion Beam in order to fully understand this behavior. Even if the global permittivity and dielectric loss do not increase dramatically, it is possible that some agglomerates appear locally inside the film and contribute to decrease the interfacial polarization and consequently the density of trapped charges. This effect yields a decrease in the deflection strain amplitude.

By using the same method, measurements have been performed on different PU88-SiC@C nanocomposites. Figure 22 exhibits the strain deflection versus the square of the electric field. Even if the polyurethane host is not the same for the study, it is possible to compare the influence of the two types of nanowires, ie SiC@C and Fe_3C@C, on the actuation performances.

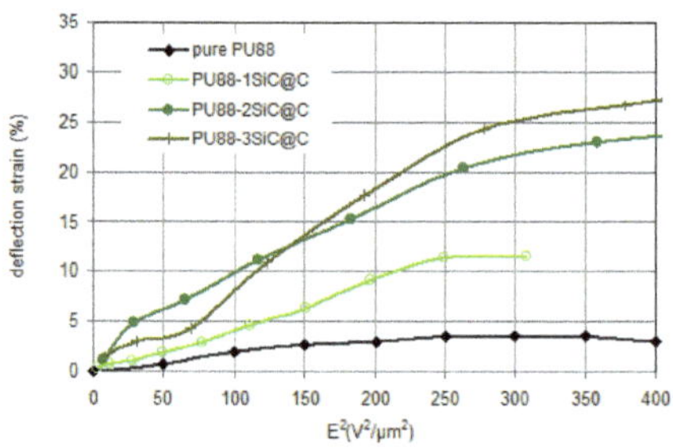

Figure 22. The deflection strain versus the square applied electrical field for PU88-SiC@C nanocomposites films

The electric field induced a deflection of PU88 films which was lower than that of the PU75 films; it is in agreement with a similar value of the relative permittivity in both cases and a lower value of Young's modulus of the PU75 films. Indeed the strain depends directly on the ratio of permittivity over Young's modulus as shown in the equation (5).

At high electrical fields, the PU88-3SiC@C shows a 25% deflection strain which is 6 times higher than for pure PU88. The gain in actuation is similar to that measured on PU75-2.5Fe3C@C. This is consistent with results presented in previous section: dielectric studies for both nanocomposites have shown an increase of the relative permittivity near 1 kHz with a gain of 1.35. Moreover, the mechanical reinforcement was also comparable for the 2 types of composite films.

4. Conclusions

Two different methods of synthesis were investigated and exploited to develop new carbon-based nanowires, SiC@SiO_2, SiC@C and Fe_3C, used as fillers in PU matrices to improve the deflection capabilities of these electroactive polymers.

Polyurethane composites filled with low amounts of SiC@C and Fe_3C@C nanowires (0 – 5 wt %) were prepared by two solvent methods and yielded to very large surface films in the range of 30-80 μm in thickness. It can be noted that the deposit way was different for the two systems: spin coating for PU88-SiC@C films and doctor blade application for PU75-Fe_3C@C, both methods being used for preparation of actuators made of polymers.

The dielectric properties were studied at different frequencies and showed an increasing of the relative permittivity of the composites with the amount of nanowires. Permittivity and conductivity results showed also that the percolation threshold was not reached whatever the studied system. The mechanical properties were performed by using a simple tensile test and it was shown that the increase in the Young's modulus induced by the nanowires was only moderate.

Based on DSC analysis, only a very slight modification of the polyurethane host was detected with the incorporation of the nanowires and no difference in the degree of hard segments/soft segments mixing was found.

The actuation capabilities were studied by measuring the strain under an applied electric field with two methods.

The first method consisted in the measure of the electric field-induced thickness strain S with a double-beam laser interferometer at low frequency (0.1 Hz). At a moderate electrical field (6V/μm), it was found a 3 fold increase of the electrostriction coefficient for only 0.5%wt of SiC@C into PU88 polymer. The polyurethane is a good candidate for electromechanical applications, it has a good ability to collect space charges due to its special structure consisting of a mixing of hard and soft segments. It was shown that the nanowires acted as charge traps and then enhanced the electrical field induced deformation.

The second method is to measure the deflection strain of a diaphragm type dielectric actuator working at high frequency (near 1kHz). The deflection strain was considerably enhanced with the incorporation of either Fe_3C@C and SiC@C nanowires into polyurethane films whatever the type of the polyurethane host. The maximum deflection strains measured at high electrical fields (18 V/μm) reached 130 % for the PU75-Fe3C@C nanocomposites and 25% for the PU88-3SiC@C. These values are considerable and at least 6 times higher than those obtained with corresponding pure polymer films.

Due to their evident large actuation capabilities, these new nanocomposites based on polyurethane and carbon-based nanowires are very promising materials for micro electromechanical systems.

Acknowledgments

Authors thank the Institut Carnot for its financial support and Veronique Perrin from LGEF (EA682 INSA Lyon) for mechanical tests and thickness strain measurements with laser interferometer. The authors also acknowledge the CTμ (Centre Technologique des Microstructures) of the Université Lyon 1 for access to the SEM.

Author details

V. Salles[1], L. Seveyrat[2*], T. Fiorido[1], L. Hu[1], J. Galineau[2], C. Eid[1], B. Guiffard[2], A. Brioude[1] and D. Guyomar[2]

*Address all correspondence to: laurence.seveyrat@insa-lyon.fr

1 Laboratoire des Multimatériaux et Interfaces, UMR 5615 CNRS, Université Lyon1, Université de Lyon, Bat. Berthollet, France

2 Laboratoire de Génie Electrique et Ferroélectricité, LGEF – INSA Lyon, France

References

[1] Bar-Cohen Y, Zhang Q. Electroactive Polymer Actuators and Sensors. MRS Bulletin 2008;33 173-181.

[2] Carpi F, Rossi D, Kornbluh R, Pelrine R, Sommer-Larsen P. Dielectric Elastomers as Electromechanical Transducers : Fundamentals, Materials, Devices, Models and Applications of an Emerging Electroactive Polymer Technology. Amsterdam: Elsevier Science; 2008.

[3] Nan CW, Shen Y, Ma J. Physical Properties of Composites Near Percolation. Annual Review of Materials Research 2010;40 131-151.

[4] Wongtimnoi K, Guiffard B, Bogner-Van de Moortèle A, Seveyrat L, Gauthier C, Cavaillé JY. Improvement of Electrostrictive Properties of a Polyether-based Polyurethane Elastomer Filled with Conductive Carbon Black. Composites Science and Technology 2011;71(6) 885-892.

[5] Park C, Kang JH, Harrison JS, Costen RC, Lowther SE. Actuating Single Wall Carbon Nanotube-Polymer Composites: Intrinsic Unimorphs. Advanced Materials 2008;20(11) 2074-2079.

[6] Zhang S, Zhang N, Huang C, Ren K, Zhang Q. Microstructure and Electromechanical Properties of Carbon Nanotubes/Poly(vinylidene fluoride-trifluoroethylene-chlorofluoroethylene) Composites. Advanced Materials 2005;17(15) 1897-1901.

[7] He F, Fan JT, Lau S. Thermal, Mechanical and Dielectric Properties of Graphite Reinforced Poly(vinylidene fluoride) Composite. Polymer Testing 2008;27(8) 964-970.

[8] Seveyrat L, Chalkha A, Guyomar D, Lebrun L. Preparation of Graphene Nanoflakes/Polymer Composites and Their Performances for Actuation and Energy Harvesting Applications. Journalf of Applied Physics 2012;111(10) 104904-104913.

[9] Guiffard B, Seveyrat L, Sebald G, Guyomar D. Enhanced Electric Field-Induced Strain in Non-percolative Carbon Nanopowder/Polyurethane Composites. Journal of Physics D: Applied Physics 2006;39(14) 3053-3057.

[10] Watanabe M, Hirai T, Suzuki M, Amaike Y. Electric Conduction in Bending Electrostriction of Polyurethanes. Applied Physics Letters 1999;74(18) 2717-2719.

[11] Guiffard B, Guyomar D, Seveyrat L, Chowanek, Bechelany, Cornu D, Miele P. Enhanced Electroactive Properties of Polyurethane Films Loaded with Carbon-coated SiC Nanowires. Journal of Physics D: Applied Physics 2009;42(5) 055503.

[12] Yimnirun R, Eury S, Sundar V, Moses PJ, Jang SJ, Newnham RE. Electrostriction Measurements on Low Permittivity Dielectric Materials. Journal of the European Ceramic Society 1999;19(6-7) 1269-1273.

[13] Pelrine R, Kornbluh R, Joseph J, Heydt R, Pei Q, Chiba S. High-field Deformation of Elastomeric Dielectrics for Actuators. Materials Science and Engineering 2000;11(2) 89-100.

[14] Guillot FM, Balizer E. Electrostrictive Effect in Polyurethanes. Journal of Applied Polymer Science 2003;89(2) 399-404.

[15] Diaconu I, David A, Dorohoi DO. An Experimental Investigation of Electroactive Polyurethane. Journal of Optoelectronics and Advanced Materials 2005;7(6) 2797-2801.

[16] Bechelany M, Brioude A, Stadelmann P, Ferro G, Cornu D, Miele P. Very Long SiC-based Coaxial Nanocables with Tunable Chemical Composition. Advanced Functional Materials 2007;17(16) 3251-3257.

[17] Huang ZM, Zhang TZ, Kotaki M, Ramakrishna S. A Review on Polymer Nanofibers by Electrospinning and Their applications in Nanocomposites. Composites Science and Technology 2003;63(15) 2223-2253.

[18] Chronakis IS. Novel Nanocomposites and Nanoceramics Based on Polymer Nanofibers Using Electrospinning Process-A Review. Journal of Materials Processing Technology 2005;167(2-3) 283-293.

[19] Shin DG, Riu DH, Kim HE. Web-type Silicon Carbide Fibers Prapared by the Electrospinning of Polycarbosilanes. Journal of Ceramic Processing Research 2008;9(2) 209-214.

[20] Eick BM, Youngblood JP. SiC Nanofibers by Pyrolysis of Electrospun Preceramic Polymers. Journal of Materials Science 2009;44(1) 160-165.

[21] Welna DT, Bender JD, Wei X, Sneddon LG, Allcock HR. Preparation of Boron-Carbide/Carbon Nanofibers from a Poly(norbornenyldecaborane) Single-source Precursor via Electrostatic Spinning. Advanced Materials 2005;17(7) 859-862.

[22] Guron MM, Wei XL, Welna D, Krogman N, Kim MJ, Allcock H, Sneddon LG. Preceramic Polymer Blends as Precursors for Boron-Carbide/Silicon-Carbide Composite Ceramics and Ceramic Fibers. Chemistry of Materials 2009;21(8) 1708-1715.

[23] Wu H, Sun Y, Lin D, Zhang R, Zhang C, Pan W. GaN Nanofibers based on Electrospinning: Facile Synthesis, Controlled Assembly, Precise Doping, and Application as High Performance UV Photodetector. Advanced Materials 2009;21(2) 227-231.

[24] Salles V, Bernard S, Brioude A, Cornu D, Miele P. A New Class of Boron Niride Fibers with Tunable Properties by Combining An Electrospinning Process and the Polymer-derived Ceramics Route. Nanoscale 2010;2 215-217.

[25] Eid C, Brioude A, Salles V, Plenet JC, Asmar R, Monteil Y, Khoury R, Khoury A, Miele P. Iron-based 1D Nanostructures by Electrospinning Process. Nanotechnology 2010;21(12) 125701.

[26] Wongtimnoi K. Polyurethanes électrostricitifs et nanocomposites : caractérisation et analyse des mécanismes de couplage électromécaniques. PhD thesis. Institut National des Sciences Appliquées de Lyon ; 2011.

[27] Martin DJ, Meijs GF, Gunatillake PA, McCarthy SJ, Renwick GM. The Effect of Average Soft Segment length on Morphology and Properties of a Serie of Polyurethane Elastomers. II. SAXS-DSC Annealing Study. Journal of Applied Polymer Science 1997;64(4) 803-817.

[28] Chen KS, Leon Yu T. Chen YS, Lin TL, Liu WJ. Soft and Hard-segment Phases Segregation of Polyester-based Polyurethane. Journal of Polymer Research 2001;8(2) 99-109.

[29] Koberstein JT, Russell TP. Simultaneous SAXS-DSC Study of Multiple Endothermic Behavior in Polyether-based Polyurethane Block Copolymers. Macromolecules 1986;19(3) 714-720.

[30] Dyre JC, Schroder TB. Universality of AC Conduction in Disordered Solids. Reviews of Modern Physics 2000;72(3) 873-892.

[31] Kremer F, Schönhals A. Broadband Dielectric Spectroscopy. Springer-Verlag; 2003.

[32] Su J, Zhang QM, Ting RY. Space-charge-enhanced Electromechanical Response in Thin-film Polyurethane Elastomers. Applied Physics Letters 1997;71(3) 386-388.

[33] Su, J, Zhang QM, Wang PC, MacDiarmid AG, Wynne KJ. Preparation and Characterization of Electrostrictive Films with Conductive Polymer Electrodes. Polymers for Advanced Technologies 1998;9(6) 317-321.

[34] Rybak A, Warde M, Beyou E, Chaumont P, Bechelany M, Brioude A, Toury B, Cornu D, Miele P, Guiffard B, Seveyrat L, Guyomar D. Synthesis of Polystyrene Coated SiC Nanowires as Fillers in a Polyurethane Matrix for Electromechanical Conversion. Nanotechnology 2010;21 145610.

[35] Galineau J, Guiffard B, Seveyrat L, Lallart M, Guyomar D. Study and Modeling of an Electrostrictive Polyurethane Diaphragm Loaded with Conductive Carbon Black. Accepted manuscript in Sensors and Actuators A: Physical http://dx.doi.org/10.1016/j.sna.2012.09.021.

[36] Fiorido T, Salles V, Seveyrat L, Hu L, Galineau J, Guiffard B, Brioude A, Guyomar D. Innovative Organic/Inorganic Nanocomposites for Actuators and Magnetic Sensors Applications. Submitted.

[37] Guyomar D, Yuse K, Cottinet PJ, Kanda M, Lebrun L. Focus on the Electrical Field-induced Strain of Electroactive Polymers and the Observed Saturation. Journal of Applied Physics 2010;108(11) 114910.

Section 4

Other Nanowires

Chapter 14

A 2-D/3-D Schrödinger-Poisson Drift-Diffusion Numerical Simulation of Radially-Symmetric Nanowire MOSFETs

Daniela Munteanu and Jean-Luc Autran

Additional information is available at the end of the chapter

1. Introduction

The phenomenal success of CMOS technology, and, then the progress of the information technology, can be attributed without any doubt to the scaling of the MOS transistor, which has been pushed during more than thirty years to increasingly levels of integration and performances. Then, MOSFETs have been fabricated always smaller, denser, faster and cheaper in order to provide ever more powerful products for digital electronics. Recently, the scaling rate has accelerated, and the MOSFET gate length is now less than 40 nm, with devices entering into the nanometer world [1]-[2]. The so-called "bulk" MOSFET is the basic and historical key-device of microelectronics: its dimensions have been reduced more than $\sim 10^3$ times during the three past decades. However, the bulk MOSFET scaling has recently encountered significant limitations, mainly related to the gate oxide (SiO_2) leakage currents [3]-[4], the large increase of parasitic short channel effects and the dramatic mobility reduction [5]-[6] due to highly doped Silicon substrates precisely used to reduce these short channel effects. Technological solutions have been proposed in order to continue to use the "bulk solution" until the 45 nm ITRS node. Most of these solutions envisage the introduction of high-permittivity gate dielectric stacks (to reduce the gate leakage, [4], [7]-[8]), midgap metal gate (to suppress the Silicon gate polydepletion-induced parasitic capacitances) and strained Silicon channel (to increase carrier mobility, [9]). However, in parallel to these efforts, alternative solutions to replace the conventional bulk MOSFET architecture have been proposed and studied in the recent literature. These options are numerous and can be classified in general according to three main directions: (i) the use of new materials in the continuity of the "bulk solution", allowing increasingly MOSFET performances due to their dielectric properties (permittivity), electrostatic immunity (SOI materials), mechanical (strain), or transport (mobility) properties; (ii) the complete change of the device architecture (e.g. multiple-gate devices, Silicon nanowires MOSFET) allowing better electrostatic control, and, as a result, intrinsic channels with higher mobilities and currents; (iii) the exploitation of certain new physical phenomena that appear at the nanometer scale, such as quantum transport, substrate orientation or modification of the material band structure in devices/wires with nanometer dimensions [2], [10].

Nanowire MOSFETs with completely surrounding gate are presently considered as one of the possible solutions for replacing bulk (single-gate) devices and continuing MOSFET scaling in the nanometer scale [11]-[12]. The main advantage of this architecture is to offer a reinforced electrostatic coupling between the conduction channel and the gate electrode, which considerably reduces short-channel effects (SCE) compared to conventional bulk devices [11]-[13]. Then, the constraints on channel doping levels can be relaxed and nanowire MOSFET devices can be designed with intrinsic (pure Si) channels. This offers considerable advantages, especially in terms of mobility and elimination of doping fluctuations. The intrinsic channel can also be beneficial with regard to source-to-drain transport due to the high probability of ballistic transport.

Nanowire MOSFETs are generally designed with very thin silicon films in order to reinforce the electrostatic control of the gate over the channel. The ultra-thin silicon film creates a sufficiently narrow rectangular potential well for inducing the quantization of carrier energy. Carriers are then confined in a rectangular quantum well having feature size close to the electron wavelength. This gives rise to a splitting of the energy levels into subbands (two-dimensional (2-D) density-of-states) [14]-[15], such that the lowest of the allowed energy levels for electrons (resp. for holes) in the well does not coincide with the bottom of the conduction band (resp. the top of the valence band). In nanowire devices the carrier energy is quantified in the two directions perpendicular to the source-to-drain axis (leading to one-dimensional (1-D) density-of-states). The total density of states in a 1-D system is less than that in a three-dimensional (3-D) (or classical) system, especially for low energies. Carriers occupying the lowest energy levels behave like quantized carriers while those lying at higher energies, which are not as tightly confined in the potential well, can behave like classical (3-D) particles with three degrees of freedom. As the surface electric field increases, the system becomes more quantized and more and more carriers become confined in the potential well. The quantum mechanical confinement considerably modifies the carrier distribution in the channel: the maximum of the inversion charge is shifted away from the interface into the Silicon film. Because of the smaller density of states in the 1-D system, the total population of carriers is smaller for the same Fermi level than in the corresponding 3-D (or classical) case. This phenomenon affect the net sheet charge of carriers in the inversion layer, thus requiring a larger gate voltage in order to populate a 1-D inversion layer to have the same number of carriers as the corresponding 3-D system. This leads to an increase of the threshold voltage of the MOSFET, which is an important issue, especially as the power supply voltages drop to lower levels. The gate capacitance and carrier mobility are also modified by quantum effects. These considerations indicate that the wave nature of carriers can no longer be neglected in very thin nanowire MOSFETs and has to be considered in simulation studies.

Modeling and simulation of nanowire MOSFETs devices are currently experiencing a growing interest due to unique capabilities: (i) simulation provides useful insights into device operation since all internal physical quantities that cannot be measured on real devices are available as outputs in simulation; (ii) the predictive capability of simulation studies makes possible the reduction of systematical experimental investigation of these new ultra-scaled devices; (iii) simulation offers the possibility to test hypothetical devices which have not yet been manufactured. Since computers are today considerably cheaper resources, simulation is becoming an indispensable tool for the device engineer, not only for the device optimization, but also for specific studies such as the investigation of new physical phenomena specific to the ultra-short channels (quantum confinement of carriers or short-channel electrostatic effects).

This chapter presents a simulation study of electrostatics and electronic transport in radially-symmetric nanowire MOSFETs by quantum drift-diffusion numerical simulation. In this chapter, we will investigate the operation of radially-symmetric nanowires using a 2-D/3-D Schrödinger/Poisson solver, called Cylmos. Various methods have been suggested to model quantum confinement effects. Among the approaches that are compatible with classical device simulators based on the drift-diffusion approach, the physically most accurate method is to include the Schrödinger equation into the self-consistent computation of the device characteristics [15]. Then, the solver described in this work provides self-consistent solutions of the 2-D Schrödinger equation and the 3-D Poisson equation in cylindrical coordinates [16]-[18], coupled with the drift-diffusion transport equation; this approach is commonly called in literature "quantum drift-diffusion". Simulated drain current versus gate voltage characteristics have been compared to data obtained from simulation with commercial code with an excellent agreement. Cylmos also provides a lot of additional information and valuable physical insights (such as the 3-D profile of electrostatic potential, classical and quantum carrier densities in the channel, the energy levels and total inversion charge) used to investigate the influence of short-channel and quantum-mechanical effects. Finally, the quantum drift-diffusion code will be used to analyze electrical parameters such as threshold voltage, drain-induced barrier lowering (DIBL) effect and subthreshold swing in radially-symmetric nanowire MOSFETs. The difference between classical (i.e., without quantum confinement) and quantum threshold voltage, as function of different film silicon thicknesses in nanowire MOSFETs will be also discussed.

The chapter is organized as follows. In section 2 we describe in detail the theoretical background of our approach and the simulation code. After the description of the simulated devices, the Schrödinger, Poisson and current continuity solvers are presented. The complete discretization of these equations and the calculation algorithm are explicitly presented. In Section 3 we compare Cylmos to results obtained using a commercial simulator, particularly in terms of output parameters of the Schrödinger module and drain current. Finally, after this validation step, we use Cylmos to analyze and discuss the operation and performance of circular nanowire-based MOSFETs, including the off-state current, threshold voltage and short channel effects.

2. Theoretical background and description of the simulation code

Cylmos is based on the numerical solving of the Poisson-Schrödinger system coupled with the drift-diffusion equation. The Poisson equation is solved on the entire 3-D structure. To solve the Schrödinger equation, the device is divided into parallel vertical slices (y-z plane) (one slice per mesh point in the x direction). The 2-D Schrödinger equation is solved in each slice to obtain the wave functions, the quantum energy levels and the charge density. The solution of the Schrödinger equation in 2-D Cartesian coordinates for the nanowires with square or circular section is straightforward to implement numerically, but it requires very extensive computation time [19]-[22]. To simplify the calculations, in radially-symmetric nanowires it is possible to reduce the size of the two Poisson and Schrödinger equations. By expressing these equations in cylindrical coordinates and using the property of cylindrical symmetry of the structure (which implies a symmetry of the potential and wave function), the Poisson equation becomes a 2 dimensional equation and the Schrödinger equation becomes one dimensional. The Cartesian coordinates (x, y, z) are converted in cylindrical coordinates (x, r, θ), as shown in Fig.1, and considering these later, the structure is symmetrical with respect to the coordinate θ. Therefore, the coordinate θ will be ignored and the Poisson equation is solved in two dimensions, r and x, on the 2-D mesh shown in Fig. 2(a). In the same way, the circular symmetry allows us to simplify the Schrödinger equation, which will be solved on a 1-D mesh along the radial direction r, as shown in Fig. 2(b). The Schrödinger equation will be solved along vertical cut-lines, perpendicular to the x axis, in each mesh-point i of the x-axis. The advantage of this transformation is that, contrary to the 3-D-Poisson/2-D-Schrödinger equations, the 2-D-Poisson/1-D-Schrödinger system is numerically less CPU consuming. Nevertheless, this system of equations is more difficult to implement due to the form of the Laplacian operator [16]. In the following, we detail the equations for a n-channel fully depleted nanowire MOSFET, but similar equations can be derived for p-channel structures.

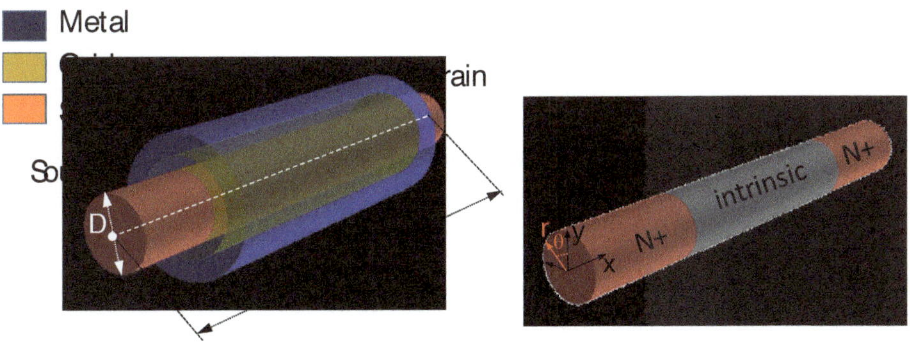

Figure 1. Schematic representation of the radially-symmetric nanowire MOS transistor simulated in this work. The main geometrical parameters are also defined.

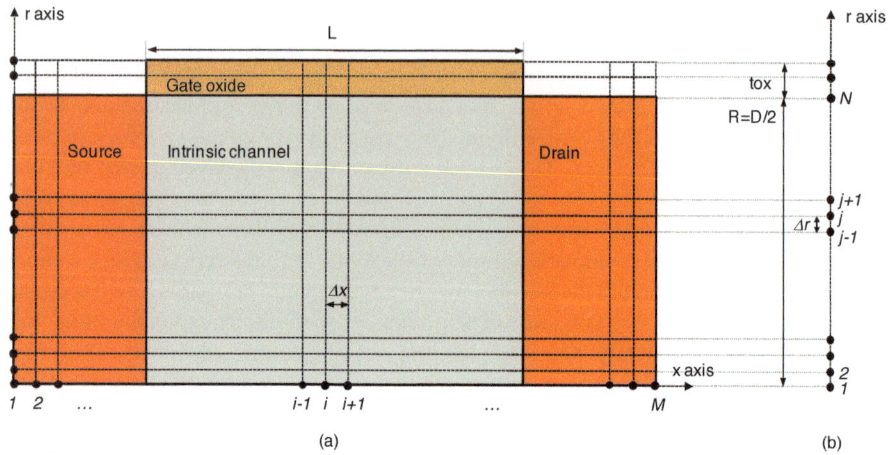

Figure 2. Schematics of the mesh used in Cylmos code and definition of mesh points in x-direction and r-direction.

The physical description of electrostatics of the nanowire device requires solving the coupled Poisson equation with the Schrödinger equation. The general Poisson equation is given by:

$$\nabla(\varepsilon \nabla V) = -\rho \tag{1}$$

where ρ is the charge density, V is the electrostatic potential and ε is the material-dependent permittivity. This is a 3-D equation which can be expressed either in Cartesian coordinates or in cylindrical coordinates. In this work we will use the cylindrical coordinates, as it will be shown in the following. The general independent of time Schrödinger equation, expressed in terms of effective mass, is given by:

$$\left\{-\frac{\hbar^2}{2}\left(\nabla_u \cdot \frac{\nabla_u}{\hat{m}^*}\right) + U(u)\right\}\Psi = E\Psi \tag{2}$$

where $\hbar$ is the reduced Planck constant, Ψ is the wave function and E the energy of the particle, $1/\hat{m}^*$ is the inverse effective mass tensor and $U(u)$ is the potential energy. The potential energy $U(u)$ is related to the electrostatic potential $V(u)$ via the relation:

$$U(u) = -qV(u) + \Delta E_C(u) \tag{3}$$

where $\Delta E_C(u)$ represents the band offset at the interface between two materials (i.e. the band offset between the conduction band of the semiconductor and the conduction band of the oxide). Similarly to the Poisson equation, the general form of the Schrödinger equation (2) will be expressed in paragraph 2.3 in cylindrical coordinates. These two equations will be solved self-consistently and the solution of this system of equations will be coupled with the courant continuity equation which is written:

$$\nabla J_n = qR + q\frac{\partial n}{\partial t} \tag{4}$$

where R is the net rate of generation-recombination and $J_n$is the electrons current density given by the following equation in the case of the drift-diffusion model:

$$J_n = -q\mu_n n\nabla\phi + qD_n\nabla n \tag{5}$$

where μ_n is where the electron mobility, D_n is the coefficient of thermal diffusion, and n is the electron density. In equation (5) the first term on the right corresponds to the drift component of the total current density and the second term to the diffusion component.

Poisson, Schrödinger and drift-diffusion equations are solved using a finite difference scheme with a uniform mesh on a domain including the channel, the source and drain regions, the gate-oxide layer and the gate electrodes (Fig. 2(a)). The electric field penetration in the source/drain and electron wave function penetration in the gate-oxide can be thus taken into account.

The general flowchart of our code is presented in Fig. 3. In a first step, the quantized energy levels as well as their associated electron wave functions and populations can be obtained from the self-consistent solving of the Schrödinger and Poisson equations. The next step concerns the continuity equation which is numerically solved to ensure a constant drain current along the device channel. The continuity equation gives the numerical solution of the quasi-Fermi level, used further in the calculation of the charge density in the Schrödinger-Poisson system. After reaching the convergence, the drain current density is finally evaluated from the drift-diffusion formalism. In the following, we will explain in detail the three modules corresponding to the solving procedures for the Poisson, Schrödinger and current continuity equations. The discretization of theses equations and the simplifying assumptions used in the calculations, assumptions mainly related to the cylindrical symmetry of the structure, will be systematically explained and justified.

2.1. Description of the simulated devices

The description of the 3-D architecture of the radially-symmetric nanowire MOSFETs considered in the simulation is presented in Fig. 1. The definition of the main geometrical parameters of this structure is also shown in Fig. 1. The structure is symmetric with intrinsic thin silicon film and highly doped source and drain (N_{SD} = 3×10^{20} cm^{-3}). A midgap metal gates (Φ_m=4.61 eV) and a 1.0-nm-thick gate-oxide have been also considered. The source is grounded, and the gate and the drain biased at V_G and V_D, respectively. In order to investigate the influence of short-channel effects and quantum mechanical confinement, different gate lengths and nanowire diameters will be simulated in section 4.

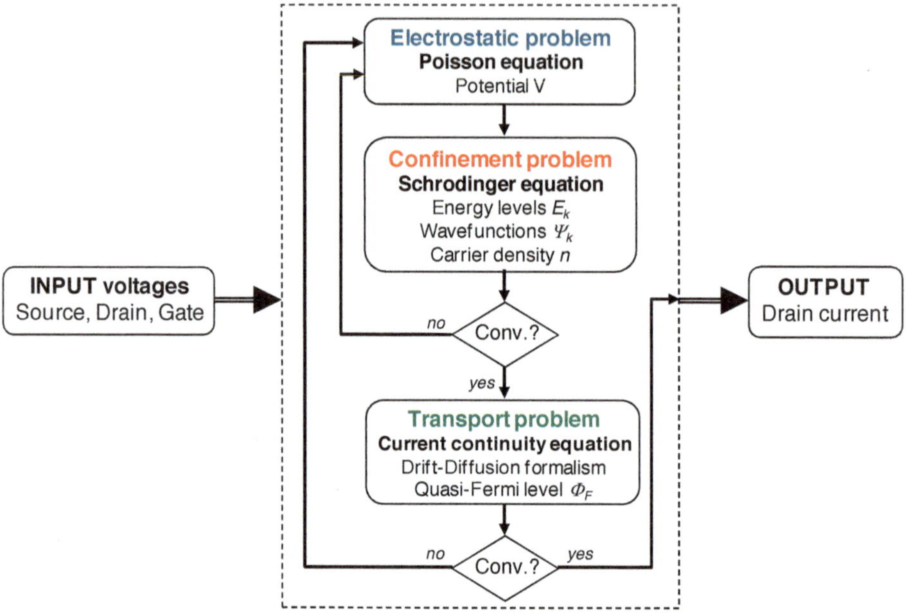

Figure 3. Flowchart of the numerical code illustrating the different problems to solve in the quantum drift-diffusion calculation of the output drain current.

2.2. Poisson solver

For a given channel geometry and bias conditions, the 2-D Poisson equation is solved (see flowchart in Fig. 3) considering a net charge density given by [23]:

$$\rho = q\left[N_{DOPING} - n\right] \tag{6}$$

where q is the absolute value of the electron charge, N_{DOPING} is the doping atom concentrations in the Si film ($N_{DOPING} = +N_D$ in source and drain regions and $N_{DOPING} = -N_A$ in the channel) and $n(x, r)$ is the electron density. This electron density is given by the Schrödinger module (see paragraph 2.3).

The Poisson equations expressed in cylindrical coordinates (x, r, θ) and ignoring the dependence in θ due to the circular symmetry of the electrostatic potential can be written under the form:

$$\frac{\partial^2 V}{\partial x^2}+\frac{1}{r}\frac{\partial V}{\partial r}+\frac{\partial^2 V}{\partial r^2}=-\frac{\rho(x,r)}{\varepsilon} \tag{7}$$

This equation is discretized using a three-point finite difference scheme applied to the grid shown in Fig. 2, as follows:

$$V_{i,j}=\left[\frac{\rho_{i,j}}{\varepsilon_j}+A_{i,j}+B_{i,j}\right]/\Sigma_{i,j} \tag{8}$$

where $A_{i,j}$, $B_{i,j}$ and $\Sigma_{i,j}$ are given by:

$$A_{i,j}=\frac{V_{i-1,j}+V_{i+1,j}}{\Delta x^2}+V_{i,j-1}\left(\frac{1}{\Delta r^2}-\frac{1}{r_j}\frac{1}{2\Delta r}\right)+V_{i,j+1}\left(\frac{1}{\Delta r^2}+\frac{1}{r_j}\frac{1}{2\Delta r}\right) \tag{9}$$

$$B_{i,j}=\frac{\varepsilon_{j+1}-\varepsilon_{j-1}}{2\Delta r}\times\frac{V_{i,j+1}-V_{i,j-1}}{2\Delta r}\times\frac{1}{\varepsilon_j} \tag{10}$$

$$\Sigma_{i,j}=2\left(\frac{1}{\Delta x^2}+\frac{1}{\Delta r^2}\right) \tag{11}$$

where the index i identifies the mesh points in x-direction (i is an integer, $i=1,..,M$), j identifies the mesh points in r-direction (j is an integer, $j=1,..,N$), Δx and Δr are the mesh sizes between two mesh points in the x-direction and r-direction, respectively (Fig. 2) and $r=j\cdot\Delta r$.Δx and Δr are considered constant in this work. The electrostatic potential calculated by eq. (8) is then used in the solving of the Schrödinger equation for the calculation of the quantum charge density (see below).

2.3. Schrödinger solver

The Schrödinger equation in cylindrical coordinates (x, r, θ) is given by:

$$-\frac{\hbar^2}{2}\left[\frac{\partial}{\partial x}\left(\frac{1}{m^*}\frac{\partial\Psi}{\partial x}\right)+\frac{\partial}{\partial r}\left(\frac{1}{m^*}\frac{\partial\Psi}{\partial r}\right)+\frac{1}{r}\frac{1}{m^*}\frac{\partial\Psi}{\partial r}+\frac{1}{r^2}\frac{\partial}{\partial\theta}\left(\frac{1}{m^*}\frac{\partial\Psi}{\partial\theta}\right)\right]+U\Psi=E\Psi \tag{12}$$

where $m^*(x, r, \theta)$ is a position dependent effective mass, $U(x, r, \theta)$ is the potential energy of the conduction-band edge. It is assumed that m^* is constant within the silicon nanowire so that it is not dependent on θ. The effective mass is also considered energy-independent (non-parabolicity effects are thus neglected).

Due to the cylindrical symmetry, we may consider wave functions of the form:

$$\Psi(x,r,\theta) = \Psi(x,r)\, e^{im\theta} \tag{13}$$

where m=0, ± 1, ± 2, ± 3,... is the angular momentum quantum number. This wave function is normalized according to:

$$2\pi \int_0^R r\left|\Psi(x,r)\right|^2 dr = 1 \tag{14}$$

where the factor 2π is due to the integration with respect to θ.

Using relation (13), equation (12) becomes:

$$-\frac{\hbar^2}{2}\left[\frac{\partial}{\partial x}\left(\frac{1}{m^*}\frac{\partial \Psi}{\partial x}\right) + \frac{\partial}{\partial r}\left(\frac{1}{m^*}\frac{\partial \Psi}{\partial r}\right) + \frac{1}{r}\frac{1}{m^*}\frac{\partial \Psi}{\partial r}\right] + \left(U + \frac{\hbar^2}{2}\frac{1}{m^*}\frac{m^2}{r^2}\right)\Psi = E\Psi \tag{15}$$

We recall here that the Schrödinger equation is solved along vertical cut-lines (1-D mesh along the radial direction), perpendicular to the x axis, in each mesh-point i of the x-axis (Fig. 2(b)). The procedure for solving the Schrödinger equation will be the same for each mesh-point i. In the following, we set a fixed mesh point i (then the dependence on x in the equation (14) vanishes); but this procedure will be repeated M times for each mesh-point of the x-axis (this will be explained in detail at the end of this paragraph).

As the Poisson equation, Schrödinger equation (14) is discretized using finite differences scheme to obtain a standard matrix eigen-value problem:

$$H\Psi = E\Psi \tag{16}$$

where H is the general Hamiltonian matrix. In practice, equation (16) is solved for each value of the angular momentum quantum number m separately. In the following, we will describe in detail the Hamiltonian H for a given m value.

To simplify the equations, we consider firstly m=0, and we will address the other cases later in this paragraph. With this condition, the discrete form of the Schrödinger equation for the ladder of states with m=0 on a mesh point j along the radial direction is written as:

$$-\frac{\hbar^2}{2}\left[\frac{\Psi_{j+1}-\Psi_j}{m^*\Delta r^2}\right] + \frac{\hbar^2}{2}\left[\frac{\Psi_j-\Psi_{j-1}}{m^*\Delta r^2}\right] - \frac{\hbar^2}{2}\frac{1}{(j-1)\Delta r}\left[\frac{\Psi_{j+1}-\Psi_{j-1}}{2m^*\Delta r}\right] + U_{i,j}\Psi_j = E\Psi_j \tag{17}$$

where the effective mass has been considered independent on r (since we simulate here a fully silicon nanowire). In U_{ij}, the index i corresponds to the cut-line on the x axis for which the Schrödinger equation is solved. The potential energy term U_{ij} is obtained using eq. (3) which is discretized as:

$$U_{i,j} = -qV_{i,j} + \Delta E_{Ci,j} \tag{18}$$

where $V_{i,j}$ is the electrostatic potential obtained from the resolution of the Poisson equation as described in the previous paragraph (Poisson module).

We explain below how the effective masses are calculated in the cylindrical silicon nanowire considered here.The conduction band of silicon has 6 equivalent minima. The variation of the energy of the conduction band near the minimum is not isotropic and constant energy surfaces are ellipsoids around each of the axes (Fig. 4). In each minimum, the electrons propagating along the axis of the ellipsoid are characterized by a longitudinal effective mass, m_l, and those that propagate in a direction perpendicular to this axis are characterized by a transverse effective mass, m_t. In the circular nanowire simulated here, the two valleys along the x direction (transport direction) have a confinement mass equal to m_t and a transport mass equal to m_l. The other four valleys following the directions y and z are anisotropic.

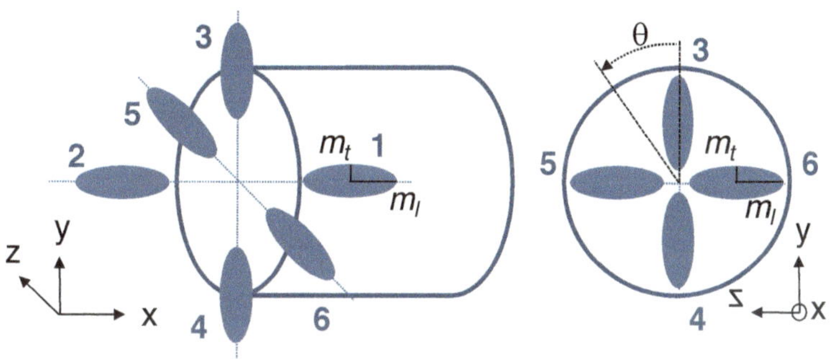

Figure 4. Schematic illustration of electron confinement in a Silicon nanowire with circular cross-section.

In order to exploit the cylindrical symmetry, we use an approximate isotropic effective mass in the radial direction. Because of this assumption, now there are only two different types of conduction band valleys: the valley along the x direction which is doubly degenerate and the valley along the radial (r) direction whose degeneracy is equal to 4. The effective masses (m_r, m_x) along radial and axial directions for the two types of valleys are given in Table 1.

Valley	Number of equivalent valleys	Radial effective mass, m_r (confinement)	Axial effective mass, m_x(transport)
1-2	2	$m_{r1} = m_t$	m_l
3-6	4	$m_{r2} = \dfrac{2\,m_t\,m_l}{m_t + m_l}$	m_t

Table 1. Definition of the effective masses along the radial and axial directions in cylindrical nanowire (m_t=0.19×m_0, m_l=0.98×m_0, m_{r2}=0.35×m_0, m_0= 9.11×10^{-31}Kg).

It is important to note that in addition to the solving in one dimension of the Schrödinger equation, which drastically reduces the computation time compared to a 2-D solving, the computation time is further reduced because of the smaller number of valleys taken into account in the calculations compared to the case where an anisotropic effective mass is considered in the radial direction.The choice of an approximate isotropic effective mass for r-direction is justified by the fact that the eigenvectors and eigenenergies resulting from the cylindrical solver are very similar to those of an exact 2-D solver taking into account the anisotropy of the effective mass [24]. In the following, the effective mass m^* used in the Schrödinger equation is equal to the radial effective mass m_r, which may have two distinct values, m_{r1} for the valleys 1 and 2 and m_{r2} for the valleys 3, 4, 5 and 6, as shown in Table 1.

Let's go back now to the solving the Schrödinger equation. Using the discrete form given by eq. (17), the Schrödinger equation for m=0 may be set in the form of a matrix equation:

$$H^0\Psi^0 = E^0\Psi^0 \tag{19}$$

where H^0(0 stands for m=0 condition) is the N×N tridiagonal hamiltonian matrix of the device (N is the total grid-point number in r-direction):

$$H^0 = \begin{bmatrix} \bullet & \times & 0 & \cdots & 0 \\ \circ & \bullet & \times & \ddots & \vdots \\ 0 & \circ & \bullet & \times & 0 \\ \vdots & \ddots & \circ & \bullet & \times \\ 0 & \cdots & 0 & \circ & \bullet \end{bmatrix} \tag{20}$$

where symbols are given by the following equations:

$$\circ = -\frac{\hbar^2}{2m_r\Delta r^2} + \frac{\hbar^2}{4m_r\Delta r^2(j-1)} \tag{21}$$

$$\bullet = \frac{\hbar^2}{m_r\Delta r^2} + U_{i,j} \tag{22}$$

$$\times = -\frac{\hbar^2}{2m_r\Delta r^2} - \frac{\hbar^2}{4m_r\Delta r^2(j-1)} \tag{23}$$

The matrix H_0 defined in (20) is a non-symmetric tridiagonal matrix and the computation of eigenvalues and eigenvectors of such matrix is numerically not optimized. However, using the method proposed in [16] by Zervos and Feiner, we can convert H^0 into a symmetric matrix S using the relation:

$$S = TH^0 \tag{24}$$

where T is an (NxN) diagonal matrix whose elements are determined as follows. Since S is a symmetric matrix, it must fill the condition:

$$S_{j,j+1} = S_{j+1,j} \tag{25}$$

Equation (24) can be rewritten in a discretized form as:

$$S_{j,j+1} = T_{j,j}H^0_{j,j+1} \tag{26}$$

Combining equations (25) and (26) we obtain the following relation for matrix T:

$$T_{j,j}H^0_{j,j+1} = T_{j+1,j+1}H^0_{j+1,j} \tag{27}$$

Then we can obtain the elements of the matrix T [16]: the first element T_{11} is set equal to 1 and the subsequent elements are obtained from (27) as:

$$T_{j+1,j+1} = T_{j,j}H^0_{j,j+1} / H^0_{j+1,j} \tag{28}$$

Matrix S is then symmetric and using equation (19) we obtain:

$$S\Psi^0 = TH^0\Psi^0 = E^0 T\Psi^0 \tag{29}$$

In equation (29) Ψ^0 and E^0 are the eigenvector and the eigenvalue of the real matrix H^0, respectively. Then, equation (29) has to be further transformed for obtaining a standard matrix eigenvalue problem. As shown in [16], we define a diagonal matrix L whose elements are the square root of the corresponding elements of T:

$$T_{j,j} = L^2_{j,j} \tag{30}$$

Multiplying eq. (29) from the left by L^{-1} we obtain:

$$L^{-1}SL^{-1}L\Psi^0 = L^{-1}LLH^0\Psi^0 = E^0L^{-1}LL\Psi^0 = E^0L\Psi^0 \tag{31}$$

We define:

$$H^N = L^{-1}SL^{-1} \tag{32}$$

and

$$\Psi^N = L\Psi^0 \tag{33}$$

Then, eq. (29) becomes:

$$H^N\Psi^N = E^0\Psi^N \tag{34}$$

which is the new eigenvalue problem. The new Hamiltonian H^N is a symmetric tridiagonal matrix whose eigenvalues are E^0 and the eigenvectors are Ψ^N. Then, we solve eq. (34) instead of eq. (19) to find eigenvalues and eigenvectors, and to finally calculate the quantum carrier density. The eigenenergies E^0 corresponding to the wave functions Ψ^N are computed using a QL algorithm [25] with implicit shifts. Next, we obtain the original wave functions Ψ^0 from eq. (33) as:

$$\Psi^0 = L^{-1}\Psi^N \tag{35}$$

So far, we explained the resolution of the Schrödinger equation for m=0, which has simplified the equations. To take into account all values of m, eq. (15) must be solved following the same procedure as that used for m=0. The discretization of equation (15) leads to a matrix relation of the same form as the relation (17):

$$H^m\Psi^m = E^m\Psi^m \tag{36}$$

In this equation, the Hamiltonian H^m is always a tridiagonal matrix which has the same elements "∘" and "×", respectively given by eqs. (21) and (23). The only term that changes is the diagonal term, "• ", which becomes:

$$\bullet = \frac{\hbar^2}{m_r\Delta r^2} + \frac{\hbar^2}{2m_r}\frac{m^2}{j^2\Delta r^2} + U_{i,j} \tag{37}$$

Equation (36) is a new eigenvalue problem to be solved separately for each m to find the ladder states for m=0, ± 1, ± 2, ± 3,... The eigenvalues and eigenvectors are determined using the same procedure as previously used for the matrix H^0. For each m, a distinct set of wave functions and energy levels is found. The energy levels obtained in this way for all values of m will be sorted and classified in ascending order, from the lowest level and to the highest level. The wave function corresponding to each classified level will be also stored. The first lowest levels (and the corresponding wave functions), in a number equal to N_{level}, will be used next to calculate the electron density, $n(r)$. In the following, we call E_k (respectively Ψ_k), these first lowest levels (respectively their corresponding wave function), where k is the subband index (k=1,..,N_{level}) which implicitly includes the angular momentum quantum number m.

The procedure described before for the solving of the Schrödinger equation is performed for both m_r=m_{r1} and m_r=m_{r2}. Two distinct sets of energy levels and wave functions are then obtained and are used in the following for the calculation of the electron density, $n(r)$.

The 1-D electron gas density is given by:

$$n(r) = \sum_{r1,r2} g_{r1,r2} \sum_{k=1}^{N_{level}} n_k \left| \Psi_k(r) \right|^2 \tag{38}$$

where $g_{r1,r2}$ is the valley degeneracy (2 for m_r=m_{r1} and 4 for m_r=m_{r2}) and n_k is the electron density per unit of length of the kth subband given by:

$$n_k = N_{1D}\, \Im_{-1/2}\left(\frac{E_{Fn} - E_k}{k_B T} \right) \tag{39}$$

where N_{1D} is the effective density of states (DOS), k_B is the Boltzmann constant, T is the temperature, E_{Fn} is the quasi-Fermi level which will be calculated in the current continuity module (by a self-consistent coupling of the solution of the Poisson-Schrödinger system to the current continuity equation) and $\Im_{-1/2}$ is the Fermi-Dirac integral of order -1/2 [26]. The 1-D density-of-states N_{1D} is calculated as:

$$N_{1D} = \frac{1}{\pi} \sqrt{\frac{2 m_{r1,r2}^{1D} k_B T}{\hbar^2}} \tag{40}$$

where $m_{r1,r2}^{1D}$ is the 1-D density of states effective mass ($m_{r1}^{1D} = m_l$for valleys 1-2 and $m_{r2}^{1D} = m_t$for valleys 3-6). Similar to energy levels and wave functions, N_{1D} and n_k are calculated for the two types of valleys described in Table 1. Then, these values are used in eq. (38) for the first sum in the right side of the equation. Note that the unity of n_k is m^{-1} which multiplied by the normalized probability density $|\Psi_k(r)|^2$(eq. 38) gives the volume electron density $n(r)$.

As we said before, the 1-D Schrödinger equation is solved (using the procedure outlined above) in a vertical cut-line in each mesh point of the x axis (which means that it will be solved M times). For each cut-line in a given mesh-point i, the electron density $n(r)$ is calculated using eq. (38), and it will then be assigned to the mesh-point point i; this makes possible to build step by step the total density $n(x, r)$. This density is injected into the module solving the 2-D Poisson equation (eqs. (6)and (8), paragraph 2.2), which will provide a new potential profile $V(x, r)$. This new potential will be introduced later in the solving of the Schrödinger equation, which will give the new carrier density $n(x, r)$ (in the same manner as described above), which in turn will be injected into Poisson's equation and so on. This procedure allows us to self-consistently solve the two Poisson and Schrödinger equations. The loop stops when the convergence criterion is reached, as shown in Fig. 3.

2.4. Current continuity module

As we explained at the beginning of Section 2, the solution of the Schrödinger-Poisson system of equations is coupled with the current continuity equation given by relation (4). In eq. (4), the current density is defined by a model of charge transport, usually obtained in a semi-classical approximation or simplifying the Boltzmann transport equation (ETB). The simplifying assumptions made on this equation lead to several different semi-classical models such as the drift-diffusion model or the hydrodynamic model. Nevertheless, whatever the model used for the calculation of the carrier density (drift-diffusion, hydrodynamic, etc), the condition of current continuity should be guaranteed. Indeed, the continuity equations describe the evolution of carriers in the silicon nanowire film (source-channel-drain) in order to maintain a constant current along the film in which there is no charge accumulation.

In this work we consider the drift-diffusion model for which the current density is given by equation (5). The charge transport in the nanowire MOSFET simulated here involves only the minority carrier conduction. Therefore, the current density used herein concerns uniquely the electrons (for an n-channel transistor, NMOSFET), but a similar equation can be used for holes in the case of a p-channel transistor (PMOSFET) and the procedure of solving the current continuity equation will be rigorously identical with that shown below for electrons. In addition, to simplify the equations, we consider a stationary regime of operation and we assume that no process of generation or recombination of carriers occurs in our simulations. With these assumptions, equation (4) becomes:

$$\nabla J_n = 0 \tag{41}$$

The current density of electrons can be rewritten as a function of the quasi-Fermi level of electrons in the silicon nanowire, Φ_F, as:

$$J_n = -q\mu_n n\nabla\phi + qD_n\nabla n = q\mu_n n\nabla\Phi_F \tag{42}$$

where the electron diffusion coefficient, D_n, is calculated from the Einstein's relation (for the non-degenerate case):

$$D_n = \mu_n \cdot \left(\frac{k_B T}{q}\right) \tag{43}$$

In this work, the electron mobility μ_n is considered constant everywhere in the silicon nanowire. Substituting equation (42) into equation (41) we obtain:

$$\nabla\left(-\mu_n n\nabla V + D_n\nabla n\right) = 0 \tag{44}$$

Using the properties of vector analysis and after various algebraic manipulations, equation (44) becomes:

$$\Delta\Phi_F + \beta\,\nabla\Phi_F\,\nabla\left(V - \Phi_F\right) = 0 \tag{45}$$

where $\beta = k_B T/q$ and V is the electrostatic potential which comes from the solution of the coupled solving of the Poisson and Schrödinger equations. Equation (45) can be rewritten:

$$\Delta\Phi_F(x,r) = -G(x,r) \tag{46}$$

where

$$G = \beta\,\nabla\Phi_F\,\nabla\left(V - \Phi_F\right) \tag{47}$$

In cylindrical coordinates, $\Delta\Phi_F(x, r)$ is given by the following expression:

$$\Delta\Phi_F(x,r) = \frac{\partial^2\Phi_F}{\partial x^2} + \frac{1}{r}\frac{\partial\Phi_F}{\partial r} + \frac{\partial^2\Phi_F}{\partial r^2} \tag{48}$$

and G is written as:

$$G = \beta\left(\frac{\partial\Phi_F}{\partial x} + \frac{\partial\Phi_F}{\partial r}\right)\left[\frac{\partial\left(V - \Phi_F\right)}{\partial x} + \frac{\partial\left(V - \Phi_F\right)}{\partial r}\right] \tag{49}$$

Equation (46) is discretized using a three-point finite difference scheme applied to the mesh shown in Fig. 2, as follows:

$$\Phi_{Fi,j} = \frac{G_{i,j} + C_{i,j}}{\Sigma_{i,j}} \tag{50}$$

where $\Sigma_{i,j}$ is given by equation (11) and $G_{i,j}$ and $C_{i,j}$ are given by:

$$G_{i,j} = \beta\,D_{i,j}\,E_{i,j} \tag{51}$$

$$C_{i,j} = \frac{\Phi_{Fi-1,j} + \Phi_{Fi+1,j}}{\Delta x^2} + \Phi_{Fi,j-1}\left(\frac{1}{\Delta r^2} - \frac{1}{r_j}\frac{1}{2\Delta r}\right) + \Phi_{Fi,j+1}\left(\frac{1}{\Delta r^2} + \frac{1}{r_j}\frac{1}{2\Delta r}\right) \tag{52}$$

$$D_{i,j} = \left[\frac{\Phi_{Fi+1,j} - \Phi_{Fi-1,j}}{2\Delta x} \cdot \frac{\left(V_{i+1,j} - \Phi_{Fi+1,j}\right) - \left(V_{i-1,j} - \Phi_{Fi-1,j}\right)}{2\Delta x} \right] \tag{53}$$

$$E_{i,j} = \left[\frac{\Phi_{Fi,j+1} - \Phi_{Fi,j+1}}{2\Delta r} \cdot \frac{\left(V_{i,j+1} - \Phi_{Fi,j+1}\right) - \left(V_{i,j-1} - \Phi_{Fi,j-1}\right)}{2\Delta r} \right] \tag{54}$$

The quasi-Fermi level calculated by eq. (50) is then injected into the system of equations Poisson-Schrödinger, where it is used for the calculation of the carrier density. This density is then used to calculate a new potential, which is then reinserted into the module solving the continuity equation that gives a new quasi-Fermi level. In its turn, this new quasi-Fermi level will be injected into the Schrödinger-Poisson and so on. The loop will stop when the convergence criterion is reached. The final values of the electron density, $n(x,r)$, and the quasi-Fermi level, $\Phi_F(x,r)$, will be used in the following for calculating the current density and the final drain current of the device.

2.5. Drain current calculation

The drain current density (per unit area) in the drift-diffusion formalism is finally evaluated from equation (42) which can be written as:

$$J_n(x,r) = q\,\mu_n\, n(x,r)\, \nabla\Phi_F(x,r) \tag{55}$$

The total current drain as a function of the applied voltages is then calculated by summing the contribution of each current line in the silicon film.

Finally, the code allows us to store the main internal key-parameters such as the carrier density, quantum energy levels, wave functions, potential, quasi-Fermi level, etc. Some of these parameters will be used for the code validation and for the analysis of the device performances in terms of threshold voltage or short-channel effects.

3. Simulation code validation

Before using the code for analyzing the operation of nanowire MOSFETs, we compared the results of the Schrödinger-Poisson module to the same data obtained from a commercial simulator (Silvaco, [24]) which can simulate cylindrical nanowires. We therefore compare in the following the results issued from Cylmos and Silvacoin terms of quantum energy levels, wave functions and drain current. Each time, we will obtain a very good agreement between the two data sets, which demonstrates the validity of our code.

3.1. Quantum energy levels and wave functions

We begin this comparison by looking in detail the internal parameters of the Schrödinger module, namely the quantum energy levels and associated wave functions. We distinguish here between the two radial effective masses, m_{r1} and m_{r2}. Thus, two sets of energy levels and wave function are compared, one set for each effective mass. The results shown in the following are obtained for a nanowire of diameter D=10 nm (i.e. R=5 nm). The simulated MOSFET is a long channel transistor, with 1-nm-thick gate oxide and an intrinsic channel. Other geometric parameters and doping levels are those described in paragraph 2.1.

Table 2 shows the comparison between the energy levels, E_k, calculated by Cylmos and those issued from Silvaco for the radial effective mass. The transistor is biased at $V_G = 0$ V and a very low polarization is applied on the drain (V_D=10 mV). As the channel is long and the drain bias is negligible, the potential does not vary much along the axial direction x. Therefore, almost identical energy levels will be obtained in any mesh point *i* of the intrinsic channel region (outside the source and drain regions). The results shown in Table 2 (as well as those reported in Figs. 5-6, Table 3 and Fig. 7) are calculated for a vertical cut-line in the middle of the channel. Table 2 shows that the energy levels for the radial effective mass calculated by Cylmos agree very well with those obtained using Silvaco.

	E_2-E_1	E_3-E_1	E_4-E_1	E_5-E_1	E_6-E_1	E_7-E_1	E_8-E_1	E_9-E_1	E_{10}-E_1
Silvaco	0.0406	0.0406	0.094	0.094	0.111	0.158	0.158	0.194	0.194
Cylmos	0.0409	0.0409	0.094	0.094	0.112	0.1601	0.1601	0.199	0.199

Table 2. Energy levels, E_k, calculated by Cylmos and issued from Silvaco for the radial effective mass m_{r1} in a circular nanowire of diameter D=10 nm. The transistor is biased at V_G=0 V and V_D=10 mV. The energy levels are calculated for a vertical cut-line in the middle of the channel.

The wave functions corresponding to the energy levels reported in Table 2 are plotted in Fig. 5 as a function of the position on the radial direction (for *k* from 1 to 6) and in Fig. 6 (for *k* from 7 to 10). The wave functions calculated by Cylmos correspond very well to those from Silvaco, particularly in the region corresponding to the silicon (0 to 5 nm). However, a rather clear difference is observed in the oxide region, due to the fact that Silvaco, contrary to Cylmos, does not take into account the penetration of wave functions in the oxide. This phenomenon is included in Cylmos, as previously stated, which leads to this discrepancy between the wave functions of Cylmos and Silvaco in the gate oxide.

The same analysis can be conducted for the energy levels and wave functions calculated for the second radial effective mass, m_{r2}. These results are shown in Table 3 and Fig. 7 for the same transistor and the same polarization as those used above. A very nice agreement is found again between the data calculated by Cylmos and those obtained from Silvaco.

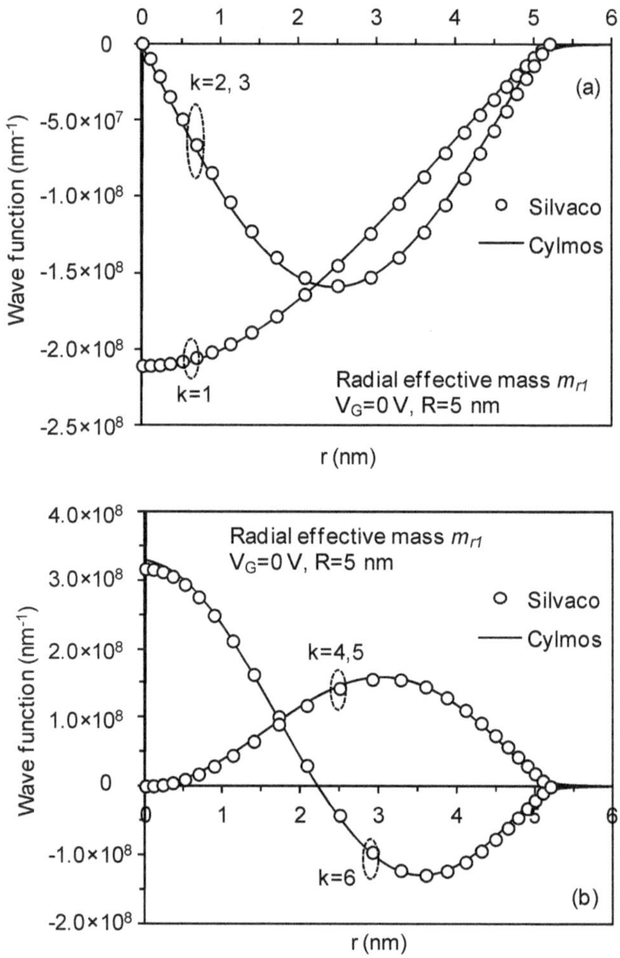

Figure 5. Wave function as a function of r calculated by Cylmos and Silvaco for the radial effective mass m_{r1} in a nanowire of diameter D=10 nm. (a) k=1,2,3; (b) k=4,5,6. Other parameters and biases are the same as reported in Table 2.

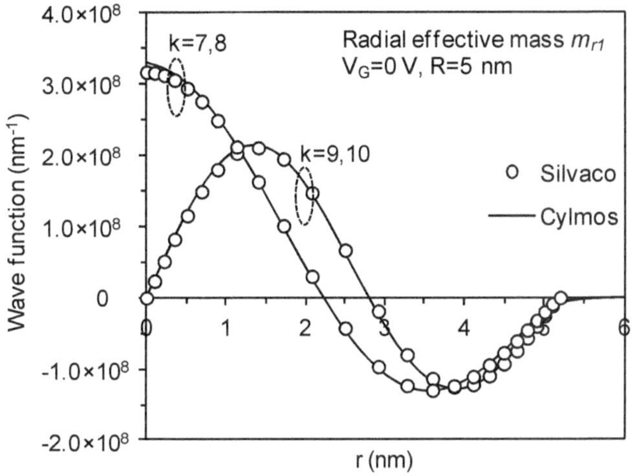

Figure 6. Wave function as a function of *r* calculated by Cylmos and Silvaco for the radial effective mass m_{r1} in a nanowire of diameter D=10 nm and k from 7 to 10. Other parameters and biases are the same as reported in Table 2.

	E_2-E_1	E_3-E_1	E_4-E_1	E_5-E_1	E_6-E_1
Silvaco	0.064	0.064	0.148	0.148	0.175
Cylmos	0.067	0.067	0.154	0.154	0.183

Table 3. Energy levels, E_k, calculated by Cylmos and those issued from Silvaco for the radial effective mass m_{r2} in a nanowire of diameter D=10 nm. Other parameters and biases are the same as reported in Table 2.

In this analysis we studied the energy levels E_k and the corresponding wave functions Ψ_k, with *k* varying between 1 and 10. We recall here that the variation of *k* includes the variation of the angular momentum quantum number, *m*. This means that each value of *k* corresponds to a value of *m*, and when *k* varies from 1 to 10, *m* can have the following values: 0, ± 1, ± 2,... We also recall that the energy levels are sorted in ascending order and that only the first N_{level} (10 in the present study) lowest levels are stored. The level *k*=1 always corresponds to *m*=0 and all energy levels with non-zero *m* are doubly degenerate, which corresponds to + *m* and-*m* or to the clockwise and counter-clockwise polarization of the wave function. This can be seen, for example in Table 2, where E_2 and E_3 levels are twofold degenerated (as well as the pairs of levels E_4-E_5, E_7-E_8 and E_9-E_{10}).

Another remark concerns the variation of the wave functions with the radial position of the axis. For *m*=0 (which corresponds to *k*=1 or *k*=6 for the case studied here), the wave function is always non zero and has zero first derivative at *r*=0; this is valid for both radial effective masses. This remark can be verified in the results presented in Figs. 5 and 7. For *m*=1, the wave function is always zero and has zero second derivative at r=0. For m>1, the wave function is always zero and has zero first derivative at r=0.

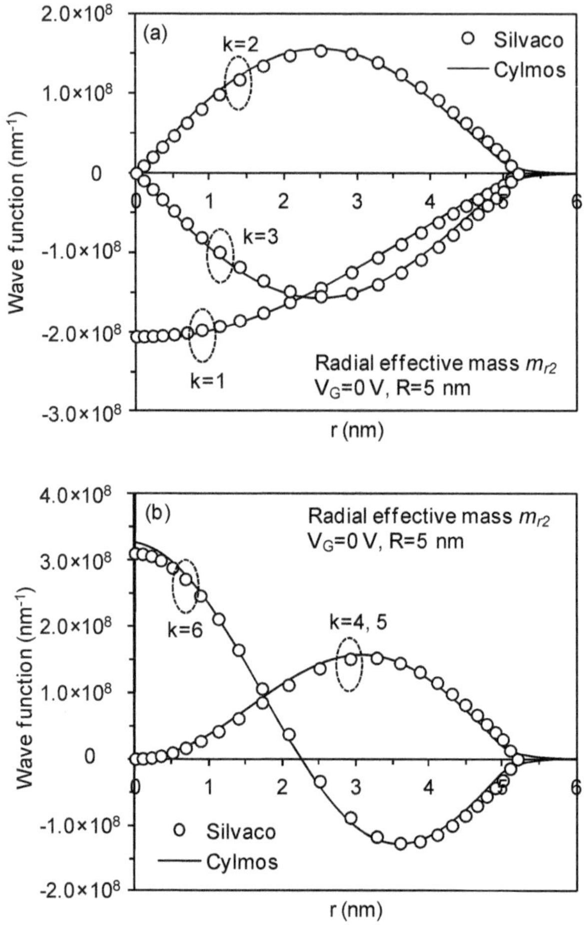

Figure 7. Wave functions as a function of r calculated by Cylmos and Silvaco for the radial effective mass m_{r2} in a nanowire of diameter D=10 nm. (a) k=1,2,3; (b) k=4,5,6. Other parameters and biases are the same as reported in Table 2.

3.2. Drain current

We also compare our code Cylmos with Silvaco in terms of drain current. Unfortunately, Silvaco does not provide the drain current, because the coupling between the Schrödinger-Poisson system of equations and the continuity equation is not yet implemented. Nevertheless, Silvaco couples the Poisson equation directly to the continuity equation (i.e. the Schrödinger equation is ignored), which makes possible the calculation of the drain current in the so-called "classical" case (this means that the quantum confinement effects are not taken into account in this approach). To facilitate comparison with Silvaco, we have implemented the same procedure in Cylmos. Thus, in our code it is possible to calculate, in addition to the "quantum" drain current (issued from the self-consistently solving of the Poisson, Schrödinger and continuity equations), the "classical" drain current (by directly coupling the Poisson equation to the continuity equation). In this "classical" case, the electron density is reevaluated using the well-known Fermi-Dirac or Boltzmann statistics:

$$n(x,r) = N_C \exp\left[-\beta_T \left(E_C(x,r) - \Psi_F(x,r)\right)\right] \tag{56}$$

where $E_C(x,r)$ is the minimum of the conduction band and N_C is the effective 3-D density-of-states for the conduction band [27].

The solving procedure is the same as explained in Section 2. The discretized Poisson equation (eq. (8)) is solved taking into account this time no more the electron density issued from the Schrödinger equation, but the "classical" electron density given by eq. (56). The potential is then calculated and is afterward injected directly into the module solving the continuity equation. The output of this module is the new quasi-Fermi level which is subsequently used in eq. (56) to calculate a new carrier density to be used in the Poisson module. In this way, a new potential is found and will be injected into the continuity equation and so on. The loop stops when the convergence criterion is reached.

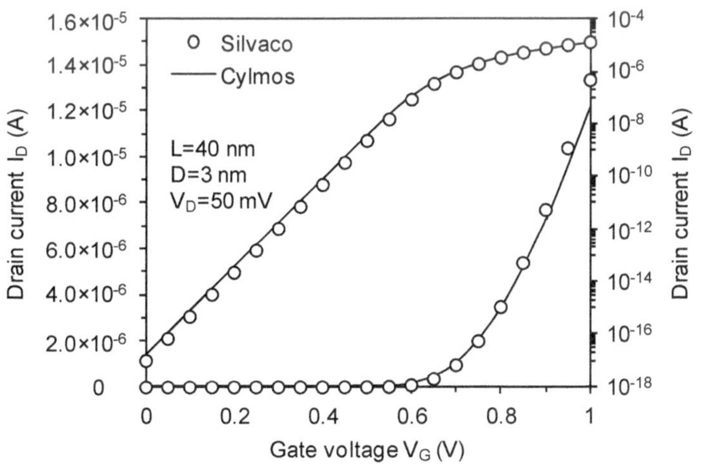

Figure 8. Drain current versus gate voltage calculated by Cylmos and Silvaco in the "classical" case (i.e., without quantum effects) for a nanowire MOSFET with L=40 nm channel length and a silicon nanowire of diameter D=3 nm.

Figure 8 shows the drain current, simulated in this way, for a long channel MOSFET (L=40 nm) with a circular nanowire with a diameter D=3 nm. The drain current is plotted as a function of the gate voltage for V_D=50 mV. This figure illustrates that a good agreement is obtained between the results of Cylmos ("classical" drain current) and those provided by Silvaco.

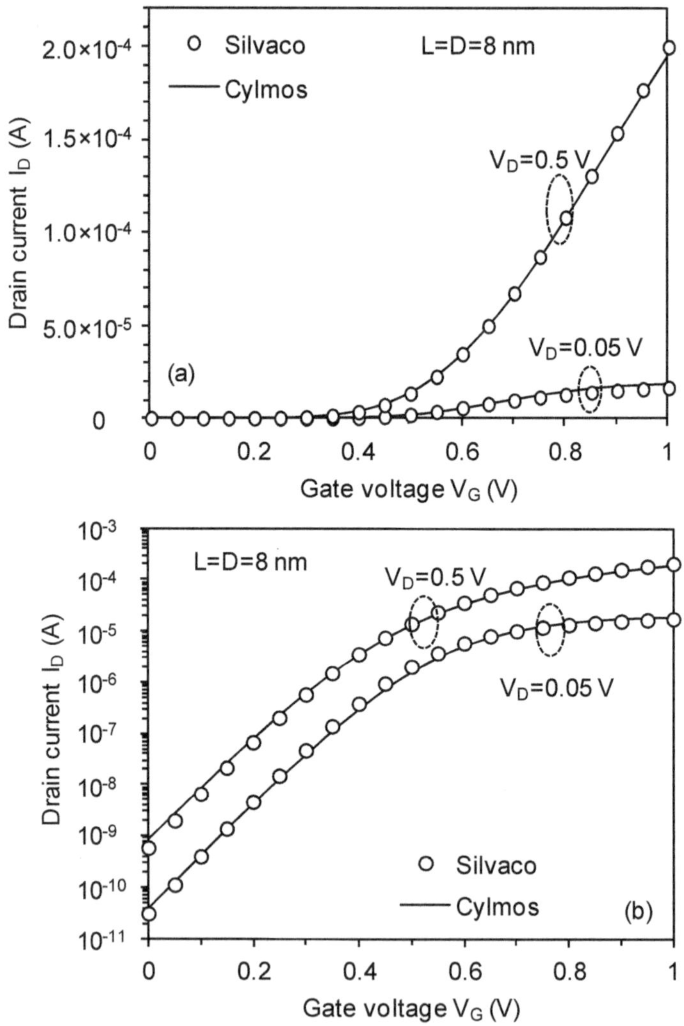

Figure 9. Drain current versus gate voltage calculated by Cylmos and Silvaco in the "classical" case (i.e. without quantum effects) for a short-channel nanowire MOSFET with L=D=8 nm.

We also tested a short-channel transistor, for which the gate length is equal to the diameter of the nanowire (L=D=8 nm). This transistor is probably provided with quite important short-channel electrostatic effects and it is necessary to verify that our code correctly reproduces these effects. In addition, the short-channel effects are considerably enhanced when the drain voltage is high, therefore it is important to test the current characteristics for several V_D. Figure 9 shows such a verification on both characteristics plotted in linear scale (for a better view on the regime above the threshold) and in logarithmic scale (to visualize the subthreshold regime). These figures show once again an excellent agreement between Cylmos and Silvaco, which means that the electrostatic effects are properly taken into account in Cylmos.

4. Simulation results

After validating Cylmos by comparison with a commercial simulator, we use in the following our code to analyze the operation and performances of circular nanowire-based MOSFETs subjected to quantum confinement. We will particularly focus on the variation of the drain current as a function of the nanowire diameter and of the gate length and we will discuss the device performances in terms of off-state current, threshold voltage and parasitic short-channel effects. These key parameters will also be compared between the classical and the quantum-mechanical case.

4.1. Drain current characteristics

We start with the variation of the drain current as a function of the nanowire diameter. Figure 10(a) plots the drain current characteristics as a function of the gate voltage, as calculated by Cylmos, for two diameters D=5 nm and D=8 nm. Data obtained in both classical and quantum cases are shown in this figure. A first remark concerns the off-state current (i.e., value of the drain current for V_G=0 V): for D=5 nm this current is lower than for D=8 nm, whatever the computation method, classical or quantum. Similarly, the current in the subthreshold regime is lower for the smallest diameter. This is due to the so-called "volume inversion" phenomenon which takes place in all multiple-gate devices in the subthreshold regime [28]-[29]. Thus, we logically find this phenomenon here in the circular nanowire, for which the gate completely surrounds the silicon film. Because of the volume inversion, the subthreshold current is directly proportional to the nanowire diameter, which implies that the off-state current decreases when the nanowire is thinned.

A second remark concerns the reduction of the drain current in the quantum case compared to the classical case, for both nanowire diameters considered here. As we explained earlier in this chapter, the existence of a strong quantum confinement of carriers leads to the reduction of the charge inversion in the channel (compared to the classic case where this effect is not taken into account). Therefore, for the same gate voltage V_G, the quantum drain current is necessarily smaller than the classical drain current, and this for all nanowire diameters. This behavior is similar to that of all other multiple-gate devices where strong quantum confinement effects (1-D or 2-D) occur [30], mainly due to the geometrical construction of these devices.

Another point concerns the difference between the current calculated in the classical case and the current calculated in the quantum case. This difference is larger for D=5 nm than for D=8 nm, as can be seen in Fig. 10(a), particularly in the subthreshold regime. The explanation comes from the large variation of quantum confinement effects with the nanowire diameter. When the nanowire is thinned, the carrier confinement is more pronounced (which also implies an increase of the allowed energy levels) and the inversion charge density in the channel reduces. This increases strongly the threshold voltage and leads to a significant decrease of the drain current.

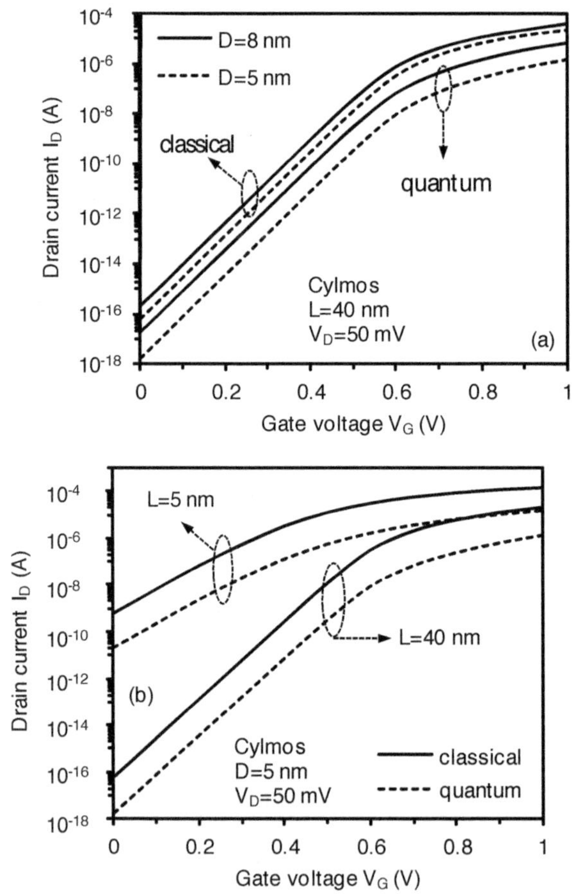

Figure 10. Drain current characteristics versus the gate voltage calculated by Cylmos: (a) L=40 nm and two nanowire diameters D=8 nm and D=5 nm; (b) D=5 nm and two channel lengths L=40 nm and L=5 nm. The drain current is calculated in both quantum and classical cases.

Finally, Fig. 10(b) shows the variation of the drain current as a function of the gate length for D=5 nm. In this figure we can observe that when the gate length is reduced (for the same nanowire diameter) the off-state current strongly increases and the subthreshold slope is much higher than in the case of a long channel (L=40 nm), reflecting a significant degradation of the device performances. This is due to the huge increase of short-channel effects, because in this ultra-short channel having a gate length equal to the nanowire diameter, the gate control on the electrostatic potential of the channel is much less effective. Thus, parasitic electrostatic effects dramatically increase and lead to this large increase in off-state current and subthreshold slope. This behavior appears for both classical or quantum cases.

4.2. Threshold voltage and short channel effects

For a more detailed analysis of these parasitic short-channel effects, we will now focus on the variations of the threshold voltage as a function of the gate length and of the nanowire diameter. The threshold voltage is extracted using the constant current method [31] from the drain current characteristics as a function of the gate voltage. In Fig. 11 we plot the variation of this threshold voltage as function of the nanowire diameter in the case of a long-channel transistor and in the case of a transistor with a channel length equal to the nanowire diameter (L=D). The results reported in this figure come from a quantum-mechanical calculation. We note that the threshold voltage of the long-channel transistor is higher than that of the short-channel transistor with L=D. This is an expected result because the short channel effects that occur in the transistor with L=D greatly reduce the threshold voltage. One can also note that when the diameter of the nanowire is reduced, the threshold voltage remains constant over a certain range of values and next begins to increase with decreasing the nanowire diameter (in both cases, long channel and short channel with L=D). This increase is due, of course, to the fact that the quantum confinement of carriers is more significant when the diameter is reduced, which induces an increase in the threshold voltage (as explained earlier). This increase is higher in the case of long channel than in the case of a short-channel. This indicates that the strong short-channel effects that exist in the channel with L=D affect the quantum confinement, which becomes weaker than in a long-channel transistor.

Another way to highlight this phenomenon is to calculate the impact of short channel effects, reflected in the SCE metric, which is equal to the difference between the threshold voltage of the short-channel transistor with L=D and the threshold voltage of the long-channel transistor (for a given diameter D). SCE is then plotted as a function of the nanowire diameter in Fig. 12. This figure shows that, after an increase and a maximum at D=8 nm, the SCE starts to decrease as the diameter reduces below 8 nm. This behavior is explained by the fact that the strong quantum confinement that occurs in these ultrafine nanowires reduces short-channel effects, which will probably lead to increased performance of these ultra-short and ultra-thin devices.

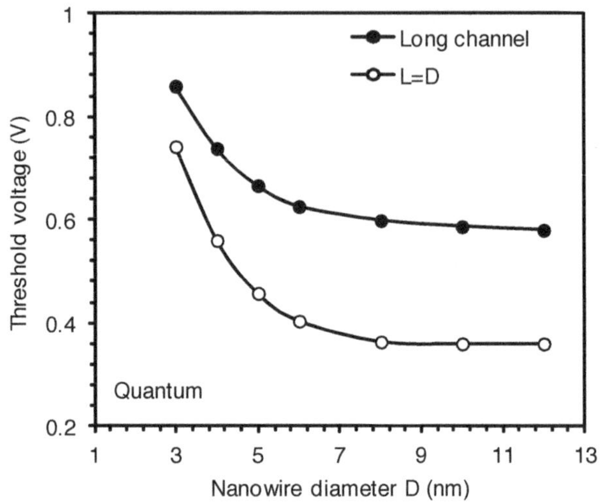

Figure 11. Threshold voltage versus the nanowire diameter extracted from quantum calculation of the drain current characteristics.

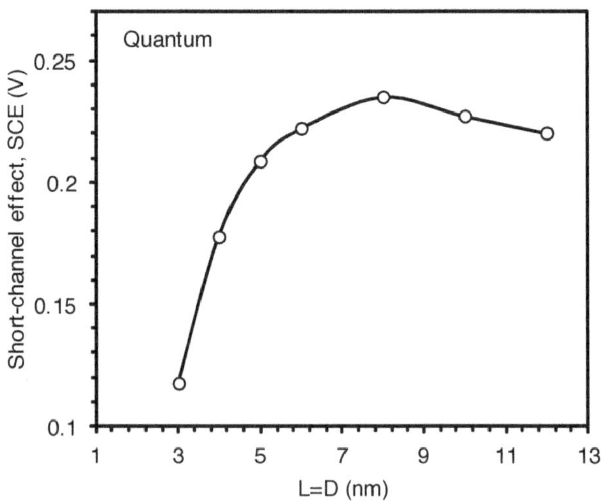

Figure 12. SCE metric versus the nanowire diameter for short-channel devices with L=D.

5. Conclusion

In conclusion, we described in this chapter a numerical solver, called Cylmos, that provides self-consistent solving of the 2-D Schrödinger equation and the 3-D Poisson equation in cylindrical coordinates, coupled with the drift-diffusion transport equation. We presented in details the discretization of these equations and the overall solving algorithm used to obtain the final drain current of the device. We also paid a particular attention to justify the approximations and simplifications used in the solving of these equations in cylindrical coordinates. A meticulous validation step was performed by comparing the results of our code with those issued from a commercial code. We demonstrated that our code properly takes into account particular phenomena specific to MOSFET devices based on circular nanowires with surrounding gate, such as volume inversion, short channel and quantum-mechanical effects. Simulated drain current versus gate voltage characteristics have also been successfully compared to data obtained from simulation with this commercial code. Our code was finally used to analyze electrical parameters such as drain current, off-state current, threshold voltage and short-channel effects in ultra-short radially-symmetric nanowire MOSFETs. We have shown that quantum-mechanical confinement is very important in circular nanowires, but the presence of this phenomenon reduces the impact of parasitic short-channel effects. This is a very encouraging result for the operation of future integrated circuits based on ultra-short and ultra-thin circular nanowires.

Author details

Daniela Munteanu and Jean-Luc Autran

CNRS &Aix-Marseille University, Marseille, France

References

[1] 2012 International Technology Roadmap for Semiconductors. Available online: http://public.itrs.net.

[2] T. Hiramoto, M. Saitoh, G. Tsutsui, "Emerging nanoscale Silicon devices taking advantage of nanostructure physics", IBM J. Res.Develop., vol. 50, no. 4/5, pp. 411-418, 2006.

[3] Y. Taur, D. Buchanan, W. Chen, D. Frank, K. Ismail, S.-H. Lo, G. Sai-Halasz, R. Viswanathan, H.-J. C. Wann, S. Wind, H.-S. Wong, "CMOS scaling into the nanometer regime", Proc. IEEE, vol. 85, pp. 486–504, 1997.

[4] E. P. Gusev, V. Narayanan, M. M. Frank, "Advanced high-k dielectric stacks with polySi and metal gates: Recent progress and current challenges", IBM J. Res. Develop., vol. 50, no. 4/5, pp. 387–410, 2006.

[5] A. Lochtefeld and D. A. Antoniadis, "On Experimental Determination of Carrier Velocity in Deeply Scaled NMOS: How Close to the Thermal Limit?", IEEE Electron Device Lett., vol. 22, pp. 95–97, 2001.

[6] M. V. Fischetti and S. E. Laux, "Long-Range Coulomb Interactions in Small Si Devices. Part I: Performance and Reliability", Journal of Applied Physics, vol. 89, no. 2, pp. 1205–1231, 2001.

[7] G. D. Wilk, R. M. Wallace, J. M. Anthony, "High-k Gate Dielectrics: Current Status and Materials Properties Considerations", Journal of Applied Physics, vol. 89, pp. 5243-5275, 2001.

[8] M. Houssa (Ed.), Fundamental and Technological Aspects of High-k Gate Dielectrics, IOP, London, 2004.

[9] K. Rim, J. L. Hoyt, J. F. Gibbons, "Transconductance Enhancement in Deep Submicron Strained-Si 12-MOSFETs", in Int. Electron Devices Meeting Tech. Dig., Dec. 1998, pp. 707–710.

[10] W. Haensch, E. J. Nowak, R. H. Dennard, P. M. Solomon, A. Bryant, O.H. Dokumaci, A. Kumar, X. Wang, J. B. Johnson, M. V. Fischetti, "Silicon CMOS devices beyond scaling", IBM J. Res Develop., vol. 50, no. 4/5, pp. 339-361, 2006.

[11] J.T. Park, J.P. Colinge, "Multiple-gate SOI MOSFETs: device design guidelines", IEEE Transactions on Electron Devices, vol. 49, no. 12, pp. 2222-2229, Dec. 2002.

[12] J. P. Colinge, M. H. Gao, A. Romano-Rodríguez, H. Maes, C. Claeys, "Silicon-on-insulator "Gate-all-around device"", in Int. Electron Devices Meeting Tech. Dig., Dec. 1990, pp. 595-598.

[13] D. Jiménez, B. Iniguez, J. Suné, L. F. Marsal, J. Pallarès, J. Roig, D. Flores, "Continuous Analytic I–V Model for Surrounding-Gate MOSFETs", IEEE Electron Device Letters, vol. 25, no. 8, pp. 571-573, 2004.

[14] S. A. Hareland, S. Jallepalli, W.-K. Shih, H. Wang, G. L. Chindalore, A. F. Tasch, C. M. Maziar, "A Physically-Based Model for Quantization Effects in Hole Inversion Layers", IEEE Trans. Electron Devices, vol. 45, no. 1, pp. 179-186, 1998.

[15] F. Stern, "Self-consistent results for n-type Si inversion layers", Phys. Rev. B, vol. 5, p. 4891-4899, 1972.

[16] M. Zervos and L.F. Feiner, "Electronic structure of piezoelectric double-barrier InAs/InP/InAs/InP/InAs (111) nanowires", Journal of Applied Physics, vol. 95, no. 1, pp. 281-291, 0021-8979, 2004.

[17] D. Munteanu and J.L. Autran, "Two-dimensional Modeling of Quantum Ballistic Transport in Ultimate Double-Gate SOI Devices", Solid State Electronics, vol. 47, no. 7, pp. 1219-1225, 2003.

[18] J. L. Autran and D. Munteanu, "Simulation of electron transport in nanoscale independent-gate DG devices using a full 2D Green's function approach", Journal of Computational and Theoretical Nanoscience, vol. 5, pp. 1120–1127, 2008.

[19] J. Martorell, H. Wu, and D. W. L. Sprung, "Systematic trends in self-consistent calculations of linear quantum wires", Phys. Rev. B, vol. 50, no. 23, pp. 17298-17308, 1994.

[20] T. Kerkhoven, A.T.Galick, U. Ravaioli, J.H. Arends, Y. Saad,"Efficient numerical simulation of electron states in quantum wires", Journal of Applied Physics, vol. 68, no. 7,pp. 3461-3469,1990.

[21] G. L. Snider, I. H. Tan, and E. L. Hu, "Electron states in mesa-etched one-dimensional quantum well wires", Journal of Applied Physics, vol. 68, no. 6, pp. 2849 – 2853, 1990.

[22] T. Kerkhoven, M. Raschke, and U. Ravaioli, "Self-consistent simulation of quantum wires in periodic heterojunction structures", Journal of Applied Physics,vol. 74, no. 2, pp. 1199 – 1204,1993.

[23] Y. Taur and T. H. Ning, Fundamentals of Modern VLSI Devices, Cambridge University Press, New York, 1998, p. 25.

[24] Atlas User's Manual, Silvaco International, Santa Clara, CA, 2010.

[25] C. Raynaud, J.-L. Autran, P. Masson, M. Bidaud and A. Poncet,"Analysis of MOS devices capacitance-voltage characteristics based on the self-consistent solution of the Schröndinger and Poisson equation", Mat. Res. Soc. Symp. Proceedings, vol. 592, pp. 159, 2000.

[26] J. S. Blakemore, "Approximations for Fermi–Dirac integrals, especially the function F1/2(ηF) used to describe electron density in a semiconductor," Solid State Electronics, vol. 25, no. 11, pp. 1067–1076, Nov. 1982.

[27] M. Moreau, D. Munteanu and J.L. Autran, "Simulation Analysis of Quantum Confinement and Short-Channel Effects in Independent Double-Gate Metal-Oxide-Semiconductor Field-Effect Transistors ", Japanese Journal of Applied Physics, vol.47, no. 9, pp. 7013-7018, 2008.

[28] Y. Taur, X. Liang, W. Wang, H. Lu, " A continuous, analytic Drain-Current model for DG MOSFETs", IEEE Electron Device Letters, vol. 25, no. 2, pp. 107-109, 2004.

[29] D. Munteanu, J.L. Autran, X. Loussier, S. Harrison, R. Cerutti, T. Skotnicki, "Quantum Short-Channel Compact Modeling of Drain-Current in Double-Gate MOSFET", Solid State Electronics, vol. 50, no. 4, pp. 630-636, 2006.

[30] D. Munteanu, J.L. Autran, V. Ferlet-Cavrois, P. Paillet, J. Baggio, K. Castellani, "3D Quantum Numerical Simulation of Single-Event Transients in Multiple-Gate Nanowire MOSFETs", IEEE Transactions on Nuclear Science, vol. 54, no. 4, pp. 994-1001, 2007.

[31] D. Munteanu, J.L. Autran, S. Harrison, K. Nehari, O. Tintori, T. Skotnicki, "Compact Model of the Quantum Short-Channel Threshold Voltage in Symmetric Double-Gate MOSFET", Molecular Simulation, vol. 31, no. 12, pp. 831–837, 2005.

Chapter 15

Cadmium Selenide-Gold (CdSe-Au) Hybrid Nanowires: Synthesis and Study of Charge Transfer

Xiaowei Teng and Xiaoxiong Jia

Additional information is available at the end of the chapter

1. Introduction

Complex hybrid nanostructures, which are assembled directly from different components without the aid from organic molecular bridges, are of major scientific and technological interest as they provide a variety of opportunities in various applications. By coupling components, each with its different functionalities, the hybrid nanostructures offer entirely novel properties that do not appear in their monocomponent counterparts. Many hybrid nanostructures, such as semiconductor/noble metal, semiconductor/semiconductor, noble metal/noble metal, noble metal/magnetic metal or metal oxide, and semiconductor/magnetic metal or metal oxide, recently were synthesized via the Polyol processes, photochemical routes, seeded growth, or site-selective lithographic approaches. However, there is a limited ability to finely control the size, morphology, phase-segregation, and position of the components in hybrid nanostructures, partially reflecting the structural complexity imposed by the random nucleation of the second component.

Among various metals, gold (Au) has been often studied in the hybrid system. The Au atom has the electron configuration of $(4f)^{14}(5d)^{10}(6s)^{1}$. Due to s-p-d hybridization, a small amount of itinerant holes existing at the d band in bulk Au has been observed. The corresponding hybridization resulted in the electron configuration as of $(4f)^{14}(5d)^{10-x}(6s)^{1+x}$. Thus, d-band holes can be generated in Au, provide various interesting properties in magnetism, catalysis and photonics [1, 2, 3].

Semiconductor nanomaterial have remarkable technological impacts on biomedical fluorescent labels, light emitting devices (LEDs), white-light laser sources, near-infrared emitters, solar cells, and tunable polarized lasers. Among various semiconductors, semiconductor-noble metal hybrid structures are of particular interest for their great potential application in catalysis and photonics. The synthesis of II-VI and III-V semiconductors with precise controlling of various shapes has been intensively studied over the past decade. CdSe, as one of the II-VI semiconductors, becomes a popular choice for researchers due to its well-studied synthetic approaches and promising applications in biological labeling and light-emitting diodes. CdSe nanostructures can be synthesized with various morphologies, such as spherical particle, rod, arrow, tear drop and tetrapod. Among these shapes, CdSe nanorods have been proved as good substrates for Au nanoparticles to grow on. Therefore, we selected CdSe nanorods as the host semiconductor to produce CdSe-Au hybrid nanorods in this work.

2. Synthesis of CdSe Nanorod

2.1. Shape Control of CdSe

The synthesis of II-VI and III-V semiconductors with precise controlling of various shapes has been intensively studied over the past decade. CdSe, as one of the II-VI semiconductors, becomes a popular choice for researchers due to its well-studied synthetic approaches and promising applications in biological labeling and light-emitting diodes. CdSe nanostructures can be synthesized with various morphologies, such as spherical particle, rod, arrow, tear drop and tetrapod. CdSe nanocrystals can be made via the injection of reactant precursors into a hot excess surfactant solvent, such as mixture of trioctylphosphine oxide (TOPO, $[CH_3(CH_2)_7]_3PO$) and stearic acid ($CH_3(CH_2)_{16}COOH$), which usually provides a fast nucleation of CdSe seeds, followed by slow growth. Two injecting methods are mostly adopted. One is a mono-injection method, from which the spherical CdSe nanoparticles are formed. In this method, the temporal separation between nucleation and growth helps to achieve relatively narrow size distribution of nanoparticles. The other is a multiple-injection method, from which CdSe nanorods are produced.

In the early study of CdSe synthetic methods, Murray *et al.* provided a simple route to the production of nearly monodispersed CdE (E = S, Se, Te) semiconductor nanoparticles [4]. A pre-heated (~300 °C) surfactant solvent was prepared first followed by co-injection of organometallic precursors (cadmium dimethyl, $Cd(CH_3)_2$ and Se trioctylphosphine, TOP, $[CH_3(CH_2)_7]_3P$) to produce a homogeneous nucleation using TOPO as a surfactant. The consequent growth at a moderate temperatures range (230~260 °C) resulted in the formation of uniform CdSe nanoparticles with diameters ranging from 1.5 to 11.5 nm.

Peng et al. also developed Murray's method and first achieved the shape control of CdSe nanostructures from CdSe nanoparticles to CdSe nanorods using $Cd(CH_3)_2$ as precursor [5]. However, instead of using pure TOPO, the mixture of hexylphosphonic acid (HPA, $CH_3(CH_2)_5P(O)(OH)_2$) and TOPO was used to serve as surfactant and reaction solvent. The function of HPA was to maintain the growth of CdSe nanorods in a controllable manner. Peng et al. also introduced multiple injections of organometallic precursors in the synthesis, which replenished monomers during the growth period of CdSe and resulted in the anisotropic growth of CdSe nanorods. By adopting HPA as surfactant and using multiple-injection method, CdSe nanorods were successfully produced in their studies. It was indicated that the growth of hexagonal crystal structure (wurtzite) of CdSe was highly anisotropic. Some growth conditions favored crystallization along the <001> axis, usually called the c-axis of wurtzite structure, and resulted in the formation of the CdSe nanorods. However, the initial injection temperature adopted in this method was in the range from 340 °C to 360 °C. Also, this multiple-injection method is relatively complex to execute compared to the mono-injection method adopted in the synthesis of CdSe nanoparticles.

Shieh et al. further explored Peng's multiple-injection approach in the synthesis of CdSe nanorods using cadmium mono-oxide (CdO) as precursor, and n-tetradecylphosphonic acid (TDPA, $CH_3(CH_2)_{13}P(O)(OH)_2$) and TOPO as surfactant [6]. They adopted a lower injection temperature (260 °C). However, they had to preheat the Se precursors to 120 °C every time before the injection.

From above discussion we can extract three important factors in the synthesis of CdSe nanocrystals:

1. In the synthesis of CdSe nanoparticles, $Cd(Ac)_2$ can be used as a precursor, avoiding high risk $Cd(CH_3)_2$;

2. a simple mono-injection method could be carried out using HPA as surfactant, avoiding the traditional co-injection method;

3. in the synthesis of nanorods, multiple-injection method is commonly used to synthesize nanorods.

In this work, we developed a mono-injection method based on traditional approach used in the synthesis of CdSe nanocrystals. We successfully achieved the shape control of CdSe nanostructures from nanoparticles to nanorods by simply adjusting the molar ratios between Se and Cd precursors. The growth models were proposed for different morphologies of CdSe and verified by the followed experimental results. The synthesized CdSe nanorods are utilized for the growth of Au nanoparticles which will be interpreted in details in the next section.

2.2. Synthesis of CdSe

In this work, we used $Cd(Ac)_2$ as the Cd precursor, a mixture of TOPO, octadecylamine (ODA, $CH_3(CH_2)_{17}NH_2$) and stearic acid as reaction solvent and surfactants. Se powder was dissolved in TOP to serve as injection solution for the growth of CdSe nanocrystals. The injection temperature was 310 °C and the growth time period for CdSe nanocrystals was 25 min. In a typical synthesis of CdSe, a mixture of $Cd(Ac)_2$ (90%, Strem Chemicals, 0.1132 g or 0.15 mmol), stearic acid (98%, Alfa Aesar, 0.0853 g or 0.3 mmol), TOPO (98%, Alfa Aesar, 2.5 g or 6.3 mmol) and ODA (90%, Acros, 2.5 g or 8.3 mmol) were added into a 50 ml three-neck round-bottom flask and then heated up via a heating mantle under vigorous stirring and argon flow protection. During the heating process, the initial white solid mixture turned into a transparent colorless solution. When the solution temperature reached 310 °C, a Se solution made by dissolving 0.18, 0.25, 0.75 or 1.0 mmol Se powder (99%, Strem Chemicals) into 1.79, 2.02, 5.38 or 7.17 mmol TOP (90%, Alfa Aesar) was quickly injected into the flask (<< 1 s). A dark purple solution formed immediately. After heating for an additional 25 min, the reaction was stopped by the removal of the heating mantle. After the solution was cooled to around 60 °C, the reaction solution was equally distributed into two 14 ml plastic centrifugation tubes. The final precipitation of CdSe was collected by washing it with ethanol and chloroform via a high speed centrifugation process.

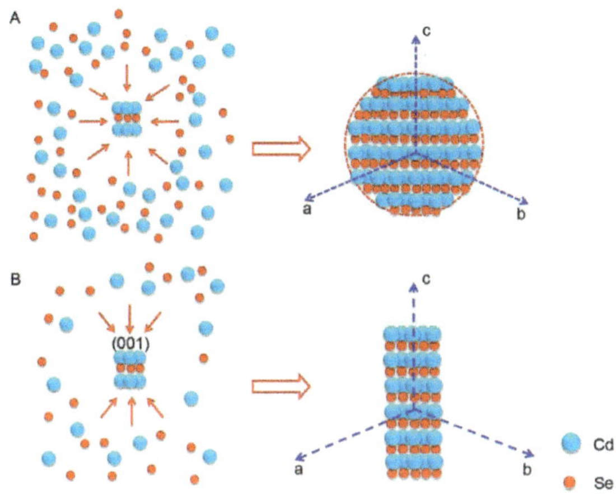

Figure 1. (A) Isotropic growth of CdSe nanoparticles. (B) Anisotropic growth of CdSe nanorods. (001) is the crystal surface perpendicular to the c-axis in the CdSe structure.

During the synthesis of CdSe, $Cd(Ac)_2$ first decomposed to form Cd atoms at high temperatures. These Cd atoms or clusters (aggregation of a few atoms) didn't nucleate before reaching the supersaturation. The formation of CdSe nanocrystals was triggered by injecting the Se/TOP solution into the flask. After injection, the Cd monomers reacted with Se to form CdSe monomers immediately, judging by the fast color changes upon the injection of Se into reaction solution. Once the concentration of CdSe reached supersaturation level, the CdSe nuclei (seeds) formed in the solution. In the meantime, CdSe monomers continued to deposit on CdSe nuclei, and resulted in the further growth of CdSe nanostructures. The continuous growth of CdSe nanostructures depleted the CdSe monomers in the solution, which would eventually decrease to the equilibrium concentration (the lowest concentration of CdSe monomer in solution). In the synthesis of CdSe with the molar ratio of Se/Cd=5, both nanorods and nanoparticles were observed. It is known that the formation of CdSe nanorods could be attributed to the elongation of c-axis in CdSe crystal structure. Namely, if the c-axis has a faster growth rate comparing to other axes in CdSe crystal structure, CdSe nanorods will be formed via an anisotropic growth along c-axis; if all the axes have a similar growth rate, CdSe nanoparticles will form via an isotropic growth. The growth rates of different axes can be also affected by the concentration of monomers in the solution. If the concentration of monomers is very high, all the axes will have a similar growth rate, CdSe nanoparticles will be formed; if the concentration of the monomers is low, CdSe nanorods will be formed via a dominant growth along c-axis. The corresponding scheme for the isotropic and anisotropic growth of CdSe nanocrystals is shown in Figure 1.

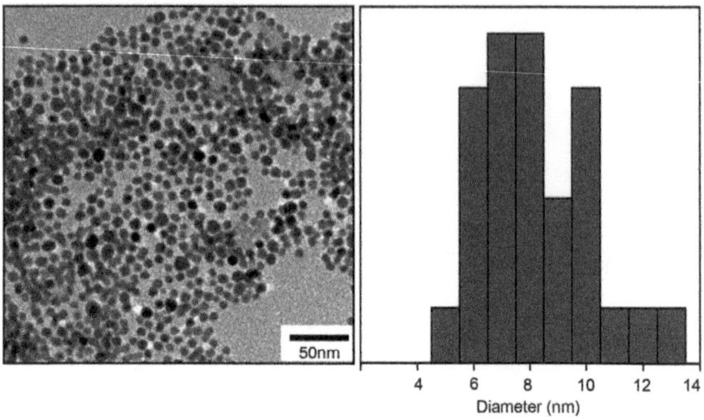

Figure 2. TEM images of CdSe nanoparticles and the corresponding sizing histogram. These CdSe nanoparticles were synthesized at 310 °C for 25 min using $Cd(Ac)_2$ (0.15 mmol), $CH_3(CH_2)_{16}COOH$ (0.3 mmol), TOPO (2.5 g), ODA (2.5 g), Se (1 mmol) and TOP (7.17 mmol). The molar ratio of Se/Cd was 6.7. The average diameter of CdSe nanoparticles was 7.7 ± 1.9 nm.

In order to verify the effect of CdSe concentration on the morphologies of CdSe as proposed in Figure 1, a series of experiments was conducted using different Se/Cd molar ratios. To simplify the experimental procedure, we kept the amount of $Cd(Ac)_2$ constant, and only adjusted the amount of Se to change the molar ratios of Se/Cd in the syntheses. If the model we proposed is valid, high amount of Se will lead to the formation of CdSe nanoparticles via the isotropic growth; while fewer amount of Se will lead to the formation of CdSe nanorods by encouraging the growth along c-axis. The TEM images of resulting CdSe nanostructures are shown in Figure 2 and 3. When Se/Cd molar ratio was equal to 6.7, CdSe crystals grow isotropically during the reaction due to the abundant existence of monomers in the solution (Figure 2). When the Se/Cd molar ratios decreased to 1.7 and 1.2 in the syntheses, the CdSe nanorod systems were synthesized (Figures 3). The observations are well consistent with the growth models we proposed. Moreover, the experimental procedure presenting in this work is more easily to conduct. This method allows the synthesis to conduct at a moderate temperature. Also by simply changing the molar ratios between Se and Cd, different shaped CdSe nanocrystals can be successfully synthesized, which provides a new thought to control the shape of CdSe nanocrystals.

In summary, a simple mono-injection method was adopted in the synthesis of CdSe nanocrystals. CdSe nanoparticles and nanorods were produced by adjusting the molar ratios between Se and Cd precursors. We proposed a plausible growth model in which the concentration of monomers in the solution played an important role in determining the final morphologies of CdSe nanocrystals. The model can be briefly described as high concentrations of CdSe monomer lead to an isotropic growth of CdSe nanoparticles; while low concentrations of CdSe monomer lead to an anisotropic growth of CdSe nanorods. A series of experiments using different Se/Cd molar ratios (Se/Cd = 6.7, 1.7 and 1.2) was carried out to verify this growth model.

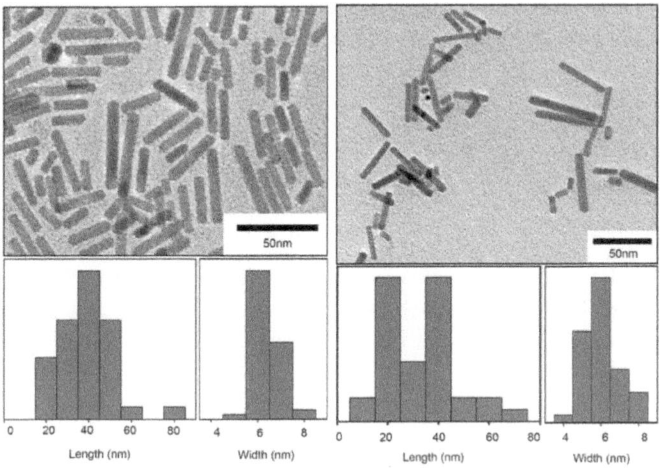

Figure 3. TEM images of CdSe nanorods and the corresponding sizing histograms. (Left) These CdSe nanorods were synthesized at 310 °C for 25 min using $Cd(Ac)_2$ (0.15 mmol), $CH_3(CH_2)_{16}COOH$ (0.3 mmol), TOPO (2.5 g), ODA (2.5 g), Se (0.25 mmol) and TOP (2.02 mmol). The molar ratio of Se/Cd was 1.7. The length of the rod was 33.4 ± 13.0 nm, the width was 5.9 ± 0.5 nm. (Right) These CdSe nanorods were synthesized at 310 °C for 25 min using $Cd(Ac)_2$ (0.15 mmol), $CH_3(CH_2)_{16}COOH$ (0.3 mmol), TOPO (2.5 g), ODA (2.5 g), Se (0.15 mmol) and TOP (1.79 mmol). The molar ratio of Se/Cd was 1.2. The length of the rod was 27.0 ± 14.4 nm, the width was 5.6 ± 0.9 nm.

3. Synthesis of CdSe-Au Nanorod

3.1. Material Synthesis

The growth of metals on semiconductors as a means of increasing functionalities is an important subject in material science. A preparation of such hybrids usually involves the injection of a metal salt into the semiconductor nanocrystal solution. Mokari *et al.* first achieved a position-selective growth of Au nanoparticles onto CdSe nanorods by gradually adding gold chloride ($AuCl_3$) solution into CdSe nanorods solution at room temperature in the presence of dodecyldimethylammonium bromide (DDAB, $[CH_3(CH_2)_{11}]_2N(CH_3)_2Br$,) and dodecylamine ($CH_3(CH_2)_{11}NH_2$) [7]. They claimed that amine reduced $AuCl_3$ to form CdSe/Au hybrid with Au nanoparticles deposited on the tips of CdSe nanorods. Furthermore, by controlling the reaction times, the sizes of Au nanoparticles could be tuned from 2 to 4 nm. The most compelling aspect of Mokari's work was the high selective growth onto the rod's tips.

They attributed this position-selective growth to higher surface energies of the tips, and more accessible sites due to imperfect passivation of surfactants on the tips of CdSe nanorods. In their follow-up work, Mokari *et al.* further studied the growth process of Au nanoparticles on CdSe nanorods [8]. At long reaction times, the concentration of $AuCl_3$ in solution dropped to a critical value and the reaction transitioned from a growth state to a ripening state, resulting in the disappearance of small Au nanoparticles along the surface of nanorod and the growth of large Au nanoparticles on the tips of nanorod. From the above discussion we can conclude two important factors to achieve the growth of Au nanoparticles on CdSe nanorods:

1. the growth of Au nanoparticles onto CdSe nanorods can be realized via a room-temperature reaction in the presence of DDAB and dodecylamine;
2. longer reaction time yields larger Au nanoparticles due to the ripening process.

Since in the ideal Au-based DMS, the growth of small Au nanoparticles has to achieve a uniform distribution all over the surfaces of CdSe nanorods, not only on some particular sites on CdSe nanorods, an alternative method to achieve the uniform growth of Au nanoparticles all over the surfaces of CdSe nanorods is required.

The Au precursor stock solution was prepared by dissolving 12 mg (0.04 mmol) of $AuCl_3$ (99.99%, Alfa Aesar), 15 mg (0.05 mmol) of 1-dodecyltrimethylammonium bromide (DTAB, $C_{12}H_{25}(CH_3)_3NBr$, 99%, Alfa Aesar) in 2 ml of chloroform at room temperature. The color of the $AuCl_3$ solution changed to yellow from orange after 5 min sonication. 0.15 mmol of as-synthesized CdSe nanorods with 30 mg TOPO ($C_{24}H_{51}OP$, 98%, Alfa Aesar) and 30 mg Octadecylamine (ODA, $CH_3(CH_2)_{17}NH_2$, 90%, Acros) were dissolved in 12 ml of chloroform in a 50 ml three-neck round-bottom flask under vigorous stirring and argon protection. Au precursor stock solution (0.167 ml) containing 1 mg (0.003 mmol) of $AuCl_3$ was injected into the reaction flask within 10 seconds. The reaction was carried out under room temperature. The TEM samples of CdSe-Au hybrid nanorods were prepared by taking 0.3 ml reaction solution from the flask at several time intervals after the initial injection. After 5, 7, 10, 30 or 60 min, the samples were collected into a 14 ml centrifuge tube and centrifuged for 5 min to get the final products.

The galvanic replacement reaction usually involves a redox process. The driving force of this process is the reduction potential (E°) difference between two reactants, with one reactant acting as the oxidizer and the other reactant as the reducer. Since E° (Au^{3+}/Au^0) has higher reduction potential than E° (Se^{2-}/Se^{4+}), the following reaction would happen in the solution:

$$2AuCl_3 + CdSe \rightarrow 2Au + Cd^{2+} + Se^{4+} + 6Cl^-$$

After mixing with CdSe, Au^{3+} ions were reduced by Se^{2-} to form Au atoms at the surfaces of CdSe nanorods. As reaction went on, Au nuclei formed on the surfaces of CdSe nanorods, as shown in Figure 4.

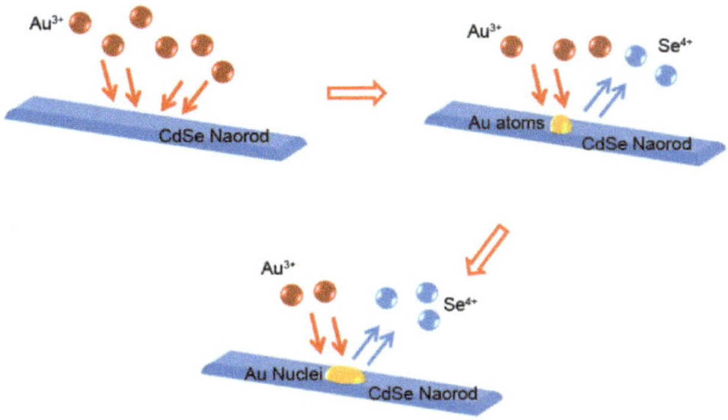

Figure 4. The galvanic replacement reactions between CdSe and $AuCl_3$ and the formation of the Au nuclei.

3.2. Time Effect

In the syntheses of CdSe-Au hybrid nanorods, the reaction times varied from 5 to 60 min to control the sizes of Au nanoparticles (See selected TEM images of CdSe-Au synthesized at 5, 10, and 30 mins in Figure 5). The CdSe nanorods used for reaction were synthesized using a Se/Cd molar ratio of 1.7. The molar numbers of CdSe nanorods and $AuCl_3$ was 0.15 mmol and 0.003 mmol. The growth of Au nanoparticle on CdSe nanorods with respect to reaction time was shown in Figure 6. From the curve we can see that in the first 10 min, the average size of Au nanoparticles increased sharply from 0 to 2.3 ± 0.7 nm; after 10 min reaction, Au nanoparticles exhibited a rather stagnant growth: sizes changed from 2.3 ± 0.7 nm at 10 min to 2.4 ± 0.9 nm at 60 min. The time-dependent growth of Au nanoparticles can be attributed to the change of concentration of $AuCl_3$ in the solution at different reaction times. The formation of Au nanoparticles was triggered by the injection of $AuCl_3$. At the initial time point (t = 0), there was no Au nanoparticle observed. Once the $AuCl_3$ was injected into the solution, the reaction between Au^{3+} and Se^{2-} occurred subsequently at the surfaces of the CdSe nanorods. This reaction led to the formation of Au atoms on the surfaces of CdSe nanorods. In the first 10 min, the Au nanoparticles kept growing at a relatively rapid growth rate due to the abundant $AuCl_3$ in the solution. As more Au nanoparticles formed in the solution, the concentration of Au^{3+} decreased until reaching the equilibrium. The growth rate of Au nanoparticles slowed down and ultimately suppressed by the depletion of the Au^{3+}, and the solution was considered to approach the equilibrium state. At an equilibrium state, the growth of Au nanoparticles transitioned to ripening state.

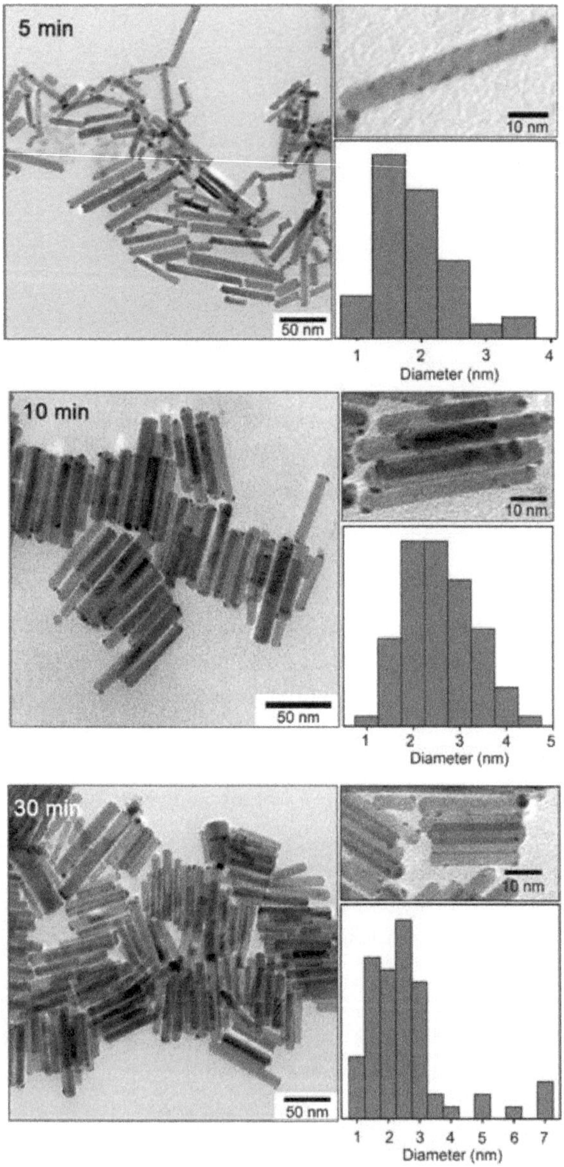

Figure 5. Original and enlarged TEM images of CdSe-Au hybrid nanorods and Au sizing histogram after 5, 10 and 30 min reactions. The average diameters of Au nanoparticles were 1.9 ± 0.8 nm, 2.3 ± 0.7 nm and 2.3 ± 1.3 nm, respectively.

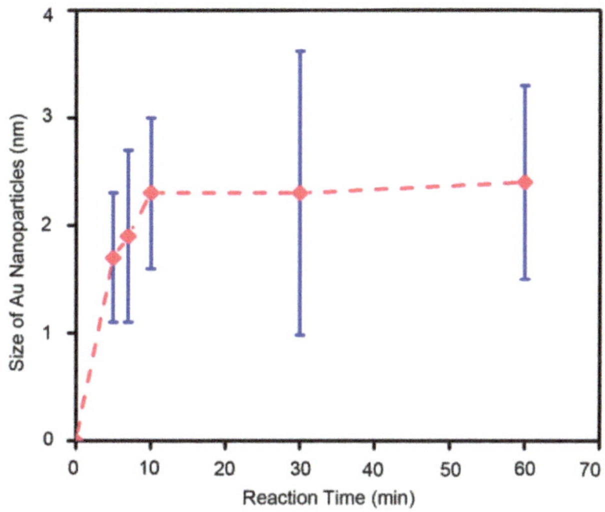

Figure 6. Changes of Au nanoparticles sizes as a function of reaction time

3.3. The Ostwald Ripening Process in CdSe-Au Solution

The Ostwald ripening process is the growth of large nanoparticles at the expense of dissolution of small ones at an equilibrium state. It is a spontaneous process because large nanoparticles are more thermodynamically stable than small ones. After the solution reached equilibrium state probably after 10 minutes reactions, large Au nanoparticles kept growing while small ones disappeared. Mokari1 *et al.* first proposed a growth mechanism for this process that small Au nanoparticles first released electrons into solution and then transferred these electrons to large Au nanoparticles through CdSe nanorods [8]. After giving out electrons, the small Au nanoparticles escaped from the surfaces of nanorods and became free Au ions in the solution. These Au ions re-deposited on large Au nanoparticles and were reduced to Au^0 (Figure 7.). The result of this process was disappearance of small Au nanoparticles and growth of large ones. However, they did not provide direct evidence to prove that small Au particles can release the electrons automatically. Also, no observation of electric current in CdSe nanorod was reported. Here, we proposed an alternative growth mechanism in which the growth of the Au nanoparticles is attributed to the Ostwald ripening process.

From a standpoint of kinetics, when the concentration of $AuCl_3$ was high, formation of small Au nanoparticles was dominant. When Au atoms/clusters monomers were depleted, the whole system transited from a fast growth state (kinetics state) to a ripening state (thermodynamics state). At the ripening state, from a standpoint of thermodynamics, small Au nanoparticles on the surfaces of CdSe nanorods dissolved into solution in the form of Au atoms (monomers) and re-deposited on the surfaces of large Au nanoparticles. As a result of this ripening process, the small Au nanoparticles dissolved to favor the growth of large particles on the CdSe nanorods. Figure 8 shows the Ostwald ripening process in CdSe-Au hybrid nanorods solution at equilibrium state. As the reaction continued, we found that Au nanoparticles tend to migrate to the tips of CdSe nanorods. After 5 minutes reactions, (Figure 6), small Au nanoparticles were found all over the surfaces of CdSe nanorods. After 30 and 60 minutes reactions, Au nanoparticles were found mostly at the tips of CdSe nanorods. This migration of Au nanoparticles can be attributed to higher surface energies on the tips of nanorod, resulting from more accessible sites on the tips where surfactant capping was weak. Therefore, the tips of nanorods are preferential growing sites for Au nanoparticles at long reaction times.

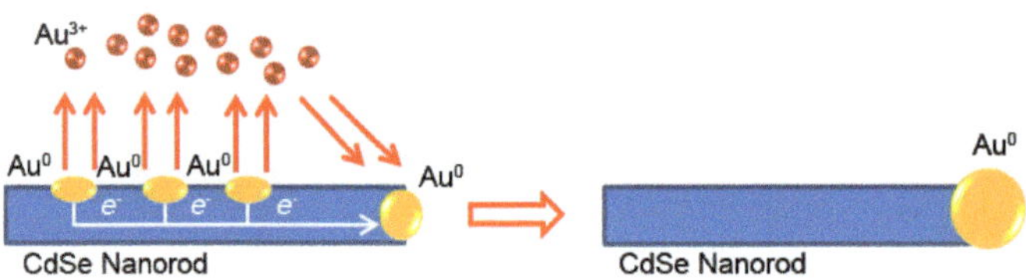

Figure 7. The growth process proposed by Mokari1 et al. Small Au nanoparticles escape to the solution as free Au ions by release their electrons. The released electrons transfer to large Au nanoparticles to reduce Au ions in the solution and results in the growth of large Au nanoparticles.

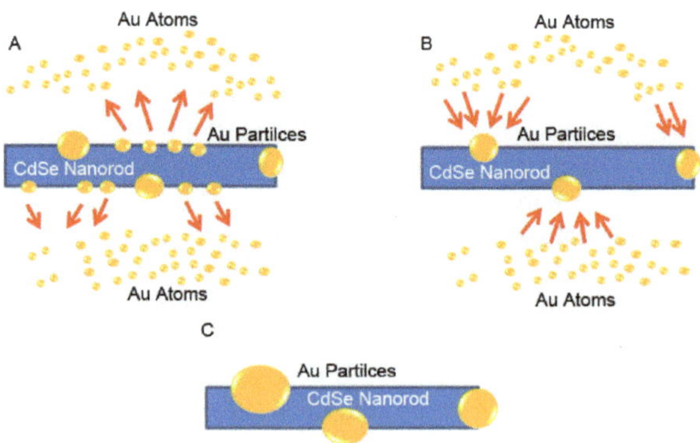

Figure 8. The Ostwald ripening process in CdSe-Au solution. (A) dissolution of small Au particles; (B) growth of large Au particles; (C) CdSe-Au hybrid nanorods with large Au particles.

3.4. In-Situ Growth of Au Nanoparticles under Electron Irradiation in TEM

A unique growth of Au nanoparticles was observed under the irradiation of the electron beams in TEM. Figure 9 shows this process. Figure 9A is the original image of CdSe-Au nanorods after 7 min reactions exposed under electron beam of low intensity, Figure 9B is the same sample exposed under electron beam of high intensity for 1 min. From Figure 9A to B we can see that as the electron beam intensity increased, the average size of Au nanoparticles experienced a clear growth from the initial 2.3 ± 0.7 nm to the final 5.0 ± 1.8 nm. This unique growth of Au nanoparticles under electron irradiation can be attributed to the size effect on the melting point of Au nanoparticles. It has been approved that the melting points of nanoparticles have great dependencies to their sizes. Typically, particles exhibit decreasing melting points with decreasing sizes. This size-dependent melting point of nanoparticles can be calculated through the Gibbs-Thomson equation:

$$T_M(d) = T_{MB}\left(1 - \frac{4\sigma_{sl}}{\rho_s d H_f}\right)$$

where T_{MB} is bulk melting temperature, σ_{sl} is solid liquid interface energy, H_f is bulk heat of fusion, ρ_s is density of solids and d is particle diameter. From this equation we can see that as particle diameter d increases, the particle melting temperature $T_M(d)$ decreases. The melting point of bulk Au metal is 1336 k (1063 °C). If the size of Au nanoparticle is 1.6 nm (the smallest Au nanoparticles we observed), the corresponding melting point is only 554 K (281 °C) (The parameters for calculation are in Table 1). Also, the operation of TEM is under vacuum, which further decreases the melting points of small Au nanoparticles. When Au nanoparticles were exposed to the electron beam in TEM, small Au nanoparticles melted easily due to their lower melting points. In the meantime, these melted Au particles were migrating towards large Au nanoparticles in order to achieve lower surface energies thermodynamically. The result of the process was the melting of small particles and the growth of large ones. The observed phenomenon also indicated that small Au nanoparticles distributed all over the surface of nanorod in the original sample. Otherwise, there were not enough growth resources for this dramatic growth of Au nanoparticles.

d/m	σ_{sl} /J m^{-2}	H_f/J Kg^{-1}	ρ_s/Kg m^{-3}	T_{MB}/K	$T_M(d)$/K	$T_M(d)$/ °C
$1.6*10^{-9}$	0.27	$6.27*10^{4}$	$1.84*10^{4}$	1336	554	281

Table 1. Parameters for Calculating the Melting Point of Au Nanoparticle

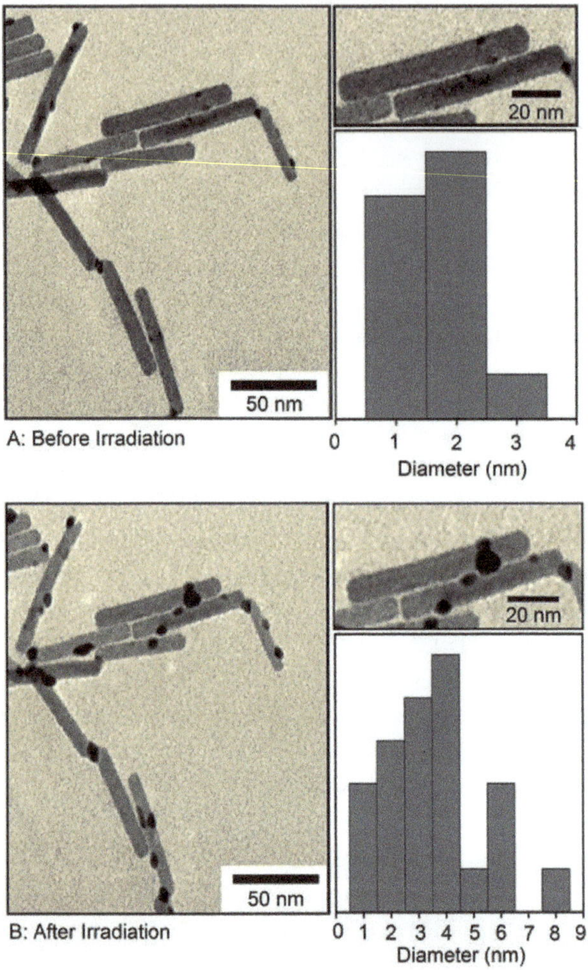

Figure 9. TEM images of CdSe-Au synthesized using 0.003 mmol $AuCl_3$ into 0.15 mmol CdSe nanorods: (A) before irradiation (B) after irradiation. The diameters of Au nanoparticles were (A) 2.3 ± 0.7 nm and (B) 5.0 ± 1.8 nm.

4. Electronic Structure Characterized by X–ray Absorption Near Edge Spectroscopy (XANES)

XANES is a powerful X–ray absorption spectroscopy technique for investigating charger transfer between distinctive elements as well as oxidation states of specific element within the nanomaterials [9]. By analyzing absorption signals in XANES, one can extract abundant chemical information of the studied material, like formal valence, coordination environment, the unoccupied band structure of a material, and fingerprint a mixture of sites/ compounds (e.g. oxidation state determination).

Since XANES is sensitive to the local electronic structure, it has been used to study the electronic redistribution or charge transfer upon alloying between various compounds. Obviously, it would also be the desirable technique to study the electronic structure of those CdSe/Au hybrid nanowires. Despite the fact that Au metal with filled d^{10} orbital usually doesn't exhibit an intense white line (the first feature in the absorption edge) at L_3 edge thresholds compared with unfilled d^9 metals (e.g. Pt), the orbital s–p–d hybridization results in a $5d^{10-x}6(sp)^{1+x}$ electron configuration, leading to a potentially noticeable increase of white line intensity. The changes of the number of holes in the d band reflect charge transfer between various CdSe-Au hybrid nanostructures with different sizes of Au components. Therefore, enhancement of the white line intensity from Au L_3 edges can thus be used as a sensor, to detect and quantify the hybrid formation. Figure 10 shows normalized Au L_3 edge XANES spectra of CdSe-Au hybrids and that of Au foil for reference. It can be seen that the strength of the Au white line feature in L_3 edge is noticeably lower than that of pure Au foil as the size of Au components decreased. Since white line features in Au L_3 edge arise from $2p_{3/2}$ to $5d_{5/2}$ dipole transitions, the observations indicate a decrease in the number of unoccupied states of d character at the Au site upon the decreasing the size of Au components. The data can be further interpreted in terms of gain of d charge in Au site. These findings dramatically highlight the intriguing structure–property relationship of nanoscaled electronic materials, and deserve further investigations.

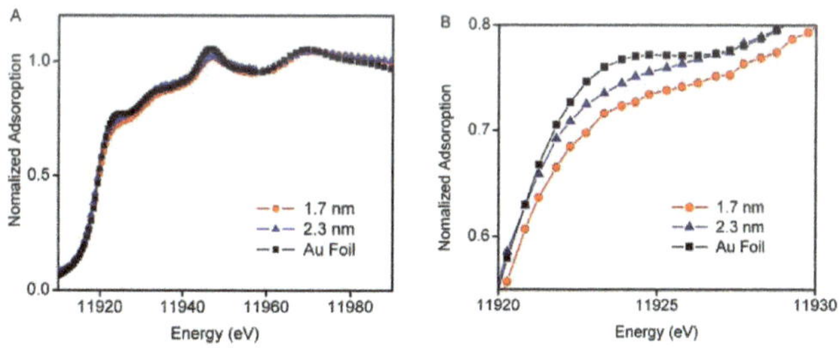

Figure 10. Normalized XANES spectra of Au in the CdSe-Au hybrid nanorods.

5. Conclusion

The target of this work was to synthesize CdSe-Au hybrid nanorods. The first step was to synthesize CdSe nanorods. A simple mono-injection method was adopted in the syntheses.

By reducing the molar ratios between Se and Cd, we have successfully achieved the shape control of CdSe nanocrystals from nanoparticles to nanorods. We found 5 to1 was a critical molar ratio between Se and Cd to form different CdSe nanostructures. When the molar ratio between Se and Cd was higher than 5 (6.7), CdSe nanoparticles were synthesized. When the molar ratio was lower than 5 (1.2 and 1.7), CdSe nanorods were synthesized. The growth models for different CdSe nanostructures were proposed and validated by the subsequent experimental results. In the model, the concentration of monomer plays an important role in the shape control of CdSe nanostructures. High concentration of monomer leads to the isotropic growth of CdSe nanoparticles. Low concentration of monomer leads to the anisotropic growth of CdSe nanorods. Comparing with traditional approach adopted in the shape control of CdSe nanostructures, the synthesis approach used in this work is more easy to conduct; also a relatively moderate reaction temperature and more favorable Cd precursor make the synthesis approach safer.

The growth of Au nanoparticles on CdSe nanorods involves galvanic replacement reactions between $AuCl_3$ and CdSe nanorods at room temperature. In the reaction, Au^{3+} was reduced to Au atoms while Se^{2-} was oxidized to Se^{4+}. By adjusting the reaction times, Au nanoparticles with diameter from 1.7 ± 0.6 nm to 2.4 ± 0.9 nm could be synthesized on the surfaces of CdSe nanorods. Before the reaction closing to equilibrium state, Au nanoparticles exhibited a fast growth rate due to a high concentration of $AuCl_3$ in solution. When the concentration of $AuCl_3$ approaching to equilibrium level, the reaction was closely related to the Ostwald ripening process in which the small Au nanoparticles dissolved into solution and the mass was transferred to increase the size of the large nanoparticle on the nanorods.

Acknowledgements

This work is supported by the University of New Hampshire. Use of the National Synchrotron Light Source, Brookhaven National Laboratory, was supported by the U.S. Department of Energy, Office of Science, Office of Basic Energy Sciences, under Contract DE-AC02-98CH10886 and beam lines X19A/X18B are partly supported by Synchrotron Catalysis Consortium under contract DE-FG02-05ER15688.

Author details

Xiaowei Teng* and Xiaoxiong Jia

*Address all correspondence to: xw.teng@unh.edu

University of New Hampshire, USA

References

[1] Saunders, A. E., Popov, I., & Banin, U. (2006). Synthesis of Hybrid CdS-Au Colloidal Nanostructures,. *J. Phys. Chem. B*, 110, 25421-25429.

[2] Crespo, P., Litran, R., Multigner, M., de la Fuente, J. M., Sanchez, J. C., Lopez, M. A., Garcia, C., Lopez, Cartes. A., Hernando, S., & Penades, A. Fernandez. (2004). Permanent magnetism, magnetic anisotropy, and hysteresis of thiol-capped gold nanoparticles,. Phys. ReV. Lett. , 93, 087204.

[3] Zhong, C. J., & Maye, M. M. (2001). Core-shell assembled nanoparticles as catalysts. *Advanced Materials*, 13, 1507-1511.

[4] Murray, C. B., Noms, D. J., & Bawendi, M. G. (1993). Synthesis and characterization of nearly monodispersed CdE (E = S, Se, Te) semiconductor Nano crystallites. *J. Am. Chem. Soc.*, 115, 8706-8715.

[5] Peng, X. G., Manna, L., Yang, W., Wickham, J., Escher, E., Kadavanich, A., & Alivisatos, A. P. (2000). Shape control of CdSe nanocrystals. *Nature*, 404, 59-61.

[6] Shieh, F., Saunders, A. E., & Korgel, B. A. (2005). General shape control of colloidal CdS, CdSe, CdTe quantum rods and quantum rod heterostructures,. *Phys. Rev. B. Lett*, 109, 8538-8542.

[7] Mokari, T., Rothenberg, E., Popov, I., Costi, R., & Banin, U. (2004). Selective Growth of Metal Tips onto Semiconductor Quantum Rods and Tetrapods,. Science. , 304, 1787-1789.

[8] Mokari, T., Sztrum, C. G., Salant, A., Rabani, E., & Uri, Banin. (2005). Formation of asymmetric one-sided metal-tipped semiconductor nanocrystal dots and rods. *Nature Mater*, 4, 855-863.

[9] Frenkel, A. (2007). Solving the 3D structure of metal nanoparticles. *Zeitschrift Fur Kristallographie*, 222, 605-611.

Chapter 16

Recent Advances in Mechanical Properties of Nanowires

Hui Li and Fengwei Sun

Additional information is available at the end of the chapter

1. Introduction

Over the recent decade, nanowires (NWs) have attracted great attention due to their small size and thus large surface area to volume ratio which results in interesting surface effect. As the devices become smaller and smaller, NWs serve as basic building blocks for future electronic and electromechanical systems, such as performing sensitive mass and force detection, acting as high frequency resonators and so on. The potential of NWs in future application has led to significant interest in experimental and theoretical characterization of the size-dependent properties of NWs. Constructing and determining NWs with better mechanical properties than the corresponding bulk materials are a challenge even though the technical equipments are improving. Meantime, another important thing is how to integrate NWs with the existing technology. Material simulations of investigating mechanical properties of NWs are still an important technique until now. This is essential to understand the inherent mechanism of the NWs' deformation. The deformation mechanism depends on several aspects including the intrinsic material properties, crystal structures, surface geometry, applied stress state, axial and surface orientation and exposed transverse surfaces. Until now many experimental techniques have been utilized to investigate mechanical properties. Scanning tunneling microscope (STM) was employed to study atomic contact of NWs and their conductance [1-3] in early experiments. Since a novel nanobridge structure of Au NWs is generated by electron-beam irradiation on a thin Au film in an ultrahigh vacuum transmission electron microscope (TEM) [4], many researches have been carried out to study the properties of this stable structure. A specific developed STM supplemented with a force sensor is used in order to reveal the process of forming thin Au NWs [5]. Atomic force microscope(AFM) has also attracted particular attention due to its high spatial resolution and force-sensing capabilities[6] and was used for bending tests of single crystal, micromachined silicon beams[7]. Recently developed in situ TEM technique can also be applied to perform bending deformation of single SiC NWs[8]. Theoretically, some new simulation methods are developed and widely used to study nanoscale materials. For FCC crystal metal NW, most simulations were performed with embedded atom method (EAM) [9-17], which reproduce exactly the experimental second-order and third-order elastic moduli as well as the phase stability. Furthermore, the modified EAM with the developed potential [18-20] is utilized to study the phase transformation of Au NWs. While for the semiconductor NWs, Tersoff many-body bond-order reactive potential and Stillinger-Weber many-body potential [21, 22] are appropriate to describe interatomic interactions[23] in Si and SiC NW. Tight-binding molecular dynamics simulations [24-26], which lies in between first principles and empirical methods is also used to study the atomic structure of Au NW. It is more accurate than empirical potential methods because it explicitly includes the electronic structure, and is much faster than first principles methods.

This review is organized as follows. In section 2, we discuss some novel mechanical properties of single crystal metal NWs caused by surface effect, strain rate, temperature and so on. In section 3, we consider the mechanical properties of semiconductor NWs such as Si, Si compound and ZnO et. al. In section 4, some other irregular structures of NWs are investigated to compare the mechanical properties with bulk and normal crystal NWs. In the last section, we discuss the enhancement of NWs which are encapsulated in carbon nanotubes during compression.

2. Single crystal metal nanowires

2.1. Side surface effect on axial deformation mode

Several researchers have demonstrated that the structures and properties of nanomaterials can be quite different from those of bulk materials by experiments or molecular dynamics (MD) simulations due to surface effect which results in different deformation modes.

In compression of the <110> NW, low energy {111} side surfaces prevented free-surface initiated slip. On the contrary, slip is observed for higher energy {100} and {110} side surfaces. For the <100> wires, distributed stacking faults appearing in the <100>/{110} wires result from simple and important rotation of the crystal orientation along its longitudinal axis, which is different from the deformation mode of twinning for the <100>/{100} wires. Some other important findings of fundamental deformation mechanism are revealed in Ref.[27].

2.2. Size effect on Young's modulus

Mechanical properties based on Au NW bending under the lateral load from an AFM tip are performed by Wu et.al [6]. Their experiments reveals that Young's modulus is essentially independent of diameter, whereas the yield strength is largest for the smallest diameter wires, with strengths up to 100 times that of bulk materials, and substantially larger than that reported for bulk nanocrystalline metals. MD simulations demonstrate that Young's modulus is independent of strain rate and cross sectional size of specimen. Neither geometry nor surface structure has an appreciable effect on the modulus[28]. However, in some other molecular statics simulations[29], the modulus values for NWs with a <110> axial orientation are observed to increase with decreasing cross-sectional thickness, while for NWs with a <100> axial orientation the modulus values decrease with decreasing cross-sectional thickness. This is attributed to the nonlinear elasticity of the NW core and dissimilar surface and core elastic modulus values[12]. The size-dependence modulus[28] is consistent with predictions based on core nonlinear elasticity, but it is difficult to separate the surface elasticity and core nonlinearity effects because of a lack of strong trends with NW size or surface orientation[29]. Meantime, deformation induced by surface stresses might also has a significant impact on the Young's modulus in bending simulations, where a strong boundary-condition dependence is also found [30]. Therefore, a nonlinear, finite deformation formulation [30] that captures changes in both bulk and surface elastic properties resulting from surface stress-induced deformation is critical to reproducing the experimentally observed boundary-condition dependence in Young's modulus of metal NWs.

2.3. Surface stress effect

2.3.1. Surface stress effects on the resonant properties

Boundary conditions determine how the NWs deform in response to surface stresses. Fixed/free NWs (cantilever) are able to contract, thus relieving the tensile surface stresses, and leading to a state of compression in the NW core. Fixed/fixed NWs are constrained such that they cannot deform axially, causing them to exist in a state of tensile stress.

The reduction in resonant frequencies for the fixed/free boundary condition and increase in resonant frequencies for the fixed/fixed boundary condition [31] indicate that those NWs are expected to be elastically softer, and stiffer, respectively, than the corresponding bulk material, agreeing with recent experimental results for both free-standing[32], and fixed/fixed NWs [33-35]. The variation in NW resonant frequencies due to surface stresses was found to be dependent on the NW aspect ratio. In contrast, no dependence of the resonant frequencies on the surface area to volume ratio was found, again agreeing with recent experimental data [36]. The variation in resonant frequency for the fixed/free case occurs due to the fact the bulk <100> material softens under compression. Therefore, the resonant frequencies of fixed/free NWs with other orientations that stiffen under compression, such as <110>, are expected to increase rather than decrease as compared to the corresponding bulk material. In the higher order resonant modes (second bending, stretch, twist), surface stresses cause the largest variation in the second bending frequencies, while the twist and stretch frequencies show little variation due to surface stresses. All higher order resonant frequencies decrease with respect to the fundamental bending frequency with increasing aspect ratio for fixed/fixed NWs due to surface stresses, while the higher order resonant frequencies increase with respect to the fundamental bending frequency for fixed/free NWs[31].

2.3.2. Surface stress induced phase transformation

Surface-stress-induced phase transformation in Au NWs is an interesting phenomenon. The emergence of the transformation is dominated by wire size, initial orientation, boundary conditions, temperature and initial cross sectional shape. For a crystal with <100> axial orientation, surface stresses alone cause Au NWs with cross-sectional area below 4 nm^2 to transform from a FCC structure to a body centered-tetragonal (BCT) structure [20]. Not only has this transformation appeared in Au NWs, but also in alloy NWs. Park has demonstrated that B2 NiAl NWs undergo a stress-induced martensitic phase transformation to a BCT phase as well by the propagation and annihilation of {101} twinning planes [37]. While for Au NWs formed along the [110] direction, quantum mechanical molecular dynamics shows that it reconstructs upon stress to form helical NWs [38]. The different final structures are probably due to temperature effect as well. The helical NW is obtained by annealing the crystal structure from 600K, while the BCT structure is relaxed to a minimum energy state at 0K with conjugate gradient method.

Some further simulations have also been performed [39] to investigate the core/shell structure in which the crystalline core is encapsulated by a glassy shell. Such core/shell heterostructures usually possess particularly attractive applications in the nanoelectromechanical system, owing to their high complexity and functionality. The elongation of the metallic NW is realized by moving the rigid atoms at one end along the axial axis. The rigid atoms are stretched 0.001nm every 1 ps. This strain rate is sufficient to dissipate excess energy to maintain a constant temperature. The Berendsen method [40] is used to control the system temperature. The motion equations are solved by a fourth-order Nordsieck-Gear predictor corrector method[41] using a time step of 1 fs. The average of the virial stress component along the axial axis during deformation is taken as the macroscopic stress of the wire. This stress measure totally differs from either engineering stress or true stress, which has been used widely in molecular simulations [42-44]. Here the low temperature (T=50K) was used to highlight the material responses upon mechanical activation, as the simulated quenching glass structure after rapid cooling is usually more prone to thermal activations than the experimental one. Continuous deformation with noticeable necking occurs to the core/shell NW instead of discrete displacement as shown in Fig.1.

The core/shell structure, which can even be spun into a single atomic chain, tends to display much larger plasticity than the crystalline one. Most importantly, during elongation, local amorphization happens in the high strained region. Generally, it is thought that the amorphization behavior in nanoscale materials is induced by the high strain rate or the entropy at a high temperature. In terms of our simulation, the strain rate of 0.006% ps^{-1} for a 16.3nm long NW, which is equivalent to an approximate stretch velocity of 1.0ms^{-1}, belongs to the range of low stretch velocity compared with previous simulation results. Therefore, we suggest that the surface amorphous layer would induce the disorder of the crystalline core during the elongating process.

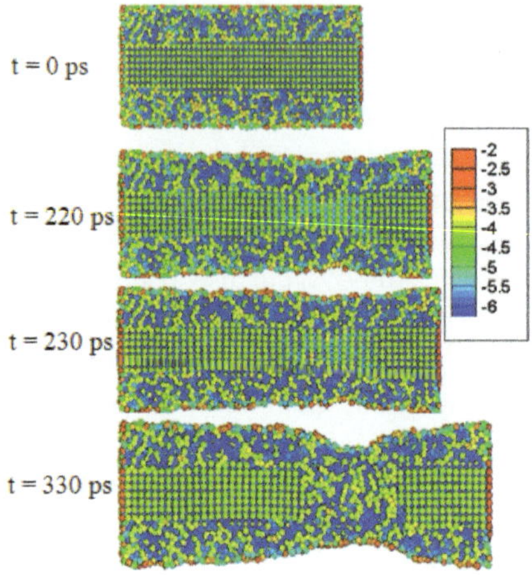

Figure 1. The stretching process of core-shelled CuZr NW, the color bar is to illustrate the contour plots of the per-atom potential energy.

To shed light on the elastic and plastic deformation mechanism in the core/shell NW, evolutions of energy and structure in three key deformation stages are provided in Fig. 2. Here, E_{per} is the total energy per atom at different stretching moment. For the sake of simplicity, an original B2 unit cell taken from the denoted region is used to track the structural transformation. Figure 2(a) shows an initial cross-section image along the axial direction. Figure 2(b) shows an incipient elongating stage. Initially, the original B2 lattice in crystalline core is cubic with a lattice parameter of 0.361 nm. When the stress is applied, a typical martensitic transformation from B2 to BCT phase is observed. That is, a set of new lattice parameters of a=0.289nm and c=0.393nm forms in the crystalline-core zone. When further elongated as shown in Fig. 2(c), the lattice in the deformed crystalline core exhibits increasingly serious distortion. Before fracture happens, atoms with high stress concentration and potential energy get rid of the constraint of the crystalline lattice and turn to be dominant. After that, the crystalline phase in the core converts into the disordered state.

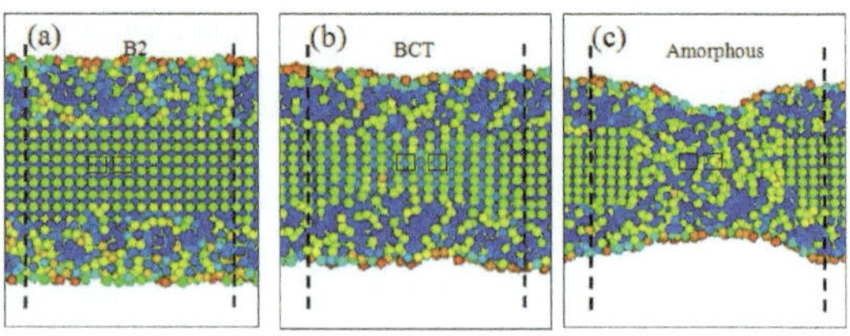

Figure 2. The structural evolution of composite nanowire in stretching, which varies from (a) the B2 state, then (b) the BCT state, to (c) amorphous state.

Only a global description of the structural evolution in the unique core/shell NW during elongation has been provided above, yet extensive analyses on those amazing phenomena are still in great need. To address this issue, three core/shell NW systems containing the same-sized crystalline core (the diameter of the crystalline core is kept at 2 nm) but different size of amorphous surface layer (the crystalline-amorphous ratios are 1 : 1, 1 : 2 and 1 : 3) are analyzed in Fig. 3. As shown in Fig. 3(a), the degree of amorphization has been employed to describe the percentage of the atoms which experiences an order-disorder transformation of the crystalline core inside the system. It is worth noting that the degree of amorphization in the core correlates closely to the crystalline-amorphous ratio. The degree of amorphization in the crystalline core rises from 18% to 40% when the crystalline-amorphous ratio changes from 1 : 1 into 1 : 3, indicating that the degree of amorphization in the core strongly depends on the crystalline-amorphous ratio. Nevertheless, the response between the degree of amorphization in the core and the crystalline-amorphous ratio is nonlinear, which should be discussed in further investigation. The corresponding total strain-stress curves with different crystalline-amorphous ratios are also plotted. As shown in Fig. 3(b), these curves indicate that the ductility of the core/shell NW is proportional to the thickness of the amorphous surface layer whereas the breaking strength keeps declining. In addition, the response of strain energy in crystalline core with strain increasing is also plotted in Fig. 3(c) which can quantitatively reflect the stress concentration exerted by the amorphous surface layer. With the amorphous surface layer thickening, the total strain energy in the crystalline core becomes higher, indicating that a thicker amorphous surface layer could introduce more stress concentration into the crystalline core and thereby convert this zone into a disordered state more easily.

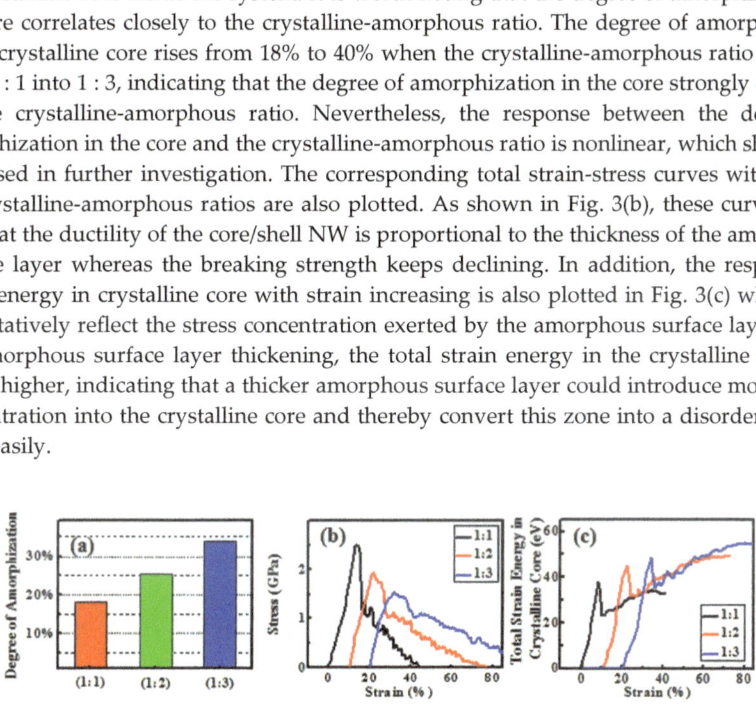

Figure 3. When increasing the surface amorphous thickness, the change curve of (a) the decrystallized degree of the internal crystalline core; (b) the total stress-strain responses; (c) and the strain energy-strain responses of the internal crystalline core. The red and blue curves in (b) and (c) are offset to a strain of 10% and 20%, respectively.

For a better understanding of the corresponding characteristics in the core/shell NW at an atomic level, real-time Honeycutt and Anderson (HA) bond-type index is introduced, which can describe and help to discern the concrete relationships between an atom and its nearest neighbors. The atoms at the high strained region in the crystalline core (ranged from 8 nm to 12 nm) are considered. Figure 4 illustrates the evolution of the BCC bond pairs, as well as the proportion of icosahedral bond pairs in the selected area of crystalline corewith strain increasing, which corresponds to the elongating process shown in Fig. 1. It is noted that when the strain increases, the total amount of BCC type bond pairs (1441 and 1661 bond pairs) remains approximately constant at first. And then both of them experience a sharp decline from 40% to about 5%. At the same time, the amount of 1541 bond pair which is related to the characteristic of defective icosahedral ordering initially keeps steady and then increases sharply. Moreover, there exists an obvious turning point in both curves which clearly reflects the BCT-disorder transformation corresponding to Fig. 2(c). That is, the BCC bond pairs may firstly convert into the 1541 bond pairs. It should be noted that the increase of the 1551 bond pairs in Fig. 4(b) happens slightly later to that of the 1541 ones. The possible reasons for this special phenomenon will be further discussed in the following.

Further studies focus on the transformation of bond-pairs at the crystalline-amorphous interface and investigate how the local amorphization happen. Figure 5 presents a supercluster consisting of over one hundred atoms taken from the core/shell NW as shown in Fig. 2(b). The red-centered cluster represents the typical BCC lattice, whereas the brown one signifies the icosahedra cluster. At the initial stage, the B2 crystalline core can convert into the BCT structure. Meanwhile, many crystal defects, such as vacancies, can be found near the crystalline-amorphous interface. With the strain increasing, partial atoms near the interface which possess a high potential energy can insert into the crystal defects and consequently result in local amorphization.

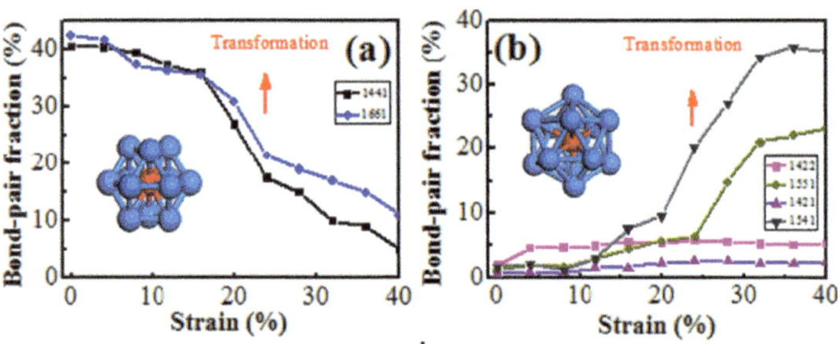

Figure 4. Variation of the main fractions of HA indices in the selected region during stretch process: (a) the decreasing bond pairs; (b) the increasing bond pairs. The insets show the typical B2 and icosahedron structures.

For example, the atom labeled 10 in Fig. 5 can insert into a BCC lattice (the corresponding atoms labels 5, 6, 8 and 9) and form 1541 bond pair. With further elongation, more and more atoms will insert into the crystalline core and the total amount of the 1541 bond pairs becomes larger. It is known that the 1551 polyhedra have a higher resistance to plastic deformation than the 1541 one owing to their higher symmetry. Thus, part of the 1541 bond pairs transforms into the 1551 ones, and as a result, the increase of the 1551 bond pairs is slightly later to that of 1541 ones as shown in fig. 4(b).

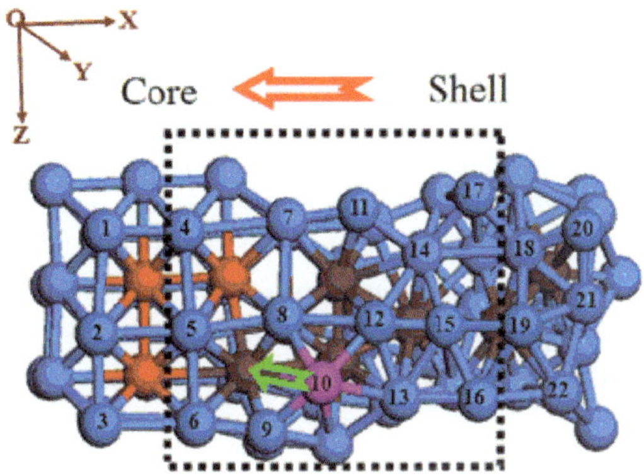

Figure 5. The evolution of bond pairs in the crystalline-amorphous interfaces. For the sake of analysis, atoms in the nearly same plane are shown.

2.4. Strain rate effect on structure of nanowire

Using molecular dynamics simulations with a many-body force field, Ikeda[45] studied the deformation of single crystal Ni and NiCu alloy NWs subjected to uniform strain rates at room temperature. For all strain rates, the Ni NWs is elastic up to 7.5% strain with a yield stress of 5.5 GPa. The crystalline phase transforms continuously to an amorphous phase at high strain rates, exhibiting a dramatic change in atomic short-range order and a near vanishing of the tetragonal shear elastic constant perpendicular to the tensile direction. This phase transformation exhibits a new mode of amorphization[45] or even a super plastic behavior [11]. Higher strain rate may result in deformation twins during plastic deformation as well. As the strain rate decreased to about 0.1% ps^{-1}, a transition of deformation mechanism from combined twining and slip to sequential propagation of slip along well defined and favorably oriented slip plane is observed [11].

2.5. Temperature effect

The mechanical behavior depends highly on the temperature of the system. Several interesting features were observed when the NW was deformed in [001] direction at different temperatures. Due to the higher crystal stability at a lower temperature, the deformation behavior of the NW was characterized by brittle slips, rupture and very low ductility [46]. At the higher temperature of 300 K, the crystal structure became less stable due to higher amplitude of vibration of the atoms about their atomic positions. A relatively stable single-walled helical substructure was formed due to the higher local vibration amplitude of the platinum atoms which enhances the ductility of the NW [46]. The calculated Young's modulus of the NW at T=300K was only about 50% that of bulk platinum, and was significantly smaller than that at T=50K.

2.6. Shape memory

Shape memory is an important property for some engineering materials at large scales. Some NWs also have characteristics of shape memory effect due to structural reorientations which are controlled by a combination of size, thermal energy, and the type of defects formed during inelastic deformation. Certain FCC NWs exhibit both shape memory and pseudoelastic behaviors in atomistic simulations. The formation of defect-free twins and surface stresses is the mechanism that controls the ability of FCC NWs of different materials to show a reversible transition between two crystal orientations during loading and thus shape memory and pseudoelasticity[14].

2.7. Superplasticity

Defect-free Au NWs show superplasticity on tensile deformation [47]. Evidences from high resolution electron microscopes indicated that the plastic deformation proceeds layer-by-layer in an atomically coherent fashion to a long distance. The stress-strain curve provides full interpretation of the deformation. After initial super-elastic deformation, the NW shows superplastic deformation induced by coherent twin propagation, completely reorientating the crystal from <110> to <100>. Uniquely well disciplined and long-propagating atomic movements deduced here are ascribed to the superb crystallinity as well as the radial confinement of the Au NWs. Other mechanical measurements in humid atmospheres suggest that salt NWs also form in ambient environments [48]. Superplastic NWs were formed by touching the NaCl(100) surface with a Au tip in a TEM. The final elongation strain of NWs reached 280%, when showered with the electron beam. More surprisingly, no dislocations were observable during the elongation due to fast diffusion.

2.8. Cold welding

General welding techniques require precise control of the heating mechanism and introduce the possibility of damage. The welding of nano-materials is likely to play an important role in the fabrication of electromechanical nano-devices. The cold welding (a process without heating) on a micrometer scale has been demonstrated, but only at macroscopic length scales and under large applied pressures. Lu et.al [49] demonstrate that single crystalline Au NWs with diameters between 3 and 10 nm can be welded together within seconds by mechanical contact alone, and under relatively low applied pressures. The nearly perfect welding structure possesses the same crystal orientation, strength and electrical conductivity as the rest of the NW. The high quality of the welds is attributed to the nanoscale sample dimensions, oriented-attachment mechanisms and mechanically assisted fast surface-atom diffusion. The cold weld technique is also applied between other kinds of metals such as gold and silver, and silver and silver.

3. Crystalline semiconductor nanowires

3.1. Silicon and silicon compound nanowires

It is generally accepted that Si NWs exhibit linear elastic behavior until fracture without any appreciable plastic deformation. However, the plasticity of Si NWs can be triggered under low strain rate inside the TEM. An electron-beam illuminating with a low current intensity would result in the bond re-forming processes, achieving the plastic deformation with a bent strain over 40% in Si NWs near the room temperature [50]. In some other tensile experiments, the Si NW also showed ultrahigh flexibility and strong toughness [51]. Large strain plasticity (LSP) of single-crystalline Si NWs, which resulted in a fourfold reduction in NW diameter before fracture occurred, was directly observed in an ultrahigh-resolution electron microscope at room temperature. The tensile plasticity was initiated by the emergence of dislocations, followed by the development of a continuous disordered lattice by the emission of dislocations and formation of disordered crystalline structures. The size-dependent fracture mechanism is attributed to the scale-related dislocation activities [52]. For ceramic materials especially at low temperature (~300 K) large strain plasticity of ceramic SiC NWs at nearly room temperature was directly observed in situ by a high-resolution TEM as dimensionality decreases. The continuous plasticity of the SiC NWs[8] is accompanied by a process of increased dislocation density at an early stage, followed by an obvious lattice distortion, which is similar with stretching Si NW.

Individual single-crystalline Si NWs were bent by forming loops or arcs with different radius with high-resolution electron microscopy. Bending-induced ripple buckling was observed and a significant strain variation along the axial direction of the compressive region was revealed. The tensile surface atomic steps and the compressive buckling are the physical origin of the asymmetric tensile-compressive properties of postelastic instabilities and the incipient plasticity. Both of the tensile surface atomic-steps and the compressive buckling initiated versatile ductile plastic dislocation events [53]. Sequential AFM manipulation-scanning protocol is used to observe large bending stress states of [112] Si NWs. It is possible to observe large bending stress states of Si NWs as their radius of curvature is progressively reduced. A slight increase in the fracture strength of Si NWs from 12 to 18 GPa as their radius decreased from 60 to 20 nm was observed in these measurements [54].

The fracture behavior of the Si NWs depends on both temperature and diameter. For NWs of diameter larger than 4 nm, cleavage fracture on the transverse (110) plane favors below the temperature 1000 K. As the temperature increases, the same NWs shear extensively on inclined {111} planes resulting in a brittle-to-ductile transition. As the diameter decreases to less than 4 nm, it fails by shear regardless of temperature. The reason is that the cleavage fracture is initiated by the nucleation of a crack, while shear failure is initiated by the nucleation of a dislocation, both from the surface [55].

Mechanical properties are extremely different at nanometer scale from those at macroscopic scale. Few crystalline nanowires have strength of over 10 GPa, however, Brambilla et al. [56]

have manufactured glass silica nanowires which are characterized by ultimate strength over 10 GPa with a top-down fabrication technique.

3.2. ZnO nanowires

The mechanical properties of ZnO are of considerable interest due to the potential applications in electromechanical devices. These mechanical properties are essential for designing, manufacturing, and operating such devices. ZnO NWs represent excellent model systems to investigate this size dependence, the ability to tune the radius over a continuous range and the manner in which their properties approach those of the bulk as a function of shape and size [57]. Young's modulus of ZnO NWs is essentially independent of diameter and close to the bulk value, whereas the ultimate strength increases for small diameter NWs with values up to 40 times that of bulk [58].

3.2.1. Size effect

A size dependence of Young's modulus in [0001] oriented ZnO NWs (NWs) with diameters ranging from 17 to 550 nm is revealed in Ref. [57]. The measured modulus for NWs with diameters smaller than about 120 nm is increasing dramatically with the decreasing diameters, and is significantly higher than that of the larger ones whose modulus tends to that of bulk ZnO. The radial elastic moduli of [0001]ZnO NWs are determined by contact resonance atomic force microscopy (CR-AFM). The radial indentation modulus and the tangential shear modulus from friction-type measurements are obtained by using realistic tip-nanowire contact geometry. Both moduli increase when the wire diameter is reduced below 80 nm. The size dependence of the elastic properties can be explained by a core-shell model that accounts for a bulk-like core and an elastically stiffer surface layer [59].

Secondly, friction type measurements are useful to directly probe the effects of surface stiffening on the shear response of the NW. Both the lateral shear modulus and radial indentation modulus increase significantly with decreasing NW diameter, which is due to a surface stiffening effect. The results could be understood by the core-shell model of the NW mentioned above, in which the elastic properties of the shell becomes predominant in the limit of small ZnO NW diameters [59].

4. Other novel structures of nanowires

4.1. Ultra thin nanowires

Computer simulations suggest that ultrathin metal wires should develop exotic, noncrystalline stable atomic structures once their diameter decreases below a critical size of the order of a few atomic spacings. The new structures, e.g. helical, spiral-structured wires, whose details depend upon the material and the wire thickness, may be dominated by icosahedral packings. The phenomenon, analogous to the appearance of icosahedral and other noncrystalline shapes in small clusters, can be rationalized in terms of surface energy anisotropy and optimal packing [60]. In some experiments, suspended Au NWs have recently been made in an ultrahigh vacuum and were imaged by the TEM [4]. Mechanisms of formation, evolution, and breaking of these atomically thin Au NWs under stress have attracted great interest.

In some MD simulations, a single chain forms and it is in agreement with experimental results. It shows how defects induce the formation of constrictions that eventually will form the one-atom chains [25]. While in the *Ab initio* calculations, the bond strength of the NW is about twice that of a bulk metallic bond. The importance of knowing bond and atom arrangement is that the total effective stiffness of the nanostructure is strongly affected by them [5].

4.2. Twinned nanowires

Simulation results reveal that with decreasing twin boundary spacing there is a transition from softening to strengthening due to a change in the dominant mechanism of plasticity deformation. Moreover, the value of critical twin boundary spacing is independent of the length of NW but depend on the diameter of the NW. The source controlled deformation mechanism in nanostructured materials will be helpful in understanding the plastic behavior of twinned NWs[61]. Some other simulations reveal the existence of a fundamental limit dividing the mechanical behavior of twinned Au NWs into either strain softening or strain hardening regimes, as a function of the NW diameter and the number of twins. In some other experiments, in-situ tensile testing of Ag NWs with diameters between 34 and 130 nm was carried out inside a SEM. Pronounced strain hardening was observed for most NWs. The strain hardening is mainly attributed to the presence of internal twin boundaries [62]. Since there are size-dependent mechanical properties in NWs, it is necessary for one to develop strategies to control the strength of these materials. The mechanical properties of Ag NWs with a unique fivefold twin structure using a lateral force AFM method were reported by Wu et.al [63]. It exhibits super elastic behavior followed by unexpected brittle failure without significant plastic deformation. However, thermal annealing resulted in a gradual transition to weaker, more ductile materials associated with the elimination of the twinned boundary structure [63].

5. Enhancement of nanowires in nano hybrid systems

The optimized Cu NWs are constrained within the hollow CNT as shown in Fig. 6. It shows the front and side views of the initial configuration of Cu NW encapsulated into CNT(6,6) and CNT(8,8). Bulk copper has a FCC structure. But at nanometer scale, randomly and fully dispersed Cu atoms form a helical or multi-shell equilibrium structure after relaxation to reach the minimum energy. This phenomenon is mainly due to the constraint effect of the template CNT and its special rolled up hexagonal network structure. Cu NW filled in CNT(6,6) (Fig.6(a)) is three-strand NW twisting with each other.

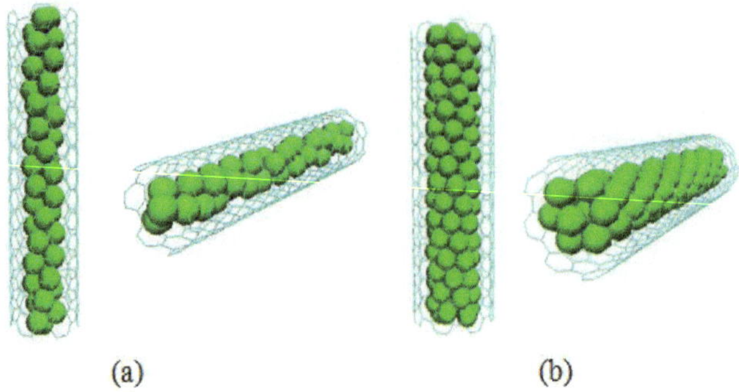

Figure 6. Front and side views of the initial configuration of Cu NW encapsulated in CNT. (a) NW@CNT(6,6), (b)NW@CNT(8,8).

In addition, another Cu NW encapsulated into CNT(8,8) has multi-shell packs that are composed of coaxial cylindrical shells (Fig.6(b)). In this case, Cu NW has a single strand chain at the center. Outside shell is formed by rows of atoms that are helically wound upwards side by side. A displacement increment of 0.0004 nm in axial direction is applied at both ends in each load step. The simulation is carried out at approximately 1 K to avoid thermal effects. Each time step used in the simulation is 1 fs. At the same time, the atoms at both ends of the composite are kept transparent to inter-atomic forces [64].

Figure 7 shows the function of the strain energy of Cu NW@CNT(16,0) with the length of 34.08Å as the time step increases. The inset is a compressive schematic, in which the widths of both fixed ends are 5Å. The constant velocity referred above is imposed on both fixed end atoms which consist of carbon and copper atoms. The strain energy per atom is determined as the difference in total energy per atom of the strained and unstrained NW@CNT. In addition, we use the strain energy of the pristine hollow CNT(16,0) under compressive load for comparison. The strain energy of CNT is increasing non-linearly without any drop before buckling. However, the strain energy of the NW@CNT undergoes a small drop atε=5.2% (Fig.8(b)) and ε=6.6% (Fig. 8(c)) because of slipping of copper atoms during the compressive process. As the strain reaches a critical point at ε=10.4% (t=3000fs), the composite undergoes abrupt buckling deformation resulting in an only 2.7% spontaneous drop which is lower than that of a pristine CNT (33.5%). This means that the composite is more stable than the corresponding pristine CNT when subject to abrupt buckling. Wang et al. [65] have investigated the stability of Ni metal atoms fully encapsulated in CNT. They found that the stability of metal filled in CNT depends much upon the amount of filling metal. When the metal atom is fully filled into CNT, it is more stable than a pristine CNT. This result is similar to some reported in our study. In our calculations, we propose that the mechanical properties of NW@CNTs depend upon the diameter, length, and carbon nanotube's chirality.

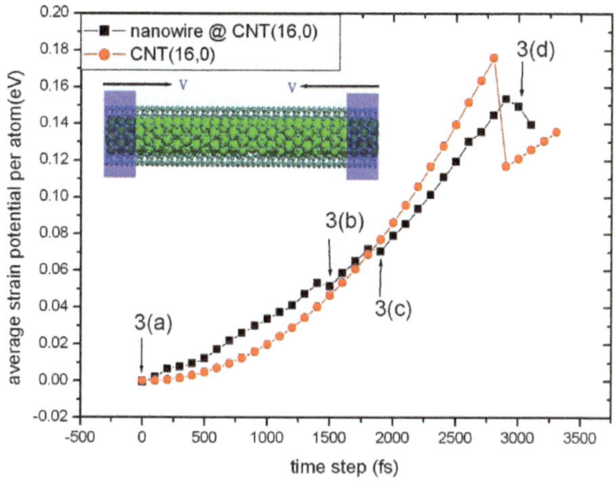

Figure 7. Average strain energy per atom for Cu NW@CNT(16,0) as a function of time step. The red line represents the strain energy of a CNT(16,0) with the same length for comparison.

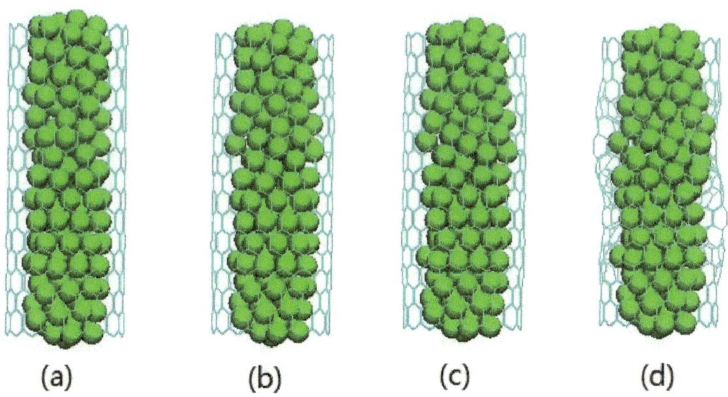

Figure 8. Morphological changes of a Cu NW@CNT(16,0) at different strains corresponding to Fig. 7. (a) Initial configuration. (b) At t=1500fs. (c) At=1900fs. (d) The buckling occurs at t=3000fs.

Table 1 illustrates the relationship between the diameter and buckling loadP_{cr} of a series of Cu NW encapsulated into armchair CNT and zigzag CNT. The lengths of the series of armchair CNTs are approximately 36.9Å, while those of zigzag CNTs are about 34Å.

NWs @armchair carbon nanotubes				
(m,n)	Diameter(Å)	Length/Diameter	Pcr(10^{-7}N)	Increase(%)
(6,6)	8.14	4.53	0.93	-23.7
(8,8)	10.85	3.40	1.28	2.4
(10,10)	13.56	2.72	1.44	14.2
(12,12)	16.27	2.27	1.50	21.9
(14,14)	18.98	1.94	1.65	35.1
(16,16)	21.70	1.70	1.81	45.9
(18,18)	24.41	1.51	2.03	59.8
(20,20)	27.12	1.33	2.29	81.7
NWs @zigzag carbon nanotubes				
(m,n)	Diameter(Å)	Length/Diameter	Pcr(10^{-7}N)	Increase(%)
(10,0)	7.83	4.35	1.12	-12.5
(12,0)	9.39	3.62	1.29	0.7
(14,0)	10.96	3.11	1.30	3.2
(16,0)	12.53	2.72	1.34	7.2
(18,0)	14.09	2.42	1.39	11.2
(20,0)	15.66	2.18	1.41	13.7
(22,0)	17.22	1.98	1.70	39.3
(24,0)	18.79	1.81	1.73	44.2

Table 1. buckling loads P_{cr} for selected Cu NW@armchair CNT and NW@zigzag CNT

When the diameter increases from 8.14 to 27.12 Å (i.e. aspect ratio decreases from 4.53 to 1.33), the buckling load P_{cr} of NW@armchair CNTs increases from 0.93 to 2.29 x10^{-7}N. Moreover, the critical buckling load P_{cr} of NW@zigzag CNTs increases from 1.12 to 1.73 x10^{-7} N as its diameter increases. In order to confirm how the insertion of Cu NW affects the pristine CNT, we calculate the buckling load P_{cr} of NW@CNT in comparison with that of the corresponding pristine CNT. The percentage increase is defined as [(critical buckling load P_{cr} of NW@CNT−P_{cr} of CNT)/P_{cr} of CNT]. As the diameter increases, the percentage increase of both NW@armchair CNTs and NW@zigzag CNTs rises in the same way. Especially for the NW@CNT (20,20) and NW@CNT(24,0), the percentage increase has reached 81.7% and 44.2%, respectively. So the insertion of Cu NW into a thicker CNT reinforces CNT more than the case with a thinner one. Another interesting point we should notice is that when the aspect ratio is 4.53 for NW@CNT(6,6), or 4.35 for NW@CNT(10,0), the percentage increase is minus.

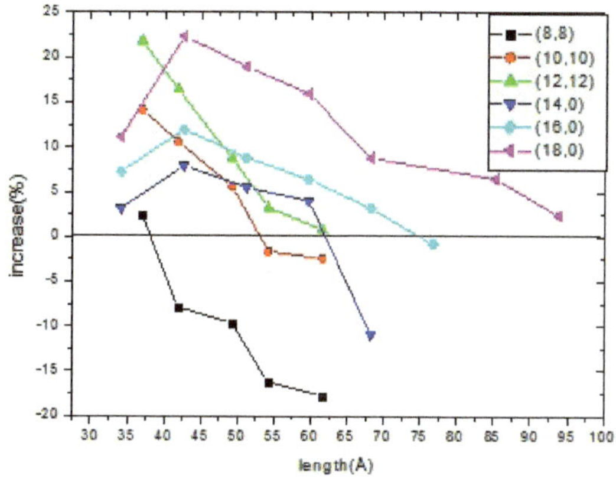

Figure 9. The percentage increase for the buckling loads P_{cr} of the composite material (NW@CNT) compared with the corresponding hollow nanotube.

To further investigate the relationship between length and reinforcing effect of NW@CNT on the critical load, we illustrate the percentage increase of NW@armchair and NW@zigzag CNT as a function of length in Fig. 9. For NW@armchair CNT, the percentage increase decreases as the length increases. There is an equilibrium value for each NW@CNT, after which the percentage increase will be minus. It means that if the length of the NW@CNT is larger than the certain value, the insertion of NW will weaken the strength of CNT. From another point of view, as the length of the composite increases, it can be viewed as one-dimension long column, which means that the stability of the solid column is not stronger than the hollow structure during compression. In continuum mechanics, the Euler formula is employed to determine the critical strain of the long beamlike buckling mode $\varepsilon_{cr}=\frac{\pi^2}{(KL\ /r)^2}$, where L is the length of the column; $r=\sqrt{I\ /A}$ is the gyration radius, where I and A are the moment of inertia and the area of cross section; and K is a constant in lieu of the effects of boundary conditions on the Euler formula. The greatest difference between CNT and NW@CNT is the area of cross section. According to the foregoing formula, the critical strain of a solid composite will be smaller than that of the hollow CNT. Furthermore, the buckling load of composite would be smaller than that of CNT.

In other words, when the length of the composite is long enough, Cu NW filled in the CNT buckles earlier than the pristine CNT at a certain strain. There will be some Cu atoms approaching the inside wall of the CNT, resulting in strong interactions along the plane perpendicular to the loading direction. As the compressive strain continues, along with the effect of van der Waals forces included, the interactions between the two different kinds of atoms will initiate the buckling of the whole composite earlier.

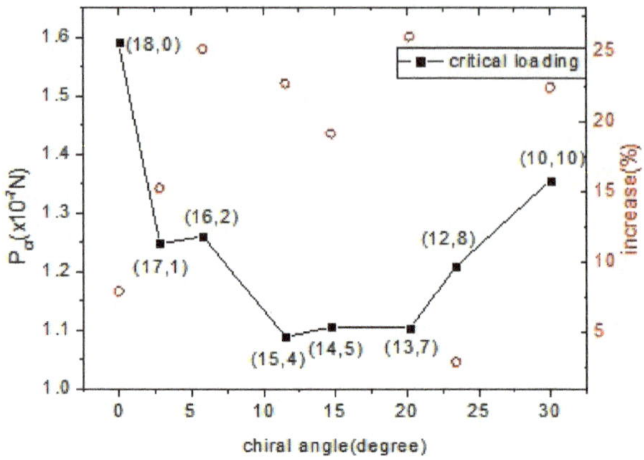

Figure 10. Critical buckling loads (black squares) and the quantity percentage increase of the composite (red circles) with approximately the same diameter and length as function of tube chirality for CNT encapsulating NWs.

Figure 10 shows both the critical loads P_{cr} and the percentage increase as functions of the nanotube's chiral angle. The chirality dependent compressive response of these NWs@CNTs with nearly the same aspect ratio (length=42 Å) is observed. The critical load decreases step-like as the chiral angle increases to about 15° and then increases to $1.36x10^{-7}N$ when the chiral angle reaches 30 degrees (armchair CNT). The critical load of zigzag CNT encapsulating NWs is the largest of all these types of composites, because there are substantial bonds which are parallel to the loading direction and they can withstand larger compressive loading before buckling. Most chiral CNTs encapsulating NWs are more sensitive than NWs@zigzag CNT in terms of percentage increase. It is seen that copper NWs can increase the critical loading of CNT (13, 7) by about 25%. The insertion of copper NWs into other CNTs with different chiral angles can also enhance the strength of the corresponding pure CNTs. So this enhancement of CNT with different chiral angles makes this kind of composite a promising material to act as building blocks.

In order to further determine the relationship between buckling load P_{cr} and composite's length, the relationship between them is plotted in Fig. 11. Euler formula for the critical buckling load is $P_{cr}=4\pi^2EI/L^2$, where L is the length. As for NW@armchair CNT with a length ranging from 35 to 65 Å (shown in Fig. 11a), the buckling loads P_{cr} for NW@CNT (8,8) and NW@CNT(10,10) fit nearly linearly with the value $1/L^2$. In addition, the similar trend is also found for NW@CNT (14,0) and NW@CNT(16,0) as shown in Fig. 11 (b).The axial Young's modulus can be defined from the continuum theory in the form of $Y=\frac{d^2E}{Vd\varepsilon^2}$, in which E is the strain energy, ε is the axial strain and V is the overall volume of the whole composite.

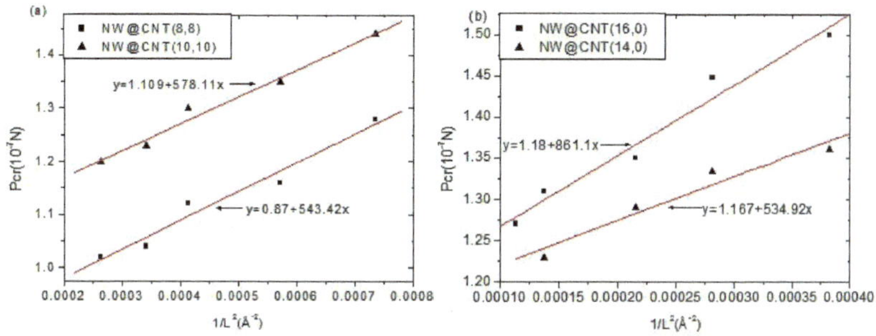

Figure 11. a) Buckling loads P_{cr} of Cu NW@CNT (8,8) and NW@CNT(10,10) verse $1/L^2$ (where L represents the length of NW@CNT), (b) Pcr of Cu NW@CNT(14,0) and NW@CNT(16,0) versus $1/L^2$.

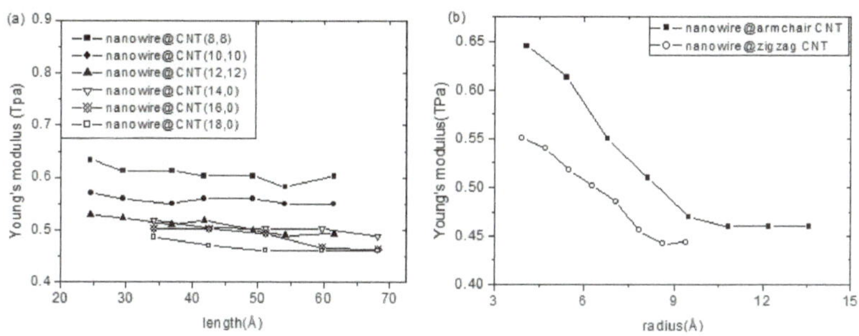

Figure 12. a) Plots depicting the effect of NW@CNT length on the Young's modulus. (b) Plots depicting the effect of NW@CNT radius on the Young's modulus.

As plotted in Fig. 12(a), an increment in length decreases the Young's modulus slightly. A decline of only 4.7% is observed in the decrement of Young's modulus of NW@CNT (8,8) as its length increases from 25 to 65 Å. And there is a similarly small decline of 5.3% in the decrement of Young's modulus of NW@CNT (14, 0) as its length increases from 35 to 70 Å. The general trend is that the Young's modulus decreases as the length of NW@CNT increases. Although there are some points that do not obey this rule, it is mainly due to the different distribution of Cu atoms filled in CNTs. As shown in Fig. 12(b), Young's modulus is sensitive to the radius of NW@CNT ranging from 8 to 14 Å. For example, it drops about 28.7% (i.e. from 0.645 to 0.46Tpa) when radius of the NW@armchair CNT increases from 4.07 to 10.85 Å. In addition, there is a 19.8% drop (from 0.551 to 0.442 TPa) for Young's modulus of NW@zigzag CNT. When the radius is sufficiently large, Young's modulus would reach a constant value 0.46TPa for NW@armchair CNT and 0.44 TPa for NW@zigzag CNT.

6. Conclusions

This paper reviews some recent development on the mechanical properties and mainly focuses on the elongation of a metallic nanowire (NW) with a core/shell structure and the enhancement of helical NWs in a hybrid system.

1. The tensile ductility of the core/shell metallic CuZr NW could be enhanced with the surface amorphous thickness increasing, whereas the breaking strength shows a downtrend instead. During elongation, the NW exhibits three successive deformation stages in crystalline core: lattice distortion, martensitic phase transformation, and local amorphization in the high strained region. Moreover, detailed analysis demonstrates that numerous 1541 bond pairs with defective icosahedral ordering form at first on the weak crystalline-amorphous interfaces. After that, part of them converts into full icosahedral ordering which has the higher resistance to plastic deformation.

2. Dispersed Cu atoms filled in CNT form the novel helical structure after the "simulated annealing" method is carried out. The mechanical properties of this novel composite reveals better than a pristine CNT. The NW@CNTs can withstand larger buckling load than the corresponding CNTs. However, the prerequisite is that the length is smaller than a certain value. On the contrary, when the length is lager than the certain value, the composite could be viewed as a long one dimensional stick; the stability and buckling load would be smaller than that of the corresponding CNT. The critical buckling load also fits linearly well with Euler formula when the aspect ratio of the composite is larger than the certain value. The Young's moduli of NW@CNTs with different diameters have been studied. We find Young's modulus of NW@CNT is more sensitive to its radius than its length. It declines to a constant value as the radius increases.

Acknowledgements

We would like to acknowledge the support from the National Basic Research Program of China (Grant No.2012CB825702). This work is also supported by the National Natural Science Foundation of China (Grant No. 50971081, 51271100). This work is also funded by the grants from Shandong Province in China (Nos. 50625101, ZR2009FM043). We also thank the support from Shandong University (No2009JQ014).

Author details

Hui Li[1*] and Fengwei Sun[2]

*Address all correspondence to: lihuilmy@hotmail.com

1 Key Laboratory for Liquid-Solid Structural Evolution and Processing of Materials, Ministry of Education, Shandong University, China

2 Department of Materials Science and Engineering, Delft University of Technology, The Netherlands

References

[1] Pascual, J. I., Mendez, J., Gomezherrero, J., Baro, A. M., Garcia, N., & Binh, V. T. (1993). Quantum Contact in Gold Nanostructures by Scanning-tunneling-microscopy. *Physical Review Letters*, 71(12), 1852-5.

[2] Agrait, N., Rodrigo, J. G., & Vieira, S. (1993). Conductance Steps and Quantization in Atomic-size Contacts. *Physical Review B*, 47(18), 12345-8.

[3] Olesen, L., Laegsgaard, E., Stensgaard, I., Besenbacher, F., Schiotz, J., Stoltze, P., et al. (1994). Quantized Conductance in an Atom-sized Point-contact. *Physical Review Letters*, 72(14), 2251-4.

[4] Kondo, Y., & Takayanagi, K. (1997). Gold Nanobridge Stabilized by Surface Structure. *Physical Review Letters*, 79, 3455-8.

[5] Rubio-Bollinger, G., Bahn, S., Agraït, N., Jacobsen, K., & Vieira, S. (2001). Mechanical Properties and Formation Mechanisms of a Wire of Single Gold Atoms. *Physical Review Letters*, 87(2), 026101.

[6] Wu, B., Heidelberg, A., & Boland, J. J. (2005). Mechanical properties of ultrahigh-strength gold nanowires. *Nature Materials*, 4(7), 525-9.

[7] Hoffmann, S., Utke, l., Moser, B., Michler, J., Christiansen, S. H., Schmidt, V., et al. (2006). Measurement of the Bending Strength of Vapor-Liquid-Solid Grown Silicon Nanowires. *Nano Letters*, 6, 622-5.

[8] Han, X. D., Zhang, Y. F., Zheng, K., Zhang, X. N., Zhang, Z., Hao, Y. J., et al. (2007). Low-Temperature in Situ Large Strain Plasticity of Ceramic SiC Nanowires and Its Atomic-Scale Mechanism. *Nano Letters*, 7, 452-7.

[9] Chantasiriwan, S., & Milstein, F. (1998). Embedded-atom models of 12 cubic metals incorporating second- and third-order elastic-moduli data. *Physical Review B*, 58(10), 5996-6005.

[10] Branício, P. S., & Rino, P. J. (2000). Large deformation and amorphization of Ni nanowires under uniaxial strain:A molecular dynamics study. *Physical Review B*, 62, 16950-5.

[11] Liang, W., & Zhou, M. (2003). Size and Strain Rate Effects in Tensile Deformation of Cu Nanowires. *Nanotechnology*, 2, 452-5.

[12] Liang, H., Upmanyu, M., & Huang, H. (2005). Size-dependent elasticity of nanowires: Nonlinear effects. *Physical Review B R*, 71, 24, 241403(R).

[13] Lin, S. J., Ju, P. S., & Lee, J. W. (2005). Mechanical behavior of gold nanowires with a multishell helical structure. *Physical Review B*, 72(8), 085448.

[14] Park, H., Gall, K., & Zimmerman, J. (2005). Shape Memory and Pseudoelasticity in Metal Nanowires. *Physical Review Letters*, 95(25), 255504.

[15] Park, H., & Zimmerman, J. (2005). Modeling inelasticity and failure in gold nanowires. *Physical Review B*, 72(5), 054106.

[16] Diao, J., Gall, K., Dunn, M. L., & Zimmerman, J. A. (2006). Atomistic simulations of the yielding of gold nanowires. *Acta Materialia*, 54(3), 643-53.

[17] Ji, C., & Park, H. S. (2006). Geometric effects on the inelastic deformation of metal nanowires. *Applied Physics Letters*, 89(18), 181916.

[18] Baskes, M. I., Angelo, J. E., & Bisson, C. L. (1994). Atomistic Calculations of Composite Interfaces. *Modelling and Simulation in Materials Science and Engineering*, 2(3A), 505-18.

[19] Baskes, M. I. (1992). Modified Embedded-atom Potentials for Cubic Materials and Impurities. *Physical Review B*, 46(5), 2727-42.

[20] Diao, J., Gall, K., & Dunn, M. L. (2003). Surface-stress-induced phase transformation in metal nanowires. *Nature Materials*, 2(10), 656-60.

[21] Stillinger, F. H., & Weber, T. A. (1985). Computer-simulation of Local Order in Condensed Phases of Silicon. *Physical Review B*, 31(8), 5262-71.

[22] Wang, Z. G., Li, J. B., Gao, F., & Weber, W. J. (2010). Tensile and compressive mechanical behavior of twinned silicon carbide nanowires. *Acta Materialia*, 58(6), 1963-71.

[23] Makeev, M., Srivastava, D., & Menon, M. (2006). Silicon carbide nanowires under external loads: An atomistic simulation study. *Physical Review B*, 74(16), 165303.

[24] Mehl, M. J., & Papaconstantopoulos, D. A. (1996). Applications of a tight-binding total-energy method for transition and noble metals: Elastic constants, vacancies, and surfaces of monatomic metals. *Physical Review B*, 54(7), 4519-30.

[25] Da Silva, E., da Silva, A., & Fazzio, A. (2001). How Do Gold Nanowires Break? *Physical Review Letters*, 87(25), 256102.

[26] Da Silva, E., Novaes, F., da Silva, A., & Fazzio, A. (2004). Theoretical study of the formation, evolution, and breaking of gold nanowires. *Physical Review B*, 69(11), 115411.

[27] Park, H. S., Gall, K., & Zimmerman, J. A. (2006). Deformation of FCC nanowires by twinning and slip. *Journal of the Mechanics and Physics of Solids*, 54(9), 1862-81.

[28] McDowell, M. T., Leach, A. M., & Gall, K. (2008). On The Elastic Modulus of Metallic Nanowires. *Nano Letters*, 8, 3613-8.

[29] McDowell, M. T., Leach, A. M., & Gall, K. (2008). Bending and tensile deformation of metallic nanowires. *Modelling and Simulation in Materials Science and Engineering*, 16(4), 045003.

[30] Yun, G., & Park, H. (2009). Surface stress effects on the bending properties of fcc metal nanowires. *Physical Review B*, 79(19), 195421.

[31] Park, H., & Klein, P. (2008). Surface stress effects on the resonant properties of metal nanowires: The importance of finite deformation kinematics and the impact of the residual surface stress. *Journal of the Mechanics and Physics of Solids*, 56(11), 3144-66.

[32] Petrova, H., Perez-Juste, J., Zhang, Z. Y., Zhang, J., Kosel, T., & Hartland, G. V. (2006). Crystal structure dependence of the elastic constants of gold nanorods. *J Mater Chem*, 16(40), 3957-63.

[33] Cuenot, S., Fretigny, C., Demoustier-Champagne, S., & Nysten, B. (2004). Surface tension effect on the mechanical properties of nanomaterials measured by atomic force microscopy. *Physical Review B*, 69(16), 165410.

[34] Jing, G. Y., Duan, H. L., Sun, X. M., Zhang, Z. S., Xu, J., Li, Y. D., et al. (2006). Surface effects on elastic properties of silver nanowires: Contact atomic-force microscopy. *Physical Review B*, 73(23), 235409.

[35] Husain, A., Hone, J., Postma, H. W. C., Huang, X. M. H., Drake, T., Barbic, M., et al. (2003). Nanowire-based very-high-frequency electromechanical resonator. *Applied Physics Letters*, 83(6), 1240-2.

[36] Verbridge, S. S., Shapiro, D. F., Craighead, H. G., & Parpia, J. M. (2007). Macroscopic tuning of nanomechanics: Substrate bending for reversible control of frequency and quality factor of nanostring resonators. *Nano Letters*, 7(6), 1728-35.

[37] Park, H. S. (2006). Stress-Induced Martensitic Phase Transformation in Intermetallic Nickel Aluminum Nanowires. *Nano Letters*, 6(5), 958-62.

[38] Amorim, E., & da, Silva. E. (2008). Helical [110] Gold Nanowires Make Longer Linear Atomic Chains. *Physical Review Letters*, 101(12), 125502.

[39] Zhang, K., Si, P. C., Li, H., Li, Y. F., Yu, H. Q., & Jiang, Y. Y. (2012). Elongation behavior and local amorphization of metallic nanowire with glassy shell and crystalline core. *EPL (Europhysics Letters)*, 97(2), 26005.

[40] Berendsen, H. J. C., Postma, J. P. M., van Gunsteren, W. F., Di Nola, A., & Haak, J. R. (1984). Molecular dynamics with coupling to an external bath. *Journal of Chemical Physics*, 81, 3684.

[41] Allen, M. P., & Tildesley, L. J. (1987). Computer Simulation of Liquids. *(Oxford University Press, Oxford).*

[42] Wei, Y., Bower, A. F., & Gao, H. (2010). Analytical model and molecular dynamics simulations of the size dependence of flow stress in amorphous intermetallic nanowires at temperatures near the glass transition. *Physical Review B*, 81(12), 125402.

[43] Koh, S. J. A., Lee, H. P., Lu, C., & Cheng, Q. H. (2005). Molecular dynamics simulation of a solid platinum nanowire under uniaxial tensile strain: Temperature and strain-rate effects. *Physical Review B*, 72(8), 085414.

[44] Zhang, K., Si, P. C., Li, H., Li, Y. F., Jiang, Y. Y., Zhang, S. L., et al. (2012). Plastic heterogeneity in nanoscale metallic glass. *Physica E: Low-dimensional Systems and Nanostructures*, 44(7-8), 1461-6.

[45] Ikeda, H., Qi, Y., C,, agin. T., Samwer, K., Johnson, W. L., & Goddard, I. I. I. W. A. (1999). Strain Rate Induced Amorphization in Metallic Nanowires. *Physical Review Letters*, 82, 2900-3.

[46] Koh, S., Lee, H., Lu, C., & Cheng, Q. (2005). Molecular dynamics simulation of a solid platinum nanowire under uniaxial tensile strain: Temperature and strain-rate effects. *Physical Review B*, 72(8), 085414.

[47] Seo, H. J., Yoo, Y., Park, Y. N., Yoon, W. S., Lee, H., Han, S., et al. (2011). Superplastic Deformation of Defect-Free Au Nanowires via Coherent Twin Propagation. *Nano Letters*, 11(8), 3499-502.

[48] Moore, N. W., Luo, J., Huang, J. Y., Mao, S. X., & Houston, J. E. (2009). Superplastic Nanowires Pulled from the Surface of Common Salt. *Nano Letters*, 9, 2295-9.

[49] Lu, Y., Huang, J. Y., Wang, C., Sun, S., & Lou, J. (2010). Cold welding of ultrathin gold nanowires. *Nature Nanotechnology*, 5, 218-24.

[50] Dai, S., Zhao, J., Xie, L., Cai, Y., Wang, N., & Zhu, J. (2012). Electron-Beam-Induced Elastic-Plastic Transition in Si Nanowires. *Nano Letters*, 12(5), 2379-85.

[51] Hsin, L. C., Mai, W., Gu, Y., Gao, Y., Huang, T. C., Liu, Y., et al. (2008). Elastic Properties and Buckling of Silicon Nanowires. *Advanced Materials*, 20(20), 3919-23.

[52] Han, X. D., Zheng, K., Zhang, Y. F., Zhang, X. N., Zhang, Z., & Wang, Z. L. (2007). Low-Temperature In Situ Large-Strain Plasticity of Silicon Nanowires. *Advanced Materials*, 19(16), 2112-8.

[53] Zheng, K., Han, X., Wang, L., Zhang, Y., Yue, Y., Qin, Y., et al. (2009). Atomic Mechanisms Governing the Elastic Limit and the Incipient Plasticity of Bending Si Nanowires. *Nano Letters*, 9, 2471-6.

[54] Stan, G., Krylyuk, S., Davydov, A. V., Levin, I., & Cook, R. F. (2012). Ultimate Bending Strength of Si Nanowires. *Nano Letters*, 12(5), 2599-604.

[55] Kang, K., & Cai, W. (2010). Size and temperature effects on the fracture mechanisms of silicon nanowires: Molecular dynamics simulations. *International Journal of Plasticity*, 26(9), 1387-401.

[56] Brambilla, G., & Payne, D. N. (2009). The Ultimate Strength of Glass Silica Nanowires. *Nano Letters*, 9, 831-5.

[57] Chen, C., Shi, Y., Zhang, Y., Zhu, J., & Yan, Y. (2006). Size Dependence of Young's Modulus in ZnO Nanowires. *Physical Review Letters*, 96(7), 075505.

[58] Wen, B., Sader, J., & Boland, J. (2008). Mechanical Properties of ZnO Nanowires. *Physical Review Letters*, 101(17), 175502.

[59] Stan, G., Ciobanu, C. V., Parthangal, P. M., & Cook, R. F. (2007). Diameter-Dependent Radial and Tangential Elastic Moduli of ZnO Nanowires. *Nano Letters*, 7, 3691-7.

[60] Gülseren, O., Ercolessi, F., & Tosatti, E. (1998). Noncrystalline Structures of Ultrathin Unsupported Nanowires. *Physical Review Letters*, 80, 3775-8.

[61] Guo, X., & Xia, Y. (2011). Repulsive force vs. source number: Competing mechanisms in the yield of twinned gold nanowires of finite length. *Acta Materialia*, 59(6), 2350-7.

[62] Zhu, Y., Qin, Q., Xu, F., Fan, F., Ding, Y., Zhang, T., et al. (2012). Size effects on elasticity, yielding, and fracture of silver nanowires: In situ experiments. *Physical Review B*, 85(4), 045443.

[63] Wu, B., Heidelberg, A., Boland, J. J., Sader, J. E., Sun, X., & Li, Y. (2006). Microstructure-Hardened Silver Nanowires. *Nano Letters*, 6, 468-72.

[64] Sun, F. W., Li, H., & Liew, K. M. (2010). Compressive mechanical properties of carbon nanotubes encapsulating helical copper nanowires. *Carbon*, 48(5), 1586-91.

[65] Wang, L., Zhang, H. W., Zhang, Z. Q., Zheng, Y. G., & Wang, J. B. (2007). Buckling behaviors of single-walled carbon nanotubes filled with metal atoms. *Applied Physics Letters*, 91(5), 051122.

Chapter 17

How and Why Alumina Matrix Architecture Influence the Shape and Composition of Nanowires Grown by AC Deposition?

Arūnas Jagminas

Additional information is available at the end of the chapter

1. Introduction

According to description, nanowires are one-dimensional materials with size ≤100 nm in two out of directions from which various architectures can be organized for recent devices offering new and sometimes unique oppurtunities. Among numerous methods can be applied to date for densely packed nanowire (*nw*) arrays formation the template-assisted electrochemical deposition is attributed to the most widespread allowing simple control on the geometrical, morphological and crystallographic properties of various nanowire arrays in an independent manner. Note that in case of formation extremely thin and densely packed nanowires, demonstrating a significant improvement of their thermoelectric, photovoltaic, catalytic and optical properties, due to huge surface size and quantum-size effects, this possibility becomes crucial (Bejenari et al.2011). Besides uniformity in wire diameter, morphology and composition, the crystallinity and crystallographic orientation also strongly influence the properties of metallic and semiconductor nanowires and their arrays (Lan et al.2009; Yan et al. 2010). However, most works to date have reported the growth of metallic and semiconductor nanowire arrays inside the alumina and polycarbonate (PC) templates with pores larger than 40 nm, especially in case of polymeric templates, and only few address the impact of pore diameter and deposition regime on the peculiarities of extremely thin nanowires growth. For example, variables of the morphology, surface roughness and crystalline orientation of the Bi_2Te_3 nanowires with the PC membrane channel diameter decrease from 150 nm down to ~10 nm have been right now reported (Pitcht et al. 2012) demonstrating the possibility of obtaining the nanowire arrays with preferential growth of either {205}, {015}, or {110} planes perpendicular to the nanowire axis from a same composition of deposition solution enabling us to tune their figure of merit and to improve the device performance.

In this chapter, the influence of alumina template barrier layer thickness on the electrochemical growth of copper and cobalt nanowires is demonstrated. Our results obtained investigating the peculiarities of bismuth selenide electrodeposition by alternating current (*AC*) treatment in alumina templates varied in pore diameter within 10 to 100 nm range are presented in subsection 5.1 showing, for the first time, a strong dependency of formed *nws* composition, morphology and their optical properties on the diameter of pores (Ø_{pore}).

2. Filling of porous alumina templates

Porous oxide films (alumina) prepared via anodization of high pure and smooth aluminum surface in an aqueous solution of sulfuric, oxalic, and o-phosphoric acid at proper concentration, temperature, and voltage are typical templates for fabrication arrays of various nanowires in quite uniform diameter and spacing with well-defined product dimensions at packing density of 10^9 - 10^{11} species/cm^2 (Masuda & Fukuda 1995; Jassensky et al. 1998; Li et al. 1998; Nielsch et al. 2002). To date high-ordered structure alumina with broad range of pore diameters as large as 300 nm (Quin et al. 2005) and as small as 5 nm (AlMawlawi et al. 1991) can be fabricated in unlimited size area. The pore diameter, cell size, and barrier-layer thickness positioned at the bottom of alumina pores (Fig. 1) are all linearly dependent on the anodizing voltage (O'Sullivan & Wood 1970), while the depth of pores can simple be controlled by anodizing time (Metzger et al. 2000). According to the final applications, the thickness of alumina templates is usually limited to 20-30 μm, but thicker templates (Römer & Steinem 2004), as well as very thin (Kokonou et al. 2007), are sometimes required for uniform filling with various *nw* arrays. Till now, direct current (*DC*), alternating current (*AC*) and pulsed current depositions have been developed for filling of alumina pores by various materials. For *DC* depositions, enabling substantial control over composition and crystallinity of *nws*, the detachment of alumina film from the substrate, opening the pore bottoms via voltage decrease and chemical etching and conductive layer formation are usually required. To remove the barrier-layer only at the bottom of alumina nanochannels or perforate it, various etching solutions (Zheng et al. 2002) and different methods (Mardilovich et al. 1995) have been proposed during last decade. Also, the sputtering of gold (Yoo & Lee 2004), silver (Sun et al. 2005) and platinum (Panet al. 2005) layer onto the back-side of perforated template as a conducting and well-adhesive layer have been applied. On the other hand, there always exists the possibility of filling alumina pores by *AC* modes, leaving intact the barrier-layer at the bottom of pores because anodic aluminum oxides conduct preferentially in cathodic direction. According to some opinions (Goad & Moskovits 1978; Clebny et al. 1993), *AC* electrolysis is an ideal method for deposition of metals and semiconductors, starting from the bottom of pores. Moreover, *AC* depositions through the rectifying barrier-layer require fewer processing steps and are more amenable to scall-up but currently provide far less control over the structure and the material deposited (Gerein & Haler 2005). As a result, different groups using this method (Preston & Moskovits 1993; Sheasby & Cook 1974) have observed interrupted growth of various polycrystalline materials and just partial depositions, namely only in a fraction of pores. Note that for alumina templates formed at higher anodizing voltages and consequently having a thicker barrier-layer, some degree of this layer thinning is essential to enable deposition even under *AC* treatment. Furthermore, the chemical composition of alumina films differs depending on the composition of anodizing solution as well as anodizing conditions due to incorporation of acid anions and water molecules into the outer part of alumina cells. For example, the sulfuric acid alumina films contained high amount of anion species (12-14 wt.% sulfate), while the phosphate and oxalate contents in corresponding alumina films are respectively 6-8 and 2-4 wt.% (Thompson 1997). Note that incorporated anion species produce a negative surface charge of the pore walls influencing the pore filling process by one or another material. Therefore, the hydrophobic/hydrophilic pore wall properties could play a significant role for nanowires growth in the precursor solution.

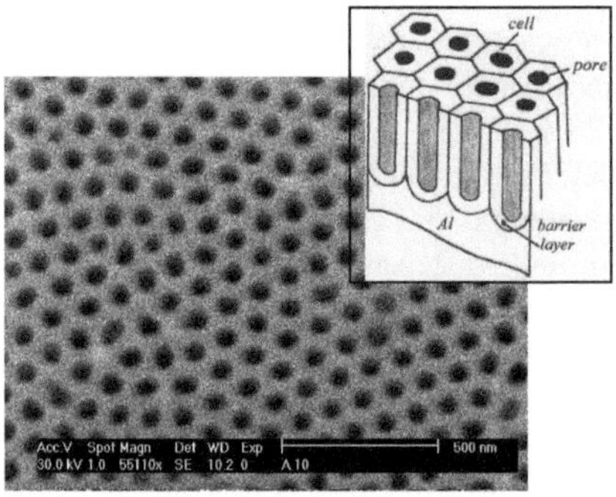

Figure 1. Typical top-side SEM view and a schematic diagram of porous alumina template.

3. The influence of porous alumina barrier-layer

Potentially nanowires in the alumina template pores can be synthesized by adsorbing and decomposing precursor species, high-pressure injection of a melt, electro less and electro deposition routes. However, the dominant synthesis technique in this area remains*AC* deposition of metals and semiconductors copying exactly the pores configuration. However, the optimal *AC* electrolysis conditions differ for various solutions and various templates whereas under others the following phenomena as the alumina barrier-layer spalling (Sheasby & Cooke 1974), pitting corrosion (Routkevich et al. 1996) and the template peeling off from the substrate (Doughty et al. 1975) take place. Besides, the filling of alumina templates grown at higher voltages in the phosphoric or oxalic acid baths is more problematic, since at higher *AC* voltages, required in this case for metal ions discharge, the alumina barrier-layer breakdown is inevitable. A few investigations devoted to uniform growth of metallic nanowires by means of pulse or *AC* electrolysis deal in this case with requirement to decrease the thickness of the alumina barrier-layer (*d*), (Forrer et al. 2000; Xu et al. 2002; Sauer et al. 2002). However, there is much uncertainty about the optimal *d* for uniform filling of alumina pores with metal nanowires. It is still far from clear which barrier-layer thickness and *AC* voltage value will be optimal for the uniform filling of one or another alumina template for given material. Reported in Paulus et al. study (2001) *d* values for deposition of Fe, Ni, and Co nanowires range from 10 to 20 nm. According to recent study (Sausa et al. 2011), the optimal barrier layer thickness of alumina templates for homogenous and complete filling of all pores with Ni by *AC* treatment approximates 10 nm. It is likely, however, that highly uniform deposition of copper into oxalic acid grown pores by *AC* treatment is significantly more challenging than deposition into sulfuric acid grown template pores because of the different chemistry and structure of their barrier-layers (Gerein & Haber 2005). Therefore, the influence of *d*is still an open issue since the composition and properties of various alumina barrier-layers are complex and ill defined, especially after thinning through voltage decrease and chemical etching.

To shed light on this problem, in this study we focused on the use of the dependency the plots of the quantity of copper deposited within the template pores on the alumina growth and subsequent *AC* deposition conditions expressed as *AC* electrolysis and the template forming voltages ratio.

The amount of copper was determined after the complete dissolution of Cu^0 nanowires from a 4.5 cm^2 template surface in HNO_3: H_2O (1:2) solution (2 cm^3) for about 3 min. All solutions used for nanowires dissolution, sample rinsing and further double soaking in triply distilled water for 3 min were mixed together, diluted to constant volume and analyzed quantitatively using a Perkin Elmer Lambda 35 UV/Vis spectrometer. To increase absorbance detection sensitivity of copper analyte at 450 nm, 1 % sodium diethyldithiocarbamate (2.0 cm^3) was used as a complexing agent. Reproducibility of the analysis was checked by 3 repeated experiments. Standard solutions were made from 99.999 % grade copper.

The electrochemical impedance spectroscopy (EIS) was utilized to characterize the alumina barrier-layer properties upon the anodizing bath voltage decrease. The EIS spectra were recorded using a VoltaLab 80 (Radiometer Analytical, France) electrochemical system for frequencies between 1 and 10^5 Hz with ten measurements and are presented in the complex capacitance form, that is, the plots of ImY/ω versus Re Y/ω, where Y is the admittance and ω is the angular frequency [$\omega = 2\pi f$, and f is the frequency in hertz (Hz)]. These plots allow simple models of the interface to be readily inferred when the electrochemical response exhibits capacitive behaviour. The amplitude of the applied *AC* signal was set to 10 mV. The spectral data were analyzed/fitted with Z View software (Scribner Associates, South Pines, NC, USA). High-frequency stray capacitance cross-talk was eliminated from the EIS spectra as described earlier (Vanderah et al. 2003).

3.1. Peculiarities of the alumina barrier-layer thinning

It is well-documented that the thickness of the barrier-layer of alumina templates is proportional to the voltage applied in the anodizing process (Diggle et al. 1969). Going through the voltage drop, at the end of anodizing, one could expect the decrease in the barrier-layer thickness due to the field-assisted ejection of Al^{+3} from the oxide lattice to the solution (Harkness & Young 1966). However, this process can proceed only in some pores if a large voltage decrement is applied in a single step (Furneaux et al. 1989) causing the increased heterogeneity in the physical properties of the films. Therefore, one might expect the concomitant changes in the EIS spectra reflecting these phenomena.

Figure 2 displays a set of the EIS spectra obtained for the alumina films of different anodization end-voltage values. As seen, all the spectra possess a typical semicircular shape, which suggest simple capacitive behavior (Vanderah et al. 2003). This indicates the dielectric film that can be modeled as an equivalent circuit of a series of R_sC elements, where R_s is the solution resistance and C is the capacitance of the alumina barrier-layer. Though being similar in shape the spectra in Fig.2 show several trends and differences. First, as the final anodizing voltage decreases the diameter of the semi-circles increases, signaling about the increase in capacitance of the barrier layer. Second, the plots of ImY/ω versus Re Y/ω of the films obtained using high end-voltage values (10 - 40 V) tend to approach the limit:$\lim_{\omega \to 0} \mathrm{Re} Y/\omega \to 0$ at the low frequency edge of the EIS spectra indicating nearly perfect capacitive behavior. While, the films obtained using low final anodizing voltages do not. This observation suggests that the low end-voltage films exhibit significant micro-heterogeneity of the physical properties, while the opposite is true for the high end-voltage films. Taking into account these observations, for fitting the experimental EIS spectra data to the R_sC model, we replaced the capacitance with the constant phase element (*CPE*) as Macdonald (1987) suggested. The *CPE* reflects the deviation of alumina impedance from the ideal behavior. In the presence of a *CPE*, the film impedance exhibits a law frequency dependence: $Z_{film} = (1/C')^{\alpha} (i\omega)^{-\alpha}$, where C' is the constant *CPE*'s coefficient, α is the *CPE*'s exponent and i is the imaginary unit. The results of the fitting to this model are summarized in Table 1.As seen from Table 1, the numerical values of C' and α are consistent with the qualitative features of the Re Y/ω vs. ImY/ω spectra, i.e., the gradual lowering of the anodizing end-voltage yields to the increase in the *CPE*'s coefficient and concomitant decrease in the CPE's exponent value.

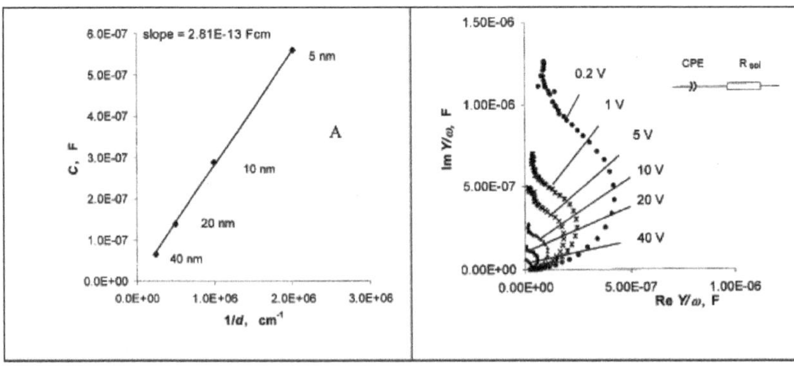

Figure 2. A) *C versus* 1/*d* plots of alumina in the range of the final anodizing voltages from 40 to 5 V calculated according to the impedance spectra of alumina layers grown in 0.2 mol/L $H_2C_2O_4$ at 40 V for 70 min with the subsequent lowering of the anodizing voltage in a step-by-step manner to $U_{a,fin}$ (V): (1) 40 (no lowering), (2) 20, (3) 10, (4) 5, (5) 1, (6) 0.2 V, shown as ImY/ω vs. Re Y/ω plots. Inset shows the elements of the equivalent model circuit.

Final anodizing voltage, V (approximate thickness of the barrier in nm)	***C′, μF***	***α***
40 (40)	0.065 ± 0.001	± 0.002
20 (20)	0.139 ± 0.002	± 0.002
10 (10)	0.288 ± 0.004	± 0.003
5 (5)	0.560 ± 0.006	± 0.004
1 (1)	0.762 ± 0.010	± 0.005
0.2 (0.2)	1.440 ± 0.032	0.943 ± 0.005

Table 1. EIS fitting data for alumina films grown in 0.2 mol/L oxalic acid bath at 18 °C and different end-voltage values. The electrode surface geometric area is 0.33 cm^2. The standard error is indicated next to the values in columns.

It is well documented, that aluminum anodizing voltage and the thickness of alumina barrier-layer are related through the equation $\delta_b = \kappa U_{a,fin}$ where κ is the "anodizing ratio" coefficient close to 1 nm V^{-1}(Diggle et al. 1969). The estimated thickness values are tabulated in Table 1, column 1(numbers in the brackets). On the other hand, the capacitance and thickness of the barrier-layer are related through the following equation:

$$C = \varepsilon_0 \varepsilon \frac{A}{d} \beta' \beta'' = \varepsilon_0 \varepsilon \frac{A}{d} \beta \tag{1}$$

where: ε_0 is the vacuum permitivitty, 8.85 10^{-14} F/cm, ε is the dielectric constant of alumina, hereinafter, the value 9.8 is accepted (Harkness & Young 1966), A is the surface area of the electrode, 0.33 cm^2, d – is the thickness of the barrier-layer and β' is the factor that accounts for the fraction of the surface occupied by the alumina pores ($0 < \beta' < 1$), and β'' is the roughness factor of the aluminum surface ($\beta'' > 1$) (Saif et al. 2002); $\beta = \beta'\beta''$. Using Eq.(1) and assuming the approximate equality $C \approx C'$ it is possible to verify whether or not the experimental capacitance (constant phase element) values follow the expected barrier-layer thickness sequence. Figure 2A shows the experimental dependence of the C vs.d^{-1}, which is linear in the range 40 to about 5 nm. The slope of the line is 2.81 10^{-13} F cm, which yields a quite realistic value of β= 0.98. However, below 5 nm there is a clear deviation from linearity (not shown). These results imply that during re-anodizing the barrier-layer thickness linearly decreases with $(U_{a,fin})^{-1}$ only down to about 5.0 nm. The further decrease in $U_{a,fin}$ results in slower reduction of d because the thickness of Al native oxide film exceeds 5 nm at room temperature (Saif et al. 2002). In other words, at $U_{a,fin}$< 5.0 V, the chemical interaction between aluminum and the ambient changes the relationship between the $U_{a,fin}$ and d. In addition, it is likely that the constant phase element decrease from 0.98 to about 0.94 reflects the enhanced fluctuations of d from pore to pore, yielding more heterogeneous distribution of the physical and chemical properties of the barrier-layer.

3.2. The m_{Cu} versus $U_v/U_{a,fin}$ plots for copper nanowires growth

Our experimental data suggest that a broad range of AC voltages can be used for deposition of copper within the alumina template pores if a suitable composition of the solution is chosen.

Unfortunately, this is applicable only for alumina grown in sulfuric acid bath where *AC* voltages U_v 6.0 to 18.0 V can be successfully applied (Fig. 3, curve 2). All attempts to fill more completely alumina templates grown in oxalic and phosphoric acid solutions varying *AC* voltage and the deposition time have failed since only limited voltages can be used in these cases. At higher *AC* voltages or somewhat longer electrolysis durations the barrier-layer breakdown of these templates was found to be inevitable (curves 1 in Figs. 4 and 5). As seen from curve 1 in Fig. 3, it is also a common practice for alumina grown in sulfuric acid bath if *d* is increased at the end of anodizing process up to about 20 nm by an increase of $U_{a,fin}$ to 20 V. On the other hand, it was established that the deposition of copper proceeds easier in the same Cu(II) solution if the barrier-layer thickness of alumina is decreased by lowering U_a. These results are in line with former findings by other investigators while depositing Au (Forrer et al. 2000), Ag (Xu et al. 2002; Sauer et al. 2002) and Fe, Ni, Co (Paulus et al. 2001). The m_{Cu} against $U_v/U_{a,fin}$ plots for alumina templates grown in oxalic and phosphoric acid solutions are shown in Figs. 4 and 5, respectively. From these plots one can see that the behavior of alumina templates in acidic Cu(II)solution manifests itself through parabolic m_{Cu}vs.$U_v/U_{a,fin}$ dependencies. Moreover, we have found that a range of *AC* voltages at which deposition of copper nanowires proceeds within the alumina pores strongly depends on the $U_{a,fin}$, decreasing with the alumina barrier-layer thickness. It should be also noted, that at *AC* voltages higher than m_{Cu}vs.$U_v/U_{a,fin}$ curve peak, $(U_v/U_{a,fin})_{peak}$, the alumina spalling and peeling off from the substrate becomes critical, especially when $U_{a,fin}>$ 20 V.Therefore, attention was focused on the range of *AC* voltages suitable for Cu^0 nanowires growth without the barrier-layer breakdown. As clearly seen from the experimental results depicted for various templates in Figs. 3 to 5, the range of voltages suitable for copper deposition shifts to the higher ratio of $U_v/U_{a,fin}$, the lower $U_{a,fin}$. On the other hand, despite the fact that a much wider range of *AC* voltages can be used for copper deposition within the pores of alumina with a quite thin barrier-layer, it seems impossible to completely fill such a matrix even at the $U_v/U_{a,fin}$ ratio 3.0 when $d< 5$ nm ($U_{a,fin}< 5$ V). This is due to a sharp decay in the current strength even during first 10-20 s of *AC* electrolysis up to a few mA cm^{-2}; characteristic feature for films with *d* lower than 5.0 nm. In these cases only imperceptible quantity of copper can be deposited within the alumina pores at the electrode edges since the gas bubbles evolved at the Al | oxide interface push away the alumina film from the substrate. The appearance of the gas bubbles at the Al | oxide interface was clearly observed by the naked eye. It is also common for alumina templates having thicker barrier-layers if prolonged *AC* electrolysis and higher *AC* voltages are used. Consequently, it seems difficult to estimate one optimal $U_v/U_{a,fin}$ for Cu^0 nanowires deposition within the pores of different templates. Evidently, the optimal *AC* voltage is lower $(U_v/U_{a,fin})_{peak}$ and the final choice is up to uniformity and completeness of the filling of the alumina pores. On the other hand, it has been found that the range of *AC* voltages suitable for copper nanowires fabrication depends also on the composition of Cu(II) solution. The m_{Cu}vs.$U_v/U_{a,fin}$ plots presented in Fig. 6 show that an increase in the solution pH widens the range of *AC* voltages suitable for copper deposition. Although the reasons of such behavior are unclear yet, we believed that this is most probably accomplished by a sharp decrease in the gas quantities evolved simultaneously with copper deposition from neutral and alkaline Cu(II) solutions at the Al | oxide boundary (Jagminas et al. 2002).

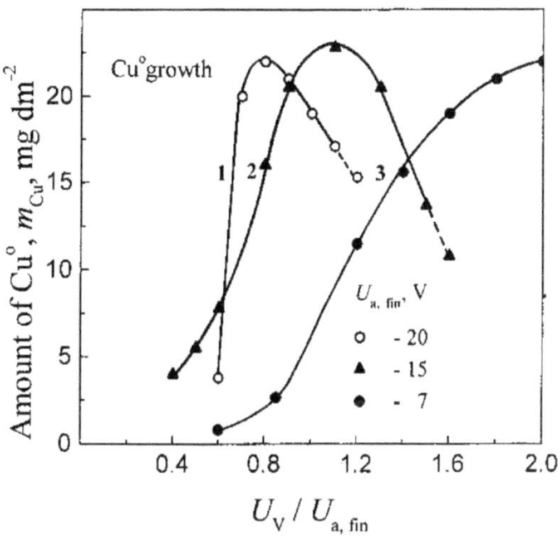

Figure 3. m_{Cu} vs. $U_V/U_{a,fin}$ plots for copper nanowires growth for 3 min within the pores of alumina with various barrier-layer thickness attained by increasing (1) or lowering (3) anodizing voltage in 1.53 mol/L sulfuric acid bath at the end of the template growth to $U_{a,fin}$ (V): (1) 20, (2) 15, (3) 7.0. An acidic solution of 0.1 $CuSO_4$ + 0.05 mol/L $MgSO_4$ + H_2SO_4 down to pH 1.50 was used for Cu^0 nanowires growth. The alumina thickness, δ = 10 μm. The dashed lines correspond to the obvious damage of alumina template.

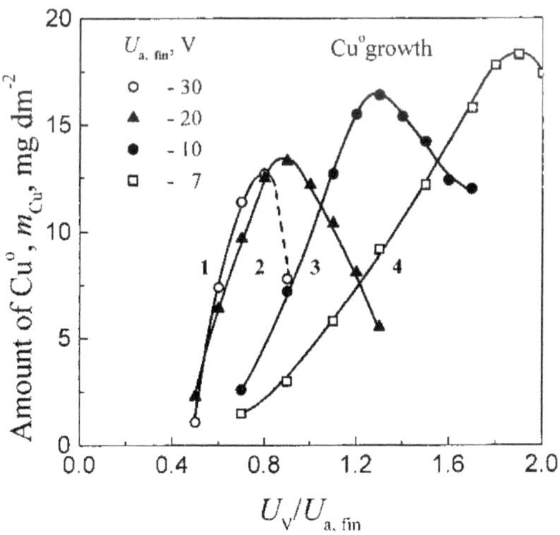

Figure 4. The same as in Fig. 3 for alumina grown in 0.2 mol/L oxalic acid bath for 70 min (δ=5.0 μm) with reduced *d* by lowering U_a 40 V to $U_{a,fin}$ (V): (1) 30, (2) 20, (3) 10, and (4) 7.0.

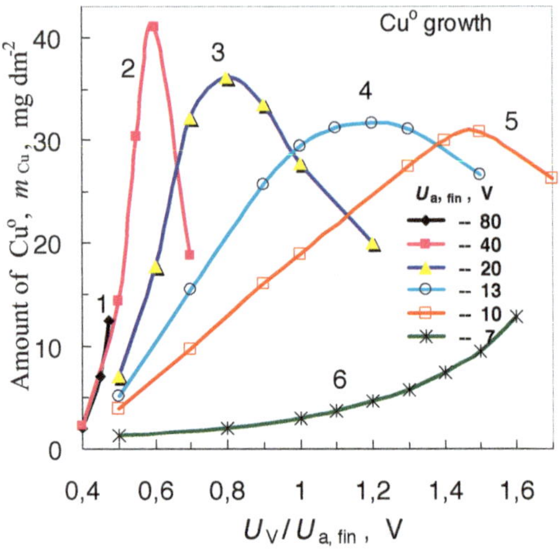

Figure 5. The same as in Fig. 3 for alumina grown in 0.4 mol/L phosphoric acid bath at 80 V for 1.5 h (δ=5.0 μm) with reduced *d* by lowering $U_a$to $U_{a,fin}$ (V): (1) 80, (2) 40, (3) 20, (4) 13, (5) 10, (6) 7.0.

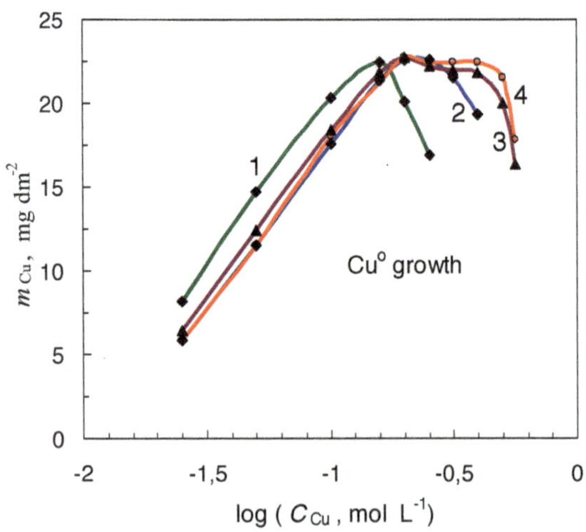

Figure 6. m_{Cu}vs.$U_V/U_{a,fin}$ plots for copper nanowires growth within the pores of alumina for 3 min *AC* electrolysis in the solution containing: 0.1 $CuSO_4$ + 0.05 mol/L $MgSO_4$ + H_2SO_4 down to pH 1.0 (1), 1.50 (2), and 6.0 (3) adjust by adding 0.2 M TEA. The alumina was grown in 0.4 mol/L phosphoric acid bath at 80 V for 1.5 h. $U_{a,fin}$ 20 V.

3.3. Concluding remarks

The above results show that in order to grow copper nanowires within the pores of alumina template obtained by Al anodizing at higher voltages the alumina barrier-layer thickness should be lowered. Using the acidic anodizing baths, a linear dependence of *d* on the final anodizing voltage, $U_{a,fin}$, is observed down to 5 V. The linearity was verified by impedance spectroscopy data, so, this technique might be used to monitor the parameters of the alumina template formation. However, further $U_{a,fin}$ decrease below 5 V results in significant barrier-layer thickness fluctuations, which are possibly due to uneven native oxide formation at the bottom of the pores. Spectrometric analysis of deposited copper content has shown that the range of AC voltage suitable for copper nanowires growth within the alumina pores sharply increases with decrease in $U_{a,fin}$ and hence *d*. The most favourable $U_{a,fin}$ range for uniform copper nanowires growth is 15 to 7.0 V.

4. Alumina template-dependant growth of cobalt nanowire arrays by *AC* deposition

4.1. Depositions into as-grown templates

In this study, different electrochemical regimes and porous alumina fabricated by aluminum anodizing in either sulphuric or oxalic acid solutions were applied for template synthesis of cobalt nanowire arrays, revealing several peculiar cases. By this way, we found that the growth of cobalt nanowires depends much stronger on the conditions for fabricating the alumina template than other metals like copper, silver or tin. For example, only sulfuric acid alumina templates can be successfully filled in the optimized deposition solutions by Co *nws* using *AC*, while the use of the same solution for uniform growth of Co *nws* in oxalic or phosphoric acid alumina pores is problematic. Therefore, in this study we focus on the investigation the peculiarities of the Co *nws* electrochemical growth using oxalic and sulfuric acid alumina templates at different regimes.

In this study, the composition of solution for Co *nws* electrosynthesis within the alumina pores was organized using $CoSO_4$, as a precursor for cobalt ions, and H_3BO_3, as a buffering ingredient, to prevent any pH variation within the alumina pores and to inhibit the formation of cobalt hydroxide species, as a result of hydrogen evolution (Zech & Landolt 2000). Furthermore, $MgSO_4$ was added as Mg^{2+} source to prevent the breakdown of alumina barrier-layer (Jagminas et al. 2003) during *AC* depositions. To adjust the pH of solutions, H_2SO_4 and triethanolamine(TEA) 1:1 solutions in water were used. From a review of patent's literature (Herrman 1972), arises that these components have been frequently used for *AC* coloring of anodized aluminum in the Co^{2+}-containing baths. To optimize the concentration of components and the solution pH, 15 μm thick sulfuric acid alumina templates were applied. All depositions in this setup were performed using 50 cm^2 specimens, a 50 Hz frequency *AC* and constant peak-to-peak voltage of 32 V, for 15 min. Observation of the alumina color uniformity and the amount of deposited cobalt were the main criteria adopted in the search of the optimal composition. Consequently, the solution composed of 0.2-0.25 $CoSO_4$, 0.65-0.7 H_3BO_3, 0.005-0.05 mol/L $MgSO_4$ and TEA, to adjust a pH at 5.5-5.75, was determined as optimum for *AC* filling of alumina templates with Co *nws*, having average $Ø_{pore}$ of 15 nm. The XRD patterns (not shown herein) of Co *nws* deposited from this solution demonstrated the polycrystalline nature of cobalt in hexagonal closely packed lattice. For any of the used *AC* deposition conditions no phase modification in the XRD patterns was observed. Figure 7 demonstrates typical variation of the amount of Co assembled inside the alumina pores (m_{Co}) with the deposition time and *AC* current frequency (*f*) used for depositions. As seen, the amount of deposited Co increases linearly during the first 10-20 min of *AC* treatment at constant voltage. However, with further processing the rate of deposition inside the alumina pores progressively decreases. Moreover, it was observed that this solution allows the growth of cobalt *nws* within a wide range of *AC* frequencies, ca. from 10 to 200 Hz, coloring the template uniformly from bronze to deep black. A further increase in frequency, up to 1000 Hz, results in a smaller amount of deposited cobalt and therefore in a lighter template color intensity. As seen from the inset of Fig. 7, the maximum amount of cobalt can be deposited using 100 Hz frequency. The bath temperature within 10 to 40°C range was found to have negligible effect on the amount of deposited cobalt as well as on the uniformity of depositions.

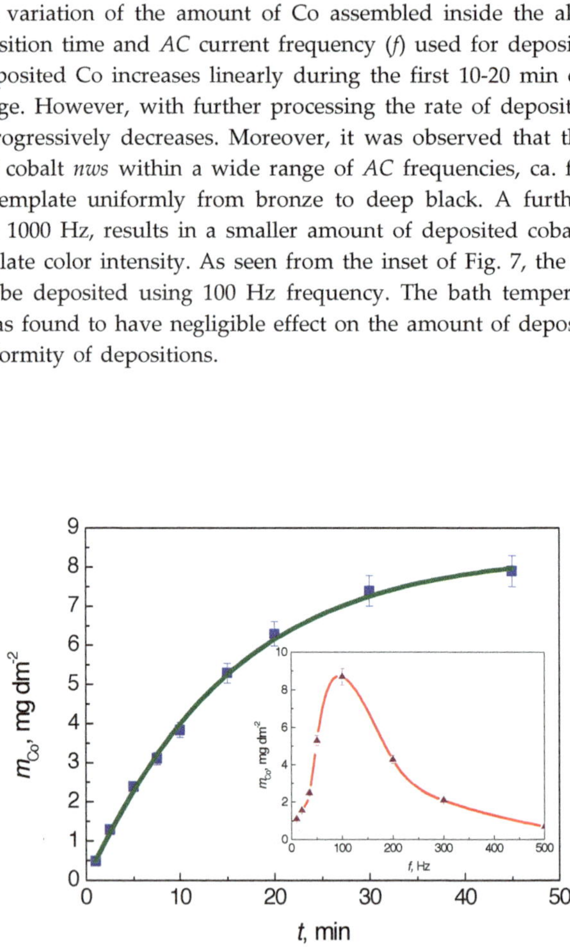

Figure 7. Variation of the amount of Co (m_{Co}) inside the alumina pores for the deposition solution made of 0.2 $CoSO_4$, 0.7 H_3BO_3 and 0.01 mol/L $MgSO_4$ and TEA up to pH 5.7, at a constant *AC* voltage U_{p-p} of 32 V, 50 Hz and 20 °C, as a function of the deposition time. $Ø_{pore}$ 15 nm; template thickness (δ_{AAO}) 15 μm. In the inset: variation of m_{Co} versus *AC* frequency (*f*) for the same deposition solution and U_{p-p} value for 15 min of deposition.

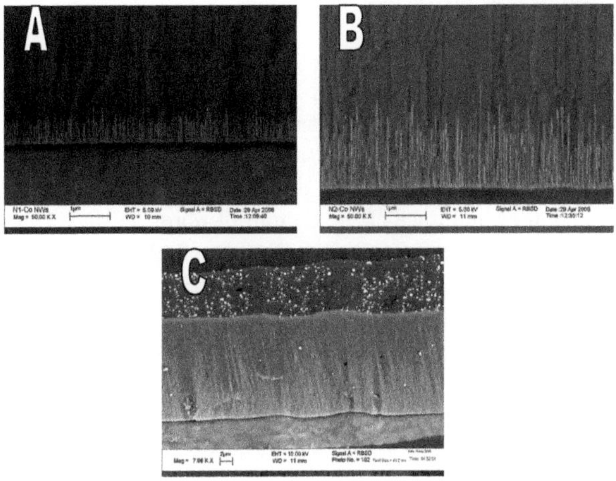

Figure 8. Cross-sectional FESEM images of sulfuric acid alumina templates grown at 25 (A, B) and 15 V (C) following the template etching in 2 mol/L H_2SO_4 for 15 min and deposition of cobalt *nw* arrays by *AC* (50 Hz) treatment in a same as in Fig. 7 solution at a constant current density of 0.3 A dm^{-2} and room temperature for: 5 (A), 15 (B) and 60 min (C). δ_{AAO} =18 μm, $Ø_{pore}$~28 (A,B) and ~20 nm (C).

In order to visualize the uniformity of Co *nws* growth by *AC* deposition, cross-sections of the alumina templates were investigated using field emission scanning electron microscopy (FESEM). Figures 8A-C show the arrangement of Co *nws* deposited inside the pores of sulfuric acid alumina templates at the same *AC* frequency (50 Hz) and peak-to-peak voltage ($U_{p\text{-}p}$) during 5, 15 and 60 min, respectively. The obtained data show quite uniform growth of Co *nws* from the bottom of almost all pores only at the onset of the process. The pores filling rate depends on *AC* voltage and of the pores diameter, $Ø_{pore}$. In case of sulfuric acid alumina templates formed at 25 V ($Ø_{pore}$ 25-30 nm after pore widening), the uniform growth corresponds to a filling rate v_{Co}~ 6 μm/h for the first 10 min of deposition at a constant $U_{p\text{-}p}$of 32 V. Processing further, v_{Co} decreases due to an increase of the template resistance, causing a reduced *AC* current. A smaller $Ø_{pore}$ results in a faster growth of cobalt *nws* under the same *AC* treatment conditions. For example, v_{Co}~ 9.3 μm/h was detected at $U_{p\text{-}p}$of32 V for alumina templates with average $Ø_{pore}$ of 15 nm. It can be observed from Fig. 8B that some cobalt *nws* grew faster than *nws* front. For prolonged *AC* treatment, this progressively leads to the formation of more and more uneven lengths of Co *nws* emerging onto the template surface in cobalt caps (Fig. 8C). The *nws* height uniformity was found to be independent on the *f*. Typical morphology of cobalt *nw* arrays assembled inside the sulfuric acid alumina pores by short-term AC deposition after template etching is presented in Figure 9. As seen, in case of 1-2 μm length of Co nws they are densely packed and quite uniform.

4.2. Depositions through a reconstructed barrier-layer

The key feature of *AC* deposition process of the densely packed Co *nws* is that only sulfuric acid alumina templates can be successfully applied. Fabrication of Co *nws* in the nanochannels of alumina template formed in the oxalic or phosphoric acid anodizing baths, however, is problematic by*AC* deposition way. One possible explanation for this effect is the adsorption and incorporation of acid anions at some depth of the alumina barrier-layer, changing the state of alumina/solution interface at the bottom of pores (surface charge, free energy, etc.) and preventing the discharge of Co^{2+} ions. Besides, highly ordered oxalic and phosphoric acid alumina templates are usually formed at higher voltages (Masuda & Fukuda 1995; Masuda et al. 1997; Li et al. 2000) and, therefore, present much thicker barrier-layers at the metal | oxide interface. To use these templates for *AC* deposition of various materials, the step-wise voltage decreasing at the end of anodizing process has been proposed (Furneaux et al. 1989) and successfully used in several works. We found herein that this is helpful also for the Co case, however, only for short-time processing. The Co *nw* array produced by long*AC* treatment, i.e. longer than 15 min, viewed nonuniform from pore to pore with some mushroomed Co fragments (Fig. 10) outgrowing from the breakdown sites of the alumina barrier-layer. The modification of alumina barrier-layers through (i) the decreasing of anodizing voltage (U_a) at the end of oxalic acid alumina growth down within 13 to 5 V, (ii) the cathodic treatment in the same anodizing bath for 3 min at various potentials and (iii) the chemical etching in the solution of sulfuric acids inhibit the uniformity of the depositions (see Fig. 11). The most uniform alumina color was obtained after decreasing U_a and chemical etching in the solution of sulfuric acid. Nevertheless, in this case the SEM cross-sectional observations of templates revealed the formation of Co *nws* tufts in random areas of template (Fig. 12A). It is worth to note that these tufts were found to arise from cobalt balls (Figs 11B and 12 C) formed at the metal/template interface.

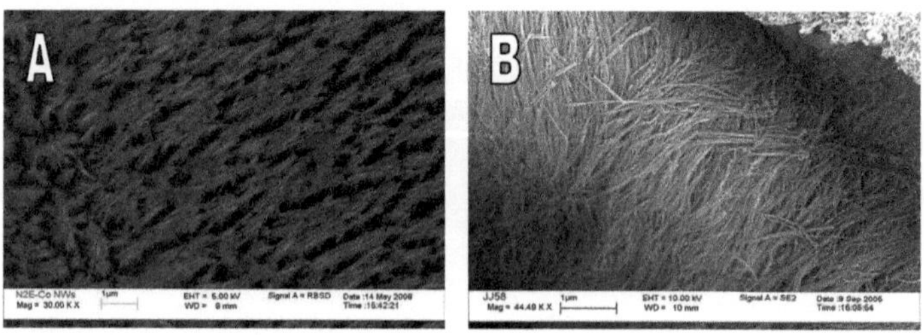

Figure 9. Top view FESEM images for fragments of Co *nw* arrays fabricated under the optimized *AC* deposition conditions in the sulfuric acid alumina template pores after the template etching in 0.5 mol/L H_3PO_4 for 70 (A) and 100 min (B).

Figure 10. Typical top view SEM images of Co 'mushroom' grown by *AC* treatment of oxalic acid alumina template in the solution of this study. Before deposition the thickness of alumina barrier- layer was reduced by lowering the anodizing voltage down to final value ($U_{a,fin}$) equal to 15 V.

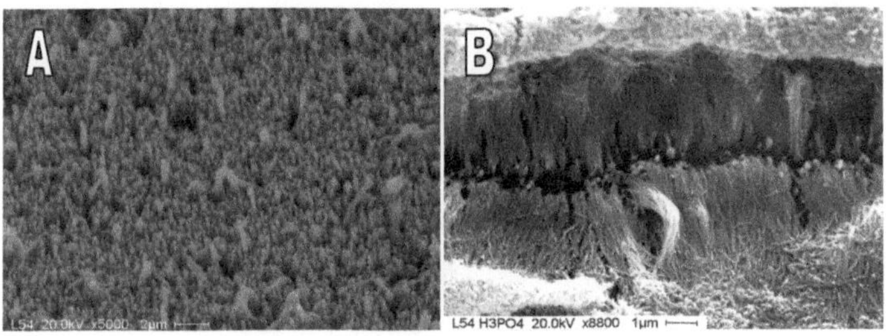

Figure 11. Top view FESEM image for fragments of Co *nw* arrays fabricated in the oxalic acid alumina pores by *AC* (50 Hz) deposition from the optimized solution at j_{ac} ~ 0.5 A dm^{-2} (U_{p-p} from 18 to 32 V) and room temperature for 10 (A) and 20 min (B). Before depositions, the barrier-layer of the as-grown templates was thinned by decreasing anodizing voltage down to 20 V and etching in 0.5 mol/L H_3PO_4 at 30 °C for 32 min. Following depositions the template was etched in the same 0.5 mol/L H_3PO_4 solution.

Variations in the conditions of cathodic treatment and chemical etching of the oxalic acid template as well as variations of the deposition potential were found to be ineffective for rod-like Co formation by *AC* deposition through the remained barrier-layer. Moreover, in case of *DC* deposition, the detachment of alumina template from the substrate even after several minutes of treatment took place.

4.3. Galvanostatic DC deposition

An alternative approach for Co *nws* deposition was further examined for oxalic acid alumina templates by a *DC* constant current density deposition, after removing or perforating the barrier-layer on the pores bottoms. In this setup, we used either an electrochemical/chemical method for the barrier layer perforation, or we detached the alumina from the substrate, removed the barrier layer and made a subsequent evaporation of Cr/Au layer, acting later as a conducting pad. The resulting Co *nws* released from the as-filled templates are shown in images A and B of Fig. 13. We found here that by applying a low current density during the entire deposition process, ca. ≤ 0.2 mA cm^{-2}, well-ordered, densely packed, continuous and highly aligned Co *nw* arrays, faithfully reproducing the shape of the pores and with height up to several tens of micrometers, can be synthesized by this way. Furthermore, after alumina dissolution these *nws* seems not to collapse and stuck together, as in the case of *AC* synthesis, implying an easier their application in future nanoelectronics and novel efficient sensors.

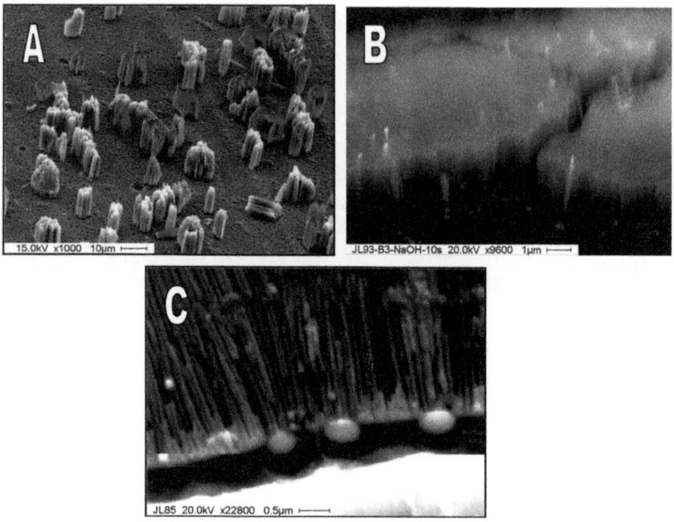

Figure 12. SEM images of reconstructed oxalic acid alumina templates after deposition of Co *nws* by *AC* treatment in the solution as in Fig. 7 at $U_{p\text{-}p}$ 32 V and room temperature for 20 min. δ_{AAO} 13 µm, $\varnothing_{pore}$~ 45 nm, τ_w 30 min in 0.5 mol/L H_3PO_4. (A) top-side view after dissolution some part of alumina; (B) *in plane* view; (C) cross-sectional view.

A typical XRD profile of a template filled with Co *nws* via *DC* deposition at a constant current density of 0.12 mA cm^{-2} for 5 hours is shown in Figure 14. Only a single peak is observable at $2\Theta = 41.59$. According to the XRD library patterns for bulk Co (PDF 89-4308), this peak corresponds to the (100) reflection of the hexagonal closely packed Co lattice. Further, some additional weak signal situated at $2\Theta = 75.89$, ascribed to hexagonal Co phase in (110) direction, can be observed. This weak feature probably inferred that Co *nws* are not single crystals but consist of oriented polycrystals with a preferred (100) growth direction perpendicular to the substrate. We note that the preferential growth of hexagonal Co phase in (100) direction is not a trivial case and differs from the growth of Co *nws* inside the alumina pores via *AC* and potentiostatic depositions reported by Kartopu et al. (2008) where the formation of Co *nw* arrays with a preferred (110) orientation has been demonstrated.

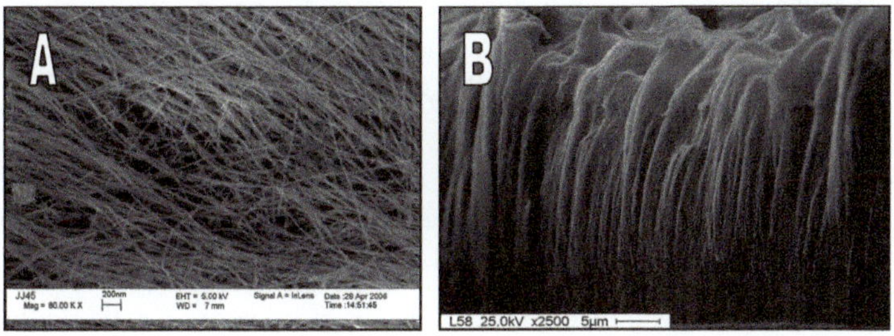

Figure 13. Top-view FESEM images of Co *nw* arrays after dissolution of oxalic acid alumina template fully (A)and in part (B). Template fabrication: anodizing 3 hours; $U_{a,fin}$ 5.0 V; τ_w 30 min in 0.5 mol/L H_3PO_4. Following the detachment of aluminafrom the substrate and a back-side evaporation of a Cr/Au layer, the deposition of Co was conducted in the solution: 0.45 $CoSO_4$, 0.7 mol/L H_3BO_3 and TEA up to pH 5.7 at a constant *DC*current density of 0.12 mA cm^{-2} for 5 hours.

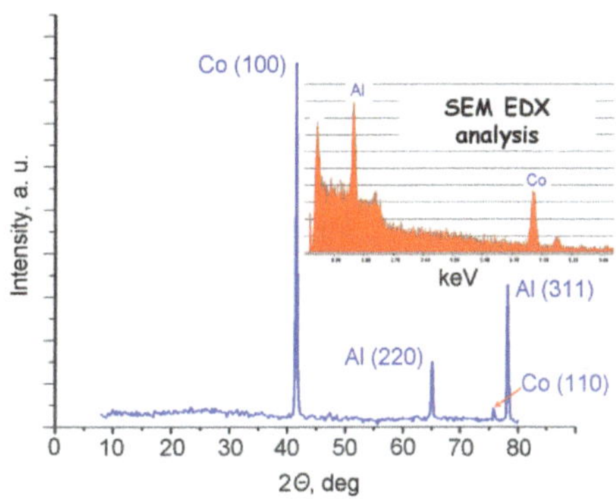

Figure 14. Typical X-ray diffraction spectrum for a Co *nws* array fabricated inside the oxalic acid alumina pores by DC deposition under the same galvanostatic conditions as in Fig. 13.

5. Compositional, structural and optical properties of bismuth selenide nws synthesized by template approach

Bulk bismuth selenide, Bi_2Se_3, is a V/VI semiconducting material with a band gap of ~ 0.35 eV and belongs to semiconductors group with a set of optical and physical properties which permits their use in photosensitive, photoelectrochemical and optoelectronic devices. In one-dimensional structures, especially when the nanometric dimensions turn down to Bohr radius ($Bi_2Se_3 r_{Bh}$~ 8 nm: Hillhouse & Tuominen 2001), these properties can be drastically enhanced due to quantum confinement effects (Li & Wang 2005). In the past decade, these effects have stimulated extremely active investigations on creating methods for fabrication of densely packed arrays of semiconductor species varied in composition, nm-scaled dimensions, shape and spacing.

Chemical deposition (Pejova & Grozdanov 2002; Gracia et al. 1997), electrodeposition (Torene et al. 1998), vapor deposition (Giani et al. 2002), reactive and compound evaporation (Augustine et al. 2005) and others (Sankapal et al. 2002) have been employed to date for the formation of Bi_2Se_3 thin films demonstrating their different optical properties and band gap values varied up to 2.3 eV.

During the last decade, several methods have also been proposed for the synthesis of nm-scaled bismuth selenide products. Yang et al. (2005) reported a solvothermal procedure for nanotubes and nanorods fabrication. Bi_2Se_3nanosheets and nanotubes have also been synthesized by hydrothermal co-reduction (Cui et al. 2004a) and nanorods(Cui et al. 2004b) and nanobelts(Xu et al. 2005) by a photochemical and sonochemical routes, respectively. The preparation of Bi_2Se_3nanocrystals has been also reported by Wang's group (2003). Recently, we have proposed a simple way for electrochemical formation of crystalline Bi_2Se_3 nanowire arrays inside the alumina template pores with an average pore diameter, $Ø_{pore}$= 40-50 nm (Jagminas et al. 2008). By this way, densely packed *nw* arrays from quite pure Bi_2Se_3 *nws* at a high aspect ratio can be successfully fabricated.

In this article, we report the results of investigation on peculiarities of bismuth selenide electrode position by*AC* treatment in different alumina templates varied in pore diameter, $Ø_{pore}$, within 10 to 100 nm range demonstrating, for the first time, a strong dependency of formed *nws* composition, morphology and their optical properties on the $Ø_{pore}$.

Porous alumina templates, 5.0 to 10.0 μm thick, were grown *via* two-step direct current (*dc*) anodizing of specimens for 0.5 to 20 hours in thermostated and vigorously stirred solutions under conditions indicated in Table 2. First anodization lasted two hours following the alumina film stripping in 0.2 M CrO_3 + 0.7 M H_3PO_4 at 60 ºC for 5 hours, while the second one – as indicated in Table 2. The barrier layer of alumina films after the second anodizing was set to about 15 nm by reduction of the anodizing voltage step-wise (2-1 V per 30-60 s) as in the study of Furneaux et al. (1989). All depositions of bismuth selenide *nws* within the alumina template pores were performed at room temperature in a glass cell where two graphite rods were used as the auxiliary and Al/alumina as a working electrode. The solution containing 0.02$Bi_2(SO_4)_3$ and 0.024 mol/L of H_2SeO_3 as the sources for Bi_xSe_y *nws* growth and two chelating agents for Bi^{3+}, e.g. 0.25 [TEA] and 0.07 mol/L [EDTA], kept at pH between 5.0 and 5.3 by addition of H_2SO_4 (1:1) was used in this study. The ratio of selenious acid to bismuth salt concentration was approximated to 1.2. This value has been recently determined by us as optimal for stoichiometric Bi_2Se_3*nws* synthesis inside the pores of oxalic acid alumina templates with $Ø_{pore}$ = 40-50 nm (Jagminas et al. 2008). For depositions, an alternating current, 50 Hz in frequency, centered at 0 V, under constant *AC* current density (j_{ac}) control of 0.3 ± 0.02 A dm^{-2} was applied. Reagents for the preparation of deposition solutions were: bismuth(III) sulphate, $Bi_2(SO_4)_3$, triethanolamine, $N(C_2H_4OH)_3$ – TEA, ethylendiaminetetracetic acid, $_2(HOOCH_2C)NCH_2CH_2N(CH_2COOH)_2$ - EDTA, and selenious acid, H_2SeO_3, of analytical reagent quality, purchased from Aldrich, and water from Milli-Q water system.

No.	Average $Ø_{pore}$, (nm)	Bath composition (M)	Anodizing conditions	Pore widening conditions
1	10	1.2 H_2SO_4	5 V *dc*; 5.00 ± 0.01 °C; 20 h	-
2	13	1.2 H_2SO_4	10 V *dc*; 10.00 ± 0.01 °C; 3 h	-
3	25-28	0.5 H_2SO_4	25 V *dc*; 3.00 ± 0.01 °C; 2 h	2 M H_2SO_4; 30 °C; 15-25 min
4	45-50	0.3 $H_2C_2O_4$	40 V *dc*; 17.00 ± 0.02 °C; 1 h	0.5 M H_3PO_4; 30 °C; 20-30 min
5	100	0.04 $H_2C_2O_4$	90 V *dc*; 1.00 ± 0.02 °C; 1.2 h	

Table 2. Summary of anodizing and post treatment variables employed for fabrication and etching of alumina templates.

To increase crystallinity of the final-products, samples were annealed for 3 hours in vacuum. The optimal annealing temperatures (T_{ann}) for various alumina templates were found in this study experimentally and approximated to: 250 °C for 100 and 50 nm, 200 °C for 28 and 13 nm and 170 °C for 10 nm $Ø_{pore}$ templates. The different values of T_{ann} were chosen due to the well-known dependency of melting and crystallization temperatures of nanomaterials on their size (Noh et al. 2007).

The alumina templates intended for XRD and UV-vis-IR investigations were separated from the electrode surface by one-side sequential etching of the electrode window in a solution of 1.5 mol/LNaOHand then in 10 wt% HCl and 0.1 mol/L $CuCl_2$ followed by thorough rinsing and drying in a nitrogen stream. X-ray diffraction studies were performed with a diffractometer D8 (Bruker AXS, Germany) equipped with a Göbel mirror (primary beam monochromator) for CuK_α radiation. A step-scan mode was used in the 2Θ range from 18 to 55° with a step of 0.04° and a counting time of 15 s per step.

A FESEM (model FESEM LEO 1530) and a FEI Helios NanoLab Workstation were used for cross-sectional observations of mechanically fractured samples and their surfaces chemically etched in 0.5 mol/LNaOH as well as *nw* arrays freed up from the templates.

To prepare TEM samples, deposited species were liberated by dissolving alumina template in 0.1 mol/L sodium hydroxide at 40 °C. The released products were then rinsed many times and finally dispersed in ethanol. At each stage, solvent exchange was carried out by centrifuging, extracting the supernatant and adding fresh solvent. Finally, free-standing nanospecies were re-dispersed in ethyl alcohol. For TEM observations, specimens were prepared by placing a drop of suspension on a Lacey carbon grid and left overnight at room temperature to evaporate the solvent. Nanostructured products were examined with a TEM microscope (model MORGAGNI 268) operating at 80 kV.

Optical properties of bismuth selenide arrays fabricated inside the alumina template pores were studied by recording the transmittance spectra within the 190 to 3150 nm wavelength range with respect to pure alumina template using a Shimadzu UV-3101PC spectrophotometer. The transmission data were manipulated for calculating the absorption coefficient dependency on the photon energy.

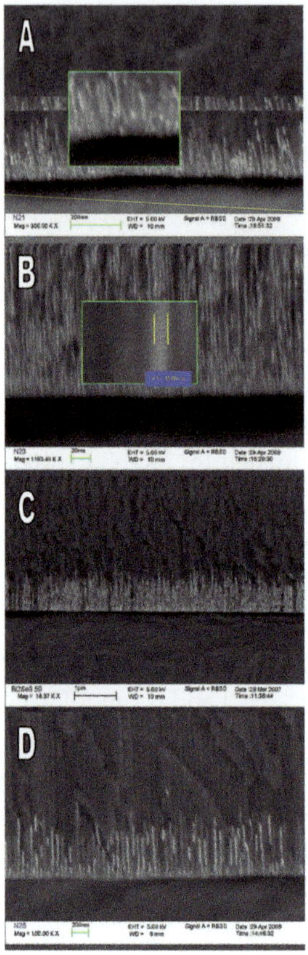

Figure 15. A typical cross-sectional FESEM view of alumina templates with average $\varnothing_{pore}$: 10 (A), 13 nm (B), 25 (C) and 50 nm (D) after *AC* treatment in the solution containing 0.02 $Bi_2(SO_4)_3$, 0.15 TEA, 0.07 EDTA and 0.024 mol/L H_2SeO_3 and H_2SO_4 to adjust pH to 5.0at j_{ac}=0.3 A dm^{-2} for 3 min.

5.1. Results

Uniform filling of alumina pores by densely packed Bi_2Se_3*nws* with an average pore diameter $\varnothing_{pore}$= 40-50 nm in the solution and conditions adopted in this study were demonstrated by us in (Jagminas et al. 2008). Figure15 presents typical FESEM images of the final products encapsulated within the alumina template pores with average diameter 10, 13, 25 and 50 nm showing that the diameters of *nws* grown inside the alumina pores by *ac* deposition are in agreement with the nominal pore diameter of templates while the height of deposited products depends on the current density, j_{ac}, deposition time, τ_{dep}, and $\varnothing_{pore}$. Under the same deposition conditions, an increase in the $\varnothing_{pore}$ resulted in filling of pores of a lower height. Furthermore, variations in $\varnothing_{pore}$ do not noticeably altered the completeness of the pore filling; for all cases most of the pores, especially at the metal/oxide interface with heights up to 1 μm, seem nicely filled. On the other hand, the dispersity of *nws* lengths increases with τ_{dep}, especially at higher j_{ac}, and *AC* voltages. Surprisingly, the pitting of alumina films, frequently observed in other solutions as a result of the alumina barrier-layer breakdown and crystallization of salts (Jagminas 2002), was not observed for the solution of this study under a wide range of deposition conditions: j_{ac} up to 0.5 A dm^{-2}, τ_{dep} up to 45 min, and $Ø_{pore}$ up to 100 nm. The influence of the solution temperature on the composition of products was also investigated here. In all cases increase in the solution temperature higher than 40 °C led to non-uniform depositions within the alumina pores of products in the lower quantity most likely due to alumina pore sealing, as could be expected.

Figure 16 shows XRD patterns of porous alumina templates with average $Ø_{pore}$ 100, 50, 25, 13 and 10 nm filled with Bi_xSe_y*Nw* arrays by *AC* electrolysis in the same optimized solution at the same constant j_{ac}, ca. 0.3 A/dm^2, for 25 min. From the patterns, it has been found that under the same electrolysis conditions Bi_2Se_3, Bi_3Se_2 or both phases of selenides can be deposited. As seen from Fig. 16a, a quite pure, Se-rich phase, Bi_2Se_3, grows when templates with $Ø_{pore} \geq 50$ nm are used. However, Bi-rich phase, Bi_3Se_2, appears to form more readily when fine structure templates with $Ø_{pore} \leq 13$ nm are employed (Fig. 16b). Moreover, the *Nws* array grown under the same conditions within extremely tiny pores, with $Ø_{pore} = 10$ nm, was found to be composed of Bi_3Se_2 and some Bi^0 inclusions while in the case of $Ø_{pore} = 28$ nm the *nws* are composed of Bi_2Se_3 with some amount of Bi_3Se_2. In the case of tiny pores ($Ø_{pore} = 13$ and 10 nm), increase in the molar ratio of the selenium and bismuth precursors ($\alpha_{Se/Bi}$) from 1.0 to 2.0 results in the formation of BiSe ($\alpha_{Se/Bi} = 1.5$) and finely grained Bi_3Se_2 ($\alpha_{Se/Bi} = 2.0$) without Bi^0 inclusions. Also, with $\alpha_{Se/Bi}$ increase a somewhat slower bath voltage growth has been determined during the deposition process at a constant *AC* current density.

The influence of the solution temperature on the composition of products was also investigated here. In all cases increase in the solution temperature higher than 40 °C led to non-uniform depositions within the alumina pores of products in the lower quantity most likely due to alumina pore sealing, as could be expected. For the same $Ø_{pore}$, however, no changes in the phase composition of deposited products have been observed within 5 to 35 °C. The compositional variations of deposited *nws* with the size of alumina template pores can be explained as follows:

It is known that Se-rich thin films of bismuth selenide, namely Bi_2Se_3, may be electrodeposited only when enough Se is present in the reaction zone. However, as it has been shown earlier by us (Jagminas et al. 2005), the discharge of SeO_3^{-2} ions from aqueous solutions of selenious acid at the bottom of tinny pores under *ac* bias is hampered. As a result, the size of *a*-Se species that can be deposited drastically decrease with $Ø_{pore}$ shrinking. Surprisingly, the increase in the *ac* current density and electrolysis time influenced the content of deposited *a*-Se in these pores only negligible. In contrast, the content of selenium deposited in wider pores, ca. 40-50 nm, under the same conditions increased many folds. Thus, the formation of Se-rich bismuth selenide nanowires in the oxalic acid alumina pores ($Ø_{pore} \geq 40$ nm) can simply be released. Again, only Bi-rich phases can be deposited in the sulfuric acid alumina templates with $Ø_{pore} \leq 13$ nm.

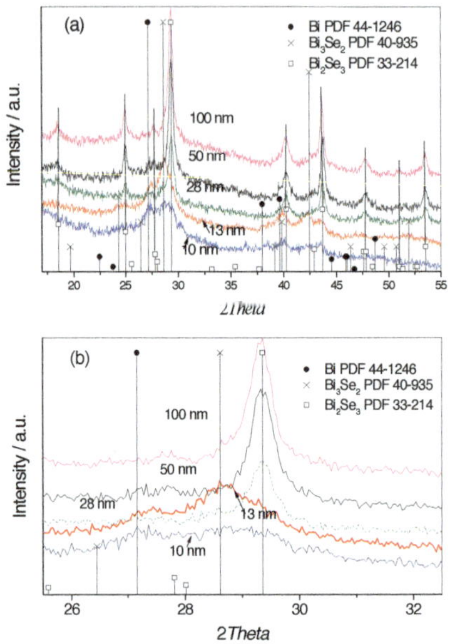

Figure 16. a) XRD patterns for various alumina templates filled by *AC* treatment at j_{ac} 0.3 A dm^{-2} for 30 min in the solution containing 0.02 $Bi_2(SO_4)_3$, 0.15 TEA, 0.07 EDTA and 0.024 mol/L H_2SeO_3 and H_2SO_4 to adjust pH to 5.0 after annealing in vacuum for 3 h. (b) The same as in part (a) within the 26↔32 2Θ range.

In this study, the optical properties of bismuth selenide *Nw* arrays fabricated within the pores of various alumina templates were investigated by recording the transmittance UV-vis-NIR spectra using the same pure alumina templates as reference. In this set-up, the thickness of alumina templates varied between 5.0 and 7.5 ± 0.5 μm and both as-grown and annealed in vacuum at 170, 200 and 250 °C templates with various $Ø_{pore}$ were studied. To achieve more precise results, the working and reference samples were anodized, post treated and annealed together. Again, the optical band-gap, E_g, for as-grown and annealed arrays was calculated using a well-known Tauc's relation:

$$\alpha(h\nu) = A\left(h\nu - E_g\right)^{n/2}, \tag{2}$$

where α is the absorption coefficient, A is a constant, $h\nu$ is the photon energy, E_g is the band gap, and n depends on the nature of transition, being equal to 1 or 3 for direct-allowed or direct-forbidden and 4 or 6 for indirect-allowed or indirect-forbidden transitions, respectively. The absorption coefficient was calculated from the transmittance spectra using a simple relation:

$$\alpha = -lnT / h_{BiSe} \tag{3}$$

The average height of bismuth selenide *nws*, h_{BiSe}, was determined from the FESEM observations of cross-sectioned templates after the optical measurements. By this way, the predominant direct band-to-band transition across the gap of different wavelengths was verified for all bismuth selenide *nw* arrays fabricated in this study. Typical plots of a^2*versus* the corresponding values of photon energy, $h\nu$, for Bi_xSe_y *nws* deposited inside the alumina pores with different $Ø_{pore}$ are given in Fig. 17. As seen, variation of α vs. $h\nu$ demonstrates a wide light absorption region from NIR to UV. In case of alumina templates with average $Ø_{pore}$ of 100 nm (curve 4) extrapolating the straight line part of the curve α^2vs. $h\nu$ to the energy-axis, the value of $E_{g,dir}$ equal to 0.4 eV was obtained for as-formed Bi_2Se_3*nw* arrays that is close to E_g= 0.35 eV of bulk bismuth selenide. Note that with decrease in the diameter of Bi_2Se_3*nws*, the absorption of higher energy light increases. In case of $Ø_{pore}$ = 25 nm, the shape of α^2vs. $h\nu$ plot implied two absorption edges, perhaps due to the deposition within such alumina pores of species composed of a Bi_2Se_3 and Bi_3Se_2 mixture, as it has been shown above by XRD investigations. For Bi_3Se_2nanoscaled products encased within the alumina template pores with $Ø_{pore}$ 10 and 13 nm, the α^2vs. $h\nu$ plots demonstrate similar shapes (see curves 1 and 2). However, we found that the straight parts both in α vs. $h\nu$ and α^2vs. $h\nu$ plots are not clear and thus numerous tangents can be extrapolated to the energy-axis from these plots indicating, for example, that E_g value for 13 nm *nws* could be between 0.9 and 1.7 eV, while for 10 nm *nws* E_g approximated to from ~2.2 to 2.7 eV. Consequently, the effective band gaps of these arrays cannot be precisely distinguished from the absorption spectra. Notice that these results are in line with the results presented in the recent publication (Sun et al. 2008) where the same problem raised analyzing the absorption spectra of CdTe quantum wires. Nevertheless, an obvious blue shift of α^2vs. $h\nu$ plot is observed for 10 nm nanowired products, e.g. when the diameter of alumina template pores approach to the Bohr radius of bismuth selenides (see *Inset* in Fig. 17).

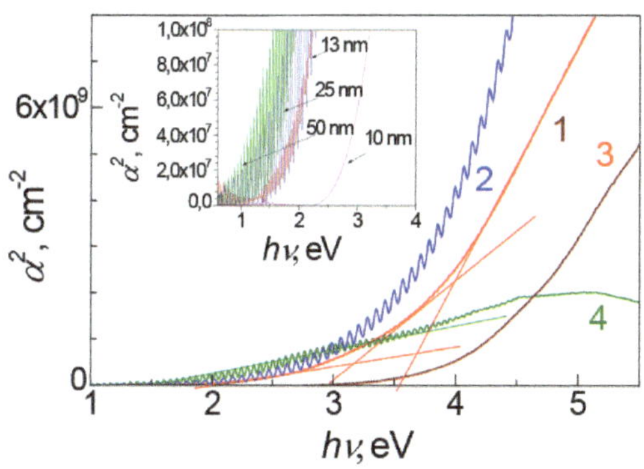

Figure 17. Variations of absorption coefficient, α^2, with photon energy, $h\nu$, for bismuth selenide *nw* arrays fabricated by *AC* deposition from the solution as in Fig. 15 at j_{ac}0.3 A dm^{-2} for 3 min (h_{BiSe} 0.75 – 1.0 μm) within the alumina template pores differing in pore diameter: (1) 10; (2) 13; (3) 25; (4)100 nm.

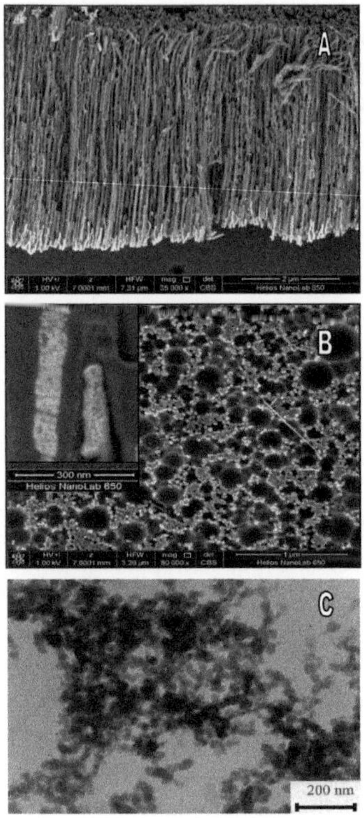

Figure 18. High resolution FESEM (A,B) and TEM (C) views of the Bi_2Se_3 product encased inside the alumina template pores by *AC* deposition as in Fig. 15 at j_{ac} 0.3 A dm^{-2} for 3.0 (A) and 30 min (B,C) before (A) and after template dissolution (B,C). $Ø_{pore}$ = 50nm. In the *Inset* high magnification (250,000×) Bi_2Se_3*Nw* FESEM fragments acquired at 1 kV using a concentric BackScatter detector are shown.

To understand absorption variables of bismuth selenide arrays fabricated herein, we further studied the morphology of products, deposited within the alumina templates with $Ø_{pore}$ 50 nm and 13 nm using modern high-resolution FESEM and TEM techniques. Shown in Figure 18 are the high resolution panoramic (A) and top-side (B) FESEM images of alumina films encased with Bi_2Se_3 species before (A) and following the template etching with a drop of 0.5 mol/LNaOH, while (C) depicts the TEM image of the same product disengaged from the template through the template dissolution and collection of the remained species by centrifugation and several washings. As seen from images B and C, Bi_2Se_3 deposited within the 50 nm pores has a granular shape. This granular structure can be also visualized from the high magnification FESEM observation of nanowired product (see *Inset*). Through the TEM observation the size of granules varied within the 17 to 40 nm range implying that 50 nm *nws* of Bi_2Se_3 are composed of weakly connected nanocrystals. A similar morphology was also observed for disengaged species of bismuth selenides deposited within 25-28 nm pores. In the case of tiny pores, however, the structure of the deposited material after dissolution of alumina matrix was found to differ significantly from the 50 nm products in that it consists of short and tinny *nw* fragments in length of from 0.2 to 1.0 μm (see Fig. 19) even after the template etching and liberation from the matrix procedures. It is some confusing knowing that in the case of $\O_{pore}$= 13 nm the deposited material is not a phase pure material but according to XRD results consists of a Bi_3Se_2 and Bi_2Se_3 mixture.

5.2. Discussion

Bi_2Se_3 is reported as a direct band gap semiconductor material. Of special note is that the band gap (E_g) values of the Bi_2Se_3 crystals determined from the optical measurements by different scientists disagree strongly both for bulk materials and nanomaterials. For example, for bulk Bi_2Se_3 Novoselova(1978) reported $E_g = 0.35$ eV, while Lide(1991) only $E_g = 0.16$ eV. In the case of Bi_2Se_3 films fabricated by electroless deposition, the presence of two edges corresponding to E_g = 0.354 eV and E_g = 1.03 eV has been reported by Bhattacharya and Pramanik(1980). Moreover, for thin Bi_2Se_3 films the band gaps as high as 2.3 eV has been reported by Pejova and Grozdanov (2002) linking such high values with a nanocrystalline film nature. The great E_g variations have been also reported for nm-scaled Bi_2Se_3 including 1.59 eV(Jiang et al. 2006) and 2.25 eV(Ota et al. 2006). Noteworthy that the reported variations in the band gap values of bismuth selenide thickfilms and nanomaterials frequently are ascribed more to the morphology and purity of this semiconductor than to the size quantization effects. Consequently our findings of high E_g of bismuth selenide species deposited within alumina pores under conditions of this study cannot be considered as surprising.

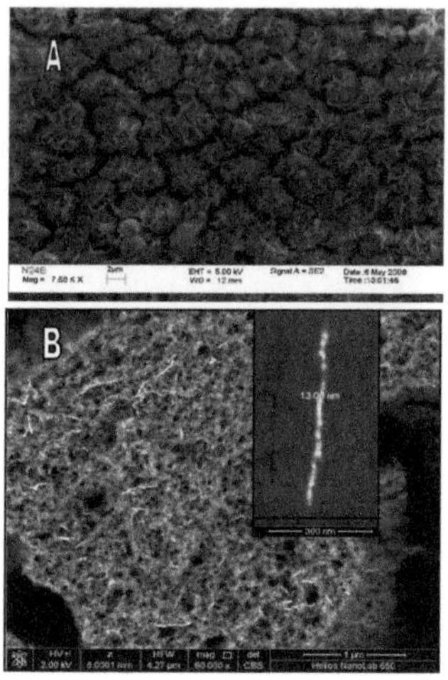

Figure 19. A top-side FESEM view of alumina surface chemically etched by a drop of 0.5 mol/LNaOH for a case when alumina template with average $\O_{pore}$= 13 nm is encased with bismuth selenide nws by AC deposition as in Fig. 15. In the Inset, 13 nm in diameter a product fragment acquired at 1 kV using a concentric BackScatter detector is shown.

5.3. Concluding remarks

We initiated this study to show for the first time that the composition and morphology of bismuth selenide deposited within the alumina template pores by means of alternating current deposition depend on the diameter of pores. Under the same electrolysis conditions *nw* arrays from Se-rich (Bi_2Se_3), Bi-rich (Bi_3Se_2) or both phases can be successfully fabricated if the templates with average $Ø_{pore} \geq 50$ nm, $Ø_{pore} = 28$ nm, and $Ø_{pore} \leq 13$ nm, respectively, are used. Also, the optical properties of bismuth selenide *nws* differing in size and phase composition, differ significantly. The band-edge absorption at 0.4 eV, characteristic to bulk E_g of Bi_2Se_3,was clearly evidenced only for bismuth selenides deposited within the alumina pores approximated to 100 nm. With decrease in $Ø_{pore}$ and diameter of bismuth selenide *nws* the blue shift of absorption edge is obvious although in the case of very tiny pores, when $Ø_{pore}$ approaches the Bohr radius, the determination of effective band gaps for deposited Bi_xSe_y *nw* arrays was found to be somewhat problematic using only the experimental transmission spectra perhaps due to at least dual composition of *nws*.

Author details

Arūnas Jagminas

State Research Institute Center for Physical Sciences and Technology, Lithuania

References

[1] Al. Mawlawi, D.; Coombs, N. Moskovits, M. (1991). Magnetic properties of Fe deposited into anodic aluminium oxide pores as a function of particle size. *J. Appl. Phys.*, Vol. 70, pp. 4421-4425.

[2] Augustine, S.; Ampili, S.; Kang, J. Ku & Mathai, E. (2005). Structural, electrical and optical properties of Bi_2Se_3 and $Bi_2Be_{(3-x)}Te_x$ thin films. *Mater. Res. Bull.*, Vol. 40, pp. 1314-1325.

[3] Bejenari, I.; Kantser, V. & Balandin, A.A. (2010). Thermoelectric properties of electrically gated bismuth teluride nanowires. *Phys. Rev. B.*, Vol. 81, pp. 075316, ISSN:

[4] Bhattacharaya, R.N. & Pramanik, P. (1980). Chemical Method for the Deposition of Thin Films of Bi_2Se_3. *J. Electrochem. Soc.*, Vol. 127, pp. 1857-1858.

[5] Clebny, J.; Doudin, B. & Anserment, J.-Ph. (1993). Nanocryst. Mater. 2, pp. 637.

[6] Cui, H.; Liu, X.; Li, X.; Wang, J.; Li, X.; Han, F.; Zhang, X. & Boughton, R.I. (2004a). Sonochemical synthesis of bismuth selenide nanobelts at room temperature. *J. Cryst. Growth* , Vol. 271, pp. 456-461.

[7] Cui, H.; Liu, H.; Li, X.; Wang, J.; Han, F. & Zhang, X. (2004b). Synthesis of Bi_2Se_3 thermoelectric nanosheets and nanotubes through hydrothermal co-reduction method. *J. Solid. State Chem.*, Vol. 177, pp. 4001-4006.

[8] Diggle, J.W.; Downie, T.C. & Goulding, C.W. (1969). Anodic oxide films on aluminium. *Chem. Rev.*, Vol. 69, pp. 365-405.

[9] Doughty, A.S.; Thompson, G.E.; Richardson, J.A. & Wood, G.C. (1975). Investigation of the electrolytic colouring of porous anodic films on aluminium using electron microscopy. *Trans. IMF*, Vol. 53, pp. 33-39.

[10] Forrer, P.; Schlottig, F.; Siegenthaler, H. & Textor, M. (2000). Electrochemical preparation and surface properties of gold nanowire arrays formed by the template technique. *J. Appl. Electrochem.*, Vol. 30, pp. 533-541.

[11] Furneaux, R.C.; Rigby, W.R. Davidson, A.P. (1989). The formation of controlled-porosity membranes from anodically oxidized aluminium. *Nature*, Vol. 337, pp. 147-149.

[12] Gerein, N.J. & Haber, J.A. (2005). Effect of ac electrodeposition conditions on the growth of high aspect ratio copper nanowires in porous aluminum oxide templates. *J. Phys. Chem. B*, Vol. 109, pp. 17372-17385.

[13] Giani, A.; Bayaz, A.Al.; Foucaran, A.; Pascal-Delannoy, F. & Boyer, A. (2002). Elaboration of Bi_2Se_3 by metalorganic chemical vapor deposition. *J. Cryst. Growth*, Vol. 236, pp. 217-220.

[14] Goad, D.G.W. Moskovits. (1978). Colloidal metal in aluminium oxide. *J. Appl. Phys.*, Vol. 49, pp. 2929-2934.

[15] Gracia, V.M.; Nair, M.T.S.; Nair, P.K. & Zingaro, R.A. (1997). Chemical deposition of bismuth selenide thin films using N,N-dimethylselenourea. *Semicond. Sci. Technol.*, Vol. 12, pp. 645-649.

[16] Harkness, A.C. & Young, L. (1966). High resistance anodic oxide films on aluminium. *Can. J. Chem.*, Vol. 44, pp 2409-2413.

[17] Hermann, E. (1972). Elektrolytisches Färben von Anodisiertem. Aluminium. Galvanotechnik *Galvanotechnik* Vol. 63, pp. 110-121.

[18] Hillhouse, H.W. & Tuominen, M.T. (2001). Modelling the thermoelectric transport properties of nanowires embeded in oriented microporous and mesoporous films. *Miroporous Mesoporous Mater.*, Vol. 47, pp. 39-50.

[19] Jagminas, A. (2002). Influence of magnesium and aluminium ions on the copper ac deposition into aluminium anodic oxide film nanotubes. *J. Appl. Electrochem.*, Vol. 32, pp. 1201-1209.

[20] Jagminas, A.; Lichušina, S.; Kurtinaitienė, M. & Selskis, A. (2003). Concentration effect of the solutions for alumina template ac filling by metal arrays. *Appl. Surf. Sci.*, Vol. 211, pp. 194-202.

[21] Jagminas, A.; Gailiūtė, I.; Niaura, G. & Giraitis, R. (2005). Template-assisted fabrication of pure Se nanocrystals in controllable dimensions. *Chemija*, Vol. 16(3/4), pp. 15

[22] Jagminas, A.; Valsiūnas, I.; Veronese, G.P.; Juškėnas, R. Rutavičius, A. (2008). Alumina template-assisted growth of bismuth selenide nanowire arrays *J. Cryst. Growth*, Vol. 310, pp. 428-433.

[23] Jessensky, O.; Müller, F. & Gösele, U. (1998). Self-organized formation of hexagonal pore structures in anodic alumina. *J. Electrochem. Soc.*, Vol. 145, pp. 3735-3740.

[24] Jiang, Y.; Zhu, Y.-J. & Chen, G.F. (2006). Synthesis of Bi_2Se_3 nanosheets by microwave heating using an ionic liquid.*J. Cryst. Growth & Design*, Vol. 6, pp. 2174-2178.

[25] Kartopu, G.; Yalcin, O.; Es-Souni, M. & Basaran, A. C. (2008). Magnetization behavior of ordered and high density Co nanowire arrays with varying aspect ratio. *J. Appl. Phys.* Vol.103, no. 093915.

[26] Kokonou, M.; Giannakopoulos, K.P. & Nassiopoulou, A. G. (2007). Few nanometer thick anodic porous alumina films on silicon with high density of vertical pores. *Thin Solid Films*, Vol. 515, pp. 3602-3606.

[27] Lan, Y.; Poudel, B.; Ma, Y.; Wang, D.; Dresselhous, M.S.; Chen, G. & Ren, Z. (2009). Interpore study of bulk nanograined thermo-electric bismuth antimony telluride. *Nano Lett.*, Vol. 9, pp. 1419-22

[28] Li, A.P.; Müller, F.; Birner, A.; Nielsch, K. & Gösele, U. (1998). Hexagonal pore arrays with a 50-420 interpore distance formed by self-organization in anodic alumina. *J. Appl. Phys.*, Vol. 84, pp. 6023-6026.

[29] Li, A.P.; Müller, F. & Gösele, U. (2000). Polycrystalline and monocrystalline pore arrays with large interpore distance in anodic alumina. *Electrochem. Solid-State Lett.*, Vol. 3, pp. 131-134.

[30] Li, J. & L.-W. Wang, (2005). Band-structure-corrected local density approximation study of semiconductor quantum dots and wires. Phys. Rev. B, Vol. 72, no. 125325

[31] Lide, D.R. Handbook of Physics and Chemistry 71st ed. Boca Raton, FL:CRC Press, 1990-1991.

[32] Masuda, H. & Fukuda, K. (1995). Ordered metal nanohole arrays made by a two-step replication of honeycomb structures of anodic alumina. *Science*, Vol. 268, pp. 1466-1468, ISSN:

[33] Masuda, H.; Hasegwa, F. & Ono, S. (1997). Self-ordering cell arrangement of anodic porous alumina formed in sulfuric acid solution. *J. Electrochem. Soc.*, Vol. 144, pp. L127-L129.

[34] Macdonald, J.R. & Franceschetti, D.R. In: Impedance Spectroscopy Emphasizing Solid Materials and Systems. R.S. MacDonald (Ed.) p. 48, John Willey and Sons, New York (1987).

[35] Mardilovich, P.P; Govyadinov, A.N.; Mukhurov, N.I.; Rzhevskii, A.M. Paterson, R. (1995). New and modified anodic alumina membranes. Part 1. Thermotreatment of anodic alumina membranes. *J. Membr. Sci.* 98, 131-142.

[36] Masuda H. Fukuda, K. (1995). Ordered metal nanohole arrays made by a two-step replication of honeycomb structures of anodic alumina. *Science*, Vol. 268, pp. 1466-1468.

[37] Metzger, R. M.; Konovalov, V. V.; Xu, M. S. T.; Zangari, G.; Xu, B.; Benakli M. & Doyle, W. D. (2000). Magnetic nanowires in hexagonally ordered pores of alumina. *IEEE Trans. Magn.*, Vol. 36(1), pp. 30-35.

[38] Nielsch, K.; Choi, J.; Schwirn, K.; Wehrspohn, R.B. Gösele, U. (2002). Self-ordering regimes of porous alumina: the 10 porosity rule. *Nano Lett.*, Vol. 2(7), pp. 677-680.

[39] Noh, K.W.; Woo, E. & Shin, K. (2007). Alteration of crystal structure of bismuth confined in cylindrical nanopores. *Chem. Phys. Lett.*, Vol. 444, pp. 130-134.

[40] Novoselova, A.B. (Ed.), Physical and Chemical Properties of Semiconductors, Handbook, Moscow, 1978.

[41] O'Sullivan, J.P. Wood, G.C. (1970). Morphology and mechanism of formation of porous anodic films on aluminium. *Proc. Royal Soc. London, A.*, Vol. 317, pp. 511-543.

[42] Ota, J. R.; Roy, P.;. Srivastava, S. K. & Popovitz-Biro, R. A simple hydrothermal method for the growth of Bi_2Se_3 nanorods. *Nanotechnology*, Vol. 17, pp. 1700-1705.

[43] Pan, H.; Sun, H.; Poh, C.; Feng, Y. Lin, J. (2005). Single-crystal growth of metallic nanowires with preferred orientation. *Nanotechnology*, Vol. 16, pp. 1559-1564.

[44] Paulus, P.M.; Luis, F.; Kröll, M.; Schmid, G. & de Jongh, L.J. (2001). Low-temperature study of the magnetization reversal and magnetic anisotropy of Fe, Ni and Co nanowires. *J. Magn. Magn. Mater.*, Vol. 224, pp. 180-196.

[45] Pejova, B. & Grozdanov, I. (2002). Chemical deposition and characterization of glassy bismuth(III) selenide thin films. *Thin Solid Films*, Vol. 408, pp. 6-10.

[46] Picht, O.; Müller, S.; Albert, I.; Rouber, M.; Lensch-Falk, J.; Medlin, D. L.; Neumann, R. & Toimil-Molares, M.E. (2012). Tuning the geometrical and crystallographic characteristics of Bi_2Te_3 nanowiresby electrodeposition in ion-track membranes. *J. Phys. Chem. C*, Vol. 116, pp. 5367-75.

[47] Qin, J.; Nogues, J.; Mikhaylova, M.; Roig, A.; Munoz, J. S. Muhammed, M. (2005). Differences in the magnetic properties of Co, Fe, and Ni 250-300 nm wide nanowires electrodeposited in amorphous anodized alumina templates. *Chem. Mater.*, Vol. 17(7), pp. 1829-1834.

[48] Routkevich, D.; Bigioni, T.; Moskovits, M. & Xu, J..M. (1996). Electrochemical fabrication of CdS nanowire arrays in porous anodic aluminum oxide templates. *J. Phys. Chem.*, Vol. 100, pp. 14037-14047.

[49] Römer, W. Steinem, C. (2004). Impedance Analysis and Single-Channel Recordings on Nano-Black Lipid Membranes Based on Porous Alumina. *Biophysical Journal,* Vol. 86, pp. 955-965.

[50] Saif, M.T.A.; Zhang, S.; Haque, A. & Hsia, K.J. (2002). Effect of native Al_2O_3 on the elastic response of nanoscale Al films. *Acta Materialia,* Vol. 50, pp. 2779-2786.

[51] Sankapal, B.P. & Lokhande, C.D. (2002). Photoelectrochemical characterization of Bi_2Se_3 thin films deposited by SILAR technique. *Mater. Chem. Phys.*, Vol. 73, pp. 151-155.

[52] Sauer, G.; Brehm, G.; Schneider, Nielsch, K.; Wehrspohn, R.B.; Choi, J.; Hofmeister, H. & Gösele, U. (2002). Highly ordered monocrystalline silver nanowire arrays. *J. Appl. Phys.*, Vol. 91, pp. 3243-3247.

[53] Sausa, C.T.; Leitao, D.C.; Proenca, M.P.; Apolinario, A.; Correia, J.G.; Ventura, J. & Aroujo, J.P. (2011). Tunning pore filling of anodic alumina templates by accurate control of the bottom barrier layer Thickness. Nanotechnology, Vol. 22, no. 315602.

[54] Sheasby, P.G. Cook, W.E. (1974). The electrolytic colouring of anodized aluminum. *Trans. IMF,* Vol. 52, pp. 103-106.

[55] Sun, J.; Wang, L.-W. & Buhro, W.E. (2008). Synthesis of cadmium telluride quantum wires and the similarity of their effective band gaps to those of equidiameter cadmium telluride quantum dots. *J. Am. Chem. Soc.*, Vol. 130, pp. 7997- 8005.

[56] Sun, X.-Y.; Xu, Fa-Q.; Li, Z.-Mu Zhang, W.-H. (2005). Cyclic voltammetry for the fabrication high dense silver nanowire arrays with the assistance of AAO template. *Mater. Chem. Phys.*, Vol. 90, pp. 69-72.

[57] Thompson, G. E.. (1997). Porous anodic alumina: fabrication, characterization and applications. *Thin Solid Films,* Vol. 297, pp. 192-201.

[58] Torane, A.P.; Lokhande, C.D.; Patil, P.S. & Bhosale, C.H. (1998). Preparation and characterization of electrodeposited Bi_2Se_3 thin films. *Mater. Chem. Phys.*, Vol. 55, pp. 51-54.

[59] Vanderah, D.J.; Valincius, G. & Meuse, C.W. (2003). Self-assembed monolayers on methyl 1-thiohexa(ethylene oxide) for the inhibition of protein adsorption. *Langmuir,* Vol. 18, pp. 4674-4680.

[60] Wang, D.; Yu, D.; Mo, M.; Liu, X. & Qian, Y. (2003). Preparation and characterization of wire-like Sb_2Se_3 and flake-like Bi_2Se_3 nanocrystals. *J. Cryst. Growth,* Vol. 253, pp. 445-451.

[61] Zech, N. & Landolt, D. (2000). The influence of boric acid and sulfate ions on the hydrogen formation in Ni-Fe plating electrolytes. *Electrochim. Acta,* Vol. 45, pp. 3461-3471.

[62] Xu, S.; Zhao, W-b.; Hong, J.-M.; Zhu, J.-J. & Chen, H.-Y. (2005). Photochemical synthesis of Bi_2Se_3 nanosphere and nanorods. *Mater. Lett.*, Vol. 59, pp. 319-321.

[63] Xu, T.; Zangari, G. Metzger, R.M. (2002). Periodic holes with 10 nm diameter produced by grazing Ar^+ milling of the barrier layer in hexagonally ordered nanoporous alumina. *Nano Lett.*, Vol. 2, pp. 37-41.

[64] Yan, X.B.; Puodel, Y.; Ma, W.S.; Liu, G.; Joshi, H.; Wang, Y.; Lan, G.; Wang, D.; Chen G. & Ren, Z.F. (2010). Experimental studies on anisotropic thermoelectric properties and structures of n-type $Bi_2Te_{2.7}Se_{0.5}$. *Nano Lett.*, Vol. 10, pp. 3373-3378.

[65] Yang, X.; Wang, X. & Zhang, Z. (2005). Synthesis and optical properties of single-crystalline bismuth selenide nanorods via a convenient route. *J. Cryst. Growth*, Vol. 276, pp. 566-570.

[66] Yoo, W.-Ch. Lee, J.-K. (2004). Field-dependent growth patterns of metals electroplated in nanoporous alumina membranes. *Adv. Mater.*, Vol. 16, pp. 1097-1101.

[67] Zheng, M.J.; Zhang, L.D.; Li, G. H. & Shen, W.Z. (2002). Fabrication and optical properties of large-scale uniform zinc oxide nanowire arrays by one-step electrochemical deposition technique. *Chem. Phys. Lett.*, Vol. 363, pp. 123-125.

Permissions

The contributors of this book come from diverse backgrounds, making this book a truly international effort. This book will bring forth new frontiers with its revolutionizing research information and detailed analysis of the nascent developments around the world.

We would like to thank Dr. Xihong Peng, for lending her expertise to make the book truly unique. She has played a crucial role in the development of this book. Without her invaluable contribution this book wouldn't have been possible. She has made vital efforts to compile up to date information on the varied aspects of this subject to make this book a valuable addition to the collection of many professionals and students.

This book was conceptualized with the vision of imparting up-to-date information and advanced data in this field. To ensure the same, a matchless editorial board was set up. Every individual on the board went through rigorous rounds of assessment to prove their worth. After which they invested a large part of their time researching and compiling the most relevant data for our readers. Conferences and sessions were held from time to time between the editorial board and the contributing authors to present the data in the most comprehensible form. The editorial team has worked tirelessly to provide valuable and valid information to help people across the globe.

Every chapter published in this book has been scrutinized by our experts. Their significance has been extensively debated. The topics covered herein carry significant findings which will fuel the growth of the discipline. They may even be implemented as practical applications or may be referred to as a beginning point for another development. Chapters in this book were first published by InTech; hereby published with permission under the Creative Commons Attribution License or equivalent.

The editorial board has been involved in producing this book since its inception. They have spent rigorous hours researching and exploring the diverse topics which have resulted in the successful publishing of this book. They have passed on their knowledge of decades through this book. To expedite this challenging task, the publisher supported the team at every step. A small team of assistant editors was also appointed to further simplify the editing procedure and attain best results for the readers.

Our editorial team has been hand-picked from every corner of the world. Their multi-ethnicity adds dynamic inputs to the discussions which result in innovative outcomes. These outcomes are then further discussed with the researchers and contributors who give their valuable feedback and opinion regarding the same. The feedback is then collaborated with the researches and they are edited in a comprehensive manner to aid the understanding of the subject.

Apart from the editorial board, the designing team has also invested a significant amount of their time in understanding the subject and creating the most relevant covers. They scrutinized every image to scout for the most suitable representation of the subject and create an appropriate cover for the book.

The publishing team has been involved in this book since its early stages. They were actively engaged in every process, be it collecting the data, connecting with the contributors or procuring relevant information. The team has been an ardent support to the editorial, designing and production team. Their endless efforts to recruit the best for this project, has resulted in the accomplishment of this book. They are a veteran in the field of academics and their pool of knowledge is as vast as their experience in printing. Their expertise and guidance has proved useful at every step. Their uncompromising quality standards have made this book an exceptional effort. Their encouragement from time to time has been an inspiration for everyone.

The publisher and the editorial board hope that this book will prove to be a valuable piece of knowledge for researchers, students, practitioners and scholars across the globe.

List of Contributors

Supab Choopun
Department of Physics and Materials Science, Faculty of Science, Chiang Mai University Chiang Mai, Thailand
ThEP Center, CHE, Bangkok, Thailand

Ekasiddh Wongrat
School of Science, University of Phayao, Phayao, Thailand

Niyom Hongsith
School of Science, University of Phayao, Phayao, Thailand
ThEP Center, CHE, Bangkok, Thailand

Usha Philipose and Gopal Sapkota
Department of Physics, University of North Texas, USA

Sheng Yun Wu
Department of Physics, National Dong Hwa University, Hualien, Taiwan, Republic of China

Sachindra Nath Das
Department of Physics, Burdwan Raj College, Burdwan, West Bengal, India
Department of Material Science and Engineering, Yonsei University, Seoul, South Korea

Junjie Xiong and Jae-Min Myoung
Department of Material Science and Engineering, Yonsei University, Seoul, South Korea

Jyoti Prakash Kar
Department of Physics, National Institute of Technology, Rourkela, Odisha, India
Department of Material Science and Engineering, Yonsei University, Seoul, South Korea

Y.Z. Long
College of Physics, Qingdao University, People's Republic of China
Key Laboratory of Photonics Materials and Technology in Universities of Shandong, Qingdao University, People's Republic of China
State Key Laboratory Cultivation Base of New Fiber Materials and Modern Textile, Qingdao University, People's Republic of China

H.D. Zhang and B. Sun
College of Physics, Qingdao University, People's Republic of China

X.L. Xu
Department of Physics, University of Science and Technology of China, People's Republic of China

M.G. Gong
College of Physics, Qingdao University, People's Republic of China
Department of Physics, University of Science and Technology of China, People's Republic of China

Takeshi Ohgai
Division of Chemistry and Materials Science, Nagasaki University, 1-14 Bunkyo-machi, Nagasaki 852-8521, Japan

Miao Zhong, Mario Kurniawan and Jean-Jacques Delaunay
School of Engineering, The University of Tokyo, Japan

Aleksandra Apostoluk and Bertrand Vilquin
INL, CNRS UMR-5270, Lyon University, France

C. Mukherjee and C. K. Maiti
Indian Institute of Technology, Kharagpur, India

J. Anaya and J. Jiménez
GdS Optronlab, Ed I+D, University of Valladolid, Valladolid, Spain

T. Rodríguez
Electronic Technology, ETSIT, Polytechnic University of Madrid, Madrid, Spain

Muhammad Y. Bashouti, Matthias Pietsch, Kasra Sardashti, Gerald Brönstrup, Sebastian W. Schmitt and Sanjay K. Srivastava
Max-Planck-Institute for the Science of Light, Erlangen, Germany

Hossam Haick
The Department of Chemical Engineering and Russell Berrie Nanotechnology Institute, Technion - Israel Institute of Technology, Haifa, Israel

Jürgen Ristein
Institue for Technical Physics, University of Erlangen-Nuremberg, Erlangen, Germany

Jordi Arbiol
Institució Catalana de Recerca i EstudisAvançats (ICREA) and Institut de Ciència de Materials de Barcelona,ICMAB-CSIC, Bellaterra, Spain

Silke Christiansen
Max-Planck-Institute for the Science of Light, Erlangen, Germany
Institute of Photonic Technology, Jena, Germany

Björn Hoffmann and Silke Christiansen
Max Planck Institute for the Science of Light, TDSU Photonic Nanostructures, Erlangen, Germany
Institute of Photonic Technology, Department Semiconductor Nanostructures, Jena, Germany

Michael Latzel, Sebastian W. Schmitt and Muhammad Y. Bashouti
Max Planck Institute for the Science of Light, TDSU Photonic Nanostructures, Erlangen, Germany

Vladimir Sivakov
Institute of Photonic Technology, Department Semiconductor Nanostructures, Jena, Germany

Jiří Dluhoš and Jaroslav Jiruse
TESCAN, a.s., Brno, Czech Republic

Marian Nowak and Marcin Jesionek
Institute of Physics, Silesian University of Technology, Poland

V. Salles, T. Fiorido, L. Hu, C. Eid and A. Brioude
Laboratoire des Multimatériaux et Interfaces, UMR 5615 CNRS, Université Lyon1, Universitéde Lyon, Bat. Berthollet, France

L. Seveyrat, J. Galineau, B. Guiffard and D. Guyomar
Laboratoire de Génie Electrique et Ferroélectricité, LGEF - INSA Lyon, France

Daniela Munteanu and Jean-Luc Autran
CNRS &Aix-Marseille University, Marseille, France

Xiaowei Teng and Xiaoxiong Jia
University of New Hampshire, USA

Hui Li
Key Laboratory for Liquid-Solid Structural Evolution and Processing of Materials, Ministry of Education, Shandong University, China

Fengwei Sun
Department of Materials Science and Engineering, Delft University of Technology, The Netherlands

Arūnas Jagminas
State Research Institute Center for Physical Sciences and Technology, Lithuania